Selected Papers of the Fourth Conference on

East Asia and Western Pacific Meteorology and Climate

WORLD SCIENTIFIC SERIES ON METEOROLOGY OF EAST ASIA

WORLD SCIENTIFIC SERIES ON
METEOROLOGY OF EAST ASIA

Vol. 1

Selected Papers of the Fourth Conference on

East Asia and Western Pacific Meteorology and Climate

Hangzhou, China October 26–28, 1999

Editors

C.-P. Chang
Naval Postgraduate School

Guoxiong Wu
Chinese Academy of Sciences

Ben Jou
National Taiwan University

C. Y. Lam
Hong Kong Observatory

World Scientific
New Jersey • London • Singapore • Hong Kong

Published by

World Scientific Publishing Co. Pte. Ltd.
P O Box 128, Farrer Road, Singapore 912805
USA office: Suite 1B, 1060 Main Street, River Edge, NJ 07661
UK office: 57 Shelton Street, Covent Garden, London WC2H 9HE

British Library Cataloguing-in-Publication Data
A catalogue record for this book is available from the British Library.

EAST ASIA AND WESTERN PACIFIC METEOROLOGY AND CLIMATE
World Scientific Series on Meteorology of East Asia Vol. 1

ISBN 981-02-4908-X

Printed in Singapore by Uto-Print

FOREWORD

In the past several decades, considerable progress has been made in the study of the meteorology of East Asia and surrounding areas. The progress arises both from the general advancement in the various fields of atmospheric sciences and the increased international exchanges and related activities that have helped to disseminate a significant amount of knowledge of the meteorology of East Asia to the world scientific community. However, the vast majority of East Asian meteorologists are not English-speaking, and countries within East Asia and surrounding regions have different languages. Thus, communications and exchanges are still hampered by language berriers. The purpose of this new Series on Meteorology of East Asia is to provide a medium to help disseminate the meteorological research in East Asia to the English-speaking world, and to facilitate the exchange among all scientists in this field. As part of the effort, World Scientific Company (WSPC) has agreed to assist in the English editing of manuscripts authored by non-English scientists. It is hoped that this book series will encourage the publication of important advances in East Asian meteorology regardless of the native language of the scientists involved, and thereby bring about increased development of East Asian meteorology and related applications.

The plan for the book series is to publish three types of volumes: selected papers from major conferences, invited review articles, and monogrpahs. Volume 1 is an example of the first type. Volume 2, which is a collection of invited reviews on the East Asian monsoon that will be published next, is of the second type. Other volumes of all three types are being planned. All fields of meteorology and related sciences are potential topics for the series. Scientists who are interested in contributing to the series, whether as an editor or an author, are encouraged to contact one of the members of the Editorial board (listed in inside cover).

I wish to thank all members of the Editorial board, from both East Asia and the U.S., for their devotion to the purpose of this endeavor; and to Professor K. K. Phua, Chairman cum Editor-in-Chief of WSPC, whose vision on the importance of meteorological sciences for the large geographic area and large population of East Asia and the surrounding regions has led to the initiation of this book series.

Chih-Pei Chang
Department of Meteorology
Naval Postgraduate School
Monterey, California

PREFACE

When the participants of the Fourth Conference on East Asia and Western Pacific Meteorology and Climate assembled in Hangzhou 26-28 October 1999, they were in effect celebrating the tenth anniversary of an endeavor that had a modest beginning in Hong Kong. The original idea for such a conference germinated in Hong Kong in 1987-88 during visits by Professors Tao Shiyan and Ding Yihui of the Chinese Academy of Sciences, Professor Ching-Yen Tsay of National Taiwan University, and Professor C.-P. Chang of Naval Postgraduate School. They noted that there were language and geographical barriers that prevented efficient exchanges among scientists studying East Asian meteorology. A bilingual conference would help to reduce these barriers. As the discussion progressed, the idea was warmly embraced by several meteorological community leaders, including Mr. Patrick Sham of Hong Kong Observatory, Dr. Richard Anthes of the University Corporation for Atmospheric Research, and the late Professor Zou Jingmeng, then President of the World Meteorological Organization.

In July 1989 the first conference in this series took place in Hong Kong. It was run with the innovative bilingual approach, which allowed both Chinese- and English-speaking scientists from various regions of East Asia and across the Pacific to come together, and share and discuss their research. It also proved to be an invaluable networking opportunity leading to a rapid growth of interaction and cooperation among a diverse group of scientists in the decade that followed. The enthusiastic participation at the fourth conference as well as the breadth and depth of the presentations were a testimony to this series of conferences coming of age.

The Conference in Hangzhou broke new ground in a number of ways. Sessions on remote sensing, atmospheric physics and chemistry as well as the impact of aircraft emissions on atmospheric composition were introduced, on top of the traditional topics like typhoons, rainstorms, monsoon and climate. Part of the conference was held concurrently with the Symposium on the Severe Rainstorm in 1998, a subject of great interest to meteorologists because of its major impact on China in the previous year. The influence of the Conference was thus amplified as it involved some 200 participants during the joint sessions.

This volume contains a collection of selected papers in English although many were originally presented in Chinese during the Conference. We hope that this would make available to the global scientific community the body of knowledge on the various aspects of the meteorology of the East Asia and Western Pacific region as represented by the transactions at the Conference.

The Conference owed its success to the able work of the Organizing Committee headed by Professor Ma Henian. It also had the benefit of financial support from the Natural Science Foundation of China and the University Corporation for Atmospheric Research, U. S. A. Many scientists served as anonymous referees to review and select the papers; they devoted much time to work with the editors in suggesting ways to improve the papers accepted for this publication. Colleagues of the Hong Kong Meteorological Society led by Mr. Y. K. Chan helped with the technical and language editing. The contributions of all these parties are gratefully acknowledged.

Finally, I wish to thank the World Scientific Publishing Company (WSPC), who extended its tradition of publishing the Proceedings of the previous conferences one step further and made this collection of selected papers the first volume of the new World Scientific Series on Meteorology of East Asia.

C Y Lam
Chairman
Hong Kong Meteorological Society

CONTENTS

~ Monsoon Observations ~

THE GLOBAL MONSOON SYSTEMS

Qing-cun Zeng
Institute of Atmospheric Physics, CAS, Beijing 100029

ABSTRACT

The global monsoon systems consist of regions with large seasonality (Zeng et al. (1994), Zeng and Zhang (1998), Xue and Zeng (1999)). It is discovered that in the lower troposphere there are one tropical monsoon region (coincident with the classical monsoon region), two subtropical monsoon regions (one each in Northern and Southern Hemisphere and occupied by the seasonal migration of subtropical high) and two middle-high latitudinal monsoon regions (associated with the storm tracks of westerlies). The division of seasons is determined by the normalized similarity (Zeng et al. (1994), Zeng and Zhang (1992), Zhang and Zeng (1998)). It is revealed that in the monsoon regions, the transitions from winter to summer and from summer to the next are abrupt.

Key words: monsoon, seasonality, season, similarity

1. SEASONALITY AND THE GLOBAL MONSOON SYSTEMS

One of the main objectives of short-range climate prediction is to predict the seasonal mean anomaly of circulation and the associated anomalies of climatic elements (e.g. precipitation and temperature), especially in monsoon regions. In classical climatology, monsoon is a regional concept summarizing all drastic seasonal variations in the tropics and some surrounding subtropical-midlatitudinal regions (e.g. East Asia and India). However, besides the classical monsoon regions there are other regions over the globe, where the seasonal variation and interannual variability are pronounced. Therefore, to extend the classical monsoon concept to include such regions could be helpful for understanding and predicting short-range climate variability. Such an attempt has been made by the calculation of seasonality (Zeng et al. (1994), Zeng and Zhang (1998), Xue and Zeng (1999)).

The seasonality, δ, of some meteorological element (e.g. the wind) is defined as the root mean square of the difference of its values between winter and summer (or January and July) divided by its annual mean (or approximated by the January and July mean). The calculation of seasonality of wind has been carried out for every 10° longitude by 10° latitude area at every level (Zeng and Zhang (1998), Xue and Zeng (1999)).

It is discovered that in the lower troposphere, there are five regions with pronounced seasonality: (a) one is coincided with the classical tropical monsoon region, (b) two in subtropical zones (one each in Northern and Southern Hemisphere) and occupied by the seasonal migration of subtropical high, and (c) two in middle-high latitudes (one in each hemisphere) and associated with the seasonal variation of westerly storm tracks. Therefore, the classical monsoon region is just one of the regions with pronounced seasonality. The (generalized) global monsoon systems can be considered as consisting of: (a) tropical monsoon, (b) subtropical monsoon, and (c) temperate-frigid monsoon (Fig. 1).

These calculations also show that the global monsoon systems possess baroclinic structure, turning into the cross-equatorial planetary monsoon patterns (different from the above mentioned) in the upper troposphere and two (one in each hemisphere) latitudinal belts associated with the seasonal variation of the night jet in the middle-high stratosphere (the maps of δ at different levels in Zeng et al. (1998), and Xue and Zeng (1999) refer).

Comparison of the seasonality (δ) maps with the inter-annual variability (IV) maps calculated by conventional method shows that they are almost coincident, except for the equatorial eastern Pacific where only IV is pronounced as a result of ENSO.

2. DIVISION OF SEASONS AND THE ABRUPT SEASONAL TRANSITION

An objective and quantitative method for dividing seasons and determining the start time, length and strength of each season is based on the (normalized) similarity R, which is the correlation coefficient between $\vec{v}'$ and $\vec{v}'_w$, departures of wind from the climatological annual mean (approximated to January and July mean) for each period (say, 10 days) and typical winter month (say, January) respectively (Zeng et al. (1994), Zeng and Zhang (1998), Zeng and Zhang (1992), Zhang and Zeng (1998)). Note that we have $\vec{v}'_w = -\vec{v}'_s$, where $\vec{v}'_s$ is the departure of wind for typical summer month (say, July). Obviously, for the typical patterns, we have R(Jan) = 1 and R (Jul) = −1, i.e. $R = 1$ and $R = -1$ correspond to the typical winter and summer respectively. We can divide the seasons as follows: $R > 0.5$ is winter, $0.5 > R > -0.5$ is spring (in the first half of the year) or autumn (in the latter half of the year), $R < -0.5$ is summer, and $R = 0$ is the typical spring (in the first half of the year) or typical autumn (in the latter half of the year).

Using the zonal component of winds and the geopotential heights, the calculations (Zeng et al. (1994), Zhang and Zeng (1998)) show that in the lower troposphere and in East Asia monsoon region: (a) $R \approx 0$ corresponds to the "south-east monsoon" near the surface from the last ten days of April to the beginning of May, or the "sky high and air brisk in autumn" period from September to October; and (b) the transitions of circulation from winter to summer and from summer to winter are abrupt and on average take place in June and October respectively (this result is in agreement with previous results of other authors) (Fig. 2).

The calculations also show that seasons in different regions and at different levels come at different times. This means that there is temporal-spatial structure of seasonal variation. It is revealed that the summer regime is first set up due to the breaking down of the night jet in the upper stratosphere and over the warm pool region in the tropical western Pacific in the lower troposphere.

ACKNOWLEDGEMENT. This paper is written based on the following references. The author is very grateful to his colleagues for cooperation and Ms. Xuan Wang for editing the paper.

REFERENCES

Xue Feng, and Zeng Qingcun, 1999: Seasonality and interannual variability of the wind fields, *Advances in Atmospheric Sciences,* **16**, 160-165.

Zeng Qingcun, and Zhang Banglin, 1992: On the seasons of general atmospheric circulation and their abrupt changes, Part I: general concept and method, *Chinese Journal of Atmospheric Sciences,* **16**, No.4, 329-336.

Zeng Qingcun, Zhang Banglin, Liang Youlin, and Zhao Sixiong, 1994: The Asian Summer Monsoon – A case study, *Proceedings of the Indian National Science Academy,* Vol.**60**, Part A, No.1, 81-96.

Zeng Qingcun, and Zhang Banglin, 1998: On the seasonal variation of atmospheric general circulation and the monsoon, *Chinese Journal of Atmospheric Sciences,* **22**, No.3, 211-220.

Zhang Banglin, and Zeng Qingcun, 1998: On the seasons of general atmospheric circulation and their abrupt changes, Part II: case study of particular year, *Chinese Journal of Atmospheric Sciences,* **22**, No.2, 103-111.

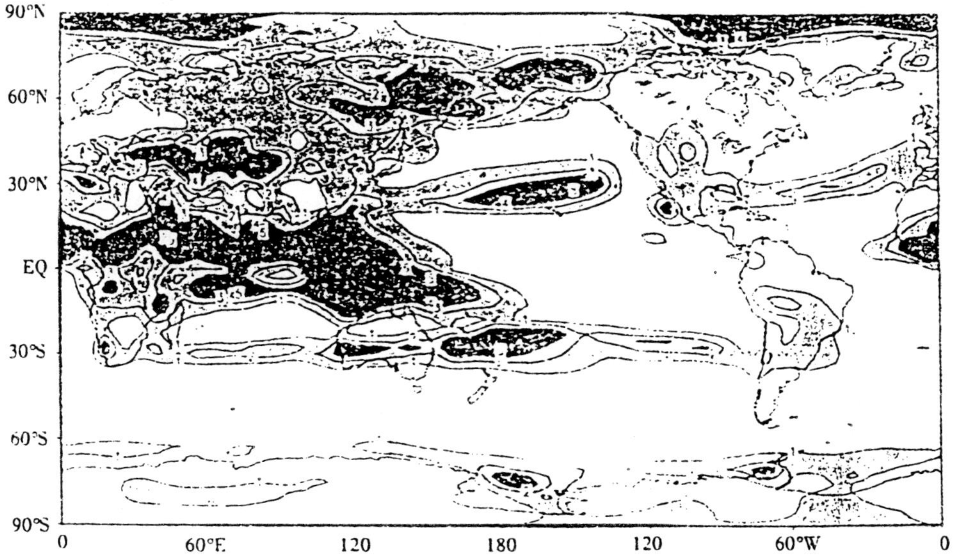

图 1. 850hPa 风场的标准化季节变率（风场为 1979～1991 年的气候平均）

Fig.1 The normalized seasonality of wind field at 850hPa (the wind is taken as climatological monthly mean for 1979-1991).

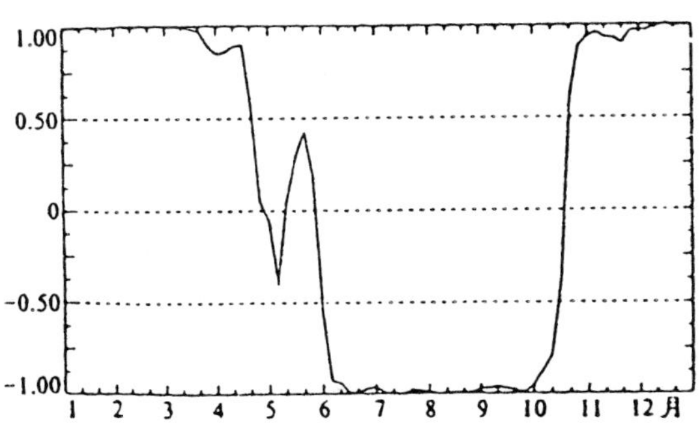

图 2.a 500hPa 位势高度场的 $R(t)$。区域为（100－120°E，20－30°N）。

Fig.2.a $R(t)$ for geopotential height of 500hPa. The region is (100-120°E, 20-30°N).

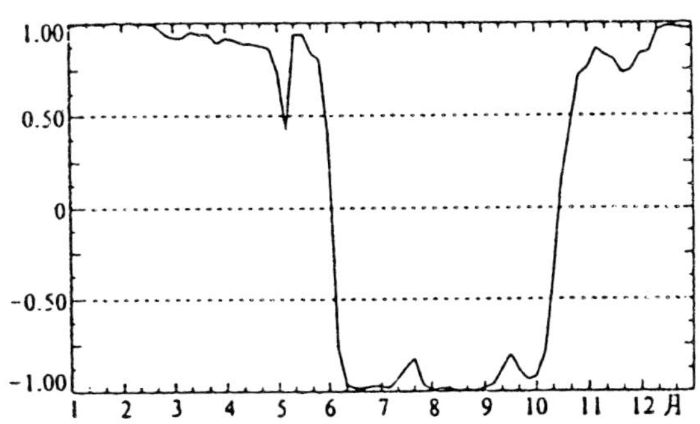

图 2.b 同图 2.a，但为 200hPa 纬向风及（100－120°E，30－40°N）。

Fig.2.b The same as Fig.2.a, but for zonal wind of 200hPa and (100-120°E, 30-40°N).

THE SCSMEX/GAME SOUNDING NETWORKS AND THE ONSET OF THE 1998 ASIAN SUMMER MONSOON

Richard H. Johnson and Paul E. Ciesielski
Department of Atmospheric Science, Colorado State University, Fort Collins, Colorado

ABSTRACT

Observations from the SCSMEX/GAME sounding domains have been used to document several features of the onset of the 1998 East Asian summer monsoon. Marked transitions of the low-level and upper-level flow occurred in the vicinity of the South China Sea (SCS) in mid-May as the summer monsoon became established over this region. Over the northern SCS, the onset of westerly monsoon flow occurred around 15 May, whereas over the southern SCS the onset was on 20 May. Deep tropospheric moistening accompanied the monsoon onset in the northern SCS as precipitation began in that region. Changes in the moisture profile in the southern SCS were more complex, although there appeared to be a transition from a tradewind-like structure (moist trade cumulus layer capped by dry air aloft) to a moistening of the mid-to-upper troposphere accompanying the onset.

Preliminary results from SCSMEX indicate that the monsoon onset over the South China Sea is rather complex and far from homogeneous over this oceanic region. A prominent mixed Rossby-gravity wave was detected at the time of onset over the southern SCS. The onset of the 1998 East Asian summer monsoon was distinctly different from the onset of South Asian or Indian monsoon.

1. INTRODUCTION

During the May-June 1998 South China Sea Monsoon Experiment (SCSMEX), a sounding network was established in the South China Sea (SCS) and surrounding region to investigate the onset and development of the East Asian monsoon. This network was contained within the larger domain of the GEWEX Asian Monsoon Experiment (GAME). Sounding sites within the SCSMEX/GAME domain are shown in Figure 1. Figure 1 contains information concerning the vertical resolution and frequency of data receipt for all of the sites during the two Intensive Observing Periods (IOPs) (5-25 May and 5-25 June). The highest sounding frequency and resolution were in the vicinity of the SCS (up to 4 launches per day and 5-hPa vertical resolution), with a mix of 1-2 and 3-4 times per day and 30-100 hPa resolution over the surrounding region. Two special sounding networks are indicated by the polygons in Figure 1, one in the northern SCS surrounding the dual-Doppler array (the TOGA radar aboard R/V *Shiyan #3* and the BMRC C-POL radar on Dongsha Island) and the other surrounding the R/V *Kexue #1* in the southern SCS. These networks were designed to determine and contrast the properties of convection in two distinct oceanic regions of the Asian monsoon.

The majority of the soundings during SCSMEX were received via the Global Telecommunication System (GTS) and decoded in real time at the Colorado State University (CSU) soundings-monitoring facility at the Hong Kong University of Science and Technology (HKUST). Preliminary sounding data have been objectively analyzed using a multi-quadric objective analysis scheme, from which gridded analyses have been prepared. In data-void regions (e.g., central Bay of Bengal and Indian

Ocean) ECMWF operational initial analysis fields have been used to supplement the soundings if no sounding data were present within a 5° radius of a grid point.

In this paper we examine several features of the East Asian monsoon onset. First, changes in the large-scale flow over East Asia accompanying the onset are documented. Next, changes in the wind and thermodynamic structure over the northern and southern SCS are investigated using Vaisala GPS sounding data from *Shiyan #3* and *Kexue #1*. In addition, evidence is presented of a prominent westward propagating, equatorial mixed Rossby-gravity wave over the longitudes of East Asia at the time of onset. Finally, the onset of the East Asian monsoon is contrasted with the onset of the South Asian or Indian monsoon.

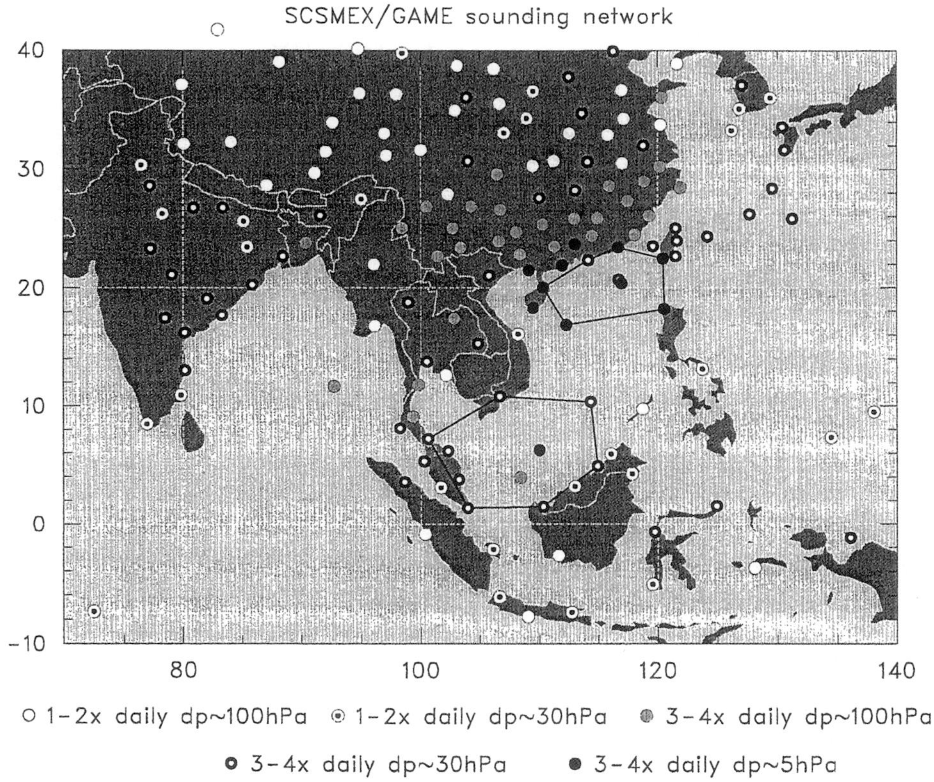

Figure 1: SCSMEX/GAME sounding network. Frequency of observations and vertical resolution are indicated by dots with various shadings. Sounding polygons in the northern and southern South China Sea are also indicated.

2. Monsoon onset: transition of the large-scale flow

The Asian summer monsoon is a major component of the Northern Hemisphere summertime circulation system. Its precipitation affects over half the world's population. The initial onset of the summer monsoon over Asia in early to mid-May occurs over the South China Sea (SCS) region (Tao and Chen 1987; Lau and Yang 1997). In early spring, convection begins a transition northward from the Indonesian "maritime continent" toward southeast Asia. The path of the monsoon rains is typically across the Borneo-Malaysia-Indo-China land bridge from mid-April until mid-May, followed by a rapid jump across the SCS during mid-to-late May. The subsequent arrival of the rains over central China and Japan marks the beginning of the Mei-Yu and Baiu rainy seasons over those respective areas.

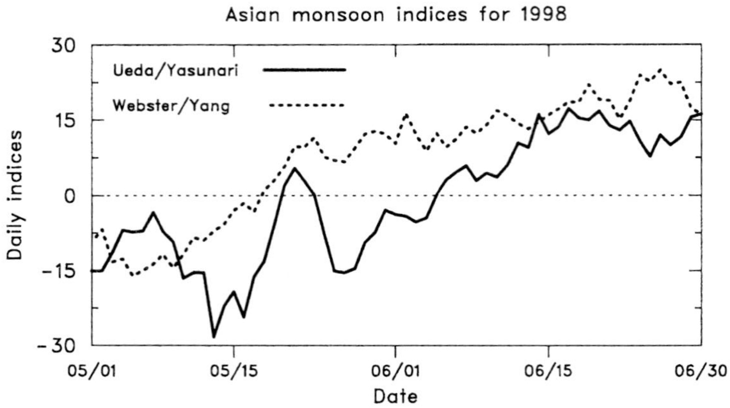

Figure 2: Time series of 1998 monsoon indices of Webster and Yang (1992) (WY) and Ueda and Yasunari (1998) (UY). Ordinate units for WY are m s^{-1} and UY decameters. Onset of the 1998 East Asian monsoon was on 20 May in the southern SCS and around 15 May in the northern SCS.

The actual timing of the monsoon onset can vary by two to three weeks and the amount of precipitation over the SCS region can also vary considerably from year to year (e.g., Lau and Yang 1997). The 1998 monsoon onset over the SCS occurred in mid-May. This timing is consistent with two monsoon indices shown in Fig. 2 from Webster and Yang (1992) (WY) and Ueda and Yasunari (1998) (UY). These indices represent a measure of the deep tropospheric shear and north-to-south temperature difference, respectively, over a large area to the south of the Tibetan Plateau. The WY index is the 850 to 200 hPa shear of the zonal wind $U_{850} - U_{200}$ in m s^{-1} computed as an average over the area 0–20°N, 90–120°E. The UY index is the 200–500 hPa thickness difference in decameters between 5 stations over the Tibetan Plateau and 5 stations over Malaysia. The two indices are roughly related through the thermal wind relation. In a very broad sense, a switch from negative to positive values (warmer upper-tropospheric temperatures over the Tibetan Plateau than the equator or a shift from westerly to easterly wind shear over India) generally corresponds to monsoon onset over the Indian Subcontinent. However, recent studies have shown that special indicies are required for the East Asian monsoon (Wang and Fan 1999; Lu and Chan 1999), which displays important differences from the monsoon over the

Indian subcontinent. In particular, the onset of the East Asian monsoon appears to be influenced by both tropical and midlatitude dynamical processes, leading Yang et al. (1999) to refer to the East Asian monsoon as a "hybrid" monsoon.

Streamline analyses at 850 and 200 hPa prior to the onset (11 May) are shown in Fig. 3. At 200 hPa the pre-onset flow was characterized by westerlies over India and China with generally northwesterly flow over the SCS. At 850 hPa a subtropical ridge extended from the western Pacific across Indo-China with east-to-southeasterly flow over the SCS. As the monsoon set in, dramatic changes in the large-scale flow occurred. Just ten days later (21 May) a 200-hPa anticyclone built northward along the Indo-China peninsula, while at 850 hPa the subtropical ridge retreated eastward and westerly low-level flow enveloped the SCS (Fig. 3). During this period, convection shifted northward along the Indo-China peninsula.

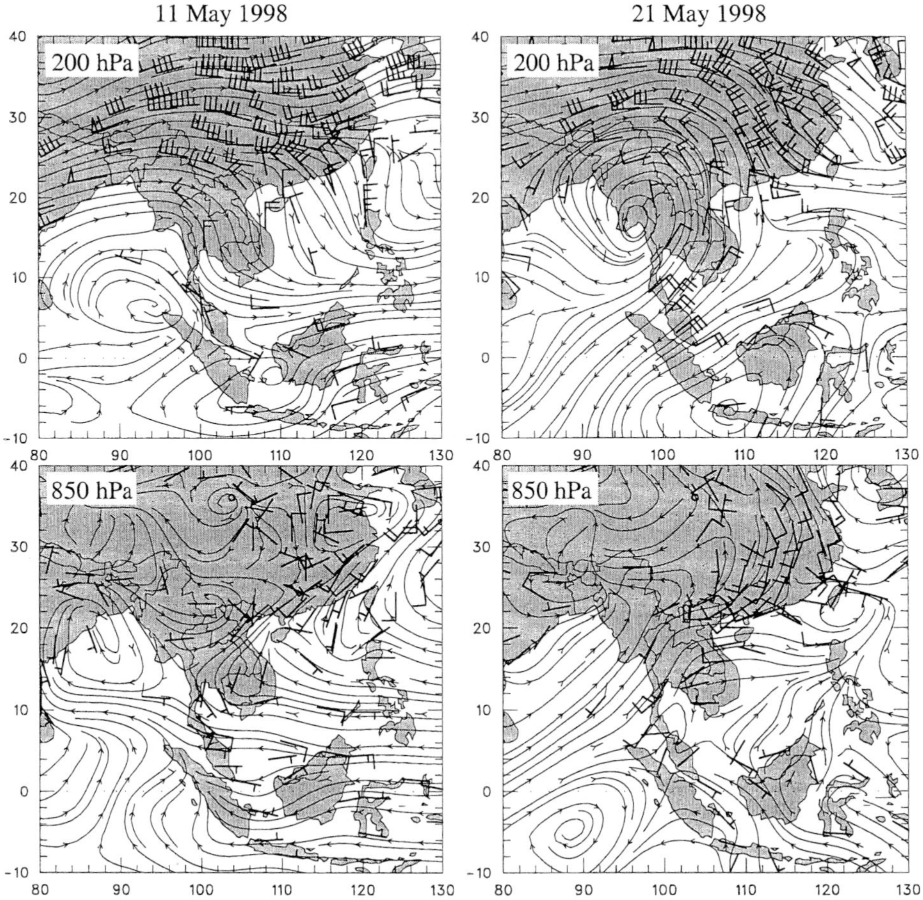

Figure 3: Streamline analyses at 200 hPa on 11 and 21 May 1998 (top panels) and at 850 hPa (bottom panels). May 11 represents the pre-onset period and 21 May the post-onset period. Wind barbs at the sounding sites are indicated.

3. Onset characteristics over the South China Sea

Figure 3 indicates marked transitions of the flow over the southern SCS between 11 and 21 May. Time series of sounding data (5-point running means of 6-hourly observations) from *Kexue #1* (6°N, 110°E) in the southern SCS (Fig. 4) indicate that in the upper troposphere the flow shifted from westerly to easterly flow around 12–13 May as the upper-level anticyclone built northward. At low levels, a shift from easterly to westerly flow occurred on 20 May. Since monsoon onset is generally linked

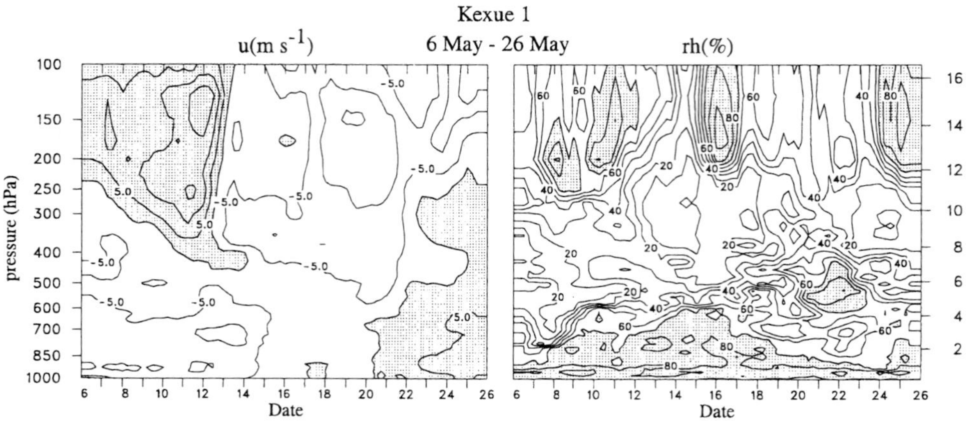

Figure 4: Time series of filtered zonal wind component (m s^{-1}) and relative humidity (%) at *Kexue #1* from 00 UTC 6 May to 00 UTC 26 May 1998. Contour interval for u is 5 m s^{-1} and for relative humidity 10%, with values greater than 0 and 70%, respectively, shaded.

to the behavior of the low-level flow, 20 May could be regarded as the date of the monsoon onset over the southern SCS.

Although very little precipitation occurred at *Kexue #1* accompanying the monsoon onset, there were several changes in the humidity profile associated with the flow reversal in that region (Fig. 4). Prior to the onset, in early May, the relative humidity (RH) structure in the easterly low-level flow resembled that of the tradewind regime, with a low-level moist region below 800–700 hPa capped by very dry air aloft. During a 10-day period prior to the monsoon onset, this moist layer deepened with time, presumably in response to the weakening subtropical ridge and corresponding weakening subsidence, which allowed the shallow cumulus clouds to grow to progressively greater depths. When the low-level westerly flow commenced around 20 May, the RH structure changed to one of moist conditions in the mid- to upper troposphere with somewhat drier conditions near 700 hPa.

The monsoon onset over the northern SCS exhibited a markedly different behavior (Fig. 5). A reversal of the zonal component of the flow in the upper troposphere at *Shiyan #3* (21°N, 117°E) did not occur until 22 May. However, a strengthening of the low-level westerlies occurred one week earlier, around 15 May (which is regarded as the onset date for the northern SCS). Unlike the southern SCS, the northern SCS

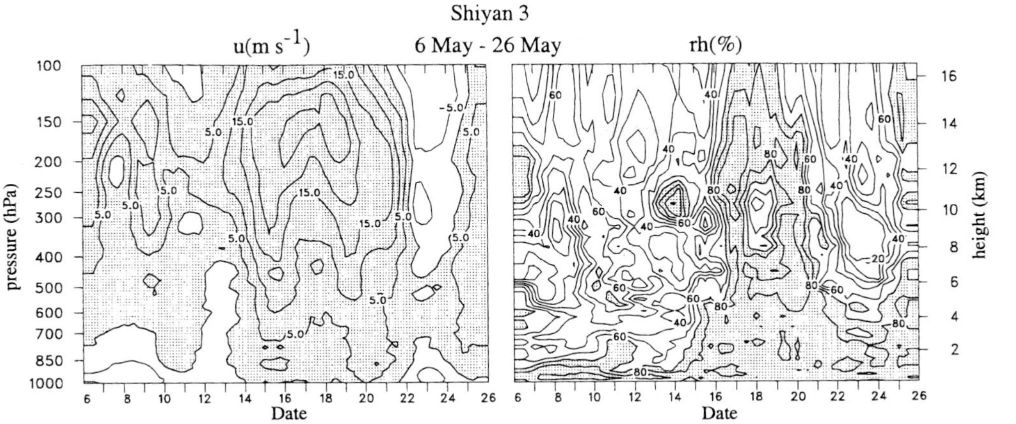

Figure 5: Time series of filtered zonal wind component (m s^{-1}) and relative humidity (%) at *Shiyan #3* from 00 UTC 6 May to 00 UTC 26 May 1998. Contour interval for u is 5 m s^{-1} and for relative humidity 10%, with values greater than 0 and 70%, respectively, shaded.

westerlies increased throughout the entire troposphere at the time of onset in this region. The westerly maximum in the upper troposphere (between 150 and 200 hPa) is associated with a midlatitude baroclinic disturbance. This coupling of the onset of the westerlies over the northern SCS with midlatitude disturbances makes the onset process in this region distinct from that of the Indian summer monsoon (Yang et al. 1999), the latter region exhibiting a behavior more like that in the southern SCS (Fig. 4). In the southern SCS and over the Indian subcontinent, the onset of low-level westerlies is accompanied by an onset of upper-tropospheric easterlies.

The monsoon onset in the northern SCS is also accompanied by a distinct change in the RH profile (Fig. 5). Relatively dry tropospheric conditions above a moist boundary layer existed prior to the onset. However, following the onset virtually the entire troposphere became moist as precipitation developed in this region.

4. Mixed Rossby-gravity wave at time of onset

A time series of the meridional wind v at *Kexue #1* (Fig. 6) indicates a prominent shift in the upper-tropospheric winds near the time of onset over the southern SCS, with a maximum amplitude in v fluctuations near 150 hPa. This disturbance has the signature of a mixed Rossby-gravity wave (Matsuno 1966). Inspection of similar time series of v from Malaysia indicates that this wave moved westward at $\sim$16 m s^{-1} and had a wavelength of $\sim$5000 km.

Recent studies of equatorially trapped waves indicate that westward-propagating mixed Rossby-gravity waves can exhibit strong vertical coupling as a result of deep convection (Wheeler et al. 2000). A sequence of 200 and 850 hPa streamline analyses from 20 to 22 May (Fig. 7) reveals a westward-moving disturbance at 200 hPa over the far southern SCS during this period and a simultaneous shift in the 850-hPa flow

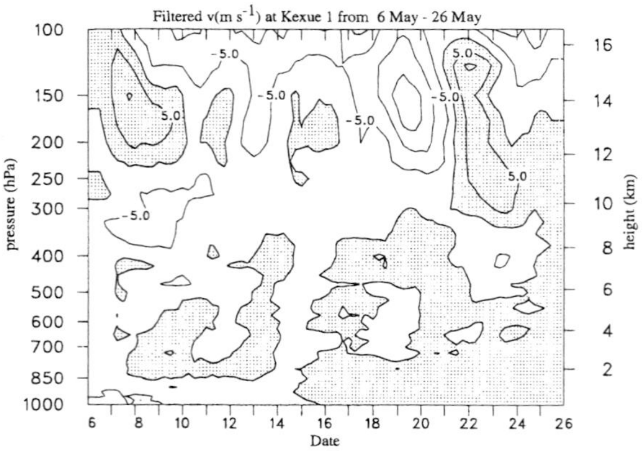

Figure 6: Time series of filtered meridional wind component v (m s^{-1}) at *Kexue #1* from 00 UTC 6 May to 00 UTC 26 May 1998. Contour interval is 5 m s^{-1}.

over the SCS from southerly to westerly. On the 19th (not shown) the 850-hPa flow over the southern SCS had been easterly. The mixed Rossby-gravity wave can best be seen as a closed circulation centered near the equator at 200 hPa passing from east of Borneo to Sumatra over the 3-day period. These flow transitions resemble those described by Wheeler et al. (2000) for vertically coupled mixed Rossby-gravity waves and suggest a possible link between the timing of the monsoon onset in the southern SCS and the occurrence of equatorial waves.

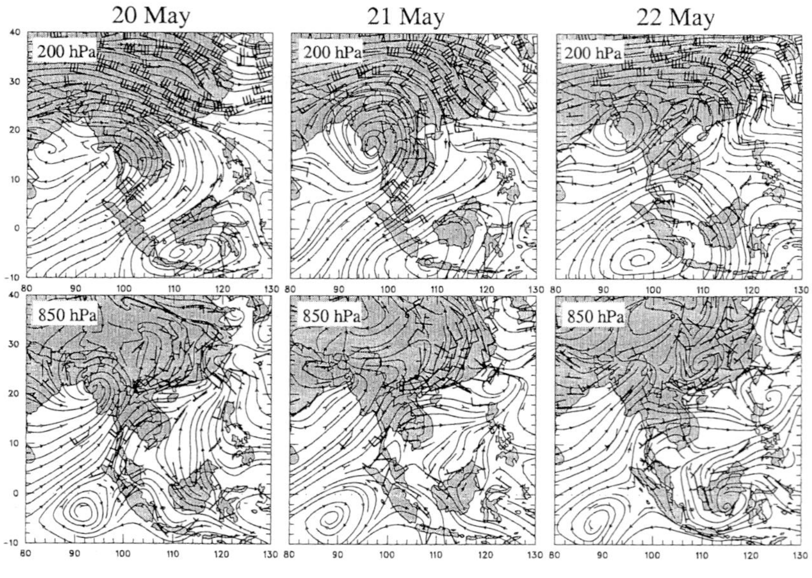

Figure 7: Streamline analyses at 200 hPa on 20–22 May 1998 (top panels) and at 850 hPa (bottom panels). Wind barbs at the sounding sites are indicated.

5. Contrasting the East Asian and Indian monsoon onsets

To examine the contrasting behavior of the onset of the East Asian and South Asian (Indian) monsoons, Hovmöller diagrams of Outgoing Longwave Radiation (OLR) and 850-hPa zonal component of the flow u have been prepared for two longitudinal bands: one over the Bay of Bengal (80–100°E) and one over the South China Sea (110–120°E).

The Hovmöller diagram of OLR and 850-hPa u for the Bay of Bengal (Fig. 8) shows an absence of high cloudiness (inferred deep convection) over the Bay until the first part of May. Beginning in the first week of May, an area of deep convection developed near the equator and moved northward reaching 15°N in mid-May.

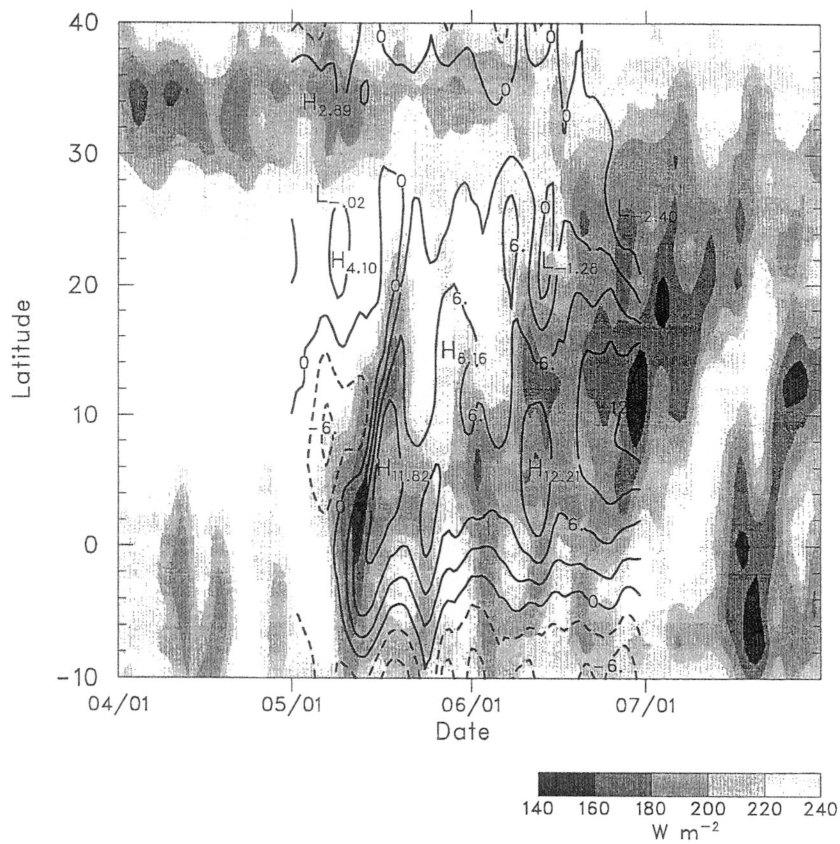

Figure 8: Hovmöller diagram of OLR (shaded; scale at lower right) and zonal wind component u (m s^{-1}; westerlies, solid; easterlies, dashed) at 850 hPa between 80 and 100°E (Bay of Bengal). Winds are shown for SCSMEX May and June 1998 periods.

Prior to the arrival of deep convection in the southern Bay of Bengal, 850-hPa winds were easterly. However, as convection developed and moved northward, westerlies ensued and accompanied the progression of the convection. Thereafter, westerly winds persisted over the Bay of Bengal, thereby signaling the onset of the summer monsoon there.

An important feature of the monsoon onset in this South Asian region is that there is only evidence of *northward propagation* of convection and wind anomalies in association with the onset. There is no evidence of southward propagating signals from the Tibetan Plateau.

A Hovmöller diagram of OLR and 850-hPa u for the longitude band of the South China Sea (110–120°E) is shown in Fig. 9. In early May a gradual northward shift of near-equatorial convection can be seen. In addition, there is an apparent southward progression of convection in mid-May along with a development of westerly flow. The shift from easterly to westerly flow, i.e., the monsoon onset, over the SCS appears to be closely connected with southward-propagating disturbances from midlatitudes. Some evidence of northward-propagating convection and westerlies from the equator can also be seen.

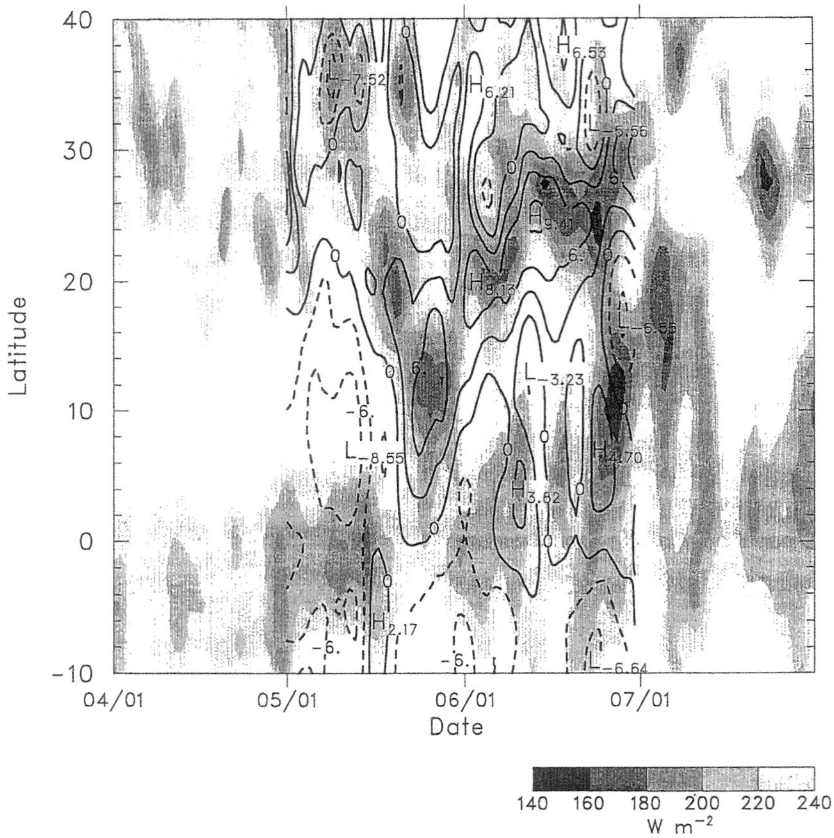

Figure 9: Same as Fig. 8, except between 110 and 120°E (South China Sea).

6. Summary and conclusions

Observations from the SCSMEX/GAME sounding domains have been used to document several features of the onset of the 1998 East Asian summer monsoon. Marked transitions of the low-level and upper-level flow occurred in the vicinity of the South China Sea (SCS) in mid-May as the summer monsoon became established over this region. Over the northern SCS, the onset of westerly monsoon flow occurred around 15 May, whereas over the southern SCS the onset was on 20 May. Deep tropospheric moistening accompanied the monsoon onset in the northern SCS as precipitation began in that region. Changes in the moisture profile in the SCS were more complex, although there appeared to be a transition from a tradewind-like structure (moist trade cumulus layer capped by dry air aloft) to a moistening of the mid-to-upper troposphere accompanying onset.

Analyses of the flow over the southern SCS indicate the passage of a prominent mixed Rossby-gravity wave at the time of monsoon onset. This disturbance had the characteristics of convectively coupled equatorial modes recently identified in the western Pacific.

The SCSMEX/GAME data show a sharp contrast of the onset of the summer monsoon of East Asia with that over the Indian subcontinent (the South Asian monsoon). It is found that the onset of the 1998 South Asian summer monsoon over the Bay of Bengal was associated with the northward progression of deep convection from the equator to Bangladesh in mid-May. No evidence of southward-propagating disturbances was seen. However, over East Asia, the monsoon onset was characterized by both southward-propagating disturbances from midlatitudes and northward-propagating equatorial convection. These findings are consistent with the Yang et al. (1999) labeling of the East Asian monsoon as a "hybrid" monsoon.

Acknowledgements

This research has been supported by the National Oceanic and Atmospheric Administration under Grant NA67RJ0152.

REFERENCES

Lau, K.-M., and S. Yang, 1997: Climatology and interannual variability of the Southeast Asian Summer Monsoon. *Adv. in Atmos. Sci.*, **14**, 141–162.

Lu, E., and J. C. L. Chan, 1999: A unified monsoon index for South China. *J. Climate*, **12**, 2378–2385.

Matsuno, T., 1966: Quasi-geostrophic motions in the equatorial area. *J. Meteor. Soc. Japan*, **44**, 25–43.

Tao, S. Y. and L. X. Chen, 1987: A review of recent research on the east Asian summer monsoon in China. *Monsoon Meteorology III*, C.-P. Chang and T. N. Krishnamurti, eds., Oxford University Press, 60–92.

Ueda, H., and T. Yasunari, 1998: Role of warming over the Tibetan Plateau in early onset of the summer monsoon over the Bay of Bengal and the South China Sea. *J. Meteor. Soc. Japan*, 76 1–12.

Wang, B., and Z. Fan, 1999: Choice of South Asian summer monsoon indices. *Bull. Amer. Meteor. Soc.*, **80**, 629–638.

Webster, P. J., and S. Yang, 1992: Monsoon and ENSO: Selectively interactive systems. *Quart. J. Roy. Meteorol. Soc.*, **118**, 877–926.

Wheeler, M., G. N. Kiladis, and P. J. Webster, 2000: Large-scale dynamical fields associated with convectively coupled equatorial waves. *J. Atmos. Sci.*, **57**, 613–640.

Yang, S., K.-M. Lau, and K. Kim, 1999: Asian monsoons: Variability, predictability, and sensitivity to external forcing. Abstracts, AGU Spring Meeting, 1-4 June, Boston, MA.

RELATIONSHIP BETWEEN THE PLANETARY-SCALE CIRCULATOIN OVER EAST ASIA AND THE INTENSITY OF THE SOUTH ASIAN SUMMER MONSOON

Jianjun Xu and Johnny C.L. Chan*

Department of Physics and Materials Science, City University of Hong Kong, Hong Kong

ABSTRACT

This paper attempts to relate the intensity of the South Asia summer monsoon (SASM) to large-scale circulation anomalies associated with the East Asia summer monsoon (EASM). By partitioning the monthly mean horizontal winds at various tropospheric levels into the vertical mean and vertical shear components, the kinetic energy of the shear flow (K_s) averaged between June and September within the box (40° – 100 °E, 0 – 20 °N) is found to be strongly related to the SASM, and is therefore defined as the south Asian summer monsoon index (SASMI) to represent the intensity of the SASM. A good correlation is found between the SASMI and the other intensity indices of the SASM.

Besides the conditions in south Asia, the value of SASMI is also closely related to the large-scale circulation anomalies associated with the EASM. When the value of the SASMI is large, positive (negative) temperature anomalies are found over the Asian landmass (Indian Ocean). This gives rise to an above-normal north-south land-sea temperature gradient. As a result, low-level southwesterlies extend all the way from the Indian sub-continent to the Philippines, which results in a monsoon trough over south China. The southerlies to the north of this trough converge with the northerlies in the midlatitudes over northern China to form a monsoon frontal trough. For small values of the SASMI, the entire atmospheric pattern apparently shifts southward.

1. INTRODUCTION

Because the characteristics of the Asian summer monsoon (ASM) differ greatly in each local region, many indices have been developed to describe the intensity and inter-annual variability of the monsoon (e.g. Shukla and Paolina 1983; Mooley and Parthasarathy 1984; Lu and Chan 1999; Goswami et al. 1999). Recently, Webster and Yang (1992) defined an index (hereafter the Webster and Yang index, or WYI) based on the vertical shear of the zonal wind over Asian monsoon area, which seems to be a good one to describe the inter-annual variation of the ASM. However, Wang and Zhen (1999) suggested that two or three indices are necessary to describe the monsoon at different locations, which would made it difficult to discuss the inter-annual variability of the entire ASM. It would therefore be beneficial to have an index that may represent more than one monsoon system, which is the main motivation of the present study.

From Gill's (1980) linear theory of thermal forcing, the vertical shear in the monsoon area should have a fundamental relationship with the total heating in the atmospheric column. Based on this hypothesis, Webster and Yang (1992) used the difference between the 850- and 200-hPa zonal winds, which is the simplest approximation of the vertical shear, to indicate the intensity of the South Asian summer monsoon (SASM). However, Xu and Wu (1999) showed from observations that the vertical shear over the Asia region is present in both the zonal and the

* *Corresponding author address*: Prof. Johnny Chan, Dept. of Physics & Materials Science, City University of Hong Kong, 83 Tat Chee Ave., Kowloon, Hong Kong, China. Email: Johnny.Chan@cityu.edu.hk

meridional flows. The entire vertical shear should therefore be represented. A convenient way to do this is to decompose the flow in the troposphere into its vertical mean and the shear flow as proposed by Wiin-Nielsen (1962). Such an approach will be adopted in this study.

Section 2 describes the datasets used and the procedure to calculate the mean and shear kinetic energy. The planetary circulation over Asia is shown in section 3 to possess a significant amount of shear kinetic energy. Based on this result, the South Asian summer monsoon index (SASMI) is defined in section 4, the interannual variability of which is examined in section 5. All the results are then summarized in section 6.

2. Data and procedure

a. Data

The present study utilizes the following datasets from the NCEP/NCAR reanalyses: monthly-mean global zonal and meridional wind components from 1958 to 1997, and monthly outgoing longwave radiation (OLR) fluxes from June 1974 to February 1978 and January 1979 to January 1998. The Indian summer rainfall indices are extracted from the homepage of the Climate Prediction Center (CPC).

b. Kinetic energy calculation

Wiin-Nielsen (1962) proposed to separate the horizontal wind components into their vertical mean (u_m, v_m) and vertical shear (u_s, v_s) (departure from the vertical mean at each level) components as follows:

$$u = u_m + u_s \quad \text{and} \quad v = v_m + v_s$$

where

$$(\;)_m = \frac{1}{p_0 - p_t}\int_{p_t}^{p_0} (\;)\, dp$$

and p_t and p_0 are 100 hPa and 1000 hPa respectively. Using the above equations, the kinetic energy and its components can be written as

$$K = \frac{1}{2g}\int_{p_t}^{p_0} (u^2 + v^2)\, dp$$

where

$$K_m = \frac{1}{2}(u_m^2 + v_m^2)$$

$$K_s = U_s + V_s$$

and $K = K_m + K_s$, K_m the kinetic energy of the vertical mean flow and K_s that of the shear flow. The latter can be divided into two components: zonal shear energy (U_s) and meridional shear energy (V_s). Wiin-Nielson (1962) have referred to K_m and K_s as the barotropic and baroclinic components respectively. In this study, the wind data at 12 levels (1000, 925, 850, 700, 600, 500, 400, 300, 250, 200, 150, 100 hPa) in the troposphere are used to calculate each of the kinetic energy components.

3. Kinetic energy distribution of the planetary circulation over Asia

The seasonal variation of K_s between May and October averaged over the period 1958 - 1997 shows two strong activity centers, one over south Asia and the other over the East Asia – North Pacific (EANP) region (upper six panels in Fig. 1). The temporal evolutions of these two centers are, however, different. The K_s over south Asia increases in June to a maximum in July and August, and then decreases in

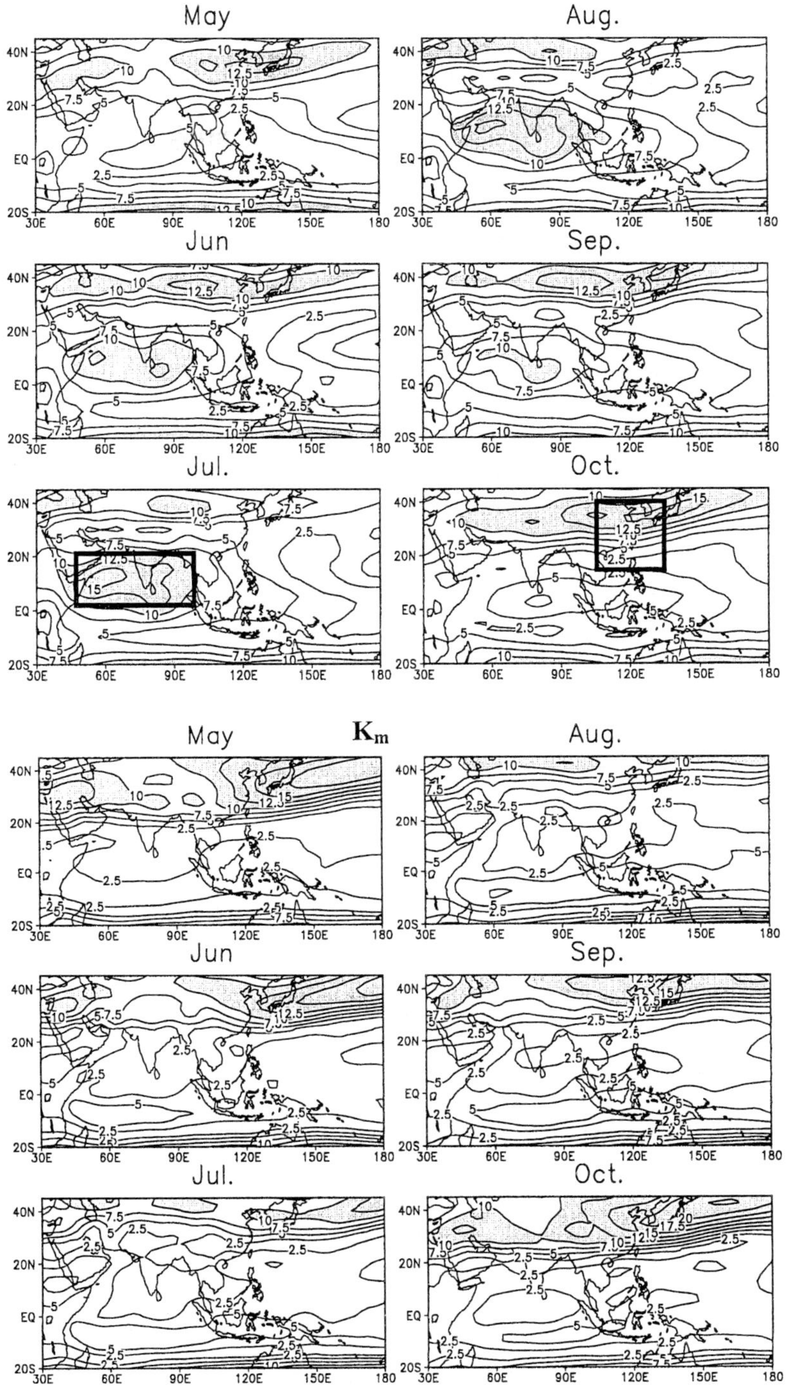

Fig. 1. Kinetic energy of the vertical shear flow (K_s) (upper six panels) and the vertical mean (K_m) (lower six panels) flow in May to October averaged between 1958-1997. The contours indicate the square root of K_s (or K_m). Unit: m s^{-1}, contour interval 2.5 m s^{-1}. Shading indicates values > 10 m s^{-1}. The two thick boxes indicate the areas of South Asia and East Asia.

September. These four months correspond to the period of the SASM. Furthermore, such a seasonal evolution is consistent with Webster's (1987) finding that the establishment of the SASM is a fast process while the retreat is a slow one. The K_s over the EANP, in contrast, is closely related to the seasonal variation of the planetary zonal winds over the Northern Hemisphere (NH), and is the weakest in July and August.

As expected, the kinetic energy of the vertical mean flow (K_m) has only one strong activity region, which is associated with the planetary zonal winds over the NH (the bottom six panels in Fig.1). The patterns are similar to those of K_s over the EANP.

The two activity centers of K_s (highlighted by the two thick-line boxes in Fig. 1) correspond to the two relatively independent monsoon systems of the Asia monsoon (Tao and Chen 1987; Lau et al 2000): the South Asian monsoon and the East Asian monsoon. The mean values of K_s over south Asia are largest during the period of June to September (Fig. 2a), but the K_m values are small throughout most of the year except in winter (Fig. 2c). Over the EANP region, the temporal variations of K_s and K_m are similar with large values in winter and small in summer (Figs. 2b and d).

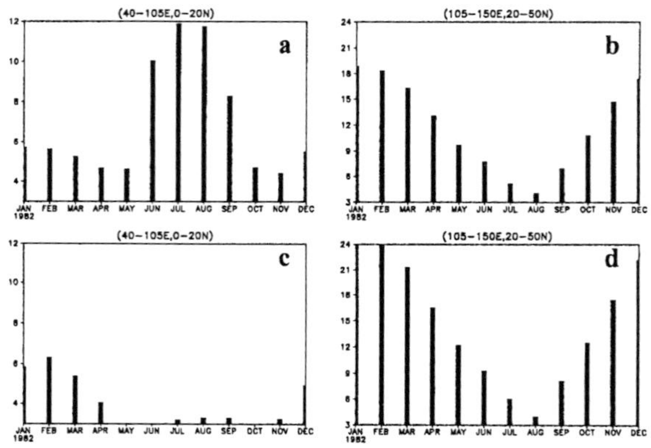

Fig. 2. Monthly mean values of the square root of K_s and K_m averaged over South Asia (40°E – 105°E, 0 – 20°N) [(a) and (c) respectively] and EANP (105°E – 150°E, 20°N – 50°N) [(b) and (d) respectively]. Unit: m s^{-1}.

It is clear that the time variation of K_s or K_m over East Asia only indicates the seasonal variation of the planetary circulation, and is not useful to describe the activity of the East Asian summer monsoon (EASM). In contrast, the kinetic energy over south Asia is dominated by K_s, which has its maximum during the development of the SASM. Therefore the K_s over south Asia should be a good indicator of SASM activity.

4. A South Asia summer monsoon index

The south Asia summer monsoon index (SASMI) can be defined as follows:

$$\text{SASMI} = \int_Q K_s \, dQ$$

where Q indicates the region over south Asia (0 – 20°N, 40°E – 100°E), which corresponds to the left thick box in Fig. 1.

The values of the SASMI are found to be very small from November to April, and those in both of May and October (the transition months) are negative (not shown). This result appears to be consistent with the monthly variation of other indices. To compare the SASMI with these indices, the standardized values of the mean SASMI for the months of June to September are computed.

The SASMI has been standardized for the whole period of January 1958 to December 1997 as follows:

$$(\mathrm{SASMI})_{s,i,j} = \frac{\mathrm{SASM}_{i,j} - \langle \mathrm{SASM} \rangle}{\sigma_{\langle \mathrm{SASM} \rangle}}$$

where $(\mathrm{SASMI})_{s,i,j}$ is the standardized value of SASM for the ith month and j^{th} year,

$$\mathrm{SASM}_{i,j} = \overline{(\mathrm{K}_s)}_{i,j} \qquad \begin{matrix} i=1, 2, 3, \dots, 12 \\ j=1958, \dots, 1997, \end{matrix}$$

where $\overline{(\ )}$ means an average over the region Q, and

$$\langle \mathrm{SASM} \rangle = \frac{\sum_{i,j} \mathrm{SASM}_{i,j}}{\mathrm{N}}$$

and N = total number of months between 1958 and 1997, and σ is the standard variance of $SASM_{i,j}$.

Table 1. Correlation coefficients among the various monsoon indices: ASMI and WYI, MHI, AIRI, CI1, CI2 based on the mean values between June and September for 1958 –1997 except for the CI1 and CI2 that cover the period of 1974-1997 (without 1978). WYI—the zonal wind difference between 850 and 200 hPa over the region (0 – 20°N, 40 – 110°E) (Webster and Yang, 1992). MHI— the meridional wind difference between 850 and 200 hPa over the region (10 – 30°N, 70 – 110°E) (Goswami et al. 1999). The AIRI—total rainfall between June and September over the Indian subcontinent (Mooley and Parthasarathy 1984). CI1 and CI2—OLR anomaly over the Indian monsoon area (8.75 – 23.75°N, 70 – 100°E) and the southeast Asian monsoon area (8.75 – 21.75°N, 115 – 140°E) respectively (Wang and Zhen, 1999). Numbers in italics indicate that the coefficients are not significant at the 0.01% level.

	ASMI	WYI	MHI	AIRI	CI1	CI2
ASMI	1.00	0.95	0.51	0.53	-0.52	-0.44
WYI	0.95	1.00	0.41	*0.34*	-0.53	-0.42
MHI	0.51	0.41	1.00	0.58	-0.77	*-0.01*
AIRI	0.53	*0.34*	0.58	1.00	-0.71	*-0.07*
CI1	-0.52	-0.53	-0.77	-0.71	1.00	*0.25*
CI2	-0.44	-0.42	*-0.01*	*-0.07*	*0.25*	1.00

The correlations between the SASMI and several other monsoon indices averaged between June and September are quite high in many cases (Table 1). In particular, as expected, the correlation with WYI reaches 0.95[1]. Note, however, that

[1] With the degrees of freedom being 40, the limiting value is 0.393 at the 0.01% significance level.

compared with the WYI, the SASMI has a higher correlation with MHI[2] and AIRI[3]. This result suggests that the SASMI is more effective in representing the interannual variation of the south Asia summer rainfall (on which AIRI is based), and in depicting the SASM circulation (which is related to the MHI). The reason is that although the distribution of U_s for the period of June to September is similar to K_s (Fig. 3), the two maxima of K_s over the western part of the Arabian Sea and to the southeast of Sri Lanka almost coincide with the two areas of maximum V_s. The V_s maximum to the southeast of the Arabian Peninsula (~ 16 $m^2 s^{-2}$) accounts for nearly 20% of K_s, and is contributed by the Somali jet, which is an important system of the SASM (Krishnamurti 1976). Webster et al (1998) also pointed out that the meridional flow in the western part of the Indian Ocean has variations similar to those in the Asia sector westerlies. Associated with the acceleration of the westerlies are pulses in the southerly cross-equatorial flow. The meridional wind shear is therefore not negligible in the SASM.

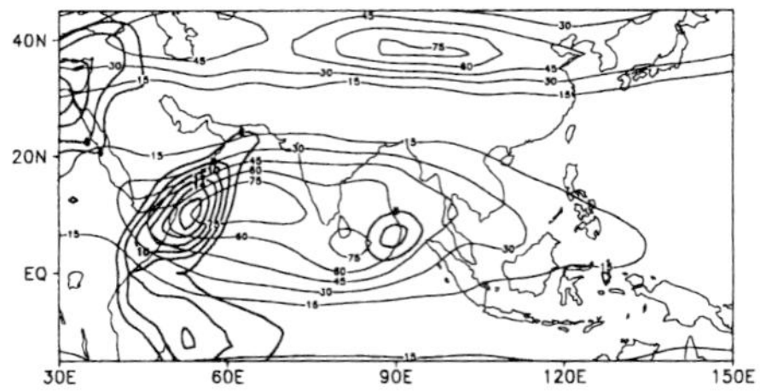

Fig. 3. Kinetic energy of the zonal shear flow (U_s) (solid, only contours >15 $m^2 s^{-2}$ plotted; interval 15 $m^2 s^{-2}$) and of the meridional shear flow (V_s) (heavy solid, only contours > 6 $m^2 s^{-2}$; interval 2 $m^2 s^{-2}$)

5. Circulation anomalies

Based on the standardized value of the mean SASMI between June and September, and using the threshold of 1.0, anomalous SASM events for the 1958-1997 period can be defined:

- strong (S) (SASMI > 1.0) : 1958, 1960, 1961, 1964, 1970, 1981, 1985
- weak (W) (SASMI < -1.0): 1972, 1974, 1976, 1979, 1987, 1996, 1997

The circulation features in these two groups of monsoon years are then studied through compositing. In the strong years, the tropical area west (east) of ~130°E between June and September is dominated by westerly (easterly) anomalies at 850 hPa (Fig. 4a). As a result, a strong-than-normal monsoon trough is found around the Philippines, southern China and extends into the Indian subcontinent. In addition, an anomalous cyclone exists over northern China. In the weak years, the anomalous circulation over the tropical monsoon area is almost opposite to that in strong years (Fig. 4b). The monsoon trough over the Indian subcontinent is replaced by an anticyclone and anomalous cyclonic flow now appears over the tropical ocean. A monsoon trough is still present over East Asia, extending from southern to central China in the southwest-northeast direction, and northern China is dominated by an anomalous anticyclone.

---

[2] MHI— the meridional wind difference between 850 and 200 hPa over the region (10 – 30°N, 70 – 110°E) (Goswami et al. 1999).

[3] AIRI—total rainfall between June and September over the Indian subcontinent (Mooley and Parthasarathy 1984).

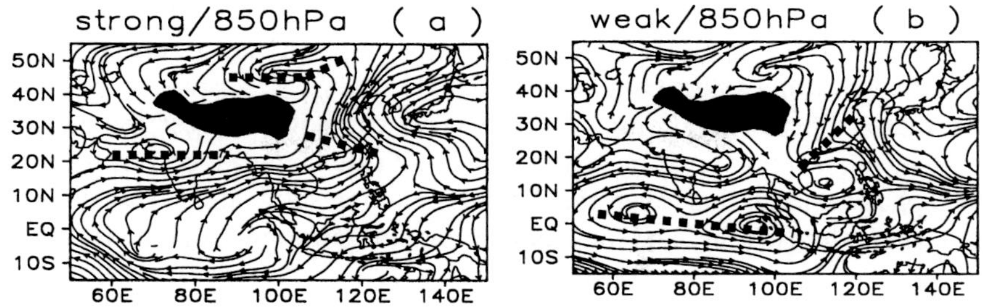

Fig. 4. Composite 850-hPa wind anomalies averaged between June and September in the strong (SASMI > 1.0) (a) and the weak (SASMI < -1.0) (b) years. The thick dashed lines indicate the position of troughs. The shading indicates the Tibetan Plateau.

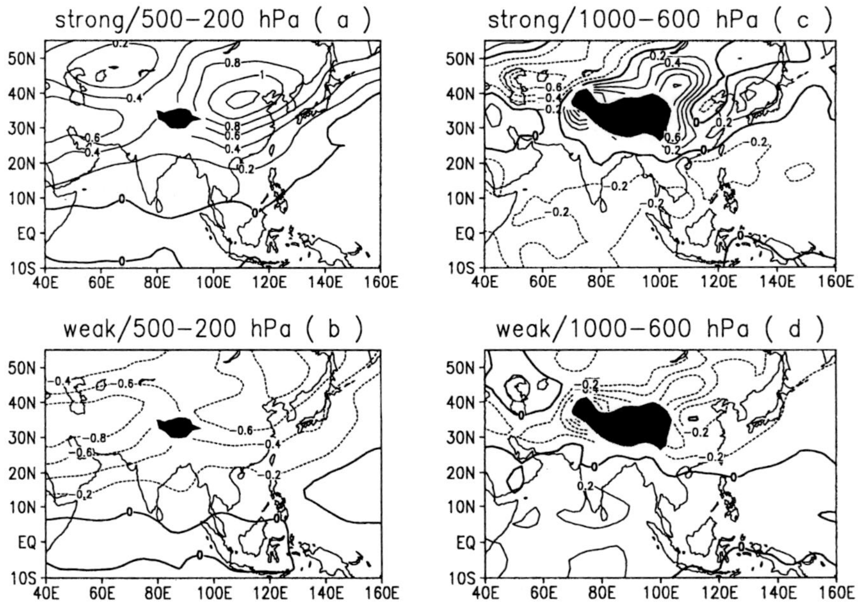

Fig. 5. Composite temperature anomalies averaged between June and September in the upper troposphere (500 – 200 hPa) in (a) strong and (b) weak and the lower troposphere (1000 – 600 hPa) in (c) strong and (d) weak ASM years. Unit: °C. The shading indicates the Tibetan Plateau. Solid-positive and dashed-negative.

In the upper tropospheric layer (500-200 hPa), the strong positive temperature anomalies dominate over the Asian landmass in the strong monsoon years (Fig. 5a). Beneath the upper warm center (1000-600 hPa), the temperature is also very high (Fig. 5c) except for a slight contraction of the positive area. In addition, negative anomalies are found over the tropical Indian Ocean. Such a pattern indicates an anomalous positive land-sea temperature gradient (temperature increasing with latitude). This therefore implies a strong vertical zonal shear, which is consistent with the results from Fig. 4. Note that the line of zero temperature anomaly lies along ~ 20°N, which just passes through the Indian subcontinent and coincides with the location of the Asian monsoon trough in Fig. 4a. In contrast, the temperature distributions in both the upper (Fig. 5b) and lower (Fig. 5d) layers are almost the opposite in the weak monsoon years.

To summarize, these results suggest that SASMI can not only describe the south Asian summer monsoon anomalies, but can also indicate some aspects of the large-scale atmospheric circulation over the East Asian monsoon area. The planetary circulation anomalies within and outside the Asian monsoon region are almost opposite in the strong and weak monsoon years.

6. Summary and discussion

a. Summary

Using the method of Wiin Nielsen (1962), the vertical mean and shear components of kinetic energy over Asia have been examined in this study. It is found that the seasonal variation of kinetic energy of the shear flow (K_s) over the south Asia is concurrent with the occurrence of the South Asia summer monsoon (SASM). The strongest signal of K_s is over the Asian monsoon area during the summer period. Based on this result, the K_s over the box (40° - 110°E, 0 - 20°N) is defined as the South Asia summer monsoon index (SASMI), which is used to measure the intensity of the south Asian summer monsoon. The SASMI is found to have a good correlation with other intensity indices of the SASM, especially with that proposed by Webster and Yang (1992). Furthermore, SASMI includes more of the fundamental dynamics associated with the SASM and are more consistent with observations.

The SASMI not only can describe the interannual variation of the SASM, but is also related to the of East Asian summer monsoon. The circulation anomalies are almost opposite between the strong and weak SASM. When the value of the SASMI is large, a strong positive temperature anomaly dominates the Asian landmass to give rise to a positive north-south temperature gradient. A stronger-than-normal monsoon trough forms over the Indian continent and extends eastward to the Philippines. An anomalous cyclonic flow can also be identified over northern China at the low level associated with anomalous southerlies over the midlatitudes of East Asia. The situation is almost the reverse for small values of the SASMI.

b. Discussion

Despite the fact that the SASM and EASM have their own characteristics (Lau et al 2000), the results of this study suggest that the SASM and EASM are, to a certain extract, inter-dependent. Their intensities are apparently related to the same land-sea contrast on a planetary-scale. Changes in the heating/cooling of the land and/or ocean would modify the meridional temperature gradient, which determines the strength of the southwesterlies associated with the SASM. Whether these southwesterlies can extend to the East Asia monsoon region depends on the same heating/cooling distribution. Thus, the SASMI proposed here, while primarily defined to represent the intensity of the SASM, can also be used to describe the some aspects of circulation anomalies of the EASM. This means that it may be possible to relate the SASMI to the rainfall distributions associated with the EASM, which will be the subject of a future study.

Acknowledgements

The authors would like to thank NCEP for providing the reanalysis data. This research is sponsored by the City University of Hong Kong Grant 9360017 and 7100092. The work of the first author is also partly sponsored by the National Natural Foundation of China Grant 49705062.

References

Barnett, T. P., 1983: Interaction of the monsoon and Pacific trade wind systems at interannual time scales. Part I: The equatorial zone. *Mon. Wea. Rev.*, **111**, 756-773.

Gill, A. E., 1980: Some simple solutions for heat-induced tropical circulation. *Quart. J. Royal Meteor. Soc.*, **106**, 447-462.

Goswami, B. N., V. Krishnamurthy, and H. Annamalai, 1999: A broad scale circulation index for the interannual variability of the Indian summer monsoon. *Quart. J. Royal Meteor. Soc.*, **125**, 611-633.

Huang, R. H., and W. J. Li, 1987: Influence of the heat source anomaly over the western Pacific on the subtropical high over East Asian, *Proceedings International Conference on General Circulation of East Asia, April 10-15, 1987, Chengdu, China.*

Krishnamurti, T. N., and T. Bhalme, 1976: Oscillations of a monsoon system, I, Observational aspects. *J. Atmos. Sci.*, **33**, 1937-1944.

Lau, K.-M., K.-M. Kim and S. Yang, 2000: Dynamical and boundary forcing characteristics of regional components of Asian summer monsoon. *J. Climate* (In press).

Lu, E. and J. C. L. Chan, 1999: A unified monsoon index for South China, *J. Climate,* **12**, 2375-2385.

Mooley, D. A. and B. Parthasarathy, 1984: Flutuations in All-Indian summer monsoon rainfall during 1871 – 1978. *Clim. Change*, **6**, 287-301.

Nitta, T., 1987: Convective activities in the tropical western Pacific and their impact on the Northern hemisphere summer circulation. *J. Meteor. Soc. Japan*, **65**, 373-390.

Reynolds, R. W., and T. M. Smith, 1994: Improved global sea surface temperature analyses using optimum interpolation. *J. Climate*, **7**, 929-948.

Shukla, J, and D. A. Paolina, 1983: The southern oscillation and longrange forecasting of the summer monsoon rainfall over Indian, *Mon. Wea. Rev.*, **111**, 1830-1837.

Tao, S. Y., and L. X. Chen, 1987, A review of recent research on the East Asian summer monsoon in China, *Monsoon Meteorology*, Edited by C. P. Chang and T. N. Krishnamurti (Eds), Oxford University Press, 60-92.

Wang, B. and F. Zhen, 1999: Choice of South Asian summer monsoon indices. *Bull. Amer. Meteor. Soc.*, **80**, 629-638.

Webster, P. J., 1987: The elementary monsoon, *Monsoons*, J. S. Fein and P. L. Stephens (Eds), John Wiley, New York, 3-32.

Webster, P. J., and S. Yang, 1992: Monsoon and ENSO: Selectively interactive system. *Quart. J. Royal Meteor. Soc.*, **118**, 877-926.

Webster, P. J., V.O. Magana, T. N. Palmer, J. Shukla, R. A. Tomas, M. Yanai, and T. Yasunari, 1998: Monsoons: processes, predictability, and the prospects for prediction. *J. Geophy. Res.*, **103**, C7, 14451-14510.

Wiin-Nielsen A., 1962: On transformation of kinetic energy between the vertical shear flow and the vertical mean flow in the atmosphere. *Mon. Wea. Rev.*, **90**, 311-323.

Xie, P. A., and P. A. Arkin, 1996: Analyses of global monthly precipitation using gauge observations, satellite estimates and numerical model prediction, *J. Climate*, **9**, 804-858.

Xu, J. and G. X. Wu, 1999: Dynamics of Asian summer monsoon and its maintenance mechanism. *Adv. Atmos. Sci.*, **16**, 523-536.

Yasunari, T., 1990: Impact of Indian monsoon on the coupled atmosphere/ocean system in the tropical Pacific. *Meteor. Atmos. Phys.*, **44**, 29-41.

WIND CHANGES OVER THE WESTERN PACIFIC

W. Timothy Liu, Xiaosu Xie, Wenqing Tang, and Son V. Nghiem
Jet Propulsion Laboratory, California Institute of Technology, Pasadena, CA 91109

ABSTRACT

The impacts of space-based observations of ocean surface wind vectors in our understanding of the environmental and climatic changes in East Asia and the Western Pacific are demonstrated. The daily global coverage and the high spatial resolution of the data open up new studies in typhoons, monsoons, and their subsequent flooding over land. Information on data access and future satellite missions is also discussed.

1. INTRODUCTION

The history of China has been swayed by winds over the Western Pacific. Winds carried the Eunuch Admiral, Zheng He, and his fleet to the Indian Ocean and Africa in the 15th century, dazzling the maritime countries with the riches and power of the Ming Dynasty. A century later, European traders, following the Portuguese Admiral Vasco da Gama, rode the winds and gave the Ching Empire the ultimate humiliation with extraterritoriality and colonies. Through millennia, steady winds have brought rain to nourish the agrarian economy of the Yangtze and Pearl deltas, but the vagaries of wind and rain have also devastated livelihoods and governments. The consequences of ocean winds may be well observed over land, but their breeding grounds over the oceans are not adequately monitored. Observations of marine winds, with adequate spatial and temporal scales to monitor and predict their environmental and climatic effects, can be achieved only from the vantage point of space. A few examples to demonstrate the impact of the most recent space-based wind observations are given in this brief report.

2. GLOBAL WIND

Fig. 1 is based on preliminary data derived from the scatterometer, SeaWinds, on the QuikSCAT mission of the National Aeronautics and Space Administration (NASA) [Graf et al., 1998]. It was launched in June 1999 to replace the NASA Scatterometer (NSCAT). A scatterometer sends microwave pulses to the surface of the Earth and measures the backscatter power from the surface roughness. The roughness may describe characteristics of polar ice or vegetation over land. Over the ocean, which covers more than three-quarters of the Earth's surface, the backscatter is largely due to small (centimetre) waves on the surface. The idea of remote sensing of ocean surface winds was based on the belief that these surface ripples are in equilibrium with the local wind stress. The backscatter power depends not only on the magnitude of the wind stress but also on the wind direction relative to the direction of the radar beam. The capability of measuring both wind speed and direction is the major uniqueness of the scatterometer.

In Fig. 1, the strong (violet color) trade winds blow steadily from the cooler subtropical oceans to the warm water of the Intertropical Convergence Zone (ITCZ) located just north of the equator. The air rises over the warm water of the ITCZ and sinks in the subtropics at the horse latitudes, forming the Hadley Circulation. Both the convergence area at the ITCZ and the divergence area at the horse latitudes are indicated by low wind speed of blue color. In the midlatitudes, the strong baroclincity caused by temperature gradient generates storms (yellow spirals), particularly in the southern oceans, which are in winter. The eastern North Pacific is dominated by a persistent high-pressure system, whose anticyclonic (clockwise) flow creates strong winds blowing parallel to the coast of Canada and the United States. This creates strong ocean upwelling and a few degrees of cold temperature anomalies along the coasts of the US. Typhoon Paul is observed south of Japan, and tropical depression Rachael is forming southeast of Taiwan. Tropical depression Eugene is visible as close circulation in the eastern Pacific off Central America.

The ocean is largely covered by clouds; the clouds prevent observations of the sea surface by instruments measuring at infrared and visible frequencies. Because the atmosphere is largely transparent to microwaves, SeaWinds was able to cover 93% of the global oceans, under both clear and cloudy conditions, in a single day. Objective interpolation is used to produced the uniformly gridded map [Tang and Liu, 1996]. Weather satellites, despite their frequent coverage, can view only part of the world's oceans. SeaWinds, on the other hand, provides in Fig. 1 a synoptic view of tropical cyclones in both the eastern and western Pacific. Spaceborne scatterometers have the capability of synoptic global coverage and high spatial information [see also Liu et al., 1998]. Atlas et al. [1999] showed NSCAT winds improve significantly the forecasts and analyses of global numerical weather prediction models. Chang et al. [1999] took advantage of the high spatial information of NSCAT winds to improve regional precipitation forecasts in Taiwan. The normalized backscatter coefficients over land are related to land vegetation, and those over Antarctica provide information of sea-ice extent and morphology [Long and Drinkwater, 1999].

3. Typhoon

The scatterometer was not designed to study tropical cyclones. The relation between radar backscatter and wind vector under strong wind and high precipitation conditions is not well established. Because of the strong economic and environmental impact, however, scatterometer data have been applied in studying and monitoring tropical cyclones since the time of Seasat [e.g., Hawkins and Black, 1983]. Hsu and Liu [1996] used winds from the scatterometer on the European Remote Sensing (ERS) satellite to derive pressure field, and thus to improve the estimation of location and intensity of tropical cyclones. NSCAT winds were also used by Liu et al. [1997] to study the extratropical transition of a typhoon.

The scatterometers on ERS satellites scan a 480-km swath on one side of the spacecraft and is able to cover only 40% of the global ocean daily; they may miss a large portion of the tropical cyclones. NSCAT scanned two 600-km swaths, one on each side of the spacecraft, covering 77% of the global oceans daily. NSCAT provides significant improvement over ERS scatterometers, both in coverage and spatial resolution (25 km for NSCAT versus 50 km for ERS), but the wide nadir gap between two swaths may still prevent the full coverage of a typhoon. Fig. 2 shows that Typhoon Olga is fully covered by the 1800-km continuous swath of SeaWinds. Over the ocean, in situ observations in a tropical cyclone are extremely sparse, and conventional satellite data provide only cloud images at the top of the storm. Katsaros et al. [2000] demonstrated that scatterometer winds could detect close circulation associated with tropical storms well ahead of the detection by the cloud images of weather satellite used by the National Hurricane center.

Tropical cyclones are devastating when accompanied by strong winds and heavy precipitation. Coincident measurements of wind and precipitation are important in understanding the structure of the storm and to predict its path. The 25-km resolution of SeaWinds data provides good details of surface winds, which feed moisture to fuel the severe storm. QuikSCAT and the Tropical Rain Measuring Mission (TRMM) provide the only opportunity to observe both the dynamic and hydrologic parameters before landfall. The coincident measurements of surface winds and precipitation reveal the interplay between dynamics and the hydrologic balance of the storm (Fig. 2). Liu et al. [2000] change drastically the diabatic heating profiles and surface fresh water flux in Hurricane Floyd computed from three-dimensional output of a high resolution numerical weather prediction model (Eta), by simply replacing the wind divergence at the first level with the divergence computed from SeaWinds data. Comparison with the precipitation profiles measured by the precipitation radar on TRMM show that SeaWinds data improve significantly the computation of hydrologic parameters in Hurricane Floyd. The SeaWinds data they used were specially produced for the strong wind and high precipitation conditions in a tropical cyclone and have a spatial resolution of 12.5 km.

4. Flood

Scatterometers are capable of monitoring not only the ocean winds, which feed moisture towards land, but the consequent flooding of the land. Both scattering and reflection from the surface contribute to the backscatter energy received by the scatterometer. Over dry land, scattering dominates and the horizontal polarization return (H) is smaller than the vertical polarization return (V). When the land gets flooded, however, the reflectivity of the surface increases greatly, and H becomes larger than V. The difference between reflectivities in H and V increases with incident angle, up to 80°. NSCAT measured H and V at incident angles varying from 18° to 59°; it is difficult to collect H and V pairs at fixed incident angles. SeaWinds measures at fixed incident angles, H at 46° and V at 54°, with a large difference in reflectivity. The SeaWinds observations, therefore, are more conducive to monitor flooding than NSCAT. The continuous wide swath of QuikSCAT also is superior to NSCAT in covering flooding extent.

The ratio of the vertical to the horizontal backscatter coefficients over flat terrain is a good indication of flooding extent, and it is mapped in Fig. 3. QuikSCAT observations coincide with extensive floods in Asia caused by strong monsoons and tropical cyclones. The maps in Fig. 3 show that the flooding in the Yangtze Valley of China (30°N), represented by blue patches, is clearly visible in July, following strong summer monsoons. The flooding observed in Anhui, Zhejiang, and Jiangsu provinces agrees with reports by the International Federation of Red Cross. The flooding recedes in September, with the retreat of the summer monsoon and the advance of the winter monsoon. The intensity of the flooding increases again in October, after the landfall of a typhoon. The flood in Bangladesh and east India did not start until August but continues to strengthen through October. The most severe floods in Vietnam, Thailand, and Burma occurred in August and September, at the peak of summer monsoons. By the end of the year, all the flooding disappears.

5. Monsoon and El Niño

Monsoons are the seasonal change of wind forced by continent-ocean temperature contrast. Liu and Xie [1999] used 11 years of latent heat flux (LHF) and 5 years of wind stress curl (WSC) to examine the changes of surface forcing following the seasonal changes of monsoons, in the South China Sea (SCS). Data of the Special Sensor Microwave Imager (SSMI) were used to estimate LHF [Liu, et al., 1994] and the winds from ERS scatterometers were used to compute WSC. They also examined the oceanic responses in the forms of sea level changes (SLC) observed by the altimeter on Topex/Poseidon and sea surface temperature (SST) observed by the Advanced Very High Resolution Radiometer. In this study, a longer time-series of ocean surface winds derived by combining the wind speeds measured by SSMI and all available wind direction data through variational analysis [Atlas et al., 1996; Liu et al., 1996] is used to cover major El

Niño Southern Oscillation (ENSO) events. Fig. 4 shows the preliminary results on how the monsoon driven air-sea interaction in the SCS is related to the interannual ENSO events.

The winter composites of anomalous cold events (La Niña) in Fig. 4 show that cold temperature anomalies prevail in the equatorial central to eastern Pacific and warm temperature anomalies occur in the tropical western Pacific, extending to high latitudes in the central Pacific. Associated with the temperature pattern, anomalous easterly winds dominate the equatorial central Pacific and an anomalous cyclonic gyre centered in the Philippines. The stronger than normal winter monsoon from the northeast in SCS associated with the atmospheric cyclonic gyre causes positive WSC anomalies in the central part of the marginal seas (but negative at the Chinese coast), similar to the demonstrations by Liu and Xie [1999]. The positive CWS anomalies are collocated with negative SST and SLC anomalies shown in Fig. 4 (upper and center panels) and is in consistent with a simple Ekman pumping scenario in the semi-enclosed SCS.

The anomalous cooling is not caused by increase in LHF, which is lower than normal because of the increase in moisture advected from the warm western Pacific and cooler local SST. The oceanic responses to dynamic and thermal forcing in SCS on monthly time-scale are discussed and demonstrated in more details by Liu and Xie [1999]. The anomalous cooling in the marginal sea may represent a negative feedback to anomalous warming in the tropical western Pacific, but the effects of local anomalies in the marginal seas on the larger scales anomalies in the western Pacific have not been clearly revealed. A reversed scenario is found during El Nino, and further results will be reported in the future.

6. Data

SeaWinds data in standard formats for scientific application are distributed by the Physical Oceanography Distributed Active Archive Center at the Jet Propulsion Laboratory in a delayed mode (weeks after observations). The National Oceanic and Atmospheric Administration is distributing near-real time (a few hours after observations) data to operational weather forecast agencies. A data system, Seaflux, has been established, the initial phases of which are to produce, display, and disseminate, 12 hourly wind maps (uniformly gridded data by objective analysis), within one day after observations. The surface winds are also combined with hydrologic parameters from the TRMM to monitor evolving weather systems and natural hazard in real time. Ocean–atmosphere fluxes of momentum, heat, and water will be produced at later stages. Seaflux is at website http://airsea-www.jpl.nasa.gov/seaflux.

7. Future

QuikSCAT was launched to reduce the data gap caused by the unexpected early demise of the Japanese spacecraft Midori (ADEOS-1). As originally planned, NSCAT was launched on ADEOS-1 in 1996, to be followed by two scatterometers, on ADEOS-2 and ADEOS-3, to provide more than a decade of ocean surface wind vectors. At present, another SeaWinds is planned on ADEOS-2, scheduled for launch in 2001. QuikSCAT will fly in tandem with ADEOS-2; they will be in complementary orbits to optimize the coverage by the scatterometers. The two SeaWinds are expected to demonstrate the importance of high wavenumber and high frequency forcing of the ocean. NASA is planning to supply an improved scatterometer to the Global Change Observation Mission (ADEOS-3), paving the way for an operational scatterometer system. The scatterometer will have strong scientific synergism with the ocean dynamic topography measured by Topex/Poseidon and JASON; with hydrologic parameters from TRMM and Advanced Microwave Scanning Radiometer (AMSR) on ADEOS-2 and Aqua; and with ocean color by the Sea-Viewing Wide-Field-of-View Sensor (SeaWifs), the Moderate-Resolution Imaging Spectroradiometer (MODIS) on EOS, and Global Imager (GLI) on ADEOS-2.

The Ming emperor might have missed the opportunity to sustain the strength provided by superior naval technology. But, Chinese scientists of the 21st century should not let the wind of space technology pass them by. Vanguard technology feeds scientific innovation and leadership. If Chinese scientists will venture beyond well established methodologies and operational data to use a new generation of spacebased research sensors, they will put themselves not only in the forefront of environmental and climate research, but in a position to provide direction and thrust for the budding space program in China.

Acknowledgments

This study was performed at the Jet Propulsion Laboratory, California Institute of Technology, under contract with the National Aeronautics and Space Administration (NASA). It was jointly supported by the SeaWinds Project, TRMM, and the Physical Oceanography Programs of NASA, and the Joint NOAA-NASA Program of Enhanced Data for Climate Changes. David Long generously made available spatially enhanced backscatter data over land and ice.

References

Atlas, R.S., R.N. Hoffman, S.C. Bloom, J.C. Jusem and J. Ardizzone, 1996: A multiyear global surface wind velocity dataset using SSM/I wind observations. Bull. Amer. Meteor. Soc., 77, 869-882.

Atlas, R., S.C. Bloom, R.N. Hoffman, E. Brin, J. Ardizzone, J. Terry, D. Bungato, and J.C. Jusem, 1999: Geophysical validation of NSCAT winds using atmospheric data and analyses. *J. Geophys. Res., 104*, 11,405-11,424.

Chang, C.P., S. C. Lin, C. S. Liou, and W. T. Liu, 1999: An experiment using NSCAT winds in the numerical prediction of tropical mesoscale rainfall systems under the influence of terrain. *Geophys. Res. Lett., 26*, 311-314.

Graf, J., C. Sasaki, C. Winn, W.T. Liu, W. Tsai, M. Freilich, and D. Long, 1998: NASA Scatterometer Experiment. Asta Astronautica, 43, 397-407.

Hawkins, J.D., and P.G. Black, 1983: SEASAT scatterometer detection of gale force wind near tropical cyclones, J. Geophys. Res., 88, 1674-1682.

Hsu, C.S. and W.T. Liu, 1996: Wind and pressure fields near Tropical Cyclone Oliver derived from scatterometer observations. *J. Geophys. Res., 101*, 17,021-17,027.

Katsaros, K.B., E.B. Forde, P. Chang, and W.T. Liu, 2000: QuikSCAT facilitates early identification of tropical depression in 1999 hurricane season. Geophys. Res. Lett., submitted.

Liu, W.T., H. Hu, and S.Yueh, 2000: QuikSCAT and Tropical Rain Measuring Mission reveal the interplay between dynamic and hydrologic parameters in Hurricane Floyd. EOS Trans. of AGU, accepted.

Liu, W.T., W. Tang, and R. Atlas, 1996: Responses of the tropical Pacific to wind forcing as observed by spaceborne sensors and simulated by model. *J. Geophys. Res.*, 101, 16,345-16,359

Liu, W.T., W. Tang, and R. S. Dunbar, 1997: NASA Scatterometer Observes the Extratropical Transition of Pacific Typhoon. *Eos, Trans. of Amer. Geophys. Soc., 78*, 237 & 240.

Liu, W.T., W. Tang, and P.S. Polito, 1998: NASA Scatterometer provides global ocean-surface wind fields with more structures than numerical weather prediction. *Geophys. Res. Lett.,25*, 761-764

Liu, W.T., and X. Xie, 1999: Spacebased observations of the oceanic responses to seasonal changes of south Asian monsoons. *Geophys. Res. Lett.*, 126, 1473-1476

Liu, W.T., A. Zheng, and J. Bishop, 1994: Evaporation and solar irradiance as regulators of the seasonal and interannual variabilities of sea surface temperature. J. Geophys. Res., 99, 12623-12637.

Long, D.G., and M.R. Drinkwater, 1999: Cryosphere Applications of NSCAT data. IEEE Trans Geos. Remote Sensing. 37, 1671-1684.

Tang, W., and W.T. Liu, 1996: *Objective Interpolation of Scatterometer Winds.* JPL Publication 96-19, Jet Propulsion Laboratory, Pasadena, 16 pp.

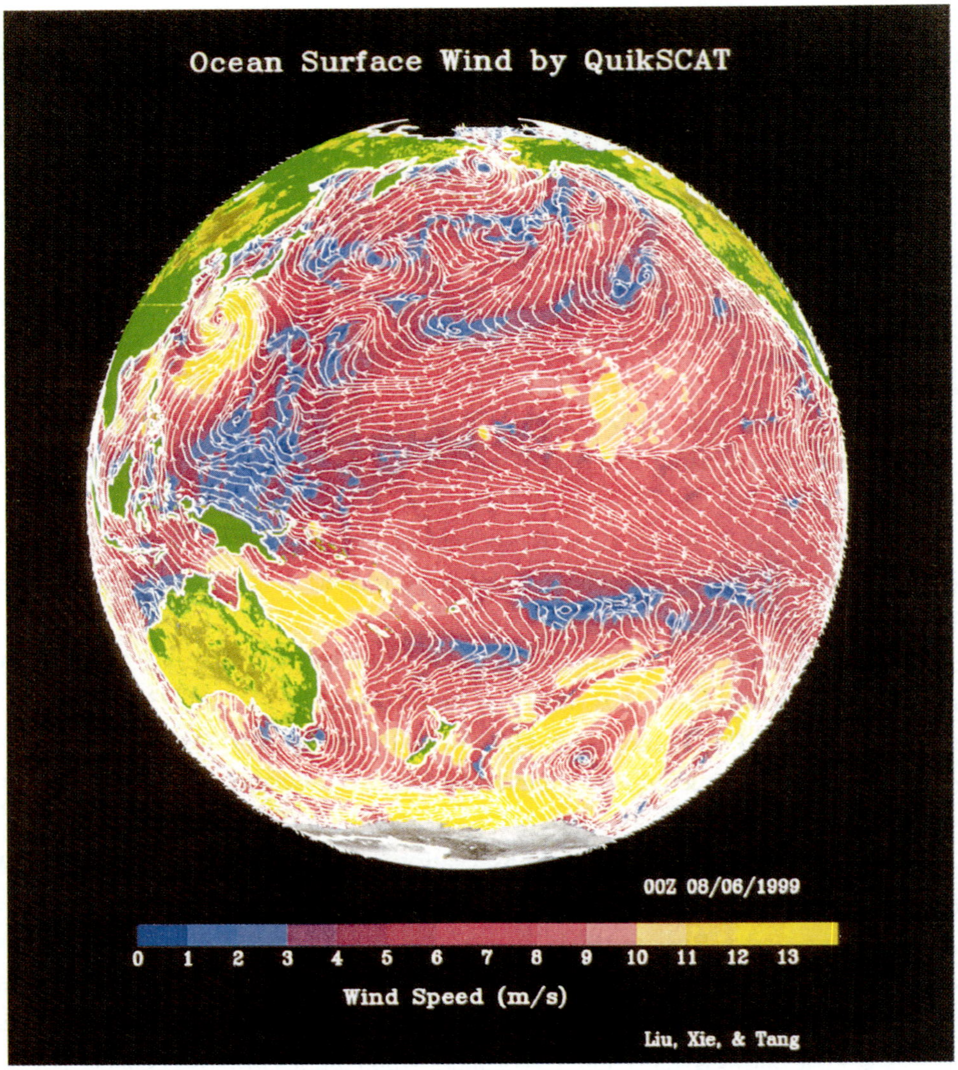

Fig. 1. Over the ocean, white streamlines indicating wind direction are superimposed on the color image of wind speed at 0Z for 6 August 1999, derived from objective interpolation. Typical average backscatter coefficients over land and Antarctica are also added. The data are all based on observations by SeaWinds on QuikSCAT.

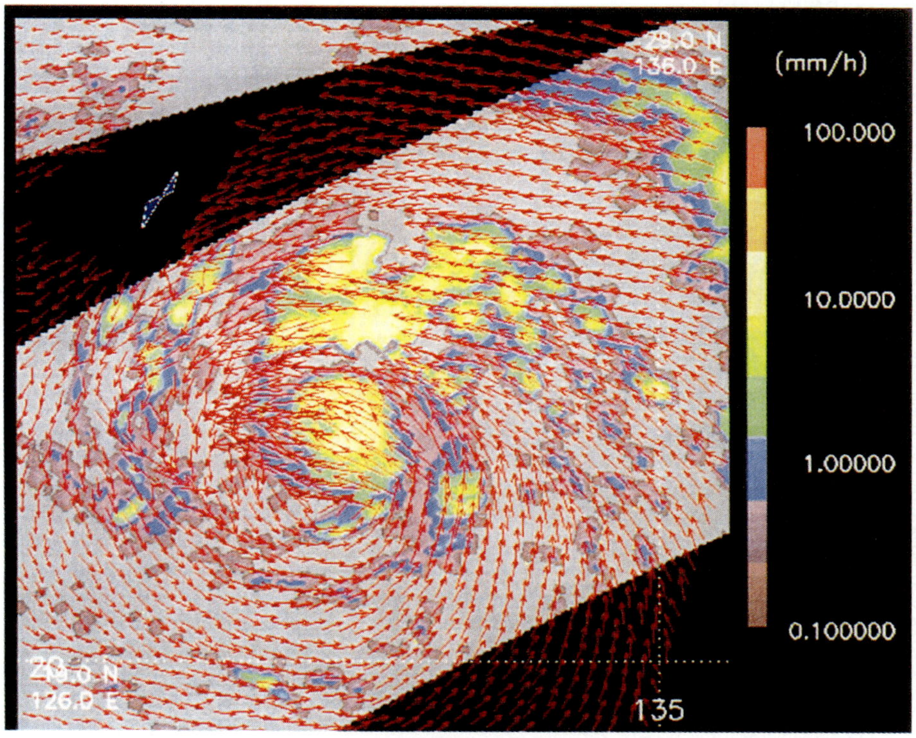

Fig. 2. Typhoon Olga is revealed by wind vectors (red arrows) from SeaWinds and surface precipitation (color image) from TRMM Microwave Imager on 31 July 1999, along the ground-tracks of QuikSCAT and TRMM, which are approximately three hours apart.

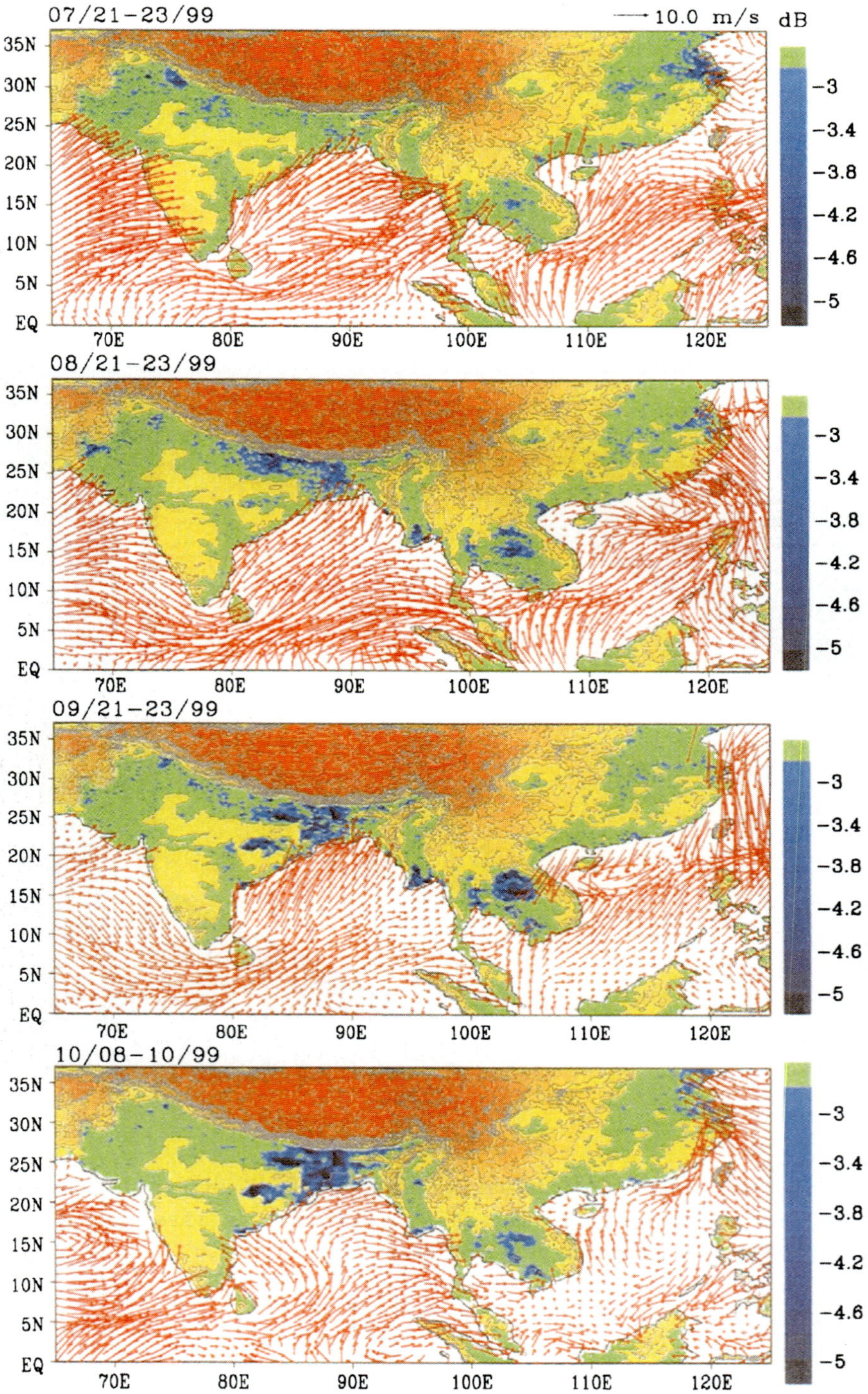

Fig. 3. Three-day average vector winds over ocean and flood indexes over land derived from SeaWinds on QuikSCAT observations, for four months in 1999. Topographic map fro areas over 50-m elevation is plotted over flood indexes.

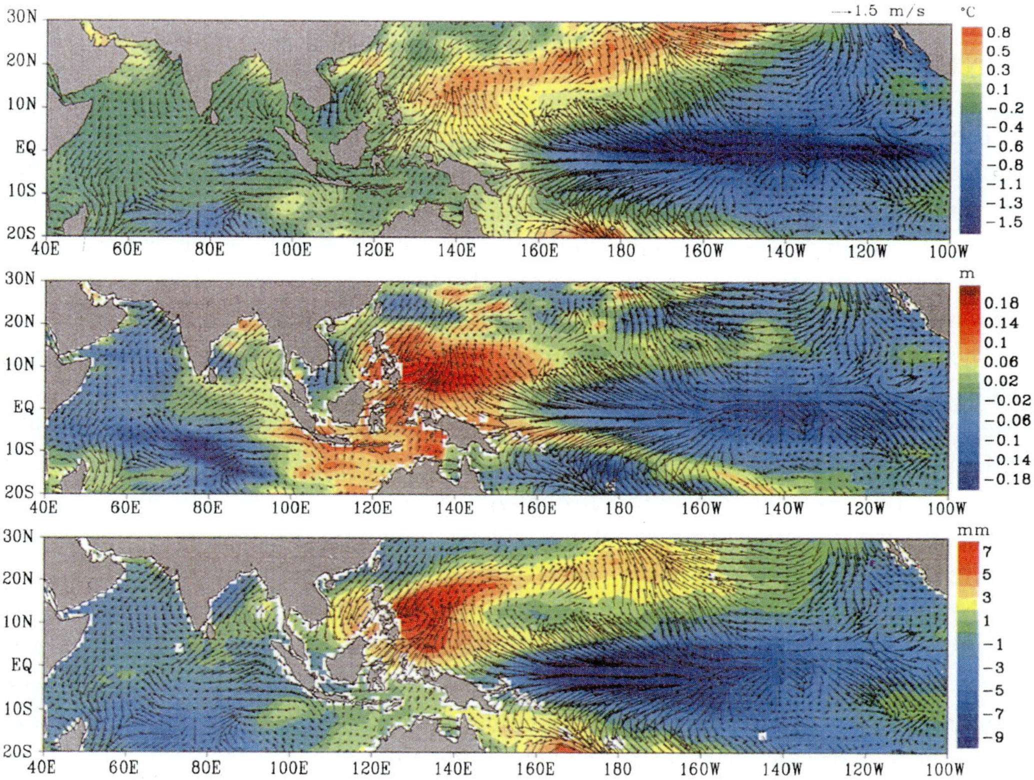

Fig. 4. Winter (December, January, and February) composites of anomalous cold events (La Niña). Dark arrows representing wind anomalies from SSMI are superimposed on color images of the anomalies of sea surface temperatures from A VHRR (upper), sea level changes from Topex/Poseidon (center), and precipitable water from SSMI (lower). Sea level anomalies are based on the 1998/99 event. The 1995/96 and 1988/89 events were added to compile water vapor and wind composites. The 1984/85 event was added to the other three to produce the sea surface temperature composite.

INTERANNUAL VARIABILITY OF THE TIMING OF THE SOUTH CHINA SEA SUMMER MONSOON ONSET AND ITS POSSIBLE MECHANISMS

Mong-Ming Lu[1] and Yun-Lan Chen
Central Weather Bureau, Taipei
Fax: +886 2 23491089
Tel: +886 2 23491320
[1]E-Mail: lu@rdc.cwb.gov.tw

1. INTRODUCTION

The onset of the South China Sea (SCS) summer monsoon is characterized by a sudden shift of the ITCZ from the equatorial region to north of 10° N (Lau et al. 1999). After that, the atmospheric low-level circulation over the SCS changes from anticyclonic to cyclonic. The onset is remarkable in the aspect that it marks the commencement of the summer regime of the East Asian Monsoon (EAM) system. In this study, the EAM system means the portion of the huge Asian Monsoon system in the east of 110°E.

The evolution of the EAM system can be represented by north-south movement of the western Pacific subtropical high (WPSH). The persistent ridge of the WPSH is the foot of the subsidence leg of the ITCZ. Therefore, the seasonal latitudinal movement of the WPSH is primarily controlled by the annual march of the sun. However, we find that in some years the seasonal transition substantially deviated from the normal cycles. To understand such deviation is essential for understanding and predicting the inter-annual variability of the EAM system.

The inter-annual variations of the SCS summer monsoon onset have not been well studied yet. Xie et al. (1998) suggested that the onset timing is sensitive to SST zonal gradient over the SCS. During the ENSO warm years, because the Walker circulation is weaker, the SST in the western Pacific is colder than normal. The SST zonal gradient over the SCS is weaker than normal and the onset timing can be delayed. During the years with warmer western Pacific SST, the onset timing can be earlier. Chang and Chen (1995) suggested that the intrusion of mid-latitude fronts into the SCS can trigger the onset of the SCS summer monsoon. He et al. (1999) showed that the onset of the SCS summer monsoon occurs when the southeastward and northeastward movements of extra-tropical and tropical convection systems merge over the SCS.

The purpose of this study is to identify possible mechanisms responsible for the significant variations of the onset timing of the SCS summer monsoon. The data used includes 42-year (1958-1999) NCEP/NCAR Reanalysis, NCEP SST and 25-year (1974-1999) OLR data sets. In Section 2, we present the contrasts in low-level flow anomalies of the early and late onset years. In Section 3, we propose a hypothesis for possible mechanisms causing the inter-annual variations of SCS monsoon onset. Concluding remarks are given in the last section.

2. EARLY VS. LATE ONSET YEARS

The onset dates of the SCS summer monsoon are determined based on the index VOR850 following three criteria. VOR850 is the average 850 hPa vorticity over the area of 10° N to 20° N, 110° E to 120° E. The three criteria are as follows. First, the onset date must be later than April 1. Second, the onset day must be the first day of continuous 5 days of positive VOR850. Third, within the period of the 45 days after the onset, the days with negative VOR850 cannot be more than 30 days. The onset dates are listed in Table 1. The earliest onset date is April 20, in 1996, and the latest is July 3, in 1959. The dates can be classified into early, normal and late groups according the distance between the dates. Sharp ratios between the sample size of the extreme groups and the normal group are necessary, since only the extreme cases can be well separated from the others. The years belong to the group of early onset are 1966, 1971, 1985, 1996 and 1999, and the late onset are 1959, 1968, 1977, 1992 and 1993.

The contrasts between early and late onset years can be clearly seen in low-level (850 hPa) flows, upper-level jet streams, and deep convection. Here only the low-level flow will be discussed because of the page limit. Fig 1a-h are the composite anomalous flows for the early and late onset years, from March to June. The anomalies are computed with respect to the average winds of the 32 normal onset years. In March (Fig 1a), we see clear anomalous cyclonic circulation over the SCS and the Bay of Bengal. Anomalous anticyclonic circulation appears over the west North Pacific accompanying with the cyclonic circulation over Siberia and the Sea of Okhotsk. The area from Siberia to southeast China is covered by strong northerly anomalies in April (Fig 1b). The anomalous northerlies are part of the northeast-southwest oriented anomalous cyclonic circulation. The area from the Bay of Bengal to the west North Pacific is also under the cyclonic anomalies. South to the cyclonic circulation, we see another cyclonic (clockwise) circulation extending from the Timor Sea to the western Pacific. The anomalous circulation over the SCS remains cyclonic in May and June (Figs 1c and d). In contrast to the early onset years, the SCS is covered by anomalous anticyclonic circulation from April to June in late onset years (Figs 1e-h). Strong southerlies are observed over Siberia and the Sea of Okhotsk in March (Fig 1e). The coastal area from the Yellow Sea to the SCS is covered by anomalous northerlies blowing from the continent to the ocean (Fig 1f). In addition, we see anomalous anticyclonic circulation over the area from the SCS to the western Pacific near 160°E and 20°N. The persistent anomalous anticyclone over the SCS can be clearly identified in both May and June (Figs 1g and h).

3. POSSIBLE MECHANISMS

(a) Surface Forcing

Atmospheric low-level circulation can be directly affected by surface forcing. In this study, we have examined the contrast in surface forcing, circulation anomalies, inter-hemispheric wave interactions and tropical convection. A hypothesis for possible mechanisms causing the extreme variations of SCS summer monsoon onset timing is proposed after the investigation. We first examined the global temperature anomalies at

1000 hPa. The land sea contrast near the area of eastern China, Korea and Japan in the early onset years is found to be sharper than in the late onset years. The contrasts between early and late onset years are particular clear in March. It is suspected that the onset timing may be particularly sensitive to the surface condition in March.

The characteristics of SST in the early and late onset years are investigated with the usage of EOF analysis. Fig 2a shows that the most distinguished area with warm anomalies is the land bridge from Australia to Southeast Asia. The first EOF (Fig 2b) represents the pattern that explains the largest portion (35.9%) of the seasonal variations of SST. Clear meridional structures appear over all oceans. The temporal structure of the first EOF depicted in Fig 2c shows a fundamental period of two years with the phase transition occurring in early summer. Comparing Figs 2a and b, using Fig 2c as a reference, we find the dominant anomalies such as the positive anomalies over the mid-latitude southeast and northeast Pacific and the cold anomalies over the equatorial central Pacific are mainly resulted from the anomalies before May. June is the transitional month for these anomalies. The SST anomalies over the SCS are negative before May and become positive after summer. The SCS SST anomalies are in phase with the anomalies over the Indian Ocean between Sumatra and Madagascar. The zonal gradient in the SCS is quite large. With the positive zonal gradient (value increases eastward, see Fig 2b), anomalous low-level westerlies are expected. However, it should be noted that the low-level westerlies over the SCS is but a small portion of the large-scale anomalous cyclonic circulation. The circulation anomalies are more directly related to the large-scale SST anomalies, from the Philippine Sea to Aleutian Islands, than the local SST anomalies.

For the late onset years, the annually average anomalies and EOFs of the composite SST are shown in Figs 3a and b. The annually average anomalies (Fig 3a) are in sharp contrast to the early onset years, particularly over the equatorial belt. Fig 3a shows warm anomalies over the central Pacific and the western Indian Ocean, but cold anomalies over the eastern Indian Ocean, the southeast Asian maritime continent, the western Pacific and the tropical Atlantic. The first EOF of the composite SST anomalies is shown in Fig 3b. The most remarkable feature is the sharp gradient of the SST over the mid-latitude western Pacific. Fig 3c suggests that the positive gradient (value increases poleward) reaches its largest magnitude in April and the strongest negative gradient should be in July. The contrast between mid-latitude and subtropical SST anomalies over the Pacific is sharp. The anomalies near 40°N-50°N over both the Pacific and the Atlantic are large positive in April and become negative in June. The anomalies around 20°N over the Pacific are negative in April and positive in June. Over the Atlantic, the large meridional SST gradient is also remarkable. The fundamental period of the first mode is about one year.

The associate low-level convergence, upper-level divergence and convection (not shown) are found to vary consistently with the surface forcing. In other words, over the area with warm SST, the expected low-level convergence, upper-level divergence and more active convection are observed.

(b) Inter-hemispheric Wave Interactions

The upper-level flow anomalies are drastically different between the early and late onset years. The difference is particularly striking over the equatorial Pacific. The upper-level flow over the equatorial Pacific is dominated by westerly anomalies in the early onset years, while it is dominated by easterly anomalies in the late onset years. The planetary wave kinetic energy can only penetrate through the equatorial westerly duct (Webster and Chang 1997). Strong westerly anomalies imply more active inter-hemispheric wave interactions. This relationship can be quickly confirmed by checking the composite high-pass filtered perturbation kinetic energy (PKE). The perturbations are defined as the kinetic energy of the wind perturbations deviating from the daily climatology obtained from 30-years (1961-1990) averages. The low-frequency part (period > 15days) of the perturbations are filtered out. The Hovmoller diagram of the latitudinally (10°N-10°S) averaged filtered PKE is shown in Figs 4a and b. The areas with large PKE coincide well with large westerly anomalies (not shown). The average PKE in the late onset years is much weaker than in the early onset years, except over the Atlantic. The temporal structure of the PKE over the Atlantic is rather sporadic in contrast to the PKE over the east Pacific. It is possible if the active zones of inter-hemispheric wave interactions shift eastwards, the onset time of the East Asia summer monsoon may be delayed.

4. CONCLUDING REMARKS

The extremely early and late SCS monsoon onset years are identified. Our preliminary results suggest that the inter-annual variability of the onset timing can be caused by global scale SST anomalies in which the SST anomalies near the East Asian coastal areas (the marginal seas) embed. In the springs (March and April) of the late onset years, a zonal wave number 1 type of SST anomalies appear in the Pacific with positive anomalies in the east and the negative in the west. The phase of higher latitude SST anomalies slightly shifts to the east of the equatorial anomalies. The largest gradient in both zonal and meridional directions is near the dateline. Tropical convection also shifts eastward, the upper-level southern hemisphere subtropical trough (downstream of the SPCZ) becomes deeper. The deeper upper-level trough, intruding from the southern hemisphere to the northern hemisphere, is the main cause for the abnormal equatorial easterlies. Such shift of convection and upper-level zonal wind anomalies may reduce energy exchange between the hemispheres and inhibit southward dispersion of the energy of the northern hemisphere extra-tropical transient eddies. On the other hand, we suspect that because the energy exchange between hemispheres is reduced, the subtropical highs in the southern hemisphere become weaker. The cross equatorial flow over the Indian Ocean also becomes weaker. Therefore, over the Southeast Asia, including SCS, surface temperature is abnormally warm (less intrusion of the transients). However, over the northwestern Pacific and central China, the surface temperature is cold. The configurations of the surface temperature anomalies may cause the low-level abnormal circulation over the SCS to remain in anticyclone until the later half of June.

In the spring (March) of the early onset years, again we observe a zonal wave number 1 type of the Pacific SST anomalies, but with positive anomalies in the west and the negative in the east. The phase of higher latitude SST anomalies also shifts slightly to the east of the equatorial anomalies. The surface temperature anomalies near the land bridge, from Australia northward to Southeast Asian continent are on the average larger than those in the Indian and the Pacific Oceans. With strong low-level convergence over the tropical eastern Indian Ocean, the Bay of Bengal, the Indonesian islands and the Philippine Sea, the tropical convection over these areas is active. On the other hand, the SST anomalies over the Northwestern Pacific are positive. The southerly flow over the Philippine Sea can be enhanced by the anomalous northward surface temperature gradient. Such gradient can destroy the low-level anticyclonic circulation over the SCS causing earlier summer monsoon onset.

REFERENCES

Chang, C. P., and G. T. Chen, 1995: Tropical circulations associated with southwest monsoon onset and westerly surges over the South China Sea. *Mon. Wea. Rev.*, **123**, 3254-3266.

He, J., L. Wang, and H Xu, 1999: Abrupt change in elements around 1998 SCS summer monsoon establishment with analysis of its onset. *Onset and evolution of the South China Sea Monsoon and its interaction with the ocean*, Ed. by Ding and Li, China Meteorological Press. 230-235.

Lau, K.-M., K.-M. Kim, and S. Yang, 1999: Dynamical and boundary forcing characteristics of regional components of the Asian summer monsoon. *J. Climate* (submitted).

Webster, P., and H. Chang, 1997: Atmospheric wave propagation in heterogeneous flow: basic flow controls on tropical-extratropical interaction and equatorial wave modification. *Dyn. Atmos. and Ocean*, **27**, 91-134.

Xie, A., Y. S. Chung, X. Liu, and Q. Ye ,1998: The interannual variations of the summer monsoon onset over the South China Sea. *Theor. Appl. Climatol.* **59**, 201-213.

Table 1. The onset dates of the SCS summer monsoon. See text for the definition.

YEAR	DATE	YEAR	DATE	YEAR	DATE	YEAR	DATE	YEAR	DATE
1958	5.24	1968	6.26	1978	5.24	1988	5.22	1998	5.21
1959	7.03	1969	5.27	1979	5.15	1989	5.17	1999	4.22
1960	5.28	1970	6.10	1980	5.15	1990	5.21		
1961	5.14	1971	4.26	1981	5.16	1991	6.08		
1962	5.19	1972	5.28	1982	6.08	1992	6.18		
1963	6.01	1973	6.12	1983	6.04	1993	6.19		
1964	5.15	1974	5.23	1984	6.06	1994	6.01		
1965	5.23	1975	6.07	1985	4.22	1995	5.12		
1966	5.03	1976	5.19	1986	5.14	1996	4.20		
1967	5.25	1977	7.02	1987	6.09	1997	5.18		

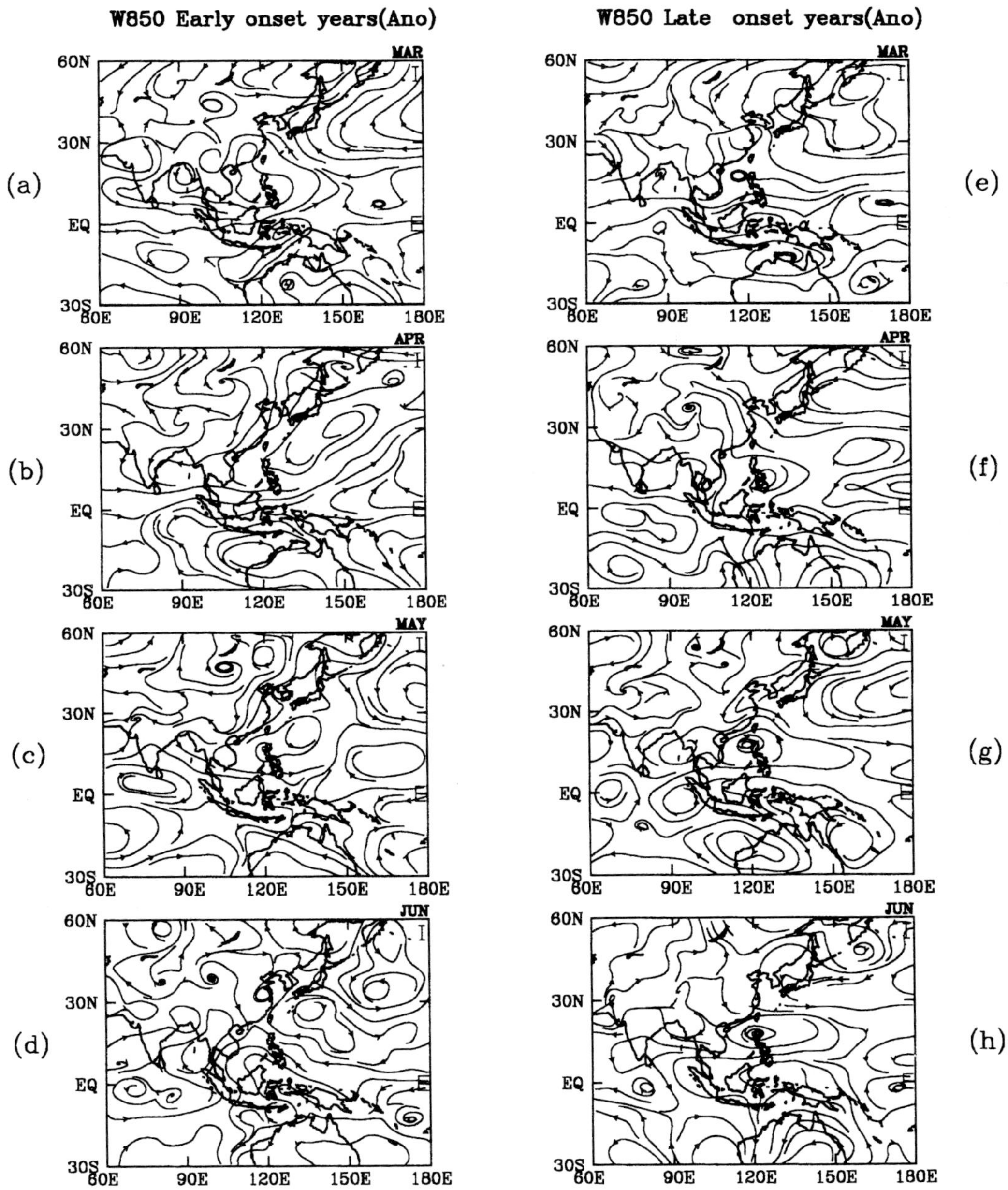

FIG 1. The composite wind anomalies with respect to the normal onset years. (a)-(d) are for the early onset years and (e)-(h) are for the late onset years. (a) and (e) are for March, (b) and (f) April, (c) and (g) May, and (d) and (h) June.

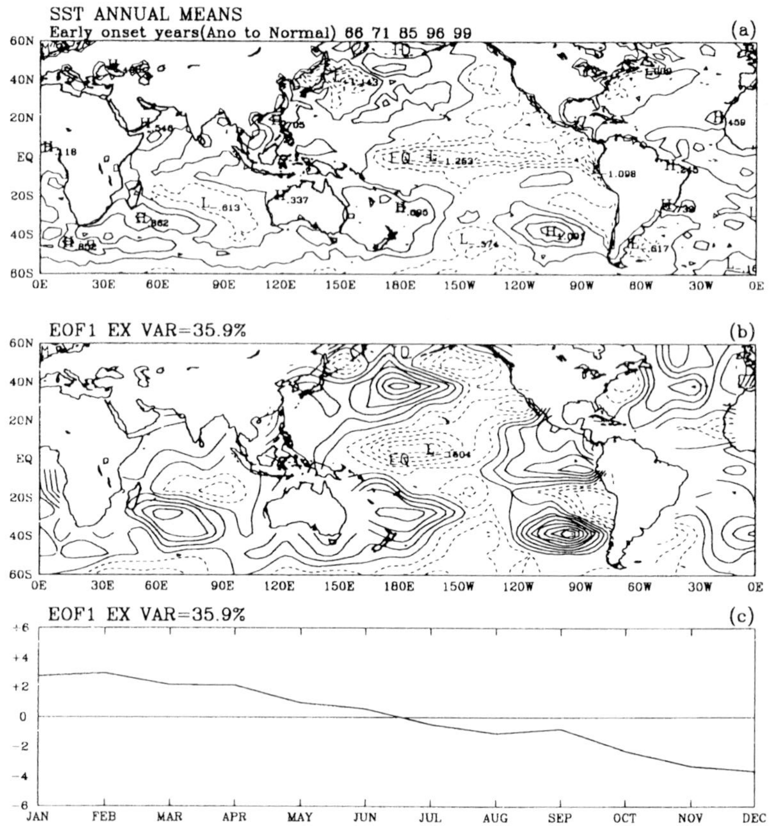

FIG 2. The composite SST anomalies for the early onset years. (a) is the annual average anomalies, (b) the spatial pattern of the first EOF mode and (c) the temporal structure of the first EOF mode. The solid lines are for positive values and the dashed lines negative.

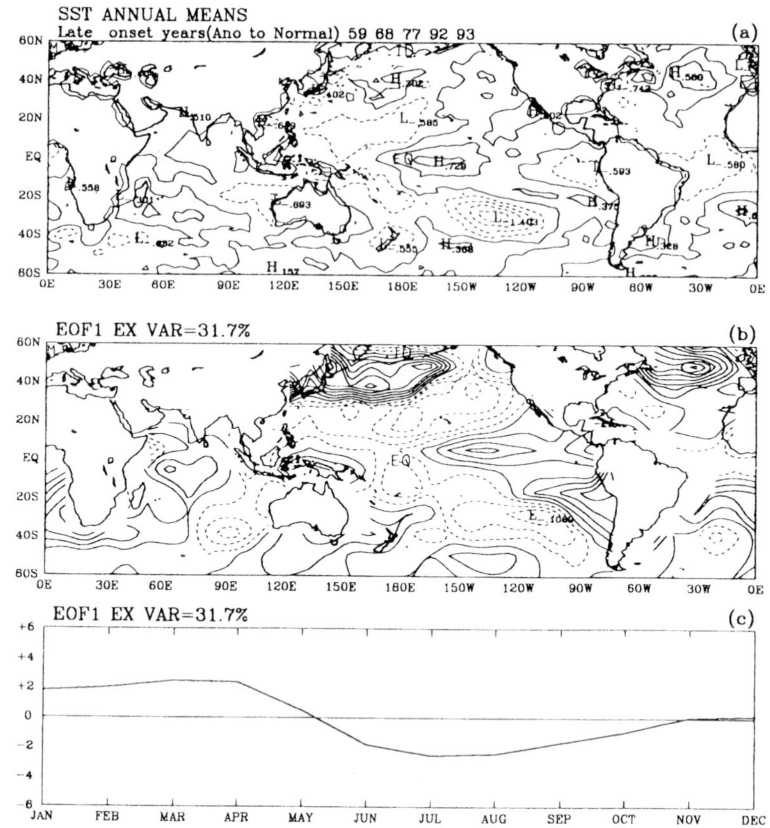

FIG 3. The same as **FIG 2** but for the late onset years.

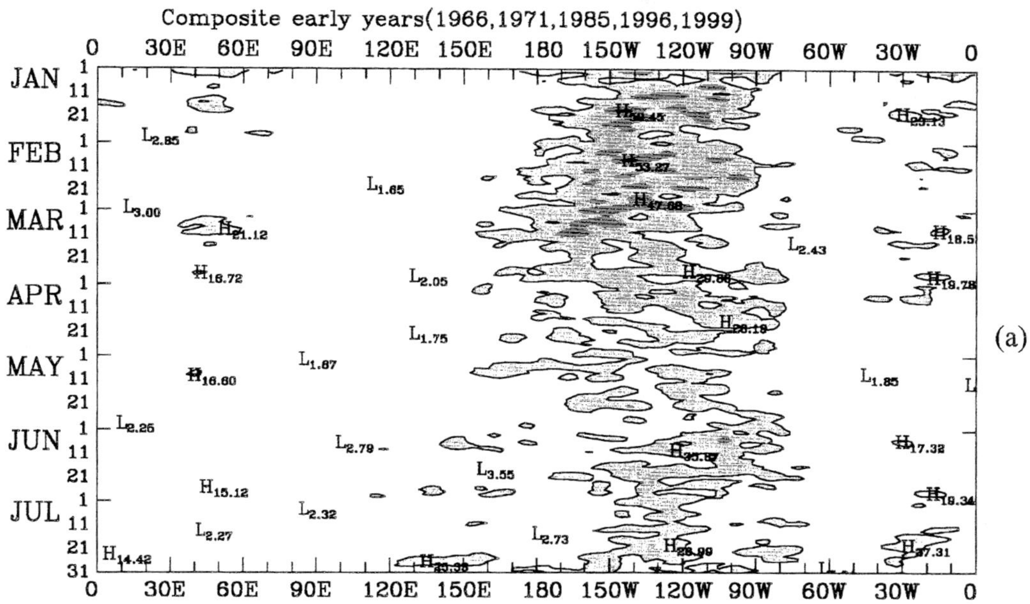

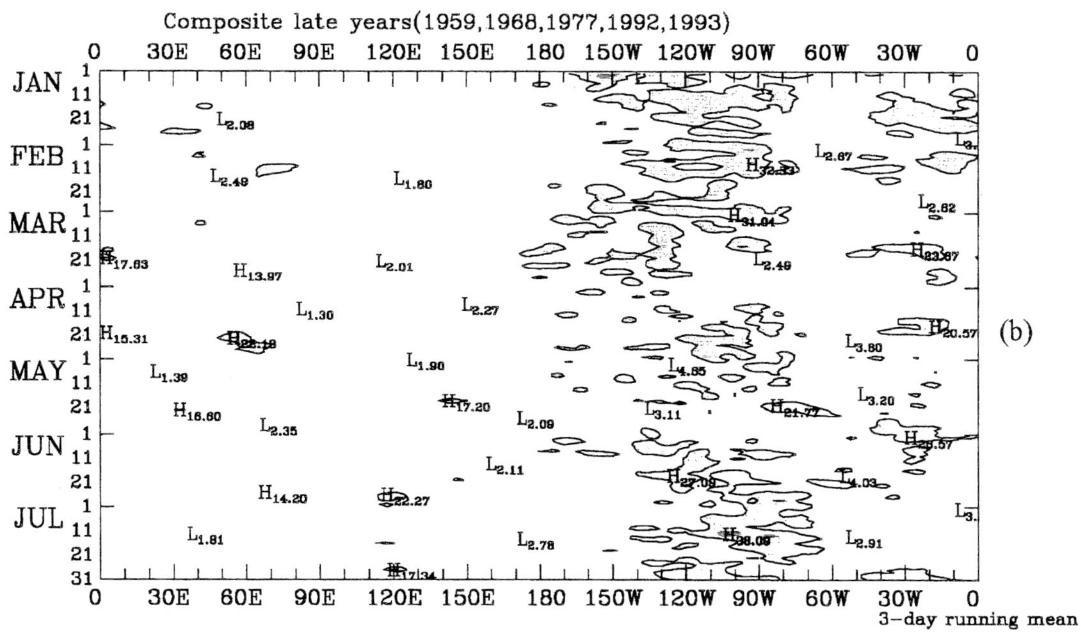

FIG 4. The Hovmoller diagram for the high-passed filtered PKE averaged over 10°N-10°S. (a) is for the early onset years and (b) the late onset years. Only values larger than 15 m^2s^{-2} are contoured and shaded. Values larger than 30 m^2s^{-2} are dark shaded.

~ Climate ~

MECHANISM OF ENSO-MONSOON INTERACTION DURING MATURE DECAY PHASES OF ENSO CYCLE

Bin Wang[#], Renguang Wu and Xiouhua Fu
International Pacific Research Center[*] and
Department of Meteorology, School of Ocean and Earth Science and Technology,
University of Hawaii

ABSTRACT

Observational evidence is presented to show an active ENSO-East Asian monsoon interaction during the mature and decay phases of ENSO. During the mature phase of major El Niño episodes, an anomalous low-level anticyclone establishes over the Philippine Sea. This anticyclone develops rapidly in late fall of the ENSO development year and persists until the following spring or early summer, causing anomalous above normal rainfall along the East Asian monsoon front stretching from southern China northeastwards to the Kuroshio extension.

Why do mature phases of ENSO have a *delayed impact* on the East Asian summer monsoon? We found that the anomalous anticyclone over the Philippine Sea is accompanied by the enhancement of the *in situ* sea surface cooling in front of the anticyclone. The persistence of the Philippine Sea anticyclone is attributed to a positive feedback between the atmospheric Rossby wave response and associated sea surface cooling through surface evaporation. This positive feedback provides a mechanism to maintain Philippine Sea anomalies until next summer against dissipations.

On the other hand, the Philippine Sea anticyclonic anomalies, which occur prior to the peaks of major warm events, induce anomalous easterlies in the equatorial western Pacific. By forcing equatorial upwelling Kelvin waves and elevating the thermocline, the easterly anomalies provide a continuing negative feedback to the eastern Pacific warming. The proposed mechanism is different from, but complementary to the delayed oscillator mechanism. The latter relies on oceanic wave adjustment process. Since the formation and maintenance of the Philippine anticyclone depend on the existence of mean circulation during the *boreal cold season*, the proposed negative feedback favors ENSO turning around in boreal winter. This may partially explain why the reversal of warming or cooling trends of ENSO cycles occurs most frequently toward the end of the calendar year.

1. INTRODUCTION

ENSO represents the strongest signal of inter-annual climate variation, while the Asian monsoon is the most energetic annual variation on the Earth. The tantalizing evidence of the

---

[#] Corresponding author address: Dr. Bin Wang, Department of Meteorology, University of Hawaii, 2525 Correa Rd., Honolulu, HI 96822, USA. E-mail: bwang@soest.hawaii.edu

[*] International Pacific Research Center (IPRC)) is sponsored in part by the Frontier Research System for Global Change.

linkage between the ENSO and Asian summer monsoon has led to the view that ENSO and monsoon should be viewed as an integrated climate system (Webster and Yang 1992). However, the precise interaction processes and the mechanisms governing the coupling remain to be further revealed and better understood.

One of the puzzling aspect of the ENSO-Asian summer monsoon interaction lies in the fact that, the mature phase of ENSO occurs most frequently in boreal winter, which is half year out of phase with the season of peak monsoon rainfall in Asia. Thus, the monsoon rainy seasons occur either in the development or the decay phases of warm or cold events when the SST anomalies are neither a maximum nor a minimum. As shown in Fig. 1, the NINO 3.4 SSTA in the summer before the mature phase of ENSO warming is about half of the mature phase SSTA, while during the summer following a mature phase of ENSO warming the NINO 3.4 is nearly normal. The weak SSTA during boreal summer is perhaps a major reason why the linkage between the Asian summer monsoon rainfall and the Pacific SST anomalies is so tantalizing.

Most of the ENSO-monsoon interaction studies have so far focussed on the South Asian summer monsoon and ENSO relationship. There is a good reason to believe that the East Asian monsoon may more directly interact with ENSO, because the western Pacific warm pool is a junction between the East Asian monsoon and ENSO. Both ENSO and East Asian monsoon have great impacts in this region and both are sensitive to the variations therein.

In the East Asia, the northern China summer rainfall tends to be weak during the development phase of a warm event, which appears in phase with the Indian summer monsoon (Huang and Wu 1989). On the other hand, the summer rainfall in the southern China tends to be above average in the year following the peak of a warm event (Ye and Huang 1996, Zhang et al. 1996). The most robust signal of the increased rainfall was found in the East Asian monsoon front (stretching from the southern China to the Kuroshio extension) during the ENSO mature winter and the following spring (Wang et al. 2000a).

In this paper, we try to address how the eastern-central Pacific warming can affect the "upstream" (with respect to the direction of the midlatitude westerlies) climate in the East Asia. In particular, since the large SST anomalies associated with ENSO occur in boreal winter, how can ENSO affect the summer rainfall in the East Asia in a "delayed" manner? We will also explore the processes through which the East Asian monsoon could feedback to ENSO evolution, in particular, the lock of mature phases of ENSO to boreal winter.

Our strategy is to focus on the most prominent basin-wide warm and cold episodes and their mature phases, because the Pacific SST and associated convection anomalies reach maximum during the mature phases and their impacts on Asian monsoon, if any, are likely most robust. Since ENSO mature phases are tightly locked to boreal winter, it is meaningful to make a composite analysis according to calendar month. The NINO 3.4 SST anomalies show seven strongest warm events (57/58, 65, 72, 82/83, 86-87, 91-92, and 97-98) and six strongest cold episodes (64, 70/71, 73/74, 75/76, 84/85, 88/89, 95/96) in the last 50 years (Fig. 1). They all occurred in the boreal winter from November to January except the 86/87 event. For clarity, we make composite scenario for the aforementioned six major warm (excluding 86/87) and six major cold episodes, which will be termed as the composite warm and cold events, respectively. Including moderate events will not change the conclusions derived here. We will emphasize only those contrasting features between composite warm and cold events that are statistically significant through a t-test at a 95% confidence level. Since the processes involved in the La Niña by large mirror those in El Niño events, we will describe the scenario associated with warm episodes only.

2. How does the Pacific warming impact East Asian climate?

Figure 2 presents composite maps of SST, surface wind anomalies around the peaks of the major El Niño for DJF season. In response to the basin-wide warming, a massive western North Pacific (WNP) anticyclone forms with one center located over the Philippine Sea. Anomalous southwesterlies prevail to the west and northwest of the Philippine Sea anticyclone, implying a weaker than normal northeasterly winter monsoon along the East Asian coast and thus a warmer than normal East Asian winter. The WNP anticyclone is a key system that bridges the eastern-central Pacific warming and the East Asian winter monsoon.

When do the wind and SST anomalies over the Philippine Sea form and how do they evolve after their formation? We present, in Fig. 3, a sequence of composite monthly mean surface wind and SST anomaly maps prior to and after the mature phase of ENSO for the major warm events. Over the WNP, the wind anomalies change suddenly from cyclonic to anticyclonic about one month prior to the warm peak. After its establishment, the WNP anticyclone rapidly intensifies and extends equatorward. Accompanying the establishment of the Philippine anticyclone, the SST in front of the anomalous anticyclone further decreases, reaching a minimum anomaly of -0.7°C, whereas the SST to the west of the Philippine Sea anticyclone further increases, inducing warming in the East Asian marginal seas including the South China Sea, East China Sea, Kuroshio currents, and Kuroshio extension.

A noticeable feature is that the wind and SST anomalies normally persist for two or even more seasons after their establishment, providing a delayed impact to the East Asian monsoon in the following spring and early summer even when the eastern Pacific warming substantially decays (the last panel in Fig. 3). Observations have shown that after El Niño, the climate in south China and Korea is warmer and wetter than normal in winter and spring (Kang and Jeong 1996, Wang and Hwu 1994, Tao and Zhang 1998). The above normal rainfall often extends into early summer in the subtropical monsoon frontal region. The devastating 1998 summer flood in Yangtze River valley is one of the examples.

3. The mechanism for the maintenance of the subtropical high anomalies

How can the anomalous subtropical high over the Philippine Sea maintain itself against damping during the decay phase of ENSO? Wang et al. (2000a) has offered an explanation. A brief summary of their explanation is provided here to make this short article self-contained. Their explanation is motivated by the observations revealed in Fig. 3. Figure 3 indicates that the maximum pressure (anticyclonic) anomalies are located about 20° longitudes to the west of the maximum cooling. Both the negative SST and positive SLP anomalies show a coherent tendency of slow eastward extension. These features suggest a coupling between the WNP wind and SST anomalies. Wang et al. (2000a) pointed out that it is the atmosphere-ocean *interaction over the western North Pacific* that provides a mechanism to maintain the Philippine Sea anticyclone and a delayed impact of ENSO on East Asian summer monsoon. More specifically, they emphasizes the following *positive feedback between the anomalous anticyclone and the sea surface cooling*.

From an atmospheric perspective, the Philippine Sea anomalous anticyclone is primarily a Rossby wave response to the suppressed convection due to *in situ* cooling of the ocean surface (Matsuno 1966). Figure 3 reveals that the low-level anticyclone is located about 20° longitude to the west of the cooling region. That is a manifestation of the Rossby waves of the lowest baroclinic mode. Numerical experiments with an atmospheric general circulation model (COLA model) and an intermediate atmospheric model (UH model) forced by anomalous SST in the western North Pacific have confirmed the above assertion (Wang et al. 2000a). On the other hand, from an oceanographic point of view, the major processes that cause the low frequency SST variations in the warm pool are the surface latent heat flux and the downward shortwave radiation flux (e.g., Godfrey and Coauthors 1995). Over the Philippine Sea, the mean winds (basic states) are dominated by northeasterly trades in *the cold season* (October – May). To the east of the anomalous anticyclone, the total wind speed and associated evaporation and entrainment cooling would be enhanced. Thus, the wind-induced SST variation would favor mixed layer cooling ahead of the anticyclone. To the west of the anomalous anticyclone, the total wind speed decreases, so does the evaporation and entrainment cooling, which induces rising of SST in the East Asian marginal seas.

The phase shift between the anomalous anticyclone and negative SSTA is a robust feature and is critical for the anticyclone-SSTA interaction. On the one hand, the anticyclonic anomalies favor lowering SST to its east. On the other hand, the negative SST anomaly located ahead (to the east) of the anomalous anticyclone would favor amplification of the anomalous anticyclone by exciting Rossby waves. Therefore, the positive feedback between the anticyclonic wind and SST through evaporation and entrainment processes in the WNP may maintain the WNP surface wind and SST anomalies against other damping processes.

5. A possible role of the Philippine Sea wind anomalies on ENSO turnabout

We discovered that *prior to* the warm peak, an anomalous high sea-level pressure (SLP) rapidly gets established in the WNP in response to the enhanced atmospheric heating in the equatorial central Pacific (ECP). A sharp increase in SLP over the Philippine Sea (120°-150°E, 10°-20°N) and abrupt development of easterly anomalies north of New Guinea (4°S-4°N, 120°-

150°E) preceded, in the majority of the strongest warm events, the warm peak by about one or two months (Fig. 4).

To better display this common feature, we present in Fig. 5 a composite tendency map for the sea level pressure and 1000hPa winds for the six warmest events. The composite is made by subtracting the sea level pressure and 1000 hPa winds at the -2 month from those at the –1 month (Lags –2 and –1 mean, respectively, two and one month prior to the warm peak in NINO 3.4 SST). Thus, the composite tendency field leads the composite warm peak by one month. Statistical significance of the composite wind tendencies between the warm and cold composites is tested using t- and F-tests. The wind anomalies over the Philippine Sea and north of New Guinea are significant at a 95% confidence level. Prior to the warm peak, anticyclonic wind anomalies over the Philippine Sea and associated easterly anomalies in the equatorial western Pacific develop abruptly. Though the timing of the maximum tendency for each event varies, the changes from –2 to –1 month with reference to the warm peaks are sufficiently common to warrant a statistically significant tendency as shown in Fig. 4. Although the wind anomalies in the equatorial central Pacific associated with the ENSO warming (cooling) are much more intense than the wind anomalies north of New Guinea, their tendencies are weak and statistically insignificant, because these wind anomalies evolve slowly, reflecting a slow coupled ocean-atmospheric mode. This suggests that the rapid establishment of the tropical wind anomalies over the Philippine Sea differs from the evolution of the slow-coupled mode.

The sudden emergence of the easterly anomalies in the EWP during the mature phase of El Niño triggers a negative feedback to ocean by stimulating equatorial upwelling Kelvin waves propagating eastward and causing decay of the surface warming. To demonstrate the excitation of the *forced* Kelvin waves, we present temporal evolution of subsurface temperature tendency in a vertical section along the equator (Fig. 6). The large temperature variations beneath the surface result primarily from vertical displacements of the thermocline, which are mainly caused by the eastward passage of the equatorial Kelvin waves. Just prior to the peak of major warm events (December 1982 and December 1991) the abrupt enhancement of easterly wind stress in the EWP (Fig. 4) raises the thermocline, triggering upwelling Kelvin waves (negative temperature tendency) propagating eastward along the flattened thermocline (Fig. 6). These upwelling Kelvin waves induce surface cooling along the equator as they journey to the east. Therefore, the sudden change of zonal winds in the EWP plays an important role in the reversal of the warming trend. To reach an opposite phase of ENSO, however, the reversed trend must be maintained sufficiently long. This is realized by the persistence of the EWP easterly anomalies after the warm peak due to the ocean-atmosphere interaction in the WNP as alluded in the previous section.

Since the formation and maintenance of the Philippine Sea wind anomalies depend on the existence of mean northeasterly trades and the central Pacific warming during the *boreal cold season*, the proposed negative feedback favors the turnabouts of ENSO cycles in boreal winter.

Observed strong warming by the end of the ENSO development year, such as in 1965, 1972, 1982, 1991, and 1997, is mostly followed by a sharp decay in the next spring and transition to a cold phase in the next summer or fall. On the other hand, some moderate warming by the end of the ENSO development year, such as in 1968, 1976, 1986, and 1993-94, was not followed by a transition to a cold phase in the next spring or summer. The difference between these two groups lies in whether the Phillipine Sea anticyclone firmly establish or not. This suggests that the Philippine Sea anticyclone may have played a significant negative feedback to the decay of the ENSO and, in particular, provided a mechanism for locking ENSO turnabout in boreal winter. This hypothesis has been further tested and demonstrated in terms of a modified Cane-Zebiak model (Wang et al. 2000b).

5 Summary and discussion

In summary, we have presented observational evidence and provided an explanation as to how mature ENSO warming can have a delayed impact on the East Asian summer monsoon. The delayed impact is caused by the persistence of the WNP subtropical anticyclonic anomalies that were excited during the mature phase of the Pacific warming. The cause of the persistence is attributed to local air-sea interaction (the interaction between atmospheric Rossby waves and local cooling) in the presence of cold season mean states, the northeasterly trades. Another key point we have made is that the sudden establishment of the WNP anticyclonic anomalies occur prior to the peak warm and therefore may provide a negative feedback to weaken the eastern Pacific warming, favoring ENSO turnabout occurring in boreal winter. Therefore, the ENSO and

East Asian monsoon exhibit active two-way interaction during the mature and decay phases of ENSO cycles.

While the positive feedback between the anticyclone and sea surface cooling may play critical roles, it remains a mystery how the anomalous subtropical high can rapidly establish over the WNP prior to the warm peak. Here, we would like to speculate possible causes responsible for the formation of the subtropical anomalies over the WNP.

First, the formation of the WNP SST and wind anomalies may depend on the strong central Pacific warming. Observations show that the Philippine Sea anticyclone failed to develop or occurred briefly but failed to persist during some moderate warm events (e.g., 1968-69, 1976-77, 1993-94) and the 1986 winter. An important feature common to these four events is the insufficient strength of the central Pacific warming toward the end of the ENSO development year (the three-month running mean NINO 3.4 SST anomalies less than one standard deviation (Fig. 1). This suggests that the *strong* warming in the central Pacific is an important precondition for the development of the Philippine Sea anticyclone.

Second, the establishment of the anticyclone over the WNP is normally rapid, often taking place within a month (Fig. 3). The rapid establishment of the Philippine Sea anticyclone and enhancement of local cooling imply that an air-sea interaction occurs "abruptly". Therefore, formation of the WNP subtropical anomalies might be triggered by atmospheric processes. The sudden rise of the pressure in the WNP may be initiated by strong synoptic or intraseasonal oscillation events. A strong cold surge or an intraseasonal disturbance, for instance, can cool ocean surfaces by one degree on a time scale of one week. This has been observed during the TOGA COARE intensive observation period over the equatorial western Pacific warm pool (e.g., Sui et al. 1997). The normal path of the cold surge to low latitudes occurs over the South China Sea. Intrusion of cold surges into the Philippine Sea and WNP requires an eastward shift of the westerly jet. The strong central Pacific warming causes such a shift in jet stream by enhancing subtropical anticyclone and eastward shifting and intensifying the Aleutian Low (Horel and Wallace 1981). Associated with the eastward shift of the Aleutian low, the pressure rises to the east of Japan. This anomalous large-scale circulation favors intrusion of cold surges into the Philippine Sea. Thus, the central Pacific warming sets up an environment, which is conducive for effective air-sea interaction in the WNP, through the *interaction between extratropical and tropical circulations*. The tropical-extratropical interaction may serve as a catalyst for the western Pacific air-sea interaction in establishing the upstream teleconnection between ENSO and East Asian monsoon. These speculations are currently tested using observations and coupled ocean-atmospheric models.

ACKNOWLEDGMENTS: This work is supported by NOAA cooperated agreement No NA67RJ0154, NSF Climate Dynamics Program (grant ATM-9613776). This is the School of Ocean and Earth Science and Technology publication No. 5211 and IPRC publication No. 47.

References

Godfrey, S. P., and Coauthors, 1995: Surface fluxes and mixed layer heat and freshwater budgets in TOGA_COARE. *Proc. Int. Scientific Conf. On the Tropical Ocean Global Atmosphere Program*, Melbourne, Australia, World Meteor. Org., 464-468.

Horel, J. D., and J. M. Wallace, 1981: Planetary-scale atmospheric phenomena associated with the Southern Oscillation. *Mon. Wea. Rev.,* **109**, 813-829.

Huang, R., and Y. Wu, 1989: The influence of ENSO on the summer climate change in China and its mechanism. *Adv. Atmos. Sci*., **6**, 21-32.

Kalnay, E., and Coauthors 1996: The NCEP/NCAR 40-year reanalysis project. *Bull. Amer. Meteor. Soc.,* **77**, 437-471.

Kang, I., and Y. Jeong, 1996: Association of interannual variations of temperature and precipitation in Seoul with principal modes of Pacific SST. *J. of Korean Meteor. Soc.*, **32**, 339-345.

Matsuno, T., 1966: Quasi-geostrophic motions in the equatorial area. *J. Meteor. Soc, Japan,* **44,** 25-43.

Sui, C.-H., X. Li, K.-M. Lau, and D. Adame, 1997: Multiscale air-sea interactions during TOGA COARE. *Mon. Wea. Rev*., **125**, 448-462.

Tao, S. and Q. Zhang, 1998: Response of the East Asian summer monsoon to ENSO events. *Sci. Atmos. Sinica*, **22**, 399-407 (in Chinese).

Wang, B., R. Wu, and X. Fu, 2000a: Pacific-East Asian Teleconnection: How does ENSO affect East Asian climate? *J. Climate*, **13**, 1517-1536.

Wang, B., R. Wu, R. Lukas, and S-I. An, 2000b: A possible mechanism for ENSO turnabouts. Submitted to *J. Climate*.

Wang, J.-T., and J.-W. Hwu, 1994: Spring time East Asia large-scale characteristics post ENSO. Part I: The spatial distribution of SST and OLR. *Atmos. Sci.*, **22**, 339-360.

Webster, J. P., and S. Yang, 1992: Monsoon and ENSO: selectively interactive systems. *Q. J. Roy. Meteorol. Soc.*, **118**, 877-926.

Woodruff, S. D., R. J. Sultz, R. L. Jenne, and P. M. Steurer, 1987: A comprehensive ocean-atmosphere data set. *Bull. Amer. Meteor. Soc.,* **68**, 1239-1250.

Ye, D. and R. Huang, 1996: *Study of the Mechanism of Drought and Flood around Changjiang and Huanghe Rivers*. Shandong Sci. & Tech. Press, 387pp (in Chinese).

Zhang, R., A. Sumi, and M. Kimoto, 1996: Impact of El Niño on the East Asian monsoon: A diagnostic study of the '86/87 and '91/92 events. *J. Meteor. Soc. Japan*, **74**, 49-62.

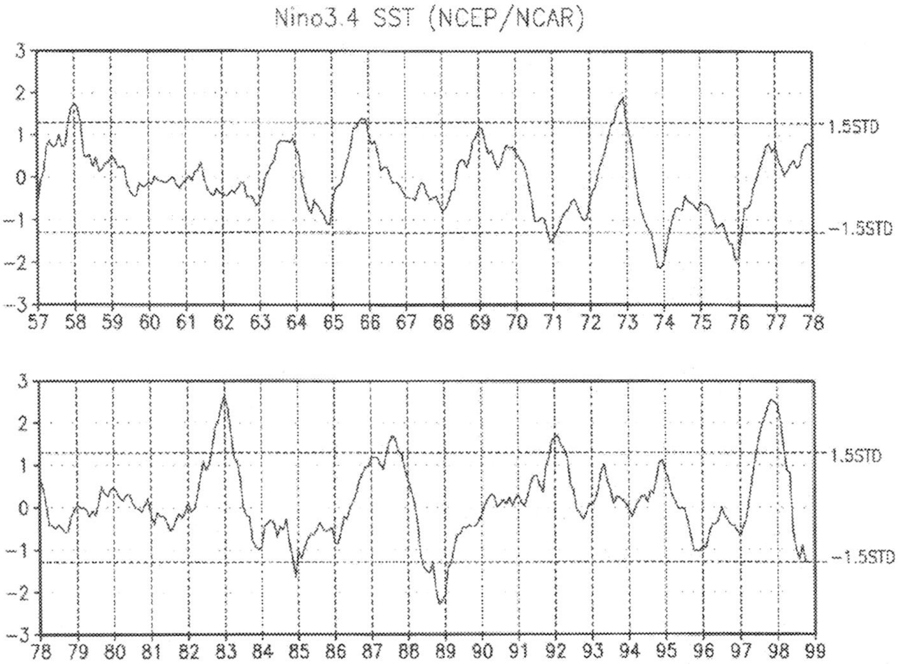

Fig. 1. Monthly mean SST anomalies averaged in NINO 3.4 region (5 °S-5°N, 170°-120°W). The SST data are derived from the National Centers for Environmental Prediction (NCEP) reanalysis (Kalney et al. 1996) except for 1957, which is derived from Comprehensive Ocean-Atmosphere Data Set (COADS) (Woodruff et al. 1987). The dashed horizontal lines indicate 1.5 times standard deviation of the SST anomalies.

Fig. 2. Composite 3-month mean surface wind and SST anomalies around the peaks of the six warmest episodes (57/58, 65, 72, 82/83, 91/92, 97/98). The shaded areas indicate regions where the difference in surface winds between the warm and cold composite are statistically significant at 99% confidence level by t-test. The contour interval for positive (negative) SST anomalies is 0.4 (0.2)°C. The vectors between the panels indicate the scale for wind anomalies.

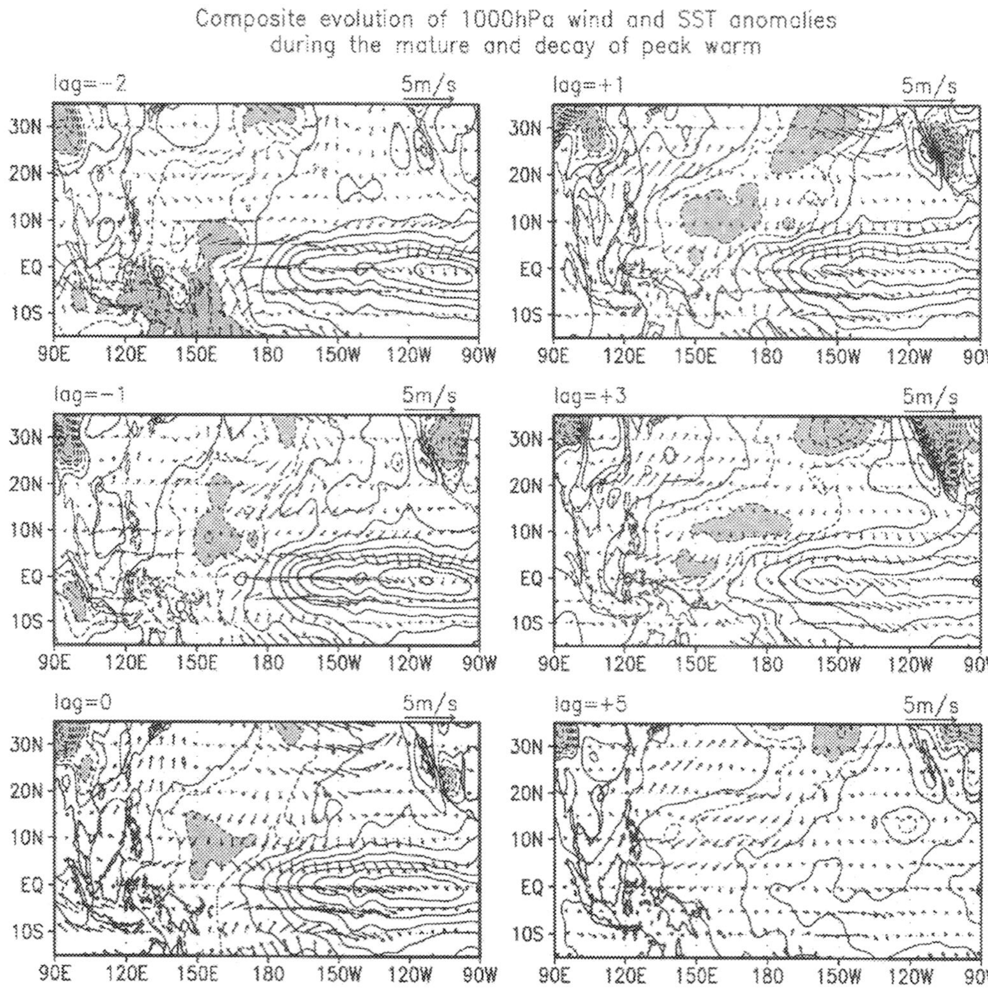

Fig. 3. Composite monthly mean anomalies of surface wind and SST for the six warmest episodes given in Fig. 2. The wind scale is given at the b ottom of the figure. The lag 0 means the time when NINO 3.4 SSTA reaches a maximum.

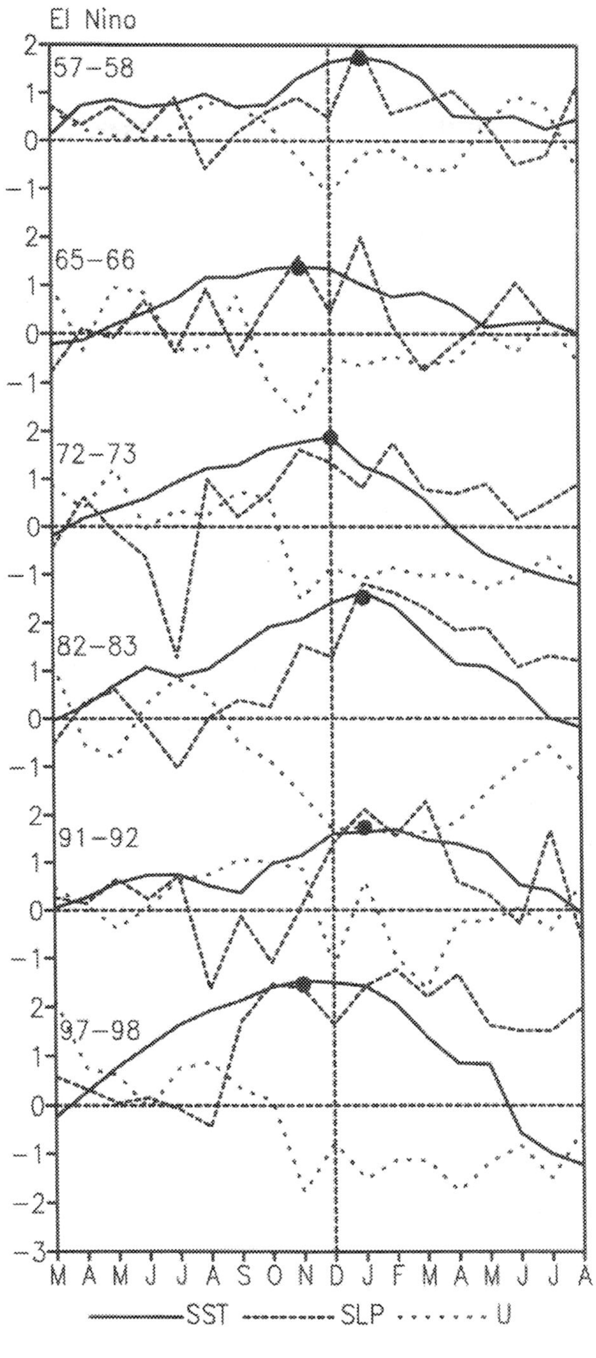

Fig. 4. Time series of sea level pressure anomalies (hPa) over the Philippine Sea (10°-20°N, 120°-150°E) (dashed) and surface zonal wind anomalies north of New Guinea (4°S-4°N, 120°-150°E) (dotted) for six strongest warm episodes during the period of 1957-1988. The abscissa denotes a time window spanning 18 months centered at November of the year during which the warm events reach a mature phase. The solid curves show NINO 3.4 SST anomalies (°C). The surface wind data are derived from the COADS (Woodruff, et al. 1987). The black dots denote the times of maximum NINO 3.4 SSTA.

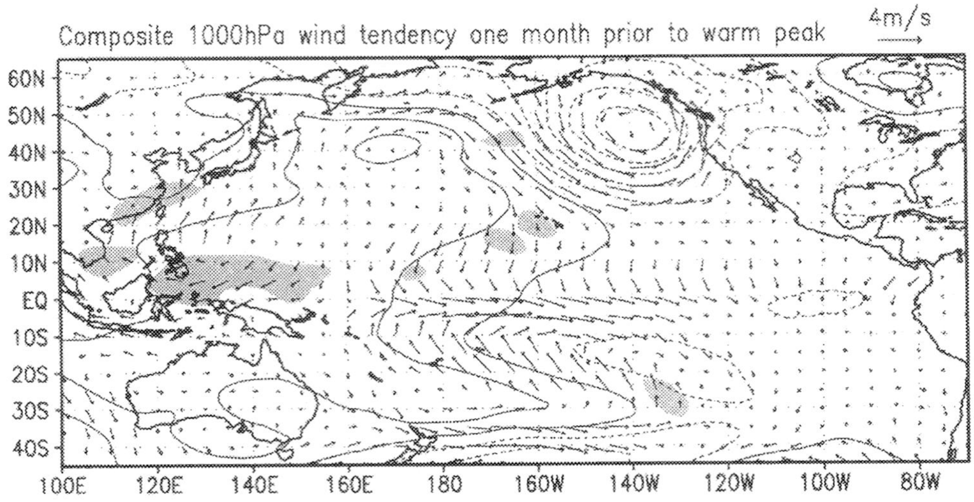

Fig. 5. Composite sea level pressure and 1000 hPa monthly mean wind tendencies at one month prior to the peak warm for the five major warm events (65, 72, 82/83, 91/92, and 97/98). The contour interval is 1 hPa. The wind scale is indicated between the two panels. Shading indicates regions where the wind tendency is significant at a 95% confidence level according to the Student t-test.

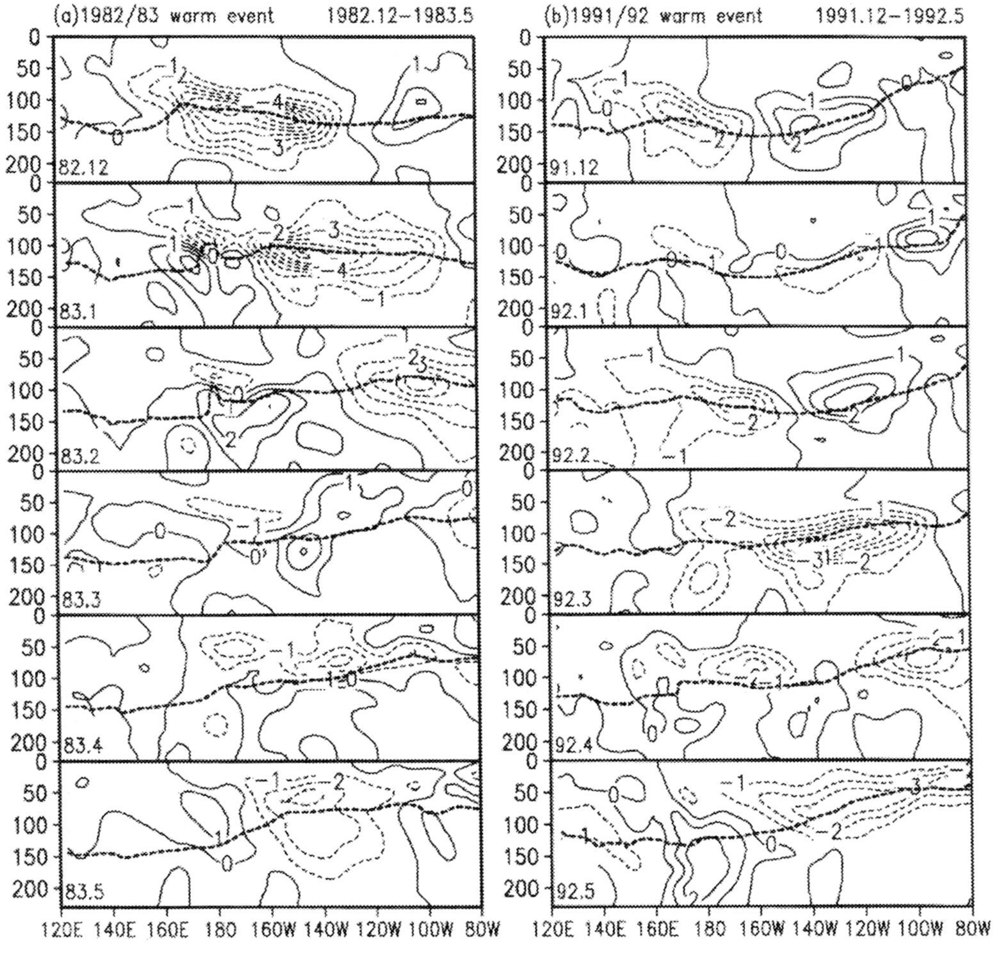

Fig. 6. Monthly mean local rates of change of subsurface ocean temperature (°C/month) in an equatorial vertical section during the turnabouts of (a) 1982-83 and (b) 1991-92 warm events. The ordinates denote the depth (m) beneath the sea surface. The thick dashed lines are the 20°C isotherm, showing the mean position of the thermocline during that month. The contour interval is 1 °C/month.

IMPACT OF LAND SURFACE PROCESSES ON SUMMER CLIMATE

Guoxiong Wu, Lan Sun, Yimin Liu, Hui Liu, Shufen Sun and Weiping Li
State Key Laboratory of Numerical Modeling for Atmospheric Sciences and Geophysical Fluid Dynamics (LASG)
Institute of Atmospheric Physics, Chinese Academy of Sciences, Beijing

ABSTRACT

Comparison experiments based upon a climate model are conducted to study the impacts of land surface processes on the mean July distributions of precipitation and subtropical anticyclone and on the seasonal variation of the Asian circulation. Results show that through changing the recycling precipitation ratio, the surface latent heat flux over land surface affects the global precipitation apparently. The latent heating of the Asian monsoon contributes to the distribution of the subtropical anticyclone over the region; and the strong surface sensible heat flux over North Africa in summer is very important for the configuration of the subtropical anticyclone over the North Atlantic Ocean. It is also shown that the surface sensible heat flux over the Tibetan Plateau is the basic factor that causes the abrupt seasonal transition from winter to summer of the circulation in the surrounding area.

1. INTRODUCTION

About two thirds of the energy of the atmospheric motion come from the energy exchange between the atmosphere and the underlying surface. The incoming solar radiation absorbed at the surface is transferred back to the atmosphere and the outer space in the form of sensible and latent heat flux and longwave radiation so that the surface energy balance is maintained. On a global and annual average, the latent heat flux accounts for about 50% of the transferred energy, the longwave radiation amounts to some 40%, whereas the sensible heat flux takes up only approximately 10% of the total released energy (Peixoto and Oort, 1991). However, over the Northern Hemisphere continents in summer the situation is quite different. Rather, the sensible heat flux of more than 100 Wm^{-2} occurs over the vast continental surfaces persistently. Such heating rate for the atmosphere is commensurate with the latent heating rate released by a precipitation of about 120 mm per month, which is no longer a small amount. Particularly the sensible heating over the Tibetan Plateau has pronounced impacts on the general circulation (Yeh et al., 1979; Wu and Zhang, 1998). The latent heat flux on the Earth's surface is transferred back to the atmosphere in the form of flux of moisture that may condense in the atmosphere and fall again as precipitation. Local precipitation comes partly from the convergence and condensation of the outer vapor and partly from the condensation of the locally evaporated vapor within the same control volume, the latter is controlled by the water content, characteristic, and vegetation type of the underlying land surface. The incorrect description for these surface attributes in numerical models will have impacts on the simulation of land surface latent heat flux and thus results in failure of the precipitation simulation.

The main goal of this study is to explore the impacts of land surface processes on the July mean global precipitation and the subtropical anticyclone over the oceans by means of numerical experiments, together with the mechanism analysis. For the present experiments, the global coupled ocean-atmosphere-land model (IAP/LASG GOALS) (Wu et al., 1997) is used. The model is composed of an OGCM (Zhang et al., 1996), a modified version of the L9R15 global spectral AGCM (Wu et al., 1996), and the SSiB land- surface model (Xue et al., 1991). The OGCM possesses 20 vertical layers ranging from the surface to the depth of 5km, and includes a 2-layer thermodynamic sea-ice model. A "monthly flux anomaly prediction- correction" scheme is used to couple the atmosphere and ocean component (Zhang et al., 1992). The SSiB model has three soil layers and one canopy layer. The implementation of the SSiB into the L9R15 AGCM gave remarkable improvements for the climate simulations (Liu and Wu, 1997). The K-distribution functions are applied in the calculation of the radiation (Shi, 1981; Wang Biao, 1996). More details of the model can be found in Wu et al. (1997). Since this paper serves to illustrate the importance of land-atmosphere interactions, the ocean component is de- coupled from the model. Instead, the prescribed seasonally monthly climatological sea surface temperature (SST) obtained from the Atmospheric Model Intercomparison Project (AMIP) data sets for the period from 1979 to 1988 is used as a boundary condition for all of the integration described below. Then the model has been integrated for ten model years for each of the experiments. By comparing the results from different experiments, the influences of land surface process on climate are evaluated. The simulation experiments involved in this study contain two runs: one with and one without the land- surface component SSiB implemented in the GOALS. There is one major difference between the two sets of simulation. In the control experiment (CON) in which the SSiB is turned off, the soil wetness is simply set to 0.25 m for all land surfaces, as was done in the bucket scheme employed in most of the AGCM's. In such models, the interactions between the atmosphere and the land surface are not taken into account. On the other hand, in the second experiment (SSiB) the component SSiB is implemented, and there include the interactive soil parameters, such as soil moisture and soil temperature. Section 2 explains why the land surface processes have impacts on the simulated precipitation. In sections 3 and 4 it is shown, respectively, how the deep convective latent heating of the Asian monsoon and the land surface sensible heat flux affect the distribution of the subtropical anticyclone over Atlantic and the Asian monsoon area. The influences of the huge sensible heat flux over the Tibetan Plateau on the abrupt seasonal transition are presented in section 5. Conclusions of this study are given in section 6.

2. LAND- SURFACE LATENT HEAT FLUX AND PRECIPITATION

The so-called recycled precipitation is defined as the precipitated water that evaporates from the land surface and condenses within the same control volume. The relative contribution of the recycled precipitation to the total precipitation within the volume is defined as the recycling precipitation ratio. It provides an indicator for the contribution of the surface evaporation to the measured precipitation in a given region. The ratio increases for relatively close regions and decreases for well open regions. Budyko (1974) proposed firstly one-dimensional approach for estimation of recycling precipitation and the process was defined along a single streamline. Then Brubaker et al. (1974), Eltahir and Bras (1994) developed a numerical model to estimate recycling precipitation consecutively. Using observational data, Yi (1995) estimated the distribution of the ratio over the Yangtze River reaches. Many of these studies involved prescribed boundary conditions and thus the results are dependent upon the choice of the boundary conditions within the computed area. Sun et al. (1998) recently proposed a new calculation scheme that can overcome this boundary limitation and provide much more objective results. In this section this new scheme is adopted. The simulated July precipitation rate in the CON and SSiB experiments are shown respectively in Figs. 1a and 1b, and their difference is presented in Fig.1c. The CON field is marked by much higher precipitation rates in most of the humid, vegetated, as well as arid regions. For example, the precipitation of more than 2 mmd^{-1} over Europe, the equatorial Africa and Asia monsoon regions are connected together and there is also much precipitation in the arid regions over the central, western, and northern North America. These results are far from the corresponding observations (see Peixoto and Oort, 1991). In the SSiB run, however, the precipitation over most of the continents has been reduced noticeably (Figs. 3b, c). Hence the distribution of the precipitation of more than 2mmd^{-1} in the monsoon region, over the equatorial Africa and Europe are disconnected. In the arid regions over the central, western, and northern North America, the precipitation rates are also reduced remarkably. This dramatic decrease in precipitation over the continental areas in the SSiB run brings the simulated precipitation much closer to the observed summer rainfall.

Fig. 1d shows the difference in the recycling precipitation ratio between the CON and SSiB runs. It is seen from this figure that, except for Europe and the western equatorial Africa, the recycling precipitation ratio is reduced over most of the continents in the SSiB experiment. In particular, the reduction is more than 20% over the central Asian, the southern North America and the Australian continents, which is closely in correspondence with the reduction in the latent heat flux over these lands in the SSiB simulation. In a word, the inaccurate precipitation distributions in the CON simulation are accompanied by the unrealistically high estimates of the calculated soil wetness, evaporation and recycling precipitation. However, such bias over the land surfaces is much reduced in the GOALS model once the interactive land- surface processes are included in the SSiB experiment. As a result, the simulated precipitation over continents is greatly improved, and become more realistic in many regions of the world.

3. VERTICALLY NON-UNIFORM DIABATIC HEATING AND SUBTROPICAL ANTICYCLONE

By employing scaling analysis to the complete form of vorticity equation (Wu and Liu, 1998), it can be shown that the vertically heterogeneous diabatic heating is the main forcing term in the equation (Wu et al., 1999). If frictional dissipation and internal vorticity forcing are neglected, then for a steady state the complete form vorticity equation can be approximated as:

$$\vec{V} \cdot \nabla \zeta + \beta v \approx \theta_z^{-1}(f + \zeta)Q_z \quad (\theta_z \neq 0). \qquad (1)$$

where ζ is the vertical vorticity, Q is diabatic heating rate, and Q_z, the vertical gradient of Q. The others are conventional. Since maximum sensible heating appears near the land surface, Q_z is negative. Consider a zonally symmetric flow field. Since the subtropical zone is an interface between mid-latitude westerlies and tropical easterlies in the lower troposphere, where $u \approx 0$, then the response of the atmosphere in the lower layers to a surface sensible heat flux of 100 Wm^{-2}, which amounts to a heating rate of several 10^{-5} Ks^{-1}, can be approximated as

$$v \approx (f+\zeta)\beta^{-1}Q_z\theta_z^{-1} < 0 \quad (\theta_z \neq 0) \tag{2}$$

and estimated to be 10 ms^{-1}. This means that, in response to a surface sensible heat flux of 100 Wm^{-2}, the equatorward winds of several ms^{-1} will be forced over the heating region in the lower layers with a thickness of several kilometers. The symmetric subtropical flow is then interrupted, and a lower layer anticyclone appears to the west of the heating region, as shown schematically in the middle panel of Fig. 2a.

In the upper layers in the subtropics, due to westerly wind shear in the vertical, zonal advection of vorticity overwhelms, and (2) can be approximated as

$$\partial\zeta/\partial x \approx (f+\zeta)u^{-1}Q_z\theta_z^{-1} < 0 \quad (\theta_z \neq 0) \tag{3}$$

For a heating scale of 10^6-10^7 m, and a westerly flow of 10^1 m, according to (3), a negative relative vorticity of the order 10^{-5} s^{-1} should appear downstream of the westerlies, resulting an upper layer subtropical anticyclone there, as is shown schematically in the upper panel of Fig. 2a. As to the surface layer, the studies of Hynes and McIntyre (1987), and Hoskins (1991) have shown that the heating generates surface cyclone (lower panel of Fig.2a).

For a strong deep convective latent heating along the subtropics, since strong extra westerlies are generated to its north and easterlies to its south, the vorticity advection across the heating region is always small and can be neglected from (1). Following a similar discussion, then, the *in situ* northerlies should be forced above the maximum heating level, while southerlies forced below that level. This then leads to the subtropical anticyclone formed to the west of the heating region in the upper layers (upper panel of Fig. 2b), but to its east in the lower layers (middle panel of Fig. 2b). Again, near the surface cyclone is produced (lower panel of Fig. 2b). To verify this hypothesis, a perpetual July experiment is developed. In this experiment, global surface is covered by ocean, the solar zenith angle is fixed at its value on July 15, and the initial fields are the zonally symmetric fields of the July means of the SSiB model climate. The rates of latent heat release at each time step and each grid-point are prescribed by using those values that were archived from the corresponding July means of the aforementioned SSiB run at the same grid-points. However, the surface sensible heat flux obtained from the surface thermal- balance consideration in the model is not allowed to heat the atmosphere. In other words, in this experiment, latent heating is the unique asymmetric and external thermal forcing. It is then referred to as LAT run. This model is integrated for 24 perpetual July months. Results from the last 20 months are retrieved for the following diagnosis. The July- mean altitude-longitude cross-sections of the zonal deviation of geopotential height at 30°N and the latent- heating rate at 20°N for the experiments LAT and SSiB, and for observation are presented in Fig. 3. By comparing LAT (a) with SSiB (b), we find that in the Asian monsoon area ranging from 60°E to the dateline, there exists a general resemblance in the deviation distribution between the two runs. In the lower layers, a high appears to the east of the heating region, whereas in the upper layers, the "South Asian High" at 200 hPa appears to the west of the heating region. It becomes clear that the general pattern of subtropical anticyclone over the Asian monsoon area can be well simulated by merely considering the impacts of latent heat release due to condensation associated with the Asian monsoon rainfall. However, in comparison with observations (Fig. 3c), the pronounced differences can be found in the LAT run, particularly over the eastern parts of the Atlantic and the Pacific, where the tropical upper tropospheric troughs (TUTT) and the surface highs are too weak. When the surface sensible heating, orography, and the annual cycle are included in the integration, as they are in the SSiB run, these differences are greatly suppressed. Results shown in Fig. 3b are fairly close to the observations shown in Fig. 3c. From these comparisons we can reach the conclusion that, the subtropical anticyclones over the western Pacific and South Asia are mainly forced by the latent heat release of the Asian monsoon rainfall, but modified to some extent by surface sensible heating and orography.

4. LAND SURFACE HEAT FLUX AND SUBTROPICAL ANTICYCLONE

To verify the impacts of land surface sensible heating on the configuration of the subtropical anticyclone, in this section the model outputs from different experiments are compared. Fig. 4 shows the simulated July mean sensible heat flux over Europe, North Atlantic and the central Asia regions in the CON and SSiB experiments, respectively. Significant differences between the two runs can be observed over the land surface. In the CON run, the sensible heat flux over the European and the North African land masses is less than 50 Wm^{-1}. On the contrary in the SSiB run over the most parts of the subtropical North Africa there exist huge sensible heat flux of more than 100 Wm^{-1}. This is because the CON run systematically overestimates the surface evaporation rates partly due to the large prescribed soil wetness over arid and semi-arid regions and partly due to the failure in accounting for vegetation stomatal resistance. Under similar meteorological forcing, this results in a large increase in the surface evaporation heat flux and a large decrease in the surface temperature over most of North Africa. The simulated surface temperature in July there is between 25 and 30℃ in the CON run (see Figs. 5a, b, Liu and Wu, 1997), much lower than the observed value of 30~35℃. This leads to a substantially reduced outgoing longwave radiation and sensible heat flux. On the contrary, in the SSiB run, the soil wetness is calculated from the model, and there is less latent heat flux in the arid region over most North Africa. Consequently, the July mean surface temperature is 30~35℃ over the region in the SSiB run (see Fig. 5c, Liu and Wu, 1997), which is in good agreement with observations. Thus, the sensible heat flux increases prominently.

Fig. 5 shows the simulated July mean sea level pressure, with the same layout as in Fig. 4. In the CON run (Fig. 5a), many unrealistic distributions are found over the North Atlantic Ocean, such as the eastward stretching of the subtropical anticyclone towards the Mediterranean Sea. However, after the SSiB is implemented into the model, these unrealistic patterns are removed, and the modeled results (Fig. 5b) become much closer to observations.

Results obtained in this section can be well explained by the theory introduced in the previous section. The huge surface sensible heating over North Africa, especially over its west, generates not only the *in situ* surface low pressure, but also remarkable negative vorticity forcing to the atmospheric circulation. Strong surface northerlies are then forced over the western coasts of Europe and North Africa, and the subtropical anticyclone is produced over North Atlantic in the SSiB experiment. It is inferred, therefore, that the configuration of the subtropical high over the North Atlantic Ocean depends strongly upon the sensible heating over the subtropical North Africa and European continents.

5. SENSIBLE HEAT-DRIVING AIR PUMP (SHAP) OVER THE TIBETAN PLATEAU AND ABRUPT SEASONAL TRANSITION

The impacts on the general circulation of the surface sensible heat flux have been demonstrated in the preceding section. When this heating is imposed on the uplifted Tibetan Plateau, such impacts should become much more significant. Yeh et al. (1979), Yanai et al. (1992), and Ye and Wu (1998) summarized the progress in this study. Using data analysis and numerical experiments, Wu et al. (1997) further found that the surface sensible heating sink in winter and strong source in summer over the Tibetan Plateau lead to the descent of 0.8×10^{-3} $hPas^{-1}$ in January and ascent of the same value in July of the aloft air column. This phenomenon is unique over the world. They also demonstrated that the atmosphere motion over the Plateau that is mainly driven by the sensible heat exchange works in a way just like an "air pump", and strongly regulates the seasonal transition from winter to summer of the general circulation in the surrounding areas. It was then defined as the "Sensible Heat-driving Air Pump" (SHAP). To study the impact of the SHAP on the seasonal variation of the regional climate, an experiment of NO sensible heat flux over the Plateau (NSH) is designed in contrast to the control experiment (CON). In this NSH experiment the sensible heat flux within the Tibetan Plateau region higher than 3km is switched off, while the others are kept the same as in the CON run. The model is then integrated over the same AMIP period. Fig. 6 shows the seasonal evolution of the westerlies along 90°E in the CON and NSH experiments. As expected, in the CON experiment (Fig. 6a), the easterlies in winter dominate near the equator and the westerly jet lies by 35°N. At the same time, the axis of the subtropical anticyclone, which is characterized by zero zonal wind in the figure, stays near 10°N. During the end of May to the beginning of June, the axis of the subtropical anticyclone moves abruptly northward from about 10°N to 27°N, and the westerly jet also shifts northward to about 45°N. The GOALS model has successfully simulated the characteristic of the abrupt transition of general circulation in spring as pointed out by Yeh et al. (1959). However, in the NSH experiment in which the SHAP works inefficiently, the northward shifting of the axis of the subtropical anticyclone from 10°N in spring to 20°N in summer is slow and smooth, lasting for about two months. Evidently, these results reveal that the SHAP is the main factor that contributes to the abrupt seasonal transition from winter to summer of the general circulation near the Plateau area. Besides generating the convergence in the lower troposphere and divergence in the upper troposphere, the elevated heating over the Plateau in the summer months produces strongly negative vorticity source over the Plateau, contributing to the formation of the South Asian High in the area. It is due to the establishment in June of the South Asian High that the axis of the aforementioned subtropical anticyclone in the upper troposphere and the easterly and westerly jets respectively to its north and south all shift northward quickly. On the contrary, in the NSH experiment, as the SHAP is weakened, the mechanism of generating the South Asian High is also weakened prominently, and the abrupt change in the seasonal transition demonstrated in Fig. 6a disappears correspondingly.

6. CONCLUSIONS

The results from the IAP/LASG GOALS model with the land- surface component SSiB implemented vis-a-vis those without the SSiB implemented show that, both the surface sensible and the latent heat flux have substantial impacts on the general circulation. The water vapor released from the underlying landmass can produce recycling precipitation. In July the recycling precipitation ratio amounts up to 10% on the continents except for the central Asia, North Africa and in the arid regions of South America, and even more than 25 % over the Plateau and the surrounding areas. Therefore, the land surface evaporation has significant influence on the rainfall climate. The Asian summer monsoon is important for the formation of the subtropical anticyclones over the western Pacific at 500 hPa and over the South Asia region at 200 hPa. Besides producing the *in situ* surface low pressure, the surface sensible heat flux over land surface can generate strong subtropical anticyclone over the ocean on the western side of the heating region through the β-effect. It is also shown that the "air pump" driven by the sensible heat flux over the Tibetan Plateau is an important factor that contributes to the abrupt seasonal transition from winter to summer of the general circulation near the Plateau area.

ACKNOWLEDGEMENT. This study is supported jointly by the "National Key Programme for Developing Basic Sciences" G1998040904(Part I) and by the Chinese Natural Science Foundation under the project numbers 49635170, 49823002 and 49825504.

REFERENCES

Brubaker, K.L. et al., 1974: Estimation of continental precipitation recycling, *J. Clim.*, **6**, 1077-1089.

Budyko, M. J., 1974: *Climate and Life*, Academic Press.

Eltahir, E.A.B., and R.L. Bras, 1994: Precipitation recycling in the Amazon Basin, *Quart. J. Roy. Meteor. Soc.* **120**, 861-880.

Haynes, P.H., and M.E. McIntyre, 1987: On the evolution of vorticity and potential vorticity in the presence of diabatic heating and frictional or other forces. *J. Atmos. Sci.*, **44**(5): 828-841.

Hoskins, B.J., 1991: Towards a PV-θ view of the general circulation. *Tellus.* **43**AB: 27-35.

Liu Hui, and Wu Guoxiong, 1997: Impacts of land surface on climate of July and oneset of summer monsoon: A study with an AGCM plus SSiB, *Adv. Atmos. Sci.*, **14**, 289-308.

Liu Yimin, Liu Ping, and Wu Guoxiong, 1998: The effect of spatially heterogeneous heating on the configuration and variation of subtropical high, II. Land surface sensible heating and the subtropical high over eastern Pacific and North America, To appear on *Acta Meteorological Sinica* (in Chinese).

Peixoto, J. P., and A.H. Oort, 1991: *Physics of Climate*, Amer. Ins. Phy. Press, New York, 520pp.

Shi Guangyu, 1981: *An Accurate Calculation and Representation of the Infrared Transmission function of the Atmospheric Constituents, Ph.D Thesis*, Dept. Sci., Tohoku Univ. Japan, 199pp.

Sun Lan, and Wu Guoxiong, 1998: Concerning the calculation of precipitation recycling ratio. Submitted to *Acta Meteor. Sinica* (in Chinese).

Wang Biao, 1996: *On the radiation transfer models in the climate modeling*, Ph.D dissertation at the Institute of Atmospheric Physics, Chinese Academy of Sciences, 92pp.

Wu Guoxiong, Li Weiping, Guo Hua, Liu Hui, Xue Jishan, and Wang Zaizhi, 1997: Sensible heat-driving air pump of the Tibetan Plateau and the Asian summer monsoon. *Collection in Memory of Dr. Zhao Jiu-Zhang*. Chinese Science Press, Beijing, 116-126.

Wu Guoxiong, Liu Hui, Zhao Yucheng, and Li Weiping, 1996: A nine-layer atmospheric general circulation model and its performance. *Adv. Atmos. Sci.*, **13**, 1-18.

Wu Guoxiong, Liu Yimin, and Liu Ping, 1999: The effect of spatially heterogeneous heating on the configuration and variation of subtropical high, I. Scale analysis, *Acta Meteor. Sinica*, **57**, 257-263 (in Chinese).

Wu Guoxiong, Zhang Xuehong, Liu Hui, Yu Yongqiang, Xin Xiangze, Kuo Yufu, Sun Shufen, Li Weiping, Wang biao, and Shi Guangyu, 1997: Global ocean-atmosphere-land system model at LASG (GOALS/LASG) and its simulation studies, *Quarterly Journal of Applied Meteorology* (supplement), 15-28 (in Chinese).

Wu Guoxiong, and Zhang Yongsheng, 1998: Tibetan Plateau forcing and the timing of the monsoon onset over South Asia and the South china Sea, *Mea. Wea. Rev.*, **126**, 913-927.

Xue Yongkang, P.J. Sellers, J.L. Kinter, and J. Shukla, 1991: A simplified biosphere model for global climate studies, *J. Clim.*, **4**, 345-364.

Yanai, M.C.F. Li, and Song Zhengshan, 1992: Seasonal heating of the Tibetan Plateau and its effects on the evolution of the Asian summer monsoon, *J. Meteor. Soc. Japan,* **70**, 319-351.

Ye Duzheng, and Guoxiong Wu, 1998: The role of the heat source of the Tibetan Plateau in the general circulation. *Meteor. Atmos. Phys.*, **67**, 181-198.

Yeh Duzheng, Tao Shiyan, and Li Maicun, 1959: The abrupt change of circulation over the Northern Hemisphere during June and October, *The Atmosphere and the Sea in Motion*, Bolin B. Ed., The Rockefeller Institute Press in Association with Oxford University Press, 249-267.

Yeh, T. C., and Gao Youxi, 1979: *Meteorology over the Tibetan Plateau*, Chinese Science Press, 278pp (in Chinese).

Yi Lan, 1995: *The water cycling and water budget in East Asian monsoon*; GEWEX GAME study, Ph.D dissertation at the Institute of Atmospheric Physics, Chinese Academy of Sciences, 156pp.

Zhang Xuehong, Bao Ning, Yu Ruchong and Wang Wanqiu, 1992: Coupling scheme experiments based on an atmospheric and oceanic GCM, *Chinese J. Atmos. Sci.*, **16**, 129-144.

Zhang Xuehong, Chen Keming, Jin Xiangze, et al., 1996: Simulation of thermohaline circulation with a twenty-layer oceanic general circulation model, *Theo. Appl. Clim.*, **55**(1-4), 65-88.

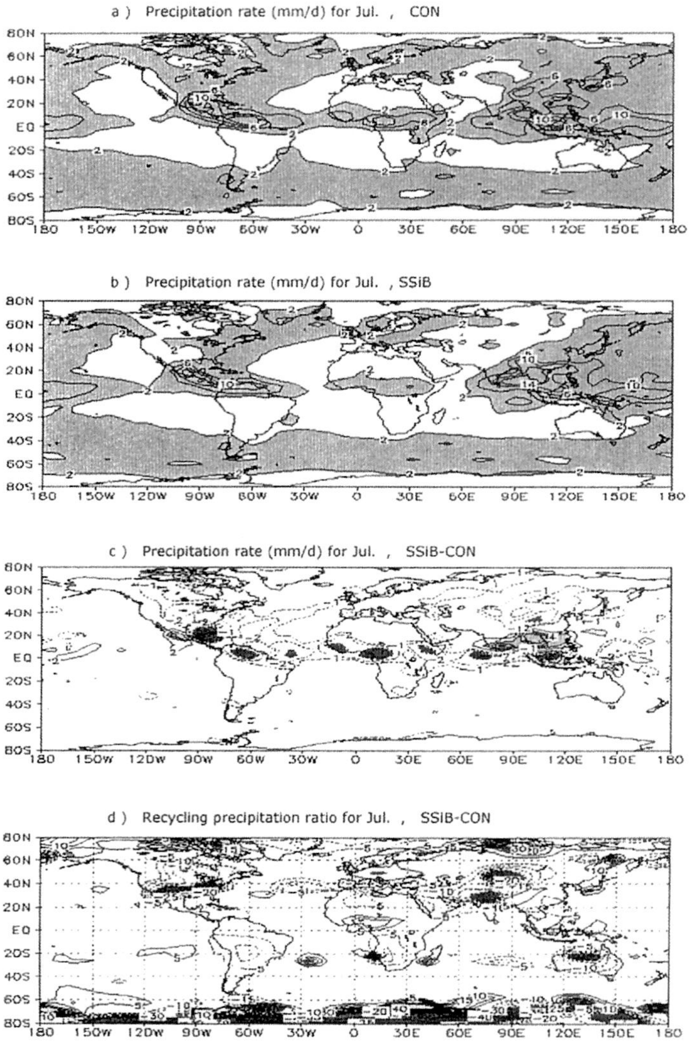

Fig. 1. Distributions of July mean precipitation (mmd^{-1}) averaged over the period from 1979 to 1988 in the CON run (a), SSiB run (b), and their difference (c), respectively, unit in mmd^{-1}. In (d) is shown the difference between the two runs of the July mean recycling precipitation ratio, unit in %.

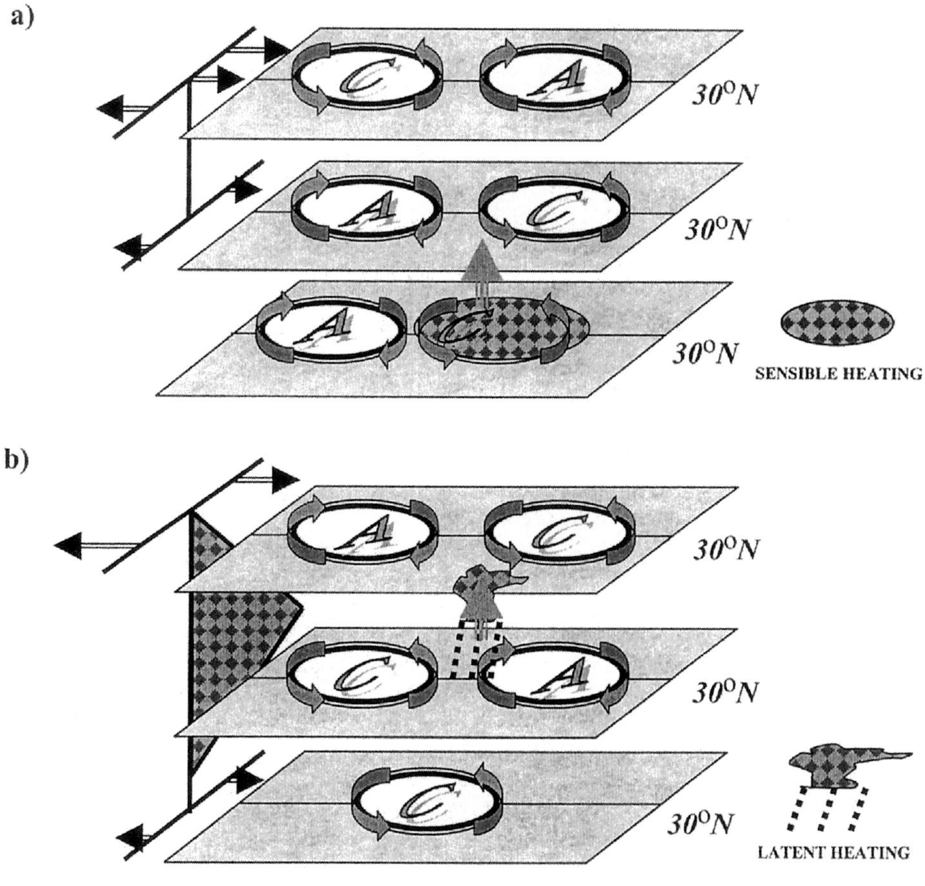

Fig. 2. (a) Schematic diagram showing the disposition of surface sensible heating and the configuration of the forced cyclones and anticyclones at upper level (top panel), lower level (middle panel), and surface (bottom panel). (b) The same as in (a), except for the deep convective heating case.

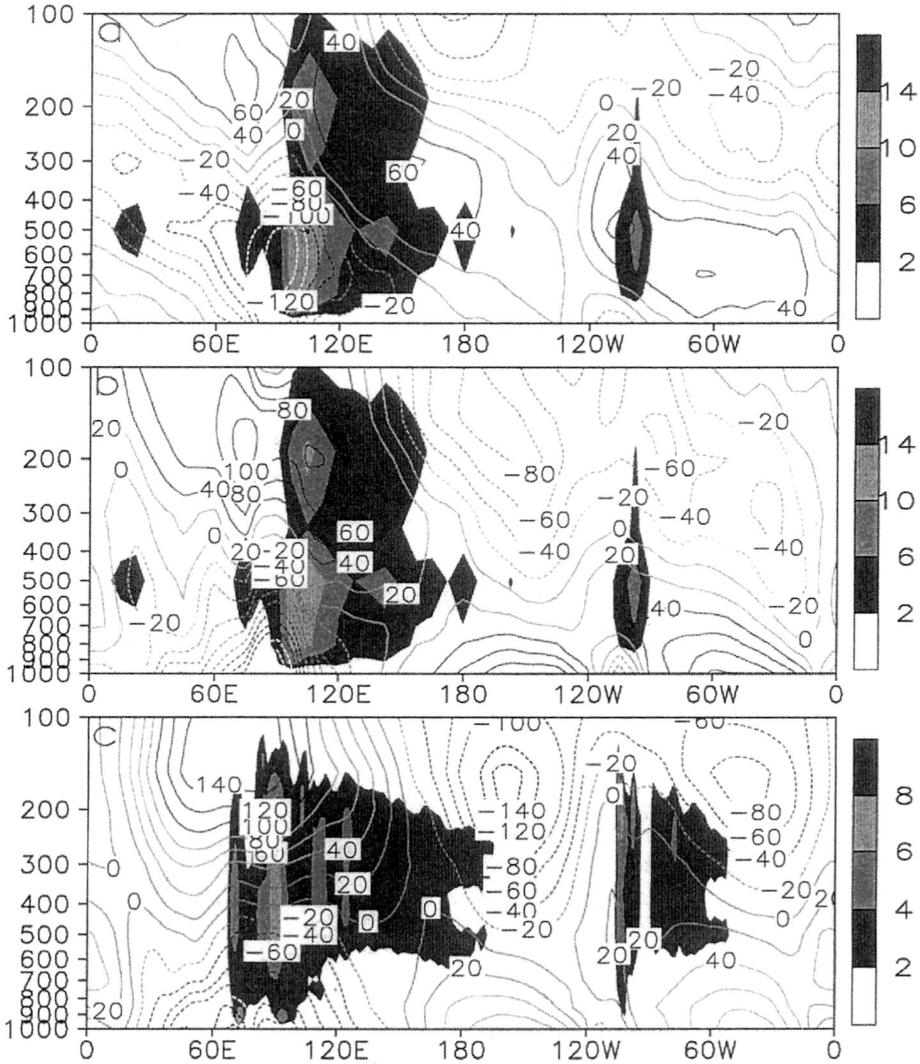

Fig. 3. Vertical cross-sections of the July means of deviation from the zonal mean of geopotential height at 30°N (contours, in gpm) and latent heating at 20°N (shaded, in Kd^{-1}) produced from experiments LAT (a) and SSiB (b), and from the NCEP/NCAR analysis of 1980-1995 (c).

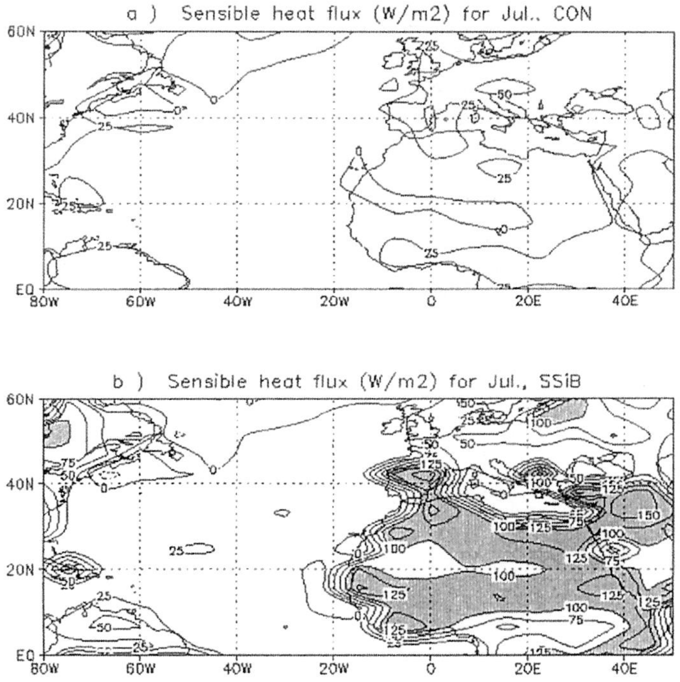

Fig. 4. Distributions of the July mean surface sensible heat flux (Wm^{-2}) over North Atlantic and Central Asia averaged over the period from 1979 to 1988 in the CON run (a) and SSiB run (b), respectively. The contour interval is 25 Wm^{-2}.

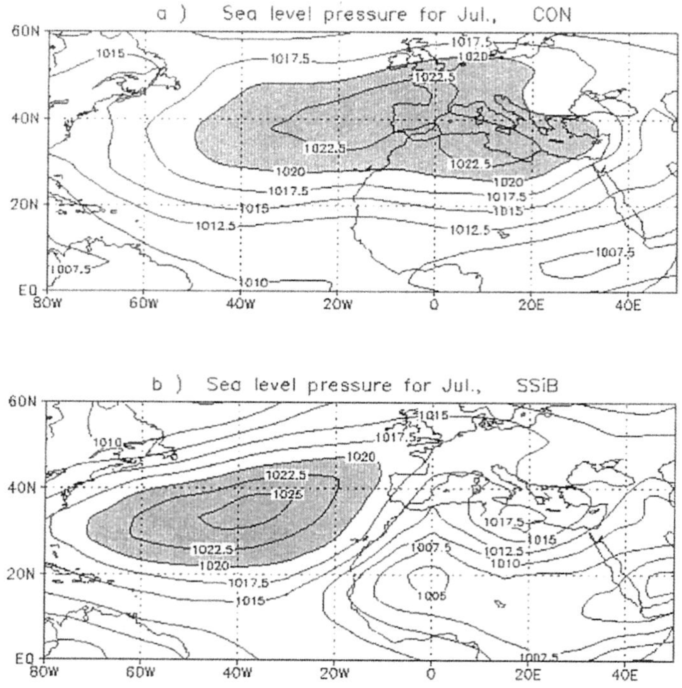

Fig. 5. As in Fig. 1, but for sea-level pressure (hPa).

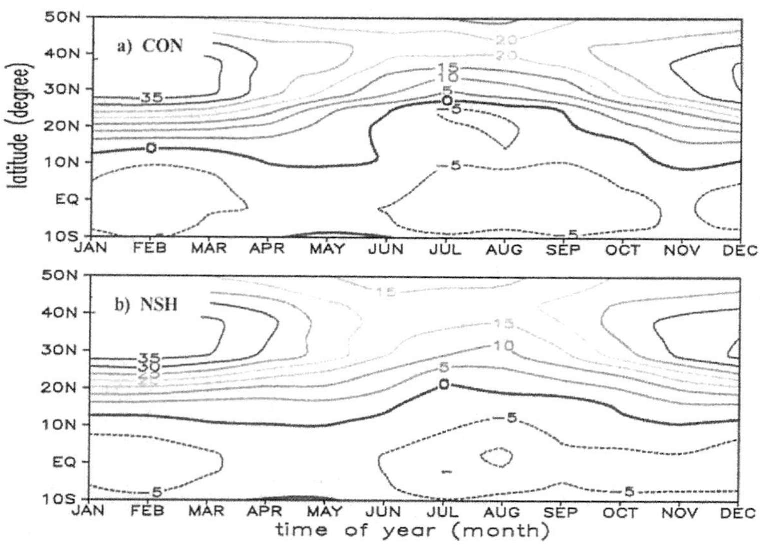

Fig. 6. The seasonal evolutions of monthly mean zonal wind (ms^{-1}) and the axis of the subtropical anticyclone (u=0) along 90°E at 300 hPa averaged over the period from 1979 to 1988 in the CON experiment (a) and in the NSH experiment (b), interval in 5.0 ms^{-1}.

IMPACT OF ENSO ON THE VARIABILITY OF THE ASIAN-AUSTRALIAN MONSOONS

Ngar-Cheung Lau
Geophysical Fluid Dynamics Laboratory/NOAA,
Princeton University, Princeton, NJ

1. INTRODUCTION

The primary goal of the present investigation is to examine the ENSO-monsoon relationship on the basis of a suite of integrations with a general circulation model (GCM). Experiments with different SST forcing scenarios are conducted to delineate the relative contributions of SST fluctuations in the tropical Pacific and those in the near-global oceans to monsoon variability, and to investigate the role of air-sea feedback processes. The simulated variability in different sub-regions of the Asian-Australian monsoon complex is documented. The dynamical basis for the linkage between SST forcing in the tropical Pacific and the monsoons in these regions is critically examined. A more detailed description of this work is provided in Lau and Nath (2000).

2. EXPERIMENTAL DESIGN

The GCM used for this investigation has been developed and maintained by the Climate Dynamics Project at the Geophysical Fluid Dynamics Laboratory (GFDL). A spectral formulation is used to represent the horizontal variations of the model fields, with a rhomboidal truncation at 30 wave numbers. The corresponding Gaussian grid has a resolution of 3.75° and approximately 2.2° in the zonal and meridional directions, respectively. Vertical variations are depicted using a sigma coordinate with 14 levels. A comprehensive set of boundary conditions and physical processes have been incorporated in this model, including ocean-continent contrast, orography, precipitation, predicted cloud cover, radiative transfer, gravity wave drag, ground hydrology and continental snow cover.

The essential characteristics of the model climatology (not shown) for the Asian-Australian sector compare well with the observations. Particularly noteworthy is the successful simulation of the large-scale flow patterns in this region in the lower and upper troposphere, and in both the summer and winter seasons. The seasonal dependence and geographical distribution of the principal precipitation centres in the model atmosphere are also in accord with observational data.

The model experiments for this study have been conducted by prescribing the temporally varying SST conditions in specific ocean sites as observed in the 46-year period from January 1950 through December 1995. The following three sets of experiments have been performed:

- *Tropical Ocean-Global Atmosphere (TOGA) runs*

  SST anomalies were inserted in the tropical Pacific between ~25°S and ~25°N only. Climatological (but seasonally varying) values were inserted in all remaining ocean gridpoints. This experiment is intended for studying the atmospheric response to SST variability which is directly associated with ENSO events in the tropical Pacific.

- *TOGA-Mixed Layer (TOGA-ML) runs*

  The domain for prescription of SST anomalies is identical to that in the TOGA runs. For all remaining ocean gridpoints, the SST conditions were predicted by a simple, motionless mixed-layer model with a constant depth of 50 m. The latter model computes the SST changes in response to variations in the heat fluxes at the air-sea interface. This experiment considers both the response to SST changes prescribed in the tropical Pacific, and the variability arising from simplified air-sea coupling outside of the tropical Pacific.

- *Global Ocean-Global Atmosphere (GOGA) runs*

  SST anomalies were inserted at all ocean gridpoints between ~40°S and ~60°N. This experiment delineates the impact of SST conditions in the near-global oceans on atmospheric variability.

For each of the three SST scenarios described above, four parallel integrations initiated from independent atmospheric conditions have been completed.

3. Precipitation anomalies during ENSO events

A composite approach has been adopted to analyze the typical atmospheric responses. The eight warm events occurring in 1957, 1965, 1969, 1972, 1976, 1982, 1987 and 1991, and the eight cold events of 1950, 1954, 1955, 1964, 1970, 1973, 1975 and 1988, have been selected for this analysis. Many previous studies have established that the most prominent ENSO events in the past several decades occurred in these years. In each of the 16 events in the above list, the spatially averaged SST anomaly in the central/eastern equatorial Pacific exceeded ± 0.4°C for a duration of more than 12 months. The years corresponding to the warm or cold events are designated as 'Year (0)'. The years following these events is designated as 'Year (1)'. Warm or cold SST anomalies typically emerge in the boreal spring and summer of Year (0), and attain maximum amplitudes a half-year later [i.e., in the boreal winter of Year(0)/Year(1)].

For each of the three experiments described in the previous section, averages were taken over the four-run ensemble means of the precipitation anomaly for the eight selected warm events to yield the 'warm composites'. Similarly, the 'cold composites' were computed by averaging over the eight cold events. The composite data were then normalized by the local standard deviation σ for the respective calendar season, as obtained from the temporal fluctuations of the four-run averages of each experiment. In Fig. 1 are shown the distributions of the difference between the normalized warm and cold composites of the precipitation in each experiment for the June-July-August season of Year (0) [hereafter abbreviated as JJA(0)], and the December-January-February season six months later [hereafter abbreviated as DJF(0/1)]. We shall henceforth refer to such patterns as the 'warm minus cold' composites.

An ubiquitous feature in all six panels of Fig. 1 is the dipolar pattern in the near-equatorial western Pacific, with below-normal precipitation in the sector between ~120°E and ~150°E, and above-normal precipitation east of ~160°E. During the JJA(0) season, the most notable precipitation signals in the Northern Hemisphere are the negative anomalies over a broad region extending from the Arabian Sea across the Indian subcontinent to the Bay of Bengal, and another dry region over the subtropical western Pacific off the Chinese coast. During the DJF(0/1) season, all three experiments yield a positive anomaly along a belt stretching from southern Japan across southern China to the northern portion of the Indochina peninsula. The other noteworthy DJF(0/1) anomalies in all three experiments are the dry conditions over much of Australia and the surrounding waters, as well as over the South China Sea-Philippine Sea region.

The TOGA experiment (with variable SST forcing confined to the tropical Pacific) is seemingly capable of reproducing many features appearing in the TOGA-ML and GOGA simulations (which include additional air-sea coupling or SST forcing elsewhere). One could infer from this finding that these common features are primarily forced by tropical Pacific SST anomalies associated with ENSO, whereas contributions due to prescribed SST forcing or air-sea interactions at maritime sites outside of the tropical Pacific only play a minor role.

4. Circulation anomalies during ENSO events

Warm minus cold composites of the zonal and meridional wind anomalies in the TOGA experiment have been constructed. The streamline patterns of the resulting circulation anomalies are displayed in Fig. 2 for the seasons of JJA(0) and DJF (0/1), and at 200 mb and 850 mb. The precipitation composites for the respective seasons are also shown in these patterns (see shading).

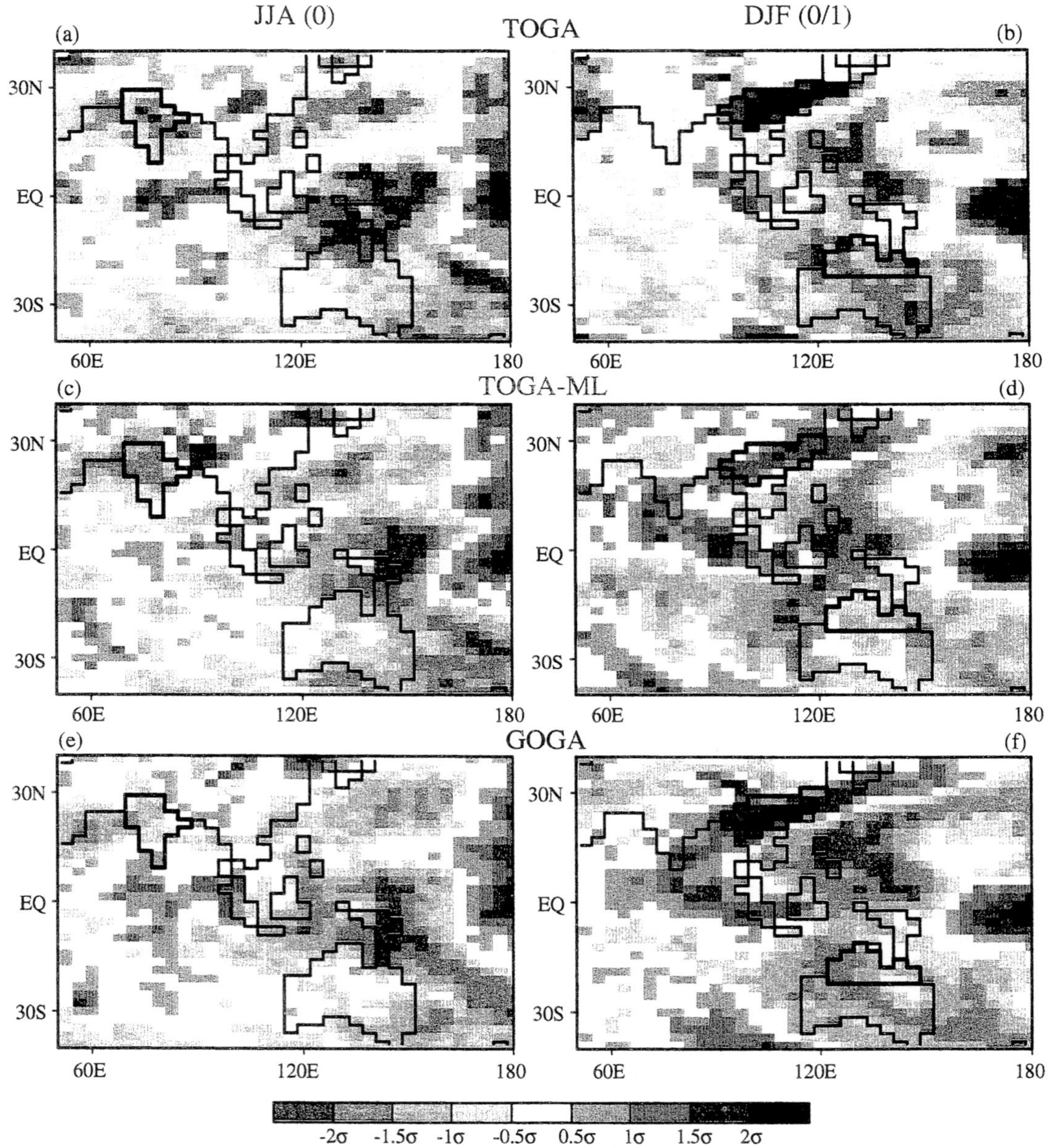

FIG. 1. Warm minus cold composites of the normalized precipitation anomalies in the (a, b) TOGA, (c, d) TOGA-ML and (e, f) GOGA experiments, for the (left panels) JJA(0) and (right panels) DJF (0/1) seasons.

The precipitation data presented in Fig. 2 are actual anomalies, whereas those in Fig. 1 have been normalized by the local standard deviation.

The 850 mb circulation pattern in the warm phase (Figs. 2c and 2d) is characterized by a pair of cyclonic centers over the North and South Pacific (labeled as 'L_N' and 'L_S', respectively), and another pair of anticyclonic anomalies in the Asian-Australian-Indian Ocean sector (denoted as 'H_N' and 'H_S'). During the northern summer, H_N is located to the northwest of the negative precipitation anomaly over the western equatorial Pacific; whereas L_N is situated to the northwest of the positive precipitation center over the central Pacific. The anomaly H_N coincides with the negative precipitation anomaly in the Asian summer monsoon region. In the southern summer, H_S (L_S) is located to the south or southwest of the negative (positive) precipitation anomalies over the equatorial Pacific.

In the DJF(0/1) season (Fig. 2d), H_N is centered over the South China Sea and the Philippines, and is accompanied by below-normal precipitation. The anomalous southwesterly flow on the northern flank of H_N is directed against the climatological off-shore monsoon circulation over eastern Asia during winter, and brings about wetter-than-normal conditions to southern China. The anomalous anticyclonic circulation associated with H_S is accompanied by dryness over northern Australia.

At the 200 mb level (Figs. 2a and 2b), the anomalous circulation pattern over the Pacific Basin is dominated by a pair of anticyclones (labeled as $\hat{H}_N$ and $\hat{H}_S$) straddling the positive equatorial precipitation anomaly. Another noteworthy feature in the upper troposphere is the cyclonic center $\hat{L}_N$ over eastern Asia. This center lies to the north or northwest of the negative precipitation anomaly over the subtropical western Pacific, and is associated with the weakening of the climatological 200 mb Tibetan Anticyclone in summer during warm ENSO events. Comparison between the upper and lower panels of Fig. 2 reveals a reversal in polarity of the near equatorial zonal wind anomaly in the lower and upper troposphere.

The placement of the cyclonic/anticyclonic centers in relation to the dipolar precipitation anomaly over the equatorial Pacific, as well as the considerable degree of symmetry of such centers in the two hemispheres about the equator, are reminiscent of the analytic solution of atmospheric flow patterns associated with tropical heat sources and sinks, as presented by Gill (1980). Specifically, the summertime anomalous anticyclone at 850 mb over South Asia (H_N in Fig. 2c),

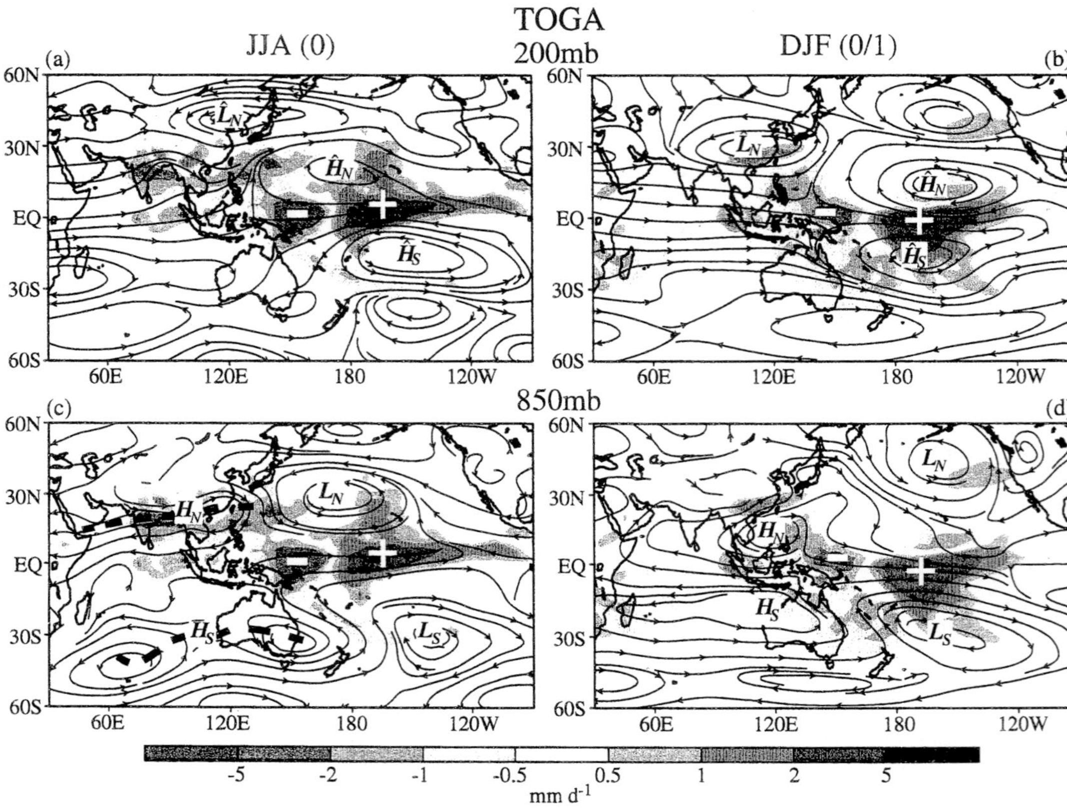

FIG. 2. Warm minus cold composites of the anomalous streamline patterns in the TOGA experiment at (a, b) 200 mb and (c, d) 850 mb for the (left panels) JJA(0) and (right panels) DJF(0/1) seasons. The labels L_N, L_S, H_N and H_S denote the 850 mb cyclonic and anticyclonic centers discussed in the text. The circulation centers at 200 mb are indicated using similar labels with the additional symbol ^. The composite precipitation anomalies for the corresponding seasons are presented using shading (see scale bar at bottom; units: mm d^{-1}).

where the monsoon rainfall is markedly reduced, may be linked to the negative diabatic heating anomaly over the western equatorial Pacific. In the DJF(0/1) season, the same heat sink may also play an important role in the formation of the anticyclone pair H_N and H_S (Fig. 2d), which modulate the winter monsoon over eastern Asia and the summer monsoon over northern Australia, respectively.

5. Ocean-atmosphere coupling processes

The difference between the responses in the TOGA-ML and TOGA runs would provide some information on the influences of air-sea coupling outside the tropical Pacific on the model atmosphere. Similarly, the difference between the results based on the GOGA and TOGA experiments would show the atmospheric impact of observed SST anomalies prescribed outside the tropical Pacific. The warm minus cold composites of surface vector wind and precipitation for the TOGA experiment have been subtracted from the corresponding composites for the TOGA-ML experiment. The results are displayed in the left panels of Fig. 3. The analogous charts as obtained by subtracting the TOGA composites from the GOGA composites are plotted in the right panels of the same figure. The patterns in Fig. 3 may be interpreted as the atmospheric responses to the SST anomalies in the Indian Ocean basin. These oceanic anomalies have been produced either through local air-sea interactions in TOGA-ML, or by direct insertion of observational data in GOGA.

The difference charts for summer (upper panels of Fig. 3) are characterized by southwesterly surface flow over the Arabian Sea and enhanced rainfall over parts of India and the Bay of Bengal. The circulation and precipitation anomalies are opposite to the remote responses of these fields to the tropical Pacific ENSO forcing as simulated in the TOGA experiment (see Fig. 2c). Hence air-sea coupling in the Indian Ocean sector results in the following negative feedback loop: during warm ENSO events: the reduced intensity of the south Asian summer monsoon leads to a heat gain in the surface waters of the Arabian Sea and Bay of Bengal through reduced ocean-to-atmosphere latent heat flux (due to below-normal surface winds) and increased solar radiation (due to less cloud amounts); the subsequent warming in these regions and the resulting anomalous large-scale spatial gradients in the SST pattern in turn force an atmospheric circulation which enhances the strength of the monsoon in this region, thus partially offsetting the (negative) monsoon anomaly that initiated this feedback loop.

The various links in the above chain of processes may be ascertained by computing warm plus cold composites of the heat flux and SST anomalies as simulated in the TOGA-ML experi-

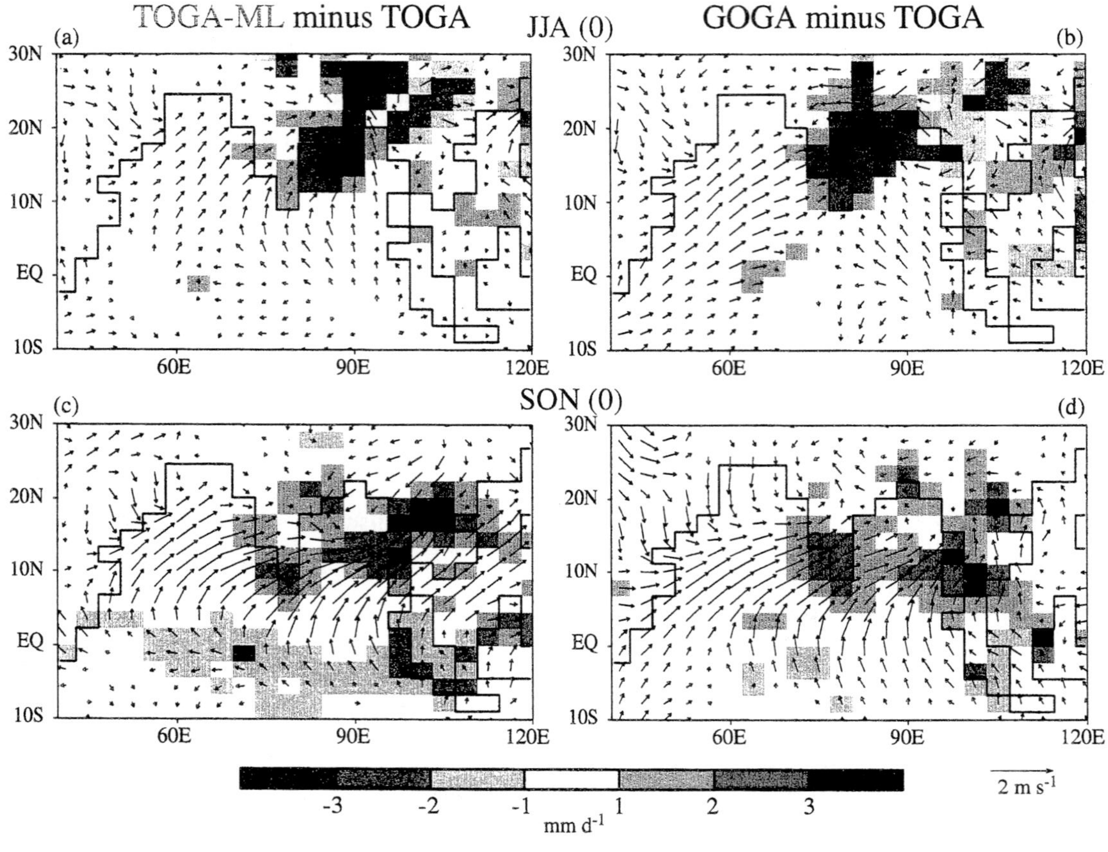

FIG. 3. Difference patterns as obtained by subtracting the warm minus cold composites for TOGA from the corresponding composites for TOGA-ML (left panels), and by subtracting the composites for TOGA from those for GOGA (right panels). Results are shown for the surface wind vector (arrow, see scale at bottom right) and precipitation (shading, see scale bar at bottom; units: mm d^{-1}) during (a, b) JJA and (c, d) SON of Year (0).

ment. These results (not shown) confirm that warm ENSO events are accompanied by positive energy input and positive SST anomalies in the northern Indian Ocean. The simulated SST anomaly pattern in the Indian Ocean sector bears some resemblance to the observed pattern used in forcing the model atmosphere in the GOGA experiment. Due to the thermal inertia of the oceanic mixed layer, the SST anomalies attain maximum amplitudes in the northern autumn, i.e., several months after the peak in the anomalous atmospheric forcing associated with changes in the Indian summer monsoon. Accordingly, the atmospheric responses in autumn [September-October-November of Year (0), or SON(0), see lower panels of Fig. 3] are stronger and more widespread than those in summer. The southwesterly flow pattern in SON(0) extends over much of the Arabian Sea, southern India and the Bay of Bengal, with positive precipitation changes over parts of India, Bay of Bengal and Indochina.

6. Discussion

A majority of observational and modeling ENSO studies have paid exclusive attention to the atmospheric circulation anomalies accompanying the *eastern* half of the dipole in condensational heating in the equatorial Pacific. The evidence presented herein indicates that the *western* portion of the dipole plays a significant role in the formation of monsoon anomalies in the Asian-Australian sector. The reality of the latter dynamical link could be further tested by comparing the solutions of mechanistic stationary models subjected to various spatial configurations of heating anomalies in the Pacific Basin.

By contrasting the results from model runs with and without air-sea coupling in the Indian Ocean (see Fig. 3), it is seen that the Indian Ocean SST perturbations arising from ENSO-related monsoon anomalies could in turn lead to atmospheric changes that oppose such monsoon anomalies. The effects of this negative feedback loop may partially account for the somewhat weaker Indian monsoon signals in TOGA-ML and GOGA (which incorporate SST changes in the Indian Ocean) than those in TOGA (with fixed Indian Ocean SST), as can be discerned from Fig. 1.

References

Gill, A.E., 1980: Some simple solutions for heat-induced tropical circulation. *Quart. J. Roy. Meteor. Soc.*, **106**, 447-462.

Lau, N.-C., and M.J. Nath, 2000: Impact of ENSO on the variability of the Asian-Australian monsoons as simulated in GCM experiments. *J. Climate*, **13**, in press.

SIMULATION AND PREDICTION OF THE REGIONAL CLIMATE AND STREAMFLOW IN TAIWAN

Ming-Chin Wu[1] , Shyh-Chin Chen[2] , Ching-Pin Tung[3]
1. Department of Atmospheric Sciences, National Taiwan University
2. Scripps Institute of Oceanography, University of California - San Diego
3. Department of Agricultural Engineering, National Taiwan University

ABSTRACT

In this study, the feasibility of applying NCEP/RSM to predict the regional climate and further with GWLF hydrological model to forecast the streamflow in Taiwan are investigated. At first, 1992 monthly rainfall in Taiwan simulated by applying NCEP/GSM/RSM were compared with the observations. It is shown that when using 25 km X 25 km grid resolution in horizontal, monthly rainfall in winter can be simulated well but tends to be over-estimated. Sensitivity runs indicated that although higher resolution models can simulate finer structure of the rainfall distribution, rainfall amount is severely over-estimated. The skills of the rainfall prediction for the other seasons are rather poor.

Monthly streamflow in Taiwan can be simulated well by applying a daily weather generating model and GWLF hydrological model. A model system consist of CCM3/ RSM climate model, daily weather generating model, and GWLF hydrological model is applied to do the assessment of climate change impacts on the regional climate and streamflow in Taiwan. Two different resolution RSM runs have been done. One is 50 km X 50 km (RSM1). The other one is 15 km X 15 km (RSM2). The CCM3/RSM model system has been run on current CO2 scenario to reproduce the present regional climate and on doubled CO2 scenario to predict the regional climate change in the future. All the climate models (CCM3, RSM1 and RSM2) predict a consistent warming trend of 1.5 to 2.3 C. However different resolution models predict rainfall and probability of rainy day vary in locations and seasons.

CCM3/GWLF predict streamflow will increase in all area while RMM1/GWLF predict the streamflow will increase in northern but decrease in central, southern, and eastern Taiwan year around. Furthermore, RSM2/GWLF predict the streamflow will increase in the southern and eastern but decrease in the northern and central Taiwan year around. Uncertainties could be caused by any links. Therefore, much more research is needed before any solid conclusion can be reached.

1. INTRODUCTION

The issue of how much will global climate change impact on regional water resources has been raised recently (Tung and Haith, 1995). The creditability of the assessment quite depends on the creditability of regional climate change prediction. Georgi and Mearns (1991) mentioned about to assess regional climate change with a full numerical model approaches. Juang and Kanamitsu (1994) developed a regional spectral model (RSM) in NCEP based on the NCEP global spectral model (NCEP/RSM). This regional model uses the same model configuration and physics of the NCEP/GSM. The only difference between the GSM and RSM is that GSM uses spherical harmonic expansion in horizontal while the RSM uses double sine-cosine instead. The difference between RSM and conventional finite-difference regional model is that all the information from large-scale model in the regional model domain will cover whole domain in RSM instead of only valid in lateral boundary. Some studies showed that RSM has skill on the regional climate simulation (Hong & Leetmaa, 1999). We will use RSM in an assessment of the global climate change impacts on the regional climate and water resources in Taiwan. Since the feasibility of applying RSM on predicting regional climate in a complex terrain area such as Taiwan has not been proved, we will study the feasibility first. Since NCEP/GSM has not shown the creditability on climate simulation yet, in the climate change assessment study we chose CCM3 instead. However, since CCM3 is not a prediction model, in the feasibility of applying RSM on the regional climate simulation study, we still use NCEP/GSM.

The other key to the numerical prediction of regional water resources is hydrological model. Tung et al. (1999) showed that with a daily weather generating model, the hydrological model, GWLF can simulate current stream flow in Taiwan well. We will use CCM3/RSM and this hydrological model system to assess the climate change impacts on the regional water resources (streamflow).

The design of experiments is to run CCM3/RSM/GWLF under current CO2 (355 ppm, 1XCO2) scenario first to reproduce current regional climate and stream flow. Then, it is to run the model system under doubled CO2 scenario (710 ppm) to predict regional climate and stream flow change in the future. Two different resolution RSM runs have been done. One is 50 km X 50 km (RSM1). The other one is 15 km X 15 km (RSM2).

2. SIMULATE REGIONAL CLIMATE IN TAIWAN

The feasibility of applying RSM on predicting regional climate in Taiwan is examined by simulating monthly climate of 1992. The initial data of simulation is taken from NCEP global analysis. This global analysis is the output from NCEP/MRF global spectral model. This global model has resolution as T126 in horizontal and 18 level sigma coordinate in vertical. Initial time was set as daily 0000 UTC. Global analysis data is used to initial global model and then make a 24 hours prediction. Then the global output is used to initialize and run the regional model. Time step of global model is 20 minutes. And that of regional model is 80 seconds. Global model calls radiation package every 3 hours, while regional model calls it every hour. Nested period is every 6 hours. Domain of regional model is chosen as centred at 121 E, 23.5 N. Two kinds of resolution runs have been done. One is in horizontal 30 sine-cosine series expansion, grids 55 X 54, and resolution 25 km X 25 km. The other one is 48 sine-cosine series expansion, 81 X 80 grids, and resolution 15 km X 15 km in horizontal.

The results of 25 km X 25 km runs indicate that RSM can simulates winter precipitation geographical distribution well but tend to over-estimate precipitation amount. Heavy precipitation is evident over where low level easterly flow impinges the topography. Maximum precipitation clusters over northeastern and eastern Taiwan in winter (eg. Fig. 1, February). This is close to observation. The 15 km X 15 km runs highlight finer structure of precipitation distribution but make the over-estimation problem severer. Distributions of monthly precipitation in other seasons are not simulated as good as those of winter (eg. Fig. 2, August). The precipitation over estimation problem is much worse than those in winter. If focus to northern Taiwan (north to 24.5 N), RSM shows good skill in autumn and winter. Table 1 lists the correlation coefficients of monthly precipitation over northern Taiwan between simulation and observation. The skills for simulated precipitation in the winter and autumn are much higher than those in the summer cases. Hong and Leetmaa (1999) did a similar study over the United States and reached a similar conclusion.

Table 1. Monthly correlation coefficients of daily precipitation over northern Taiwan in 1992 between simulation and observation.

Month	01	02	03	04	05	06	07	08	09	10	11	12
Correlation coefficient	0.385	0.279	-0.082	0.021	0.291	0.197	0.001	0.277	0.459	0.697	0.535	0.551

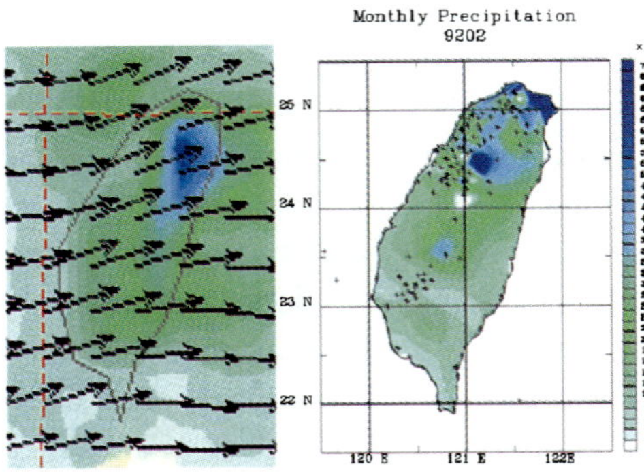

Fig. 1. (a) Moisture flux and precipitation amount of February 1992 simulated by NCEP/GSM/RSM.
(b) Precipitation observations.

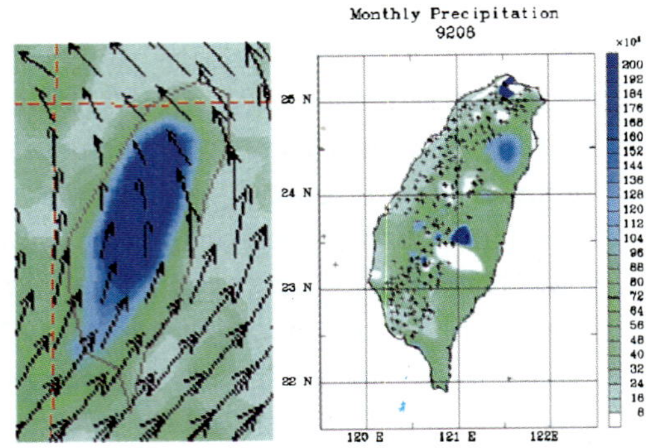

Fig. 2. (a) Moisture flux and precipitation amount of August 1992 simulated by NCEP/GSM/RSM.
(b) Precipitation observations.

3. PREDICTION OF STREAM FLOW IN TAIWAN

Regional climate simulation and prediction are further applied to assess the climate change impacts on stream flow in Taiwan. GWLF hydrological model system and CCM3/RSM1/RSM2 climate model system are applied to compute hydroclimate change after CO2 doubled (Fig. 3). Since low resolution of CCM3 (T42), Taiwan can be represented by only one grid point. The model systems predict a consistent warming trend 1.5 to 2.3 C yearly around in whole Taiwan. CCM3 predict precipitation amount and wet day probability during winter will increase in whole of Taiwan; RSM1 predict precipitation will increase in northern, southern, and eastern Taiwan, decrease in central Taiwan; RSM2 predict precipitation will increase in southern Taiwan but decrease in northern, central and eastern Taiwan. Wet day probability predicts to be decreasing in whole area by RSM1 and RSM2 except that in central Taiwan by RSM2. CCM3 predict precipitation amount and wet day probability during summer will increase in whole Taiwan. RSM1 predict precipitation amount during summer will increase in northern but decrease in central, southern, and eastern Taiwan. RSM2 predict precipitation amount during summer will increase in northern and eastern but decrease in central and southern Taiwan. RSM1 predict wet day probability during summer will increase in southern but decrease in northern, central, and eastern Taiwan. RSM2 predict wet day probability during summer will increase in southern and eastern but decrease in northern and central Taiwan. Furthermore, CCM3 predict stream flow in whole Taiwan will increase yearly around (Table 2). However, RSM1 predict stream flow will increase in northern, but decrease in central, southern, and eastern Taiwan yearly around. RSM2 predict stream flow will increase in southern and eastern but decrease in northern and central Taiwan yearly around. These results indicate that different resolution climate models predict a various trends of regional climate and streamflow.

4. CONCLUDING REMARKS

In this study, the feasibility of applying NCEP/RSM regional spectral atmospheric model to predict the regional climate and further with GWLF hydrological model to forecast stream flow in Taiwan are investigated. It is shown that the regional model with proper domain and resolution can simulate precipitation distribution well during winter season and hydrological model with hydrometeorological inputs from various resolutions climate model simulations will predict inconsistent flow trends in Taiwan. The causes of this inconsistency will be examined in the sequential study. The uncertainty of the predictions should be very large. This uncertainty of the predictions on regional climate could come from large-scale climate models and regional climate models. However, in the state of art, large-scale models could be more skillful after intensive studies in several decades. On the other hand, uncertainty from regional model still has not been clearly identified yet. The uncertainty of the predictions on stream flow should plus the uncertainty from hydrological model beyond that from regional climate model. Therefore a quantitative assessment of the climate change impacts on stream flow with confidence still has a long way to go.

(a) **DJF** (b) **JJA**

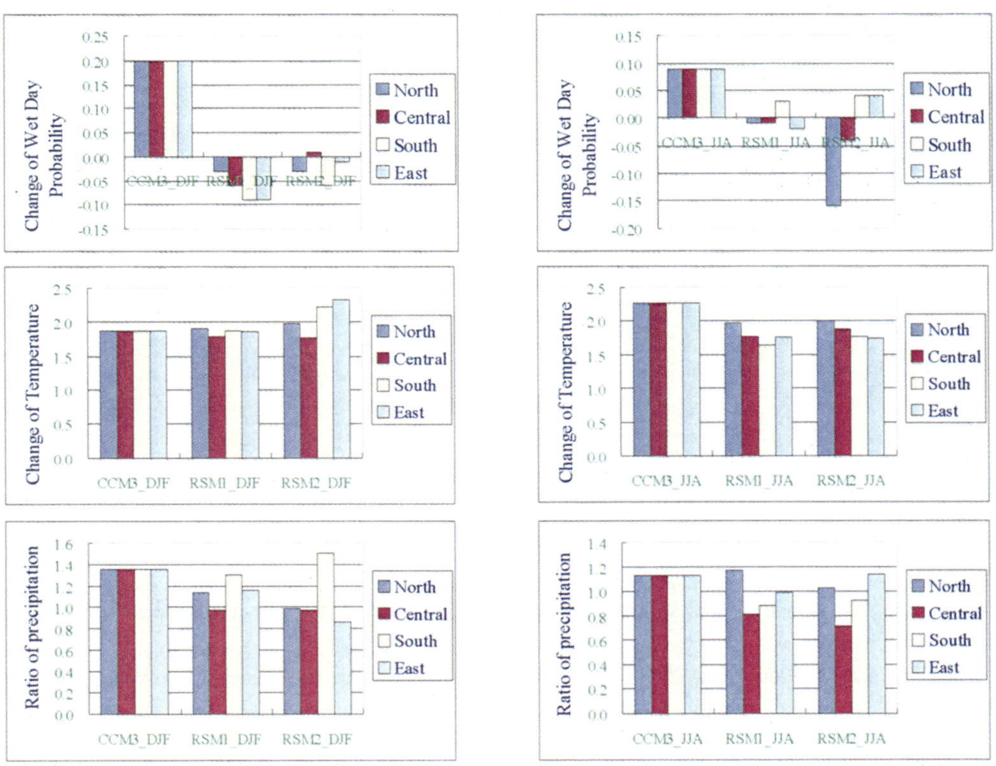

Fig. 3. Ration of Precipitation change of wet day probability, and change of temperature predict by CCM3, RSM1, and RSM2. (a) winter DJF (b) summer JJA.

Table 2. Monthly, annual, summer and winter, are stream flow hydrological model with CCM3, RSM1, and RSM3 climate model respectively.

North

	CCM3	RSM1	RSM2
1	72%	7%	-19%
2	72%	-1%	-22%
3	67%	2%	-20%
4	68%	8%	-13%
5	53%	16%	-16%
6	57%	19%	-23%
7	60%	9%	-32%
8	44%	21%	-46%
9	68%	0%	-47%
10	50%	26%	-36%
11	82%	11%	-23%
12	74%	14%	-22%
Annual	64%	11%	-28%
JJA	56%	15%	-35%
DJF	73%	7%	-20%

South

	CCM3	RSM1	RSM2
1	280%	-24%	36%
2	677%	50%	-9%
3	1045%	-10%	164%
4	1191%	-56%	267%
5	103%	-48%	21%
6	66%	-21%	3%
7	49%	-16%	-11%
8	40%	-1%	0%
9	30%	3%	6%
10	20%	-10%	0%
11	37%	-18%	4%
12	104%	-17%	3%
Annual	62%	-12%	3%
JJA	50%	-12%	1%
DJF	263%	-16%	36%

Central

	CCM3	RSM1	RSM2
1	114%	-36%	-32%
2	201%	-25%	-18%
3	120%	-43%	-7%
4	118%	-41%	-15%
5	53%	-50%	-42%
6	49%	-35%	-39%
7	50%	-36%	-29%
8	53%	-36%	-39%
9	66%	-29%	-45%
10	44%	-34%	-38%
11	68%	-32%	-36%
12	82%	-35%	-41%
Annual	80%	-37%	-32%
JJA	53%	-38%	-39%
DJF	122%	-36%	-20%

East

	CCM3	RSM1	RSM2
1	77%	-19%	14%
2	127%	-19%	11%
3	165%	-5%	10%
4	222%	-22%	11%
5	109%	-14%	61%
6	90%	-2%	63%
7	99%	-2%	58%
8	90%	-12%	59%
9	91%	-8%	61%
10	83%	-9%	65%
11	92%	-17%	18%
12	76%	-18%	19%
Annual	95%	-9%	52%
JJA	92%	-8%	61%
DJF	108%	-17%	16%

REFERENCES

Georgi, F., and L. O. Mearns, 1991: Approaches to the simulation of regional climate change: A review. *Rev. Geophys.* **29**,191-216.

Haith, D. A., and L. L. Shoevmaker, 1987: Geographized water shed loading functions fort stream flow intrients. *Water Resources Bulletin*, **23(3)**, pp.471-478.

Hong, S. Y., and A. Leetmaa, 1999: An evaluation of the NCEP RSM for regional climate modeling. *J. of Climate*, **12** 593-609.

Juang, H.-M. H., and M. Kanamitsu, 1994: The NMC nested regional spectral model. Mon. Wea. Rev., **122**, 3-26.

Tung, C. P., and D. A. Haith, 1995: Global warming effects on New York Stream flows. *Journal of Water Resources Planning and Management*, **21(2)**, pp.216-225.

Tung, C. P., N. M. Hong, and C.H. Chen, 1999: The impacts of climate change on water resources and response strategies. (personal communication)

SUMMARY OF COSMIC AND APPLICATION TO METEOROLOGY AND CLIMATE[1]

Richard A. Anthes and Ying-Hwa Kuo

University Corporation for Atmospheric Research, Boulder, Colorado

ABSTRACT

Based on the success and scientific results of the GPS/MET (Global Positioning System/Meteorology, Ware et al. 1996) project, COSMIC (Constellation Observing System for Meteorology, Ionosphere and Climate), a follow-on project for weather and climate, space weather, and geodetic science is being developed jointly by many partners. COSMIC plans to launch eight satellites in 2004. Each satellite will retrieve about 500 daily profiles of key ionospheric and atmospheric properties from the tracked GPS radio signals as they are occulted behind the Earth limb.

This paper discusses some of the applications of COSMIC data for meteorology and climate. COSMIC will provide high vertical resolution temperature, pressure and water vapour information for a variety of research studies. The COSMIC data set will allow mapping of the global atmospheric flow of water vapour that is crucial for understanding and predicting weather and climate. The data set will provide accurate geopotential heights, enable the detection of gravity waves from the upper troposphere to the stratosphere, reveal the height and shape of the tropopause globally with unprecedented accuracy, support the investigation of fronts and other baroclinic structures, and improve our understanding of tropopause-stratosphere exchange processes.

COSMIC data will complement other observing systems and improve global weather analyses, particularly over the oceans and polar regions, and numerical weather forecasts made from these analyses. Through assimilation in numerical models, COSMIC data will improve the resolution and accuracy of the global temperature, pressure and water vapour fields, and through the model's dynamical and physical adjustment mechanisms, the wind fields as well.

For climate research and monitoring COSMIC will provide an accurate global thermometer that will monitor Earth's atmosphere in all weather with unprecedented long-term stability, resolution, coverage, and accuracy. COSMIC will provide a data set for the detection of climate variability and change, the separation of natural and anthropogenic causes, the calibration of other satellite observing systems and the verification and improvement of climate models.

---

[1] Proceedings of the 4th East Asian-Western Pacific Meteorology and Climate Conference, Hangzhou, 26-28 October 1999.

1.0 Introduction

The Global Positioning System (GPS) is a satellite-based positioning system developed for navigation and positioning. The GPS today consists of 24 satellites that transmit L-band radio signals to a wide variety of users in navigation, time transfer and relative positioning and for an ever-increasing number of scientists in geodesy, atmospheric sciences, oceanography and hydrology.

Atmospheric soundings are obtained using GPS through the radio occultation technique, in which satellites in low-Earth orbit (LEO), as they rise and set relative to the GPS satellites, measure the change in frequency of the GPS dual-frequency signals. The radio occultation sounding technique was developed by scientists at Stanford University and the Jet Propulsion Laboratory (JPL) for the remote sensing of planetary atmospheres (Melbourne et al., 1994). The Doppler shifted frequency measurements are used to compute the bending angles of the radio waves, which are a function of atmospheric refractivity. The refractivity is a function of electron density in the ionosphere and temperature, pressure and water vapor in the neutral atmosphere.

A detailed discussion of the GPS radio occultation observations and data processing may be found in a number of papers, see, for example, Melbourne et al. (1994), Hocke (1997), and Kursinski et al. (1997). Therefore, we only provide a brief summary here. Each of the 24 GPS satellites continuously transmits signals at two frequencies, L1 at 1.57542 GHz (~19 cm) and L2 at 1.2276 GHz (~24.4 cm). In a GPS occultation, the GPS receiver on a LEO satellite tracks the occulted signals from a GPS satellite whenever a GPS satellite rises or sets behind the Earth (Fig. 1). Atmospheric refractivity affects the phase and amplitude of the radio waves. If we consider wave propagation in terms of rays (Fig. 1), then the effect of the atmosphere can be characterized by the total bending angle α as a function of the impact parameter, a, which are derived from the measurements of the phase and amplitude of the received signal.

The vertical profile of α is inverted into a refractivity profile by assumption of spherical symmetry and use of an Abelian transformation [Fjeldbo et al. 1971] given by:

$$ln(n(a)) = \frac{1}{\pi} \int_a^\infty \frac{\alpha(a')}{\sqrt{a'^2 - a^2}} da' \qquad (1)$$

Refractivity ($N = (n-1)\times 10^6$) is related to atmospheric pressure, P, temperature, T, water vapor partial pressure, P_w, and electron density, n_e, through

$$N = 77.6\frac{P}{T} + 3.73\times 10^5 \frac{P_w}{T^2} - 40.3\frac{n_e}{f^2} \qquad (2)$$

where f is the signal frequency. At altitudes of 90 km and above, the pressure and water vapor terms are negligible and the observed refractivity is directly proportional to the electron density in the ionosphere. In the stratosphere and the upper regions of the troposphere where T < 240K, the water vapor contribution to refractivity is negligible, and refractivity is proportional to density via the ideal gas law (after removal of the ionospheric contribution using the dual GPS frequencies). Density in turn yields pressure through the assumption of hydrostatic equilibrium and a boundary condition on pressure at an initial height (~60 km). Applying the gas law once more, density and pressure yield temperature. In the warmer regions of the troposphere, water vapor can contribute as much as 30% to refractivity and can locally dominate the vertical refractivity gradients and bending near the surface particularly at low latitudes.

At the GPS frequencies, radio waves are only minimally affected by aerosols, clouds and light precipitation (Solheim et al., 1999). The vertical resolution of variables derived from the radio occultation technique is limited by diffraction and is given roughly as the diameter of the first Fresnel zone defined as $(\lambda D)^{1/2}$ where λ is the wavelength and D is the distance from the receiver to the limb. This resolution is ~1.4 km in the ionosphere and stratosphere. In the troposphere, the large bending causes the vertical dimension of the first Fresnel zone to decrease, improving the vertical resolution to 0.2 to 1 km [Kursinski et al. 1997].

The effects of diffraction can be reduced substantially. Since the received signal has a well defined phase and amplitude, the received phase and amplitude can be used to propagate the signal backwards through the vacuum to a location closer to the Earth's limb (Gorbunov and Gurvich, 1998). The resolution can be improved to ~150 m in the stratosphere and troposphere (Mortensen and Hoeg, 1998). The back-propagation approach has the further advantage of reducing atmospheric multipath by moving the effective reception point closer to the Earth's limb thereby reducing the crossing of signal paths.

The GPS/MET program (Ware et al., 1996) was established by UCAR in 1993, jointly with the University of Arizona and JPL, to demonstrate active limb sounding of the Earth's neutral atmosphere and ionosphere using the radio occultation technique. From raw GPS/MET observations, vertical profiles of ray bending angle and refractivity were retrieved; from which ionospheric electron density, neutral atmospheric density, pressure, temperature, and moisture profiles were computed.[2] The program has been exceptionally successful, having accomplished nearly all of the "proof of concept" goals, plus a number of additional ones.

In the GPS/MET experiment, only setting occultations were observed and only about 10% of the soundings penetrated into the lowest kilometer of the atmosphere. The goal for COSMIC is to track 90% of all rising and setting occultation soundings to within one kilometer of the Earth's surface. This will be achieved with a ~10 db higher gain antenna and "open loop" tracking in the lower part of the troposphere.

The characteristics of COSMIC data and their relevance to meteorology (including weather prediction) and climate are summarized in Table 1. The set of unique characteristics will make the COSMIC data very useful as complements to other observing systems. ESA (1996), Hoeg et

[2] The techniques, algorithms and assumptions used to process GPS/MET observations are described by Hoeg et al. (1996), Kursinski et al. (1996,1997), Rocken et al. (1997) and Hocke (1997).

al. (1996) and Melbourne et al. (1994) provide excellent discussions of the potential value of radio occultation data for meteorological and climate research and operations.

Table 1: Characteristics of COSMIC Data
Limb sounding geometry complementary to ground and space nadir viewing instruments
High accuracy
High vertical resolution
Consistency of horizontal and vertical scales of observations (Lindzen and Fox-Rabinovitz, 1989)
All weather – minimally affected by clouds, aerosols, or light precipitation (Solheim et al. 1999)
Independent height and pressure
Requires no first guess sounding
Independent of radiosonde or other calibration
No instrument drift
No satellite-to-satellite bias

2.0 COSMIC and Meteorology

With a total of 4,000 soundings per day distributed over the globe, COSMIC will provide the much-needed additional data over data-sparse areas, especially over the oceans, tropical and polar regions. The high vertical resolution, accurate, and independent COSMIC data will complement data from geostationary (Menzel et al., 1998) and polar-orbiting satellites, which provide high horizontal but low vertical resolution soundings. The average horizontal spacing of the 4,000 daily soundings varies with latitude; the spacing is greatest near the poles (approximately 300 km) and least at the Equator (approximately 500 km).

2.1 Resolution of important atmospheric structures

In middle latitudes, the predominant features of interest are baroclinic waves and the associated cyclones and anticyclones. The horizontal wavelength of these waves varies from roughly 2,000 km for the shortest waves to 8,000 km for long planetary waves. With an average spacing of approximately 300-400 km poleward of 20° latitude, COSMIC data will resolve these baroclinic waves well. COSMIC data will also resolve incipient baroclinic waves, especially over the oceans, because the wavelength of maximum baroclinic instability for typical conditions of static stability is about 4,000 km (Holton, 1979).

In the tropics, COSMIC data by themselves will not resolve the detailed structure of tropical cyclones. However, they will contribute significantly to the definition of the pressure, temperature and water vapor, and through data assimilation, the wind field in the large-scale environment surrounding the storm. The typical scale of circulation affecting the motion of tropical cyclones is 3,000-5,000 km from the cyclone center (Anthes, 1982). With a horizontal spacing of 350-450 km between latitudes 10° and 30°, COSMIC soundings will resolve the environment of tropical storms and thereby contribute to improved tropical storm track forecasts.

COSMIC data will resolve vertical scales as small as 0.5 km. This vertical resolution is considerably higher than satellite-based radiometric soundings. The GPS/MET experiment has indicated that the radio occultation soundings are capable of resolving strong temperature inversions associated with upper-level fronts and the tropopause (Kuo et al., 1998). The high accuracy and vertical resolution of COSMIC data in the upper troposphere and lower stratosphere will improve the analysis of potential vorticity in the vicinity of the jet stream and tropopause, a region important in the accurate initialization of NWP models for predictions of cyclogenesis (Davis and Emanuel, 1991).

2.2 Retrieval of temperature and water vapor

As shown by (2), in the neutral atmosphere, where $n_e = 0$, the relationship between refractivity, temperature, pressure and water vapor, together with the hydrostatic equation, comprise a system of two equations with three unknowns. Where water vapor is negligible, such as in the stratosphere and upper troposphere, pressure and temperature may be obtained directly from the observed refractivity. Where water vapor is present in significant amounts, it is impossible, without additional independent data, to separate out the effects of temperature and water vapor. However, in principle it is possible to calculate either water vapor or temperature given refractivity and an independent estimate of the other variable. In practice, it is much more likely that useful estimates of water vapor can be obtained given the refractivity and an independent estimate of temperature than the other way around (Ware et al., 1996).

2.3 Upper-level fronts and the tropopause

As demonstrated by GPS/MET, COSMIC data will provide high-vertical resolution data that can contribute significantly to the observation and analysis of atmospheric fronts and the tropopause. Kuo et al. (1998) showed that GPS/MET data accurately resolved an intense upper-level front over China (Fig. 2), and that even though this front occurred in a region of dense radiosonde data, the operational analyses of NCEP and ECMWF resolved this front only in a very smooth way. Similarly, COSMIC soundings will improve the resolution of the tropopause over the world (Ware et al., 1996; Rocken et al., 1997).

2.4 Gravity waves and stratospheric dynamics

As summarized in the recent planning document for the SPARC (Stratospheric Processes and their Role in Climate) Program (WCRP, 1998), the stratosphere plays an important role in weather and climate. COSMIC data will contribute strongly to the SPARC goals of understanding stratosphere-troposphere exchange and the structure and climatology of stratospheric gravity waves and their parameterization in NWP and climate models (ESA, 1996).

Hocke (1997) and Tsuda et al. (2000) describe the use of GPS/MET data for the study of gravity waves in the middle and upper atmosphere. Kursinski et al. (1996) show a GPS/MET temperature sounding with oscillations in the stratosphere with vertical wavelengths around 3 km; these waves were in excellent agreement with waves resolved by a nearby radiosonde. The vertical oscillations shown in Figure 3 above 40 km are likely due to gravity waves.

COSMIC promises to detect gravity wave activity globally with much better temporal and spatial resolution, thus allowing studies of how gravity waves relate to wind shear, thunderstorms, cyclones and other meteorological phenomena.

2.5 Polar Meteorology

COSMIC will greatly increase the amount of atmospheric soundings in the polar regions where operational analyses often show large differences (e.g., Fig. 4). Fig. 5 shows the expected soundings from eight satellites in one day and the corresponding radiosonde network.

Weather analysis and forecasting in the polar regions are likely to be significantly enhanced by COSMIC and will have important applications in areas such as the aviation industry and in the execution of various scientific and commercial missions in these remote regions. Some examples where COSMIC can be useful to polar meteorology are:

- Improved meteorological analyses and forecasts
- Surface pressure corrections for ice balance studies with GRACE (Gravity Recovery and Climate Experiment)
- Vertical temperature profiles for ozone depletion studies
- Polar water vapor studies (assuming temperatures are known, or available through model assimilation of refractivity)
- Investigation of sudden stratospheric warming events during the polar spring
- Improved aviation weather analysis and prediction over the polar regions

3.0 COSMIC and Numerical Weather Prediction

COSMIC will generate meteorological products within three hours of observation for tests in operational numerical weather prediction (NWP). This section argues that COSMIC data, if assimilated effectively in NWP models, will improve global analyses and forecasts.

3.1 Effect on global analyses

Although the global coverage of data from the GPS/MET experiment was inadequate to make a difference operationally, the GPS/MET data provided evidence that COSMIC data will have a significant positive impact on global operational analyses. Leroy (1997) compared time-averaged analyses of geopotential heights computed from GPS/MET data with operational analyses over the same period from ECMWF and found large differences over data-sparse regions such as the South Pacific ocean. Since the differences were small over data-rich regions, this study suggests that COSMIC data will make a positive difference to operational analyses even at large scales over the oceans. Similarly, Rocken et al. (1997) found a significantly greater difference between GPS/MET refractivities and derived temperatures from NCEP analyses over data-poor regions versus data-rich regions.

Further evidence that COSMIC data will significantly modify operational analyses is shown in Fig. 6, which depicts the zonal average of temperature differences between GPS/MET and NCEP and GPS/MET and ECMWF from 500 mb to 50 mb. These comparisons show that significant differences, and therefore uncertainties, exist between objective analyses from leading world meteorological centers. Hence COSMIC data can be expected to make a significant impact on operational analyses.

On smaller scales and over data-rich regions, it is difficult to estimate in advance what the impact of COSMIC data will be. However, the improved vertical resolution of upper-level fronts and the tropopause by GPS/MET soundings compared to the operational weather analyses provide evidence that COSMIC data may have a positive impact on analyses even over data-rich regions. It is possible that the impact will be small on many days and large on a few significant days.

3.2 Initialization of NWP models using COSMIC data

COSMIC data will be available for testing in real-time in operational NWP models. It is likely that the data will be assimilated into the models using three-dimensional or four-dimensional variational techniques (3DVAR or 4DVAR). Either of these data assimilation techniques may, in principle, assimilate bending angles, refractivities, or derived temperature and water vapor data. Eyre (1994) and Zou et al. (1999) suggest that the most powerful use of COSMIC neutral atmosphere data is to assimilate the refractivity or bending angle profiles directly into numerical models. Numerical experiments have shown that assimilation of refractivity and bending angles directly into numerical models causes the model's temperature, water vapor, and wind fields to adjust toward actual atmospheric values (Zou et al., 1995; Kuo et al., 1997).

COSMIC data will provide information on the temperature, pressure and water vapor fields of the atmosphere. These variables constitute the *mass field* of the atmosphere. The type of balance between the mass and wind fields varies depending on the horizontal and vertical scales of motion of the atmospheric feature, the latitude, and whether or not significant acceleration effects are present.

In fact, the atmosphere is never in complete balance; it is continuously adjusting to small imbalances between the mass and the wind fields through the generation and propagation of inertial gravity waves. Through the data assimilation process, if one assimilates accurate observations of *either* the mass *or* the wind field into the model, the other field will adjust through the model's equations to an appropriate balance. Herein lies one of the great advantages of continuous data assimilation.

Although, in principle, either mass or wind observations may be assimilated in models and the other field will adjust appropriately (Atlas et al., 1985), the efficiency of the adjustment process varies depending on the variable being assimilated, the horizontal and vertical scale of the phenomenon, the latitude, the static stability, and the nearness to the surface of the Earth. In general, mass field (and, therefore, COSMIC data) are most useful for large horizontal and small vertical scales of motion away from the Equator where the static stability is low (relatively unstable).

Data assimilation studies have shown that if mass variables (pressure or temperature) are assimilated over time in a model, the wind field will adjust to the correct values (Atlas et al., 1985). Charney et al. (1969) illustrates this fact by showing the reduction in wind error over time given the assimilation every 12 hours of temperature data. In middle latitudes the reduction in wind error is relatively rapid. At the Equator the adjustment of the wind to the temperature data is slower in accordance with geostrophic adjustment theory, but even directly on the Equator, there is a positive impact of assimilating accurate temperature data.

In another example, numerical experiments with a primitive equation model of a tropical cyclone indicate that continuous assimilation of accurate temperature data has almost as much positive effect on the model cyclone as does the continuous assimilation of both temperature and wind data (Anthes, 1974).

In summary, we support the conclusions from the analysis of the value of GPS/MET soundings by Hoeg et al. (1996): "Atmospheric profiling using the radio occultation technique offer a major potential for improving the products from operational numerical weather prediction models."

4.0 COSMIC and Climate

Observing changes and variability in the global climate is one of the most important challenges in atmospheric sciences. Because of the many difficulties associated with making long-term in-situ measurements (for example calibration, representativeness, costs, difficulty in making temporally-continuous measurements over remote areas and throughout the entire troposphere and stratosphere), remote sensing from space is the most desirable way to observe the global atmosphere over long periods of time.

The European Space Agency (ESA, 1996) describes many of the advantages and potential contributions of radio occultation data to climate studies. The recent National Research Council (NRC, 1998) assessment of the U.S. Global Change Research Program (USGCRP) emphasizes the importance of obtaining a long-term record of global temperature change that is accurate and stable. According to this report, "..an essential requirement of the observational strategy is to establish long-term, scientifically valid, consistent records for global change studies." (NRC, 1998, p. 11). The NRC report also states (p. 26): "The importance of accuracy, continuity, calibration, documentation, and technological innovation in observations for long-term trend analysis of global change cannot be overemphasized." The characteristics of the radio occultation technique in general and COSMIC in particular include accuracy, long-term stability, no calibration, no instrument drift, relatively unaffected by clouds, precipitation and aerosols and its high vertical resolution, all at relatively low cost. The technique is in many ways the ideal "global thermometer", meeting many of the key objectives of the USGCRP as stated by the NRC and discussed recently by Goody et al. (1998).

The special nature of GPS occultation data with regard to calibration is very crucial for detecting small variations and trends in climate, since data taken in different intervals by different research groups and from different countries can be compared without concern for instrument degradation or calibration. COSMIC will deliver 4,000 soundings/day with random measurement errors of about +/-1 K. For climate studies space-time averages of COSMIC observations can be formed to reduce the random error. A one-month average over each 1% area of the Earth's surface (~1,200 soundings) will reduce the random 1 K error to 0.03°C. This is adequate to follow trends of 0.1 K per decade, which is typical of greenhouse gas or other forced climate responses. It has been shown (North and Stevens, 1998; Stevens, 1999) that stratospheric-tropospheric temperature profile information enhances the signal-to-noise ratio of climate data significantly over surface-only observations. Over much of the Earth the greenhouse-warming signal in models is strongest in the upper troposphere and lower stratosphere (Meehl et al., 2000) where the accuracy of COSMIC data will be the highest.

According to model studies and suggested by observations over the past 25 years (IPCC, 1996), the lower troposphere vapor concentrations will increase with increasing surface temperature (the water-vapor feedback effect). The situation is not so clear in the upper troposphere, where small amounts of water vapor are important for radiative heat transfer. Lindzen (1990) argues that increased surface temperature and resulting stronger tropical convection will result in increased tropical precipitation and a decrease of tropical/subtropical tropospheric moisture above 6 km. This is in contrast to model predictions of water vapor increases of 20% per degree of surface warming. Clearly it is important to resolve the differing theories and to test the performance of climate models in simulating upper troposphere water vapor. This, however, may be difficult with current observing systems as Rind (1998) points out. Radiosondes are not ideal for detecting long-term global trends, and downward-looking radiometers lack the required vertical resolution. COSMIC data will help the situation. Assuming that temperature is known at the 500 mb level from models to within 1-2 degrees, individual COSMIC profiles will measure the water vapor pressure at that height to about 0.2 mb or, assuming all errors are random, averaged over a 1% (of the area of the Earth) 1-month 1,200-sounding "climate box", to 0.006 mb. Water vapor pressure at 500-mb height varies between 0 – 6 mb, and a several percent change, especially in the tropics, should be detectable by COSMIC data with high vertical resolution. At 300 mb water vapor varies in the range of 0-0.7 mb. Following the same arguments as above, water vapor at that altitude can be measured by COSMIC to ~0.12 mb, or to a climatological average of 0.004 mb per month.

Fig. 7 shows GPS/MET derived water vapor (using ECMWF temperatures) at 300 and 500 mb for two different time periods. These results have not been validated with correlative data but look generally reasonable with the highest amounts over regions of deep convection. The East Asian monsoon shows up clearly during the June-July period, and we see the high water vapor values over Africa and South America, presumably caused by convection.

Because of their insensitivity to smoke and aerosols, COSMIC data will also be useful in studying the effect of volcanoes or large fires on climate and related global change studies. The Mt. Pinatubo eruption in 1991 caused significant changes in the global climate with warming in the stratosphere of up to 1 K and cooling in the troposphere of more than 1 K in certain regions (Fiocco et al., 1996). Both the stratospheric warming and tropospheric cooling patterns showed large variations in space – patterns that could be resolved with COSMIC data. Christopher et al. (1998) show that the biomass burning in South America significantly perturbs the regional climate. The Kuwaiti oil fires of 1991 (Husain, 1995) perturbed the weather and short-term climates of this region significantly. Accurate, high-resolution global measurements over a long period of time are usually difficult to obtain because the events often occur in remote regions, and the environment is hostile to making in-situ measurements. COSMIC can contribute to studies of perturbations to climate by similar episodic events.

Key scientific contributions of COSMIC to climate research are summarized below:

- Provide a global self-calibrating data set for climate monitoring and climate model testing
- Monitor water vapor distribution, especially in the tropics
- Study troposphere/stratosphere exchange
- Study the effects of volcanic eruptions and large fires on climate
- Stimulate development of new climate change parameters (i.e., profiles of bending angles)
- Support climate statistics and correlation studies

5.0 Summary

COSMIC data offer a number of unique advantages that complement existing observing systems and, together with these systems, offer the potential to improve understanding of global and regional atmospheric processes and to improve weather forecasts and climate projections into the future. Table 2 summarizes the most important scientific contributions to meteorology and climate that COSMIC will provide.

Table 2: Applications of COSMIC data to atmospheric sciences, including weather and climate research and operational weather prediction.

•	Provide 4,000 accurate soundings per day globally and in all weather with high vertical resolution of bending angle, refractivity and derived products such as temperature and water vapor in stratosphere and troposphere.
•	Calculate temperature directly where water vapor is negligible (e.g. stratosphere, upper troposphere).
•	Calculate water vapor directly where temperature is fairly well known independently.
•	Estimation of winds through geostrophic or gradient wind relationships in high latitudes or through assimilation of data in models.
•	Direct assimilation of bending angle or refractivity in numerical models to recover temperature, water vapor and winds.
•	Improve global analyses in real time and research mode (e.g. reanalysis) to obtain improved estimates of temperature, water vapor and winds, particularly over oceans and polar regions.
•	Improve vertical resolution of upper-level fronts and associated jet streams and tropopause, supporting process studies such as tropospheric-stratospheric exchange.
•	Improve understanding of global water cycle, climatology of water vapor.
•	Investigation of gravity waves in the stratosphere.
•	Investigation of effects of large fires and volcanoes on weather and climate.
•	Resolve the diurnal cycle of temperature globally
•	Monitor global and regional temperature change with unprecedented accuracy, vertical resolution and stability in troposphere and stratosphere for climate and global change studies.
•	Resolve discrepancies in temperature records from various sources (e.g. the Microwave Sounder Unit inferred temperature record compared to surface temperature record measured by in-situ instruments).
•	Monitor global distribution of geopotential heights for climate and global change studies.
•	Introduce bending angle and refractivity as new climate change parameters.
•	Stratospheric studies, including ozone depletion, stratospheric-tropospheric exchange and volcanic effects.
•	Improve global and regional numerical weather predictions and operational forecasts.
•	Complement existing observing systems and provide independent data for development of improved analysis systems using merged observations.
•	Provide data to calibrate or interpret other satellite-based sounding systems on GOES, POES and EOS.
•	Provide data to support research field programs (e.g. programs like TOGA-COARE, INDOEX, JASMINE, FASTEX).

Acknowledgments

We acknowledge the support of the National Science Foundation, the National Aeronautics and Space Administration, the National Oceanic and Atmospheric Administration, the U.S. Air Force, and the U.S. Navy in the United States and the National Science Council and National Space Program Office in Taiwan for their leadership and support of COSMIC.

References

Anthes, R.A., 1974: Data assimilation and initialization of hurricane prediction models. *J. Atmos. Sci.*, **31**, 702-719.

Anthes, R.A., 1982: Tropical cyclones-Their evolution, structure and effects, *Meteorological Monographs,* **19**, No. 41, American Meteorological Society, Boston, 208 pp.

Atlas, R., E. Kalnay and M. Halem, 1985: Impact of satellite temperature sounding and wind data on numerical weather prediction. *Optical Engineering*, **24**, 341-346.

Charney, J., M. Halem and R. Jastrow, 1969: Use of incomplete historical data to infer the present state of the atmosphere. *J. Atmos. Sci.,* **26**, 1160-63.

Christopher, S.A., M. Wang, J.A. Berender, R.M. Welch and S.-K. Yang, 1998: The 1985 biomass burning season in South America: Satellite remote sensing of fires, smoke and regional radiative energy budgets. *J. Appl. Met.*, **37**, 661-668.

Davis, C.A. and K.A. Emanuel, 1991: Potential vorticity diagnostic of cyclogenesis. *Mon. Wea. Rev.,* **119**, 1929-1953.

ESA, 1996: Atmospheric Profiling Mission. ESA SP-1196(7). The Nine Candidate Earth Explorer Missions. European Space Agency, ESA Publications Division, Noordwijk, The Netherlands, 58 pp.

Eyre, J., 1994: Assimilation of radio occultation measurements into a numerical weather prediction system, *Tech. Memo. 199, Eur. Cent. For Medium Range Weather Forecasts,* Reading, England.

Fiocco, G., D. Fua and G. Visconti, 1996: **The Mount Pinatubo eruption: Effects on the atmosphere and climate.** Springer, New York, 310 pp.

Fjeldbo G., A. J. Kliore, and V. R. Eshleman, 1971: The neutral atmosphere of Venus as studied with the Mariner V radio occultation experiment, *Radio Sci.*, **4**, 879-897.

Goody, R., J. Anderson and G. North, 1998: Testing climate models: An approach. *Bull. Amer. Met. Soc.,* **79**, 2541-2549.

Gorbunov, M.E., and A.S. Gurvich, 1998: Microlab-1 experiment: Multipath effects in the lower troposphere, *Jour. Geophysical Res.*, **103 (D12)**, 13819-13826.

Hocke K., 1997: Inversion of GPS meteorology data, *Ann. Geophys.*, **15**, 443-450.

Hoeg, P., A. Hauchecorne, G. Kirchengast, S. Syndergaard, B. Belloul, R. Leitinger, and W. Rothleitner, 1996: Derivation of atmospheric properties using a radio occultation technique, *Sci. Rep.,* **95-4**, Danish Met. Inst.

Holton, J.R., 1979: **An Introduction to Dynamic Meteorology**, Second Edition. Academic Press, New York, 391 pp.

Husain, T., 1995: **Kuwaiti oil fires: Regional environmental perspectives.** 1st Ed. Pergamon, Oxford, NY.

IPCC, 1996: **Climate change 1995. The science of climate change. Contribution of WGI to the second assessment report of the Intergovernmental Panel on Climate Change.** Cambridge University Press, Cambridge, UK, 572 pp.

Kuo, Y.-H., X. Zou, and W. Huang, 1997: The impact of GPS data on the prediction of an extratropical cyclone: An observing system simulation experiment, *J. Dyn. Atmos. Ocean*, 439-470.

Kuo, Y.-H., X. Zou, S.-J. Chen, Y.-R. Guo, W. Huang, R. Anthes, D. Hunt, M. Exner, C. Rocken, S. Sokolovskiy, 1998: A GPS/MET sounding through an intense upper-level front, *Bull. Amer. Met. Soc.*, **79**, 617-626.

Kursinski, E. R., G.A. Hajj, W. I. Bertiger, S. S. Leroy, T. K. Meehan, L. J. Romans, J. T. Schofield, D. J. McCleese, W. G. Melbourne, C. L. Thornton, T. P. Yunck, J. R. Eyre, and R. N. Nagatani, 1996: Initial results of radio occultation observations of Earth's atmosphere using the Global Positioning System, *Science,* **271**, 1107-1110.

Kursinski, E.R., G.A. Hajj, J.T. Schofield, R.P. Linfield and K.R. Hardy, 20 Oct. 1997: Observing the Earth's atmosphere with radio occultation measurements using the Global Positioning System, *J. Geophys. Res.*, **102**, No. D19, pp 23429-23465.

Leroy, S., 1997: The measurement of geopotential heights by GPS radio occultation, *J. Geophys. Res.,* **102**, 6971-6986.

Lindzen, R.S. and M. Fox-Rabinovitz, 1989: Consistent vertical and horizontal resolution. *Mon. Wea. Rev.*, **117**, 2575-2583.

Lindzen, R.S. 1990: Some coolness concerning global warming, *Bull. Amer. Met. Soc*,. **71**, 288.

Meehl, G.A., W. D. Collins, B. Boville, J.T. Kiehl, T.M.L. Wigley, and J.M. Arblaster, 2000: Response of the NCAR Climate System Model to increased CO2 and the role of physical processes. *J. Climate*, in press.

Melbourne, W., E. Davis, C. Duncan, G. Hajj, K. Hardy, E. Kursinski, T. Meehan, L. Young, and T. Yunck, 1994: The application of space borne GPS to atmospheric limb sounding and global change monitoring, *JPL Publ.*, **94-18**, 147 pp.

Menzel, W.P., F.C. Holt, T.J. Schmit, R.M. Aune, A. J. Schreiner, G.S. Wade and D.G. Gray, 1998: Application of GOES-8/9 soundings to weather forecasting and nowcasting. *Bull. Amer. Met. Soc.*, **79**, 2059-2077.

Mortensen, D.M., and P.Hoeg, 1998: Inversion of GPS occultation measurements using Fresnel diffraction theory, *Geophys. Res. Lett.*, **25**, 2441-2444.

North, G. R., and M. J. Stevens, 1998: Detecting climate signals in the surface temperature record, *J. Climate*, **11**, 563-577.

NRC, 1998: **Global environmental change-Research pathways for the next decade.** National Academy Press, Washington, D.C., 69 pp.

Nutter, P. A., S. L. Mullen and D. P. Baumhefner, 1998: The Impact of Initial Condition Uncertainty on Numerical Simulations of Blocking. *Mon. Wea. Rev.*, **126**, 2482–2502.

Rind, D., 1998: Just add water vapor, *Science*, **281**, 1152.

Rocken, C., R. Anthes, M. Exner, D. Hunt, S. Sokolovskiy, R. Ware, M. Gorbunov, W. Schreiner, D. Feng, B. Herman, Y. Kuo, and X. Zou, 1997: Analysis and validation of GPS/MET data in the neutral atmosphere, *J. Geophys. Res.*, **102**, pp 29849-29866.

Solheim, F.S., J. Vivekanandan, R. Ware and C. Rocken, 1999: Propagation delays induced in GPS signals by dry air, water vapor, hydrometeors and other particulates. *J. Geophys. Res. – Atmos.*, **104**, D8, 9663-9670.

Stevens, M.J., 1999: Optimal climate signal detection in four dimensions, *J. Geophys. Res. - Atmos.*, **104**, D4, pp 4089-4099.

Tsuda, T., M. Nishida, C. Rocken and R. Ware, 2000: A global morphology of gravity wave activity in the stratosphere revealed by the GPS occultation data (GPS/MET). *J. Geophys. Res.-Atmos.*, **105,** (in press).

Ware, R., M. Exner, D. Feng, M. Gorbunov, K. Hardy, B. Herman, Y. Kuo, T. Meehan, W. Melbourne, C. Rocken, W. Schreiner, S. Sokolovskiy, F. Solheim, X. Zou, R. Anthes, S. Businger, and K. Trenberth, 1996: GPS sounding of the atmosphere from low Earth orbit: Preliminary results, *Bull. Amer. Met. Soc.*, **77**, 19-40.

WCRP, 1998: Stratospheric Processes and Their Role in Climate. World Climate Research Programme Report 105, World Meteorological Organization, Geneva, 131 pp.

Zou, X. , Y.-H. Kuo, and Y.-R. Guo, 1995: Assimilation of atmospheric radio refractivity using a nonhydrostatic mesoscale model, *Mon. Wea. Rev.*, **123,** 2229-2249.

Zou, X., F. Vandenberghe, B. Wang, M.E. Gorbunov, Y.-H. Kuo, S. Sokolovskiy, J.C. Chang, J.G. Sela, and R.A. Anthes, 1999: A ray-tracing operator and its adjoint for the use of GPS/MET refraction angle measurements, *J. Geophys. Res.*, **104**, D18, 22, 301-22, 318.

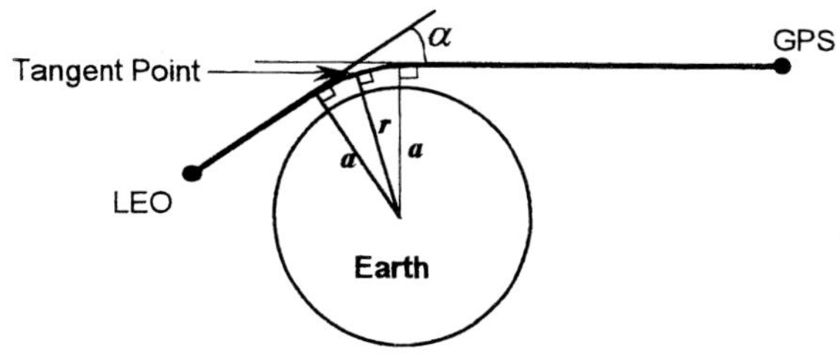

Fig. 1: Instantaneous occultation geometry for the Global Positioning System (GPS) and a low Earth orbiter (LEO) satellite. The bending angle α and impact parameter a are determined from the Doppler frequency shift, spacecraft position, and spacecraft velocity measurements.

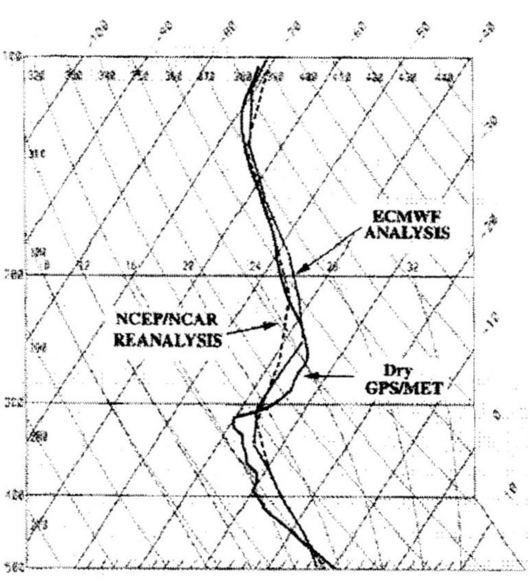

Fig. 2: Comparison of GPS/MET temperature sounding over China and the NCEP and ECMWF analyses at 00:00 UTC October 22, 1995 (Kuo et al., 1998).

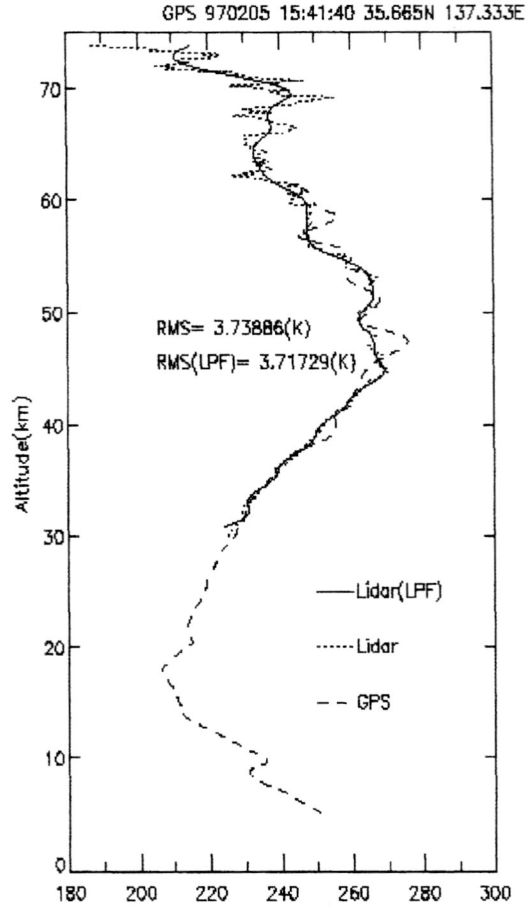

Figure 3: Comparison of temperature profiles between GPS/MET data (dashed lines) at 15:41 UTC February 5, 1997 at 35.7°N, 137.3°E and Rayleigh lidar data (dotted line) between 15:08 and 16:26 UTC at 36.1°N, 140.1°E. The lidar data smoothed with a low-pass filter with a cutoff at 3 km is also plotted as a solid line (Professor T. Tsuda, 1999, personal communication).

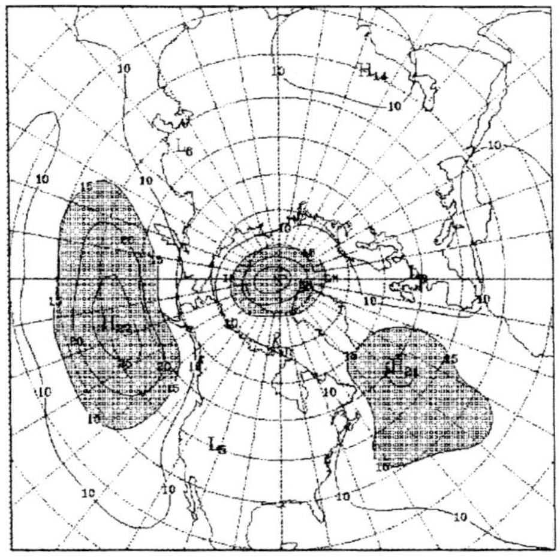

Fig. 4: Standard deviation of the difference between NCEP and ECMWF 500-mb analysis for the five winters (DJF) 1985-86 to 1989-90. Contours are 5 m with values above 15 m shaded (Nutter et al., 1998)

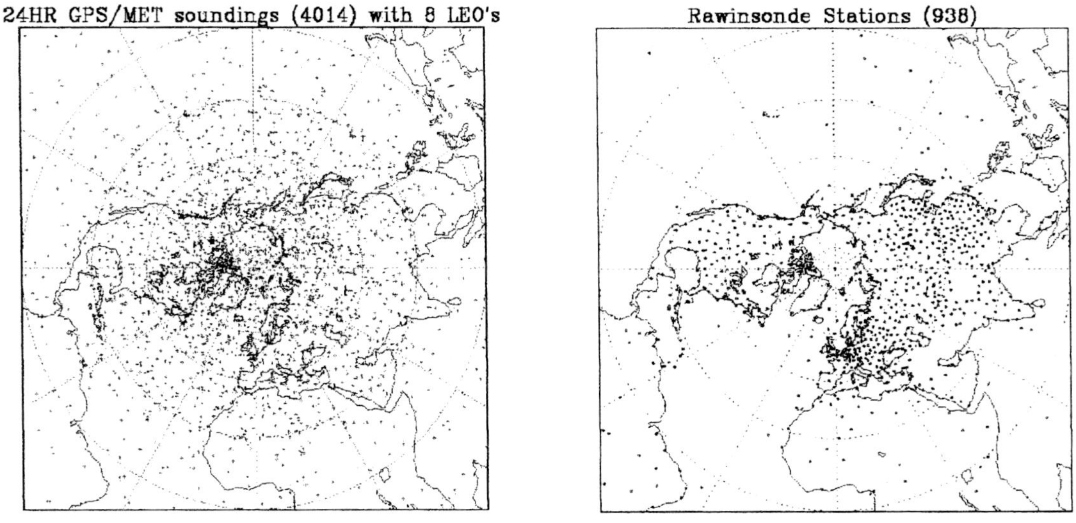

Fig. 5: Projected COSMIC soundings over the northern hemisphere every 24 hours and the present radiosonde sites.

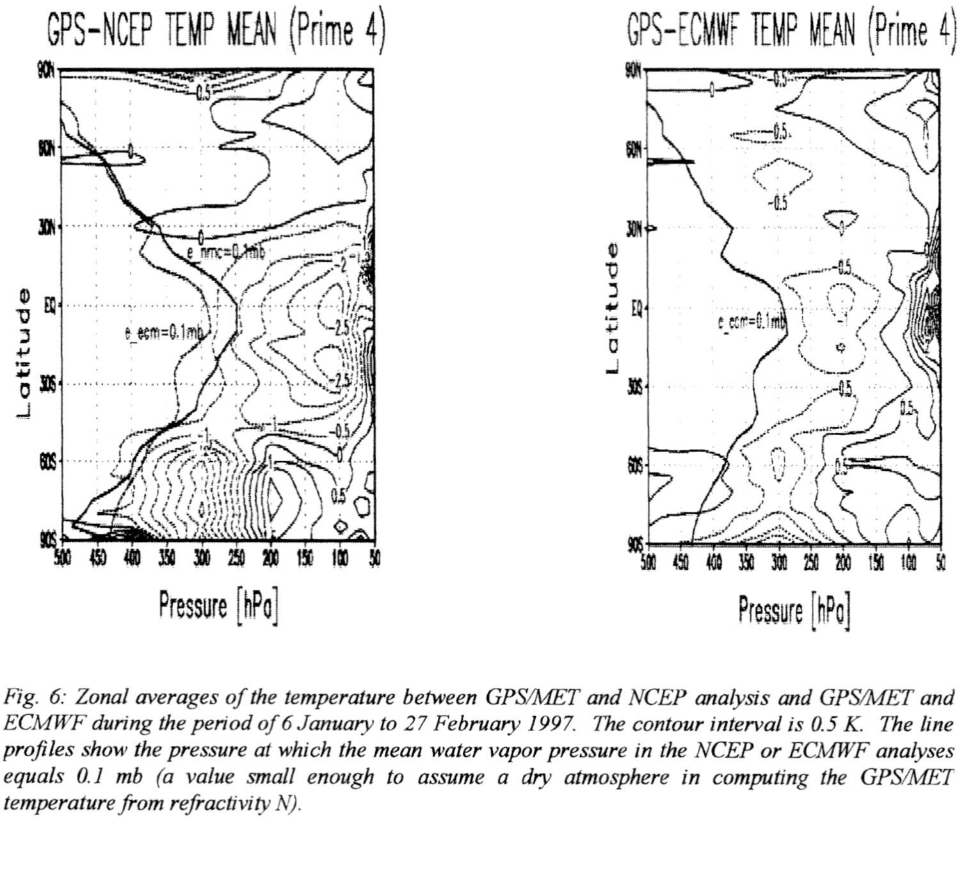

Fig. 6: Zonal averages of the temperature between GPS/MET and NCEP analysis and GPS/MET and ECMWF during the period of 6 January to 27 February 1997. The contour interval is 0.5 K. The line profiles show the pressure at which the mean water vapor pressure in the NCEP or ECMWF analyses equals 0.1 mb (a value small enough to assume a dry atmosphere in computing the GPS/MET temperature from refractivity N).

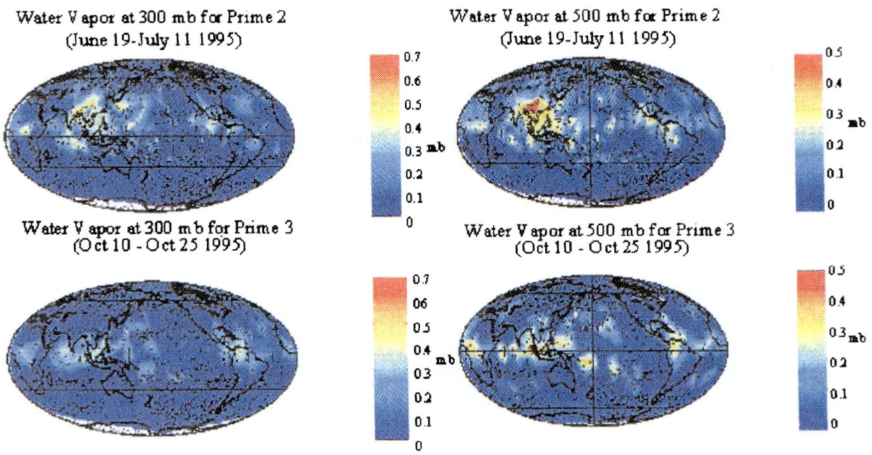

Fig. 7: Middle (500 mb) and upper (300 mb) tropospheric water vapor for period 19 June through 11 July 1995 and 10-25 October 1995 computed from GPS/MET refractivity data and ECMWF temperature data. Occultation locations are shown by black dots.

ENSO OCCURRENCE AND ANOMALIES OF SUB-SURFACE OCEAN TEMPERATURE IN THE PACIFIC WARM POOL*

Li Chongyin and Mu Mingquan
LASG, Institute of Atmospheric Physics, Academia Sinica, Beijing

ABSTRACT

Based on the data analyses and numerical simulations with a tropical Pacific coupled model, the important relationship between occurrences of El Nino/La Nina events and anomalies of sub-surface ocean temperature (SOT) in the Pacific warm pool region are studied in this paper.

The data analyses showed that most of ENSOs since 1950s are similar to the situation of 1997-98 ENSO. There have been positive anomalies of SOT in the warm pool region before the occurrence of El Nino event, but negative anomalies of SOT in the warm pool region prior to the occurrence of La Nina event. As positive (negative) anomalies of SOT propagated from warm pool region towards the equatorial eastern Pacific, the El Nino (La Nina) event would occur soon.

Eastward propagation of the anomalous SOT is closely associated with zonal wind anomaly over the equatorial western Pacific. If there is continued westerly (easterly) wind anomaly over the equatorial western Pacific, the positive (negative) anomaly of SOT in the warm pool region will propagate eastwards, and zonal wind anomaly appears a little earlier than the eastward propagation of the anomalous SOT.

Numerical simulations with a tropical Pacific atmosphere-ocean coupled model (IAP-CGCM) also showed that the anomalies of SOT in warm pool region played an important role in the occurrences of El Nino and La Nina events. These are similar to observations.

Key words: warm pool, sub-surface ocean temperature (SOT), ENSO

1. INTRODUCTION

ENSO (El Nino/Southern Oscillation) has been paid more attentions in the world, because its occurrence is always related to climate anomalies and disasters in vast areas. Previous studies have attributed ENSO to the interaction between the atmosphere and ocean in the tropical Pacific region (Bjerknes 1969; Rasmusson and Wallace 1983). There are also a lot of studies to investigate the mechanism of ENSO. Some of them emphasized the oceanic effect further more (Wyrtki 1975; Philander, Yamagata and Pacanowski 1984; McCreary and Anderson 1984; Cane and Zebiak 1985; Hirst 1986). Some focused on the atmospheric impact, particularly, the effect of the monsoon anomaly on ENSO (Li 1989, 1990, 1996; Webster and Yang 1992). Others emphasized a delayed oscillator theory in the atmosphere-ocean interaction (Suarez and Schopf 1988; Neelin 1991). Since the mid-1980s, the coupled atmosphere-ocean model (CGCM) has been made considerable headway and plaid an important role in studying ENSO (Zebiak and Cane 1987; Battisti 1988; Latif and Villwock 1990; Zhang and Levitus 1997).

But until now, a lot of questions on ENSO cycle (particularly its occurrence mechanism) have not been understood very well. It is just the purpose of the paper to investigate the relationship between the anomalies of SOT in the Pacific Ocean warm pool and the occurrence of ENSO.

* It is supported by <National Key Programme for Developing Basic Sciences>G1998040903 and NNSFC-49823002.

2. 1997 EL NINO EVENT

A very strong El Nino event occurred in 1997, with SST anomalies greater than +6°C observed in the eastern Pacific for several months. This El Nino event began in May, when positive anomalies of SST greater than +4°C were observed in the equatorial eastern Pacific, whereas the systematically positive anomalies of SST have not formed in the equatorial eastern Pacific before that.

Equatorial depth-longitude sections of ocean temperature anomalies for the different month show that there were positive anomalies of SOT (100-200 m) in the western Pacific warm pool, but negative anomalies (50-100 m) in the equatorial eastern Pacific during August-November 1996. In January 1997, the positive anomalies of SOT in the warm pool were very strong (>5°C) and have expanded to the east of 160°W. The negative anomalies of SOT in the equatorial eastern Pacific were replaced by positive anomalies along with the continued eastward expanding of the anomalous SOT. In March 1997, there was not only a strongly positive anomaly center of SOT in the western Pacific warm pool but also a positive center in the equatorial eastern Pacific. Then, the positive anomaly area of SOT moved into the equatorial eastern Pacific gradually. In May 1997, the anomalous SOT has been over +6°C in the equatorial eastern Pacific. However, there was the negative anomaly of SOT in the equatorial western Pacific. At the same time, the positive anomaly of SOT propagated gradually up to the sea surface and the positive anomaly of SST occurred in the equatorial eastern Pacific. Thus, El Nino event was excited, and the thermal regime of the equatorial Pacific Ocean was shown that there were positive anomalies (SST and SOT) in the equatorial eastern Pacific but negative anomalies in the equatorial western Pacific after the occurrence of El Nino.

The above analysis shows that the occurrence of 1997 El Nino event is closely related to the anomalous SOT in the equatorial western Pacific Ocean warm pool. The positive anomaly of SOT in the warm pool and its eastward propagation play an important role in the occurrence of the event.

3. ANALYSES WITH THE PREVIOUS DATA

In this section, we will further investigate the relationship between the anomalies of SOT in the warm pool and the occurrence of El Nino events. The temporal variations of the anomalous SOT in the warm pool (10°S-10°N, 140°E-180°) and the anomalous SST in Nino 3 region (5°S-5°N, 150°-90°W) are shown in Fig. 1 by using the JEDAC and COADS data. It is very clear that the positive anomalies of SOT in the warm pool region always exist prior to the occurrence of El Nino event, but after the outbreak of El Nino, the SOT anomalies in the warm pool region are out-of-phase to the SST anomalies in Nino 3 region. It is also shown in Fig.1 that the sub-surface warming in the warm pool region leads the outbreak of El Nino event usually by half to two years.

The sub-surface warming in the warm pool is closely related to the occurrence of El Nino event, but the eastward propagation of the positive SOT anomalies is more important. It seems that the eastward propagation of the positive SOT anomalies along the slope of the thermocline in the equatorial Pacific triggers the occurrence of the positive SST anomalies in the equatorial eastern Pacific and El Nino events. Sometimes the eastward propagation of the positive SOT anomalies is in time after warming in the warm pool region. Sometimes the eastward propagation is delayed. Therefore, there are half to two years of the time difference between warming in the warm pool and outbreak of El Nino event.

The relationship between the occurrence of El Nino and the positive SOT anomalies in the warm pool region and its eastward propagation can be obviously shown by using the time-longitude section of the SOT anomalies in the equatorial Pacific. Because the thermocline is deeper and thicker in the western Pacific than that in the equatorial eastern Pacific, the depth of the maximum SOT variation is also different. In order to represent this feature, the data should be disposed of in different depths respectively, such as from 120-160m in the equatorial western Pacific, 60-120 m in the equatorial central Pacific to 40-60m in the eastern equatorial Pacific.

As the cases analyzed, the time-longitude sections of the SOT anomalies in the equatorial Pacific are shown in Fig. 2 for 1957 and 1972 El Ninos. In 1957 El Nino case, the obvious warming in the warm pool began in April-May 1956 even though the positive anomalies of SOT were weaker in the equatorial western Pacific. The maximum anomaly of SOT was about +2.5°C in the autumn of 1956. The positive anomalies of SOT propagated into the equatorial eastern-central Pacific in the winter of 1956 and the spring of 1957, thus, the El Nino occurred in April 1957. There was about one year from obviously warming in the warm pool till outbreak of El Nino. In 1972 El Nino case, obviously warming of the warm pool began in May 1970. The strong warming was in August-October 1971 (>3°C). But until the spring of 1972, the positive anomalies of SOT just expanded into the equatorial eastern- central Pacific Ocean, and then, the El Nino occurred in May 1972. The time span from obviously warming in the warm pool to outbreak of El Nino was about two years.

We also make similar analyses of SOT anomalies in the Pacific warm pool for the composite El Nino and La Nina event because the composite method can catch the common features of the most events. We choose all of the El Nino and La Nina cases after 1955 (except 1975 La Nina, because it's quite near to 1973 case.). The nine El Ninos are 1957, 1963, 1965, 1968-69, 1972, 1976, 1982-83, 1986-87 and 1991, respectively. The four La Ninas are the 1964, 1973, 1984 and 1988 case. Their composite results are shown in Fig. 3. For the composite El Nino, the considerably positive SOT anomaly has begun to occur in the equatorial warm pool region even in the previous two years before (in Fig. 3a). After the previous January of the composite case, the positive center keeps on developing, but it still stays its origin place (the warm pool). Only after January of the onset year, the positive SOT anomaly starts to move and quickly propagate eastwards from its origin place to the equatorial eastern Pacific region. Finally, the eastward propagating SOT anomaly triggers the El Nino outbreak. The composite La Nina has the contrastive situation in the warm pool with the composite El Nino, but their signals are opposite. That means eastward propagating of the negative SOT anomaly in the equatorial warm pool is one physical mechanism for La Nina final happening.

4. ON EASTWARD PROPAGATION OF SOT ANOMALIES

The above analyses showed that the eastward propagation of the anomalous SOT in the warm pool region directly plays an important role in the occurrence of ENSO. Some studies have shown that the occurrence of El Nino is closely related to the westerly wind anomalies over the equatorial western-central Pacific. Therefore, the relationship between the eastward propagation of the anomalous SOT with the westerly wind anomaly over the equatorial western-central Pacific will be discussed in this section.

We have indicated that the anomalous winter monsoon in East Asia plays an important role in exciting ENSO through the westerly wind anomaly over the equatorial western-central Pacific (Li 1990; Li 1996). In other words, the anomalous westerly wind mainly results from the anomaly of winter monsoon in East Asia. By using the COADS data, temporal variations of the zonal wind anomaly (10°S-10°N, 140°E-160°W) and the meridional wind anomaly (20°-30°N, 120°-150°E) at the surface are shown in Fig.4. It is very clear that there is always the negative (positive) anomaly of the meridional wind over the northwestern Pacific near to the Asian continent prior to the positive (negative) anomaly of the zonal wind over the equatorial central-western Pacific. The numerical simulations with CGCM also proved that the continued strong (weak) winter monsoon in East Asia can excite the westerly (easterly) wind anomaly at the equatorial western-central Pacific and the El Nino-like (La Nina-like) phenomena (Li and Mu 1998).

Fig. 5 and Fig. 6 are the results of the nine El Nino and four La Nina cases composite, respectively. Temporal variations of the SOT anomaly in the warm pool region, the zonal wind variation at the equatorial western Pacific, the anomaly of SOT in the equatorial eastern Pacific and SSTA in Nino 3 are given in Fig. 5 (Fig. 6). It is very obvious that eastward-propagating of the positive (negative) SOT anomaly from the warm pool to the equatorial eastern Pacific is earlier than the occurrence of El Nino (La Nina). While the westerly (easterly) wind anomaly is earlier than the eastward propagation of the anomalous SOT, too. But the positive (negative) anomaly of SOT has stayed in the warm pool region (its origin place) for a quite long time before the occurrence of El Nino (La Nina).

5. NUMERICAL SIMULATION RESULTS OF A CGCM

In order to further explore the important effect of the SOT anomalies in the warm pool region on exciting ENSO, the numerical simulation results of a CGCM are analyzed. The model used in this study is a coupled atmospheric-oceanic general circulation model developed in the Institute of Atmospheric Physics/Chinese Academy of Sciences (Zhou et al., 1998). The atmospheric component is a two-level global atmospheric general circulation model formulated in σ-coordinate with the resolution of 4° in latitude and 5° in longitude (Zeng et al., 1989), which has been shown to have quite good simulating abilities for climate variations in the AMIP project. The oceanic component is a free surface tropical Pacific ocean general circulation model with the resolution of 1° in latitude, 2° in longitude and 14 unequal layers in the vertical with flat-bottom (4000 m), of which the domain is 121°E-69°W and 30°S-30°N. The detailed introduction about this OGCM was described by Zhang and Endoh in 1992. In order to control the "climate drift" in the coupled system, the linear statistical correlation is used in the model integration. The outputs are analyzed for the last forty years of the total 100 years integration that shows the model has reached equilibrium after 50 years.

The temporal variations of the simulated SSTA in Nino 3 region and the SOT anomaly in the equatorial western Pacific (140°E-180°, 6°S-6°N) are shown in Fig. 7. It can be seen that there is obviously positive anomalies of SOT prior to most of the warm events and the SOT anomalies become negative after the appearances of the warm events. On the contrary, the cold events are always corresponding to the negative anomalies of SOT in the equatorial western pacific in their earlier stage.

The composite analyses of the SOT anomalies in the equatorial Pacific for the simulated El Nino and La Nina have advantage to expose the importance of the SOT anomalies in the warm pool to trigger ENSO. In the warm case, there are positive anomalies of SOT in the warm pool region in the previous autumn and winter, and then as the eastward propagation and enhancement of the positive anomalies of SOT, the positive SSTA gradually appears in the equatorial eastern Pacific. Contrary to the warm case, the negative SSTA gradually shows in the equatorial eastern Pacific as the result of the eastward propagation and enhancement of the negative anomalies of SOT, which exists in the warm pool in the previous autumn and winter.

6. CONCLUSIONS

The close relationship between the occurrence of ENSO and the SOT anomalies in the warm pool region is studied by using data analyses and numerical simulation with a CGCM, the main results can be summarized in the following:

(1) There are continued positive (negative) anomalies of SOT in the warm pool region prior to the occurrence of El Nino (La Nina). The occurrence of ENSO is closely related to the anomalies of SOT in the warm pool region.

(2) The outbreak of El Nino (La Nina) event is directly associated with the eastward propagation of the anomalous SOT. Eastward propagating of the anomalous SOT toward the equatorial eastern Pacific and expanding to the sea surface will lead to SSTA in the equatorial eastern Pacific and the occurrence of ENSO.

(3) Even though the mechanism to cause the positive anomalies of SOT in the warm pool region has not been understood completely, it is clear that the eastward propagation of the anomalous SOT in the warm pool region is closely related to the westerly wind anomalies over the equatorial western-central Pacific, which mainly result from the anomalous East-Asian winter monsoon, and simultaneously the occurrence of the westerly wind anomalies is earlier than that of the SOT anomalies in the warm pool.

REFERENCES

Battisit, D.S., 1988: The dynamics and thermodynamics of a warm event in a coupled ocean-atmosphere model. *J. Atmos. Sci.*, **45**, 2889-2919.

Bjerknes, J., 1969: Atmospheric teleconnections from the equatorial Pacific. *Mon. Wea. Rev.*, **97**, 163-172.

Can, M.A., and S.E. Zebiak, 1985: A theory for El Nino and Southern Oscillation. *Science*, **228**, 1085-1087.

Hirst, A.C., 1986: Unstable and damped equatorial modes in simple coupled ocean-atmosphere model. *J. Atmos. Sci.*, **43**, 606-630.

Latif, M., and A. Villwock, 1990: Interannual variability as simulated in coupled ocean-atmosphere model. *J. Mar. Systems*, **1**, 51-60.

Li, C., 1989: The frequent activities of stronger aerotroughs in East Asia in wintertime and the occurrence of the

El Nino event. *Scientia Sinica (Series B)*, **32**, 976-985.

Li, C., 1990: Interaction between anomalous winter monsoon in East Asia and El Nino events. *Adv. in Atmos. Sci.*, **7,** 36-46.

Li, C., 1996: ENSO cycle and anomalies of winter monsoon in East Asia. *Workshop on El Nino, Southern Oscillation and Monsoon. ICTP, SMR/***930-18**, Trieste, 15-26 July.

Li, C., and M. Mu, 1998: Numerical simulations on anomalous winter monsoon in East Asia exciting ENSO. *Chinese J. Atmos. Sci.*, **22**, 393-403.

McCreary, J.P., and D.L.T. Anderson, 1984: A simple model of El Nino and the Southern Oscillation. *Mon. Wea. Rev.*, **112,** 934-946.

Neelin, J.D., 1991: The slow sea surface temperature mode and the fast-wave limit: Analytic theory for tropical interannual oscillations and experiments in a hybrid coupled model. *J. Atmos. Sci.*, **48**, 584-606.

Philander, S.G., T. Yamagata, and R.C. Pacanowski, 1984: Unstable air-sea interactions in the tropics. *J. Atmos. Sci.*, **41**, 604-613.

Rasmusson, E.M., and J.M. Wallace, 1983: Meteorological aspects of El Nino/Southern Oscillation. *Science,* **222**, 1195-1202.

Suarez, M.J., and P. Schopf, 1988: A delayed action oscillator for ENSO. *J. Atmos. Sci.*, **45**, 3283-3287.

Webster, P.J., and S. Yang, 1992: Monsoon and ENSO: selectively interactive systems. *Quart. J. Roy. Meteor. Soc.*, **118,** 877-926.

Wyrtki, K., 1975: El Nino - the dynamic response of the equatorial Pacific Ocean to atmospheric forcing. *J. Phys. Oceanogr.*, **5**, 572-584.

Zebiak, S.E., and M.A. Cane, 1987: A model El Nino and Southern Oscillation. *Mon. Wea. Rev.*, **115**, 2262-2278.

Zeng, Q., X. Zhang, X. Liang, C. Yuan, and S. Chen, 1989: Documentation of IAP Two-Level AGCM. *TR044*, DOE/ER/60314-H1, U.S. DOE, 383pp.

Zhang, R., and M. Endoh, 1992: A free surface general circulation model for the tropical Pacific Ocean. *J. Geophys. Res.*, **97(c7)**, 11237-11255.

Zhang, R., and S. Levitus, 1997: Interannual variability of coupled tropical Pacific Ocean-atmosphere system associated with the El Nino/Southern Oscillation. *J. Climate*, **10**, 1312-1330.

Zhou, G.Q., Q.C. Zeng, and R.H. Zhang, 1999: A coupled oceanic-atmospheric general circulation model and its preliminary results. *Progress in Natural Science*, **9**, 542-551.

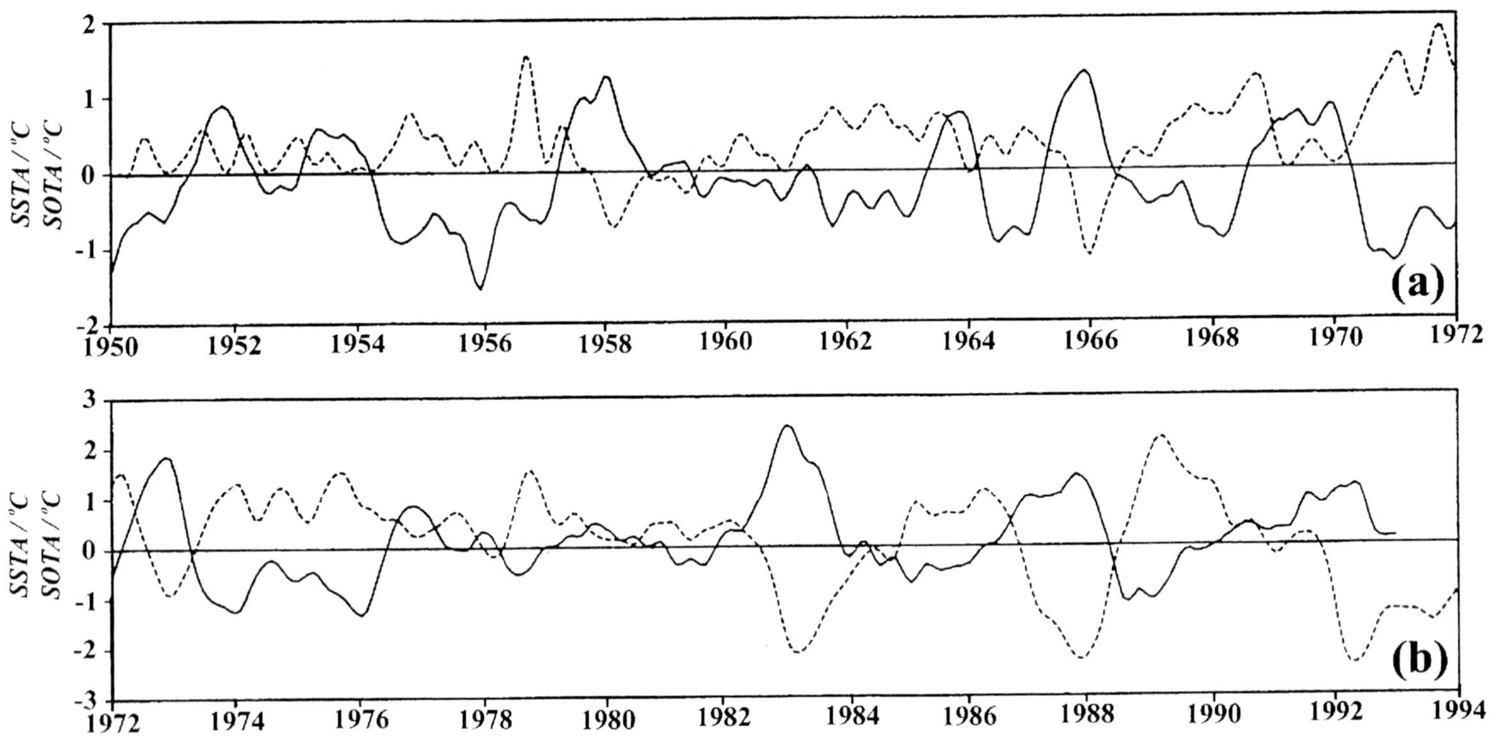

Fig. 1. Temporal variations of the SST anomalies (solid curve) in Nino 3 region (5°S-5°N, 150°-90°W) and the SOT anomalies (dashed curve, 120-160 m mean) in the warm pool region (6°S-6°N, 140°E-180°).

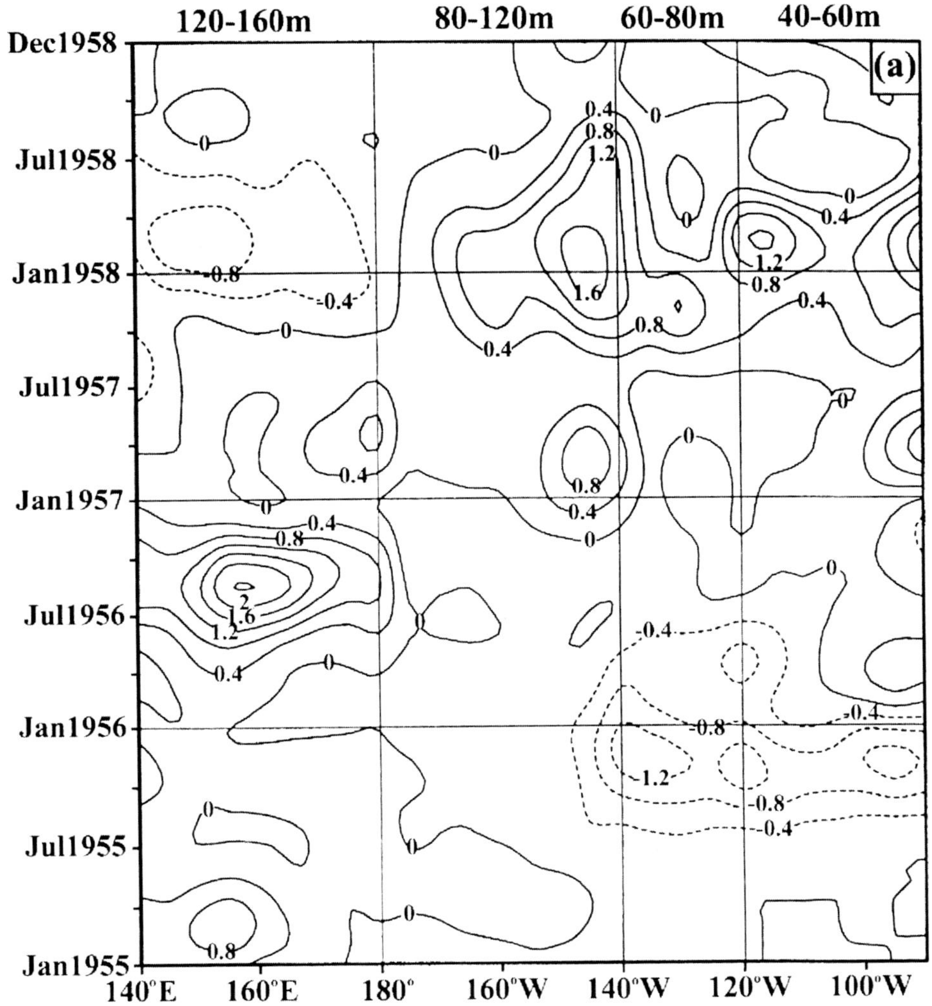
120-160m
80-120m
60-80m
40-60m
(a)
Dec1958
Jul1958
Jan1958
Jul1957
Jan1957
Jul1956
Jan1956
Jul1955
Jan1955
140°E
160°E
180°
160°W
140°W
120°W
100°W

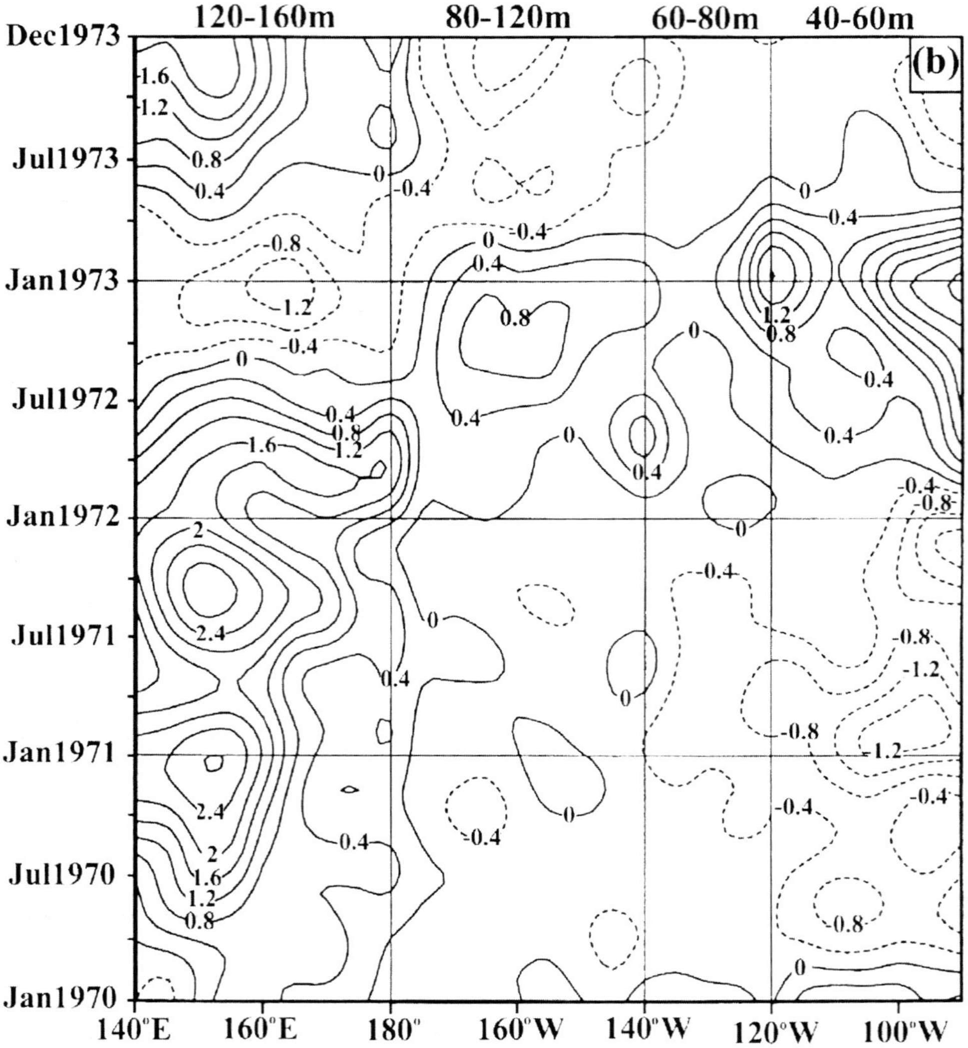

Fig. 2. Time-longitude sections of the SOT anomalies (°C) in the equatorial Pacific (10°S-10°N) along the maximum variance layer of ocean temperature for 1957 El Nino case (a) and 1972 El Nino case (b), respectively. The contour interval is 0.4 °C.

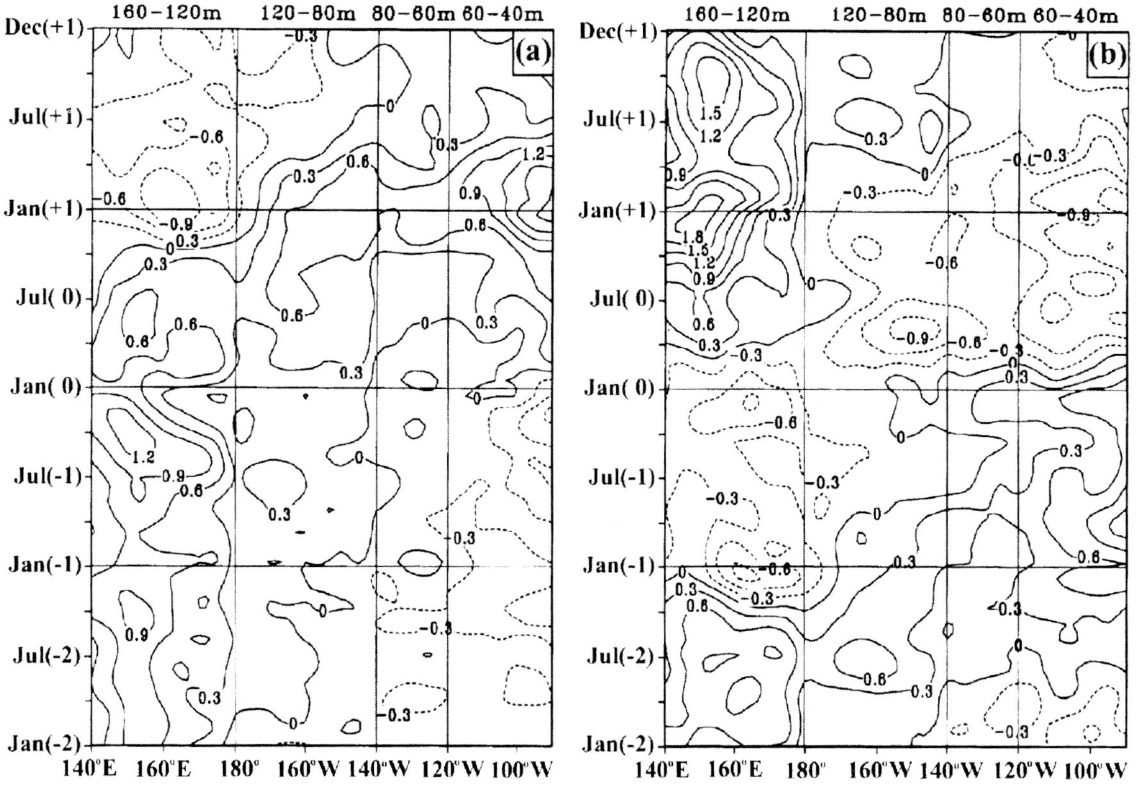

Fig. 3. Same as Fig.2, but for nine El Ninos and four La Ninas composite. The contour interval is 0.3 °C.

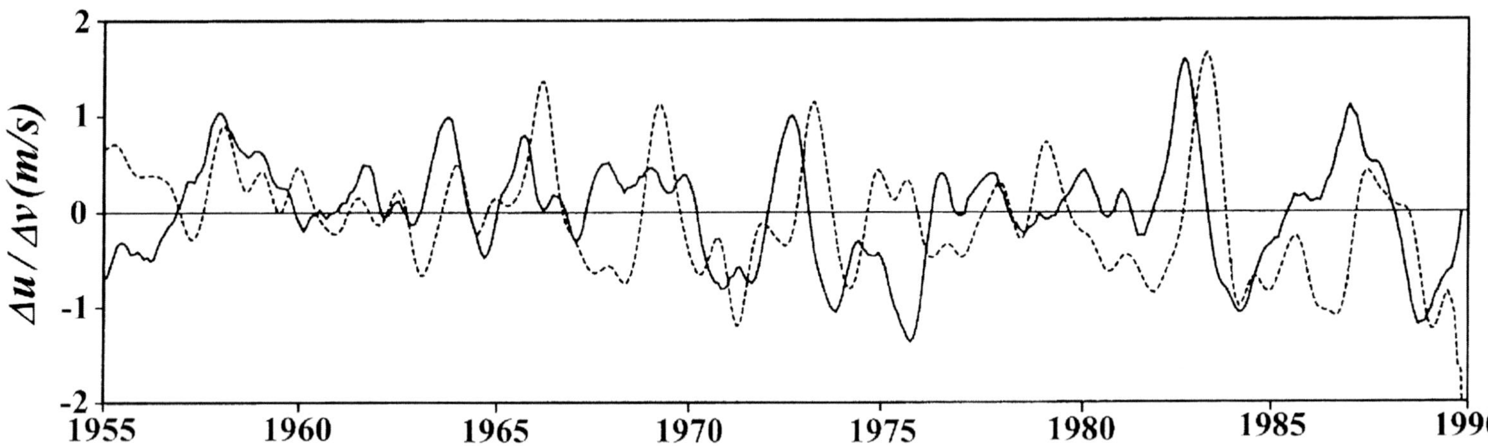

Fig. 4. Temporal variations of the surface zonal wind anomaly (solid curve) at the equatorial western-central Pacific region (10°S-10°N, 140°E-160°W) and the surface meridional wind anomaly (dashed curve) at the northwestern Pacific region (20°-30°N, 120°-150°E).

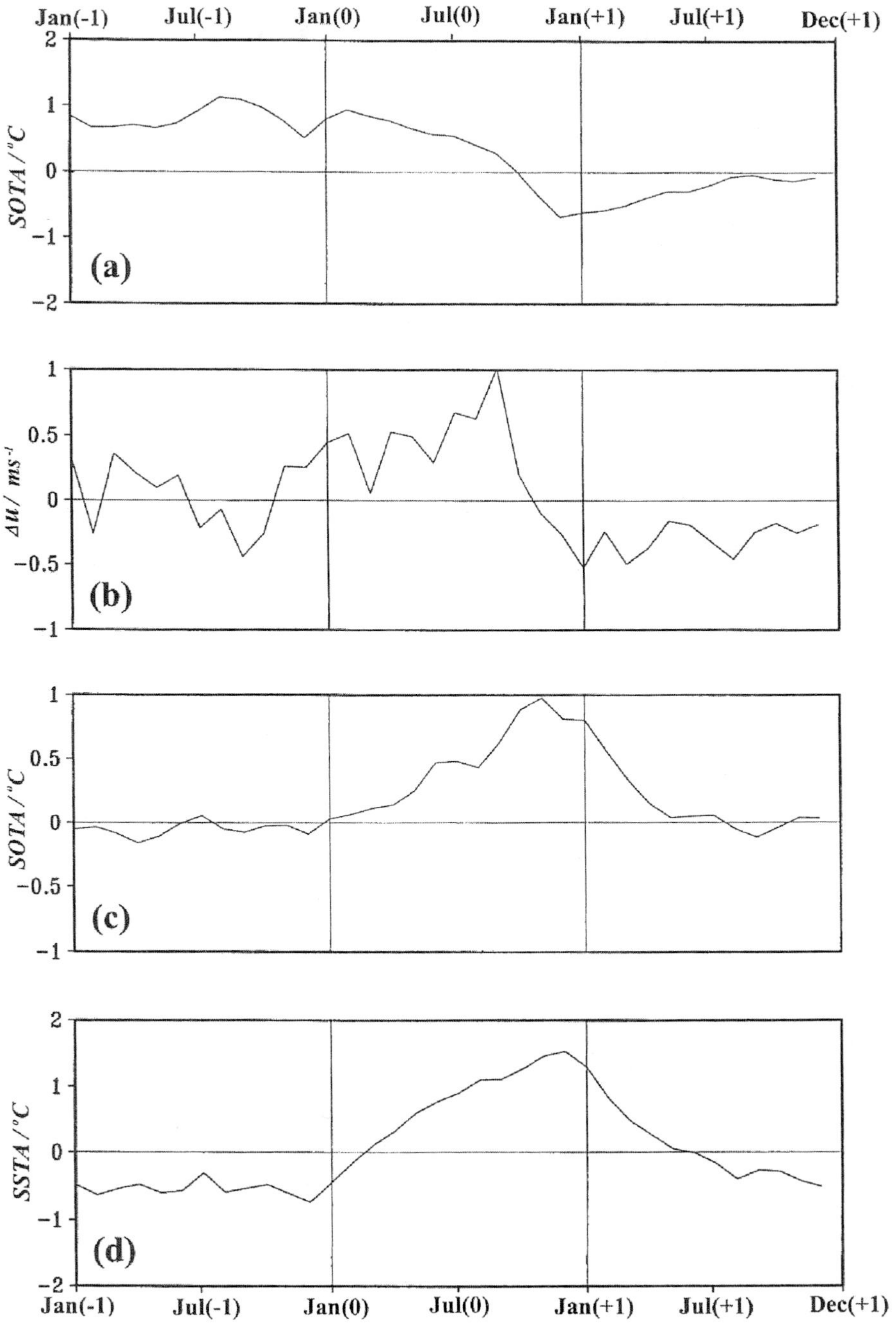

Fig. 5. Temporal variations of nine El Nino cases composite for SOT anomaly (5°S-5°N, 140°E-180°, 120-160 m mean) (a), the surface zonal wind anomaly (10°S-10°N, 120°-160°E) (b), SOT anomaly (5°S-5°N, 170°-130°W, 80-120 m mean) (c) and SSTA in Nino3 (5°S-5°N, 150°-90°W) (d), respectively.

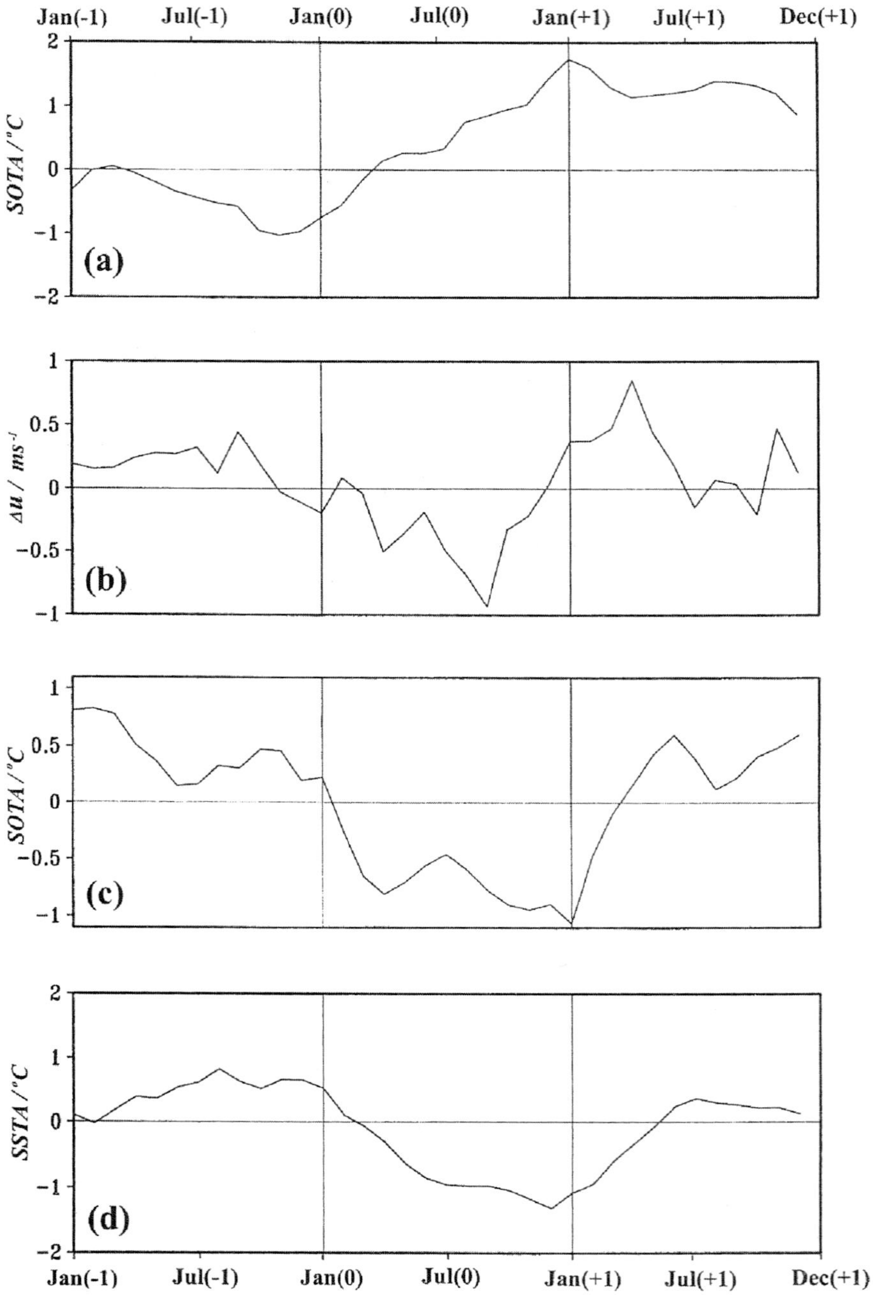

Fig. 6. Same as Fig.2, but for four La Ninas composite.

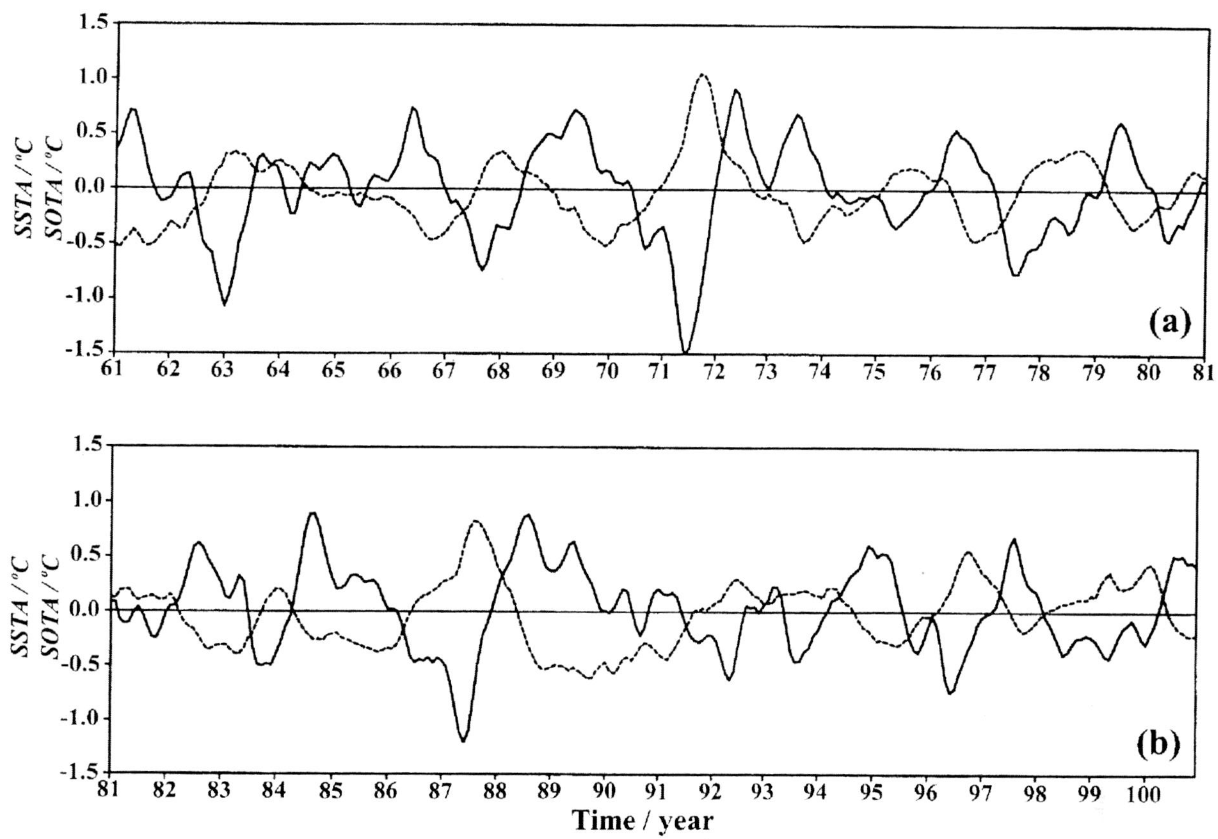

Fig. 7. Temporal variations of SSTA (solid curve) in Nino 3 region (5°S-5°N, 150°-90°W) and SOT anomaly (100-200m mean, dashed curve) in the equatorial western Pacific (6°S-6°N, 140°E-180°) simulated with a CGCM.

IMPACT OF EL NIÑO AND LA NIÑA ON PRECIPITATION AND AIR TEMPERATURE OF MACAO

Lou Mei Meng, Fong Soi Kun, Hao I Pan
Macao Meteorological and Geophysical Services

ABSTRACT

1997/98 was one of the strongest El Niño period on record, the estimated global mean surface temperature for land and marine areas averaged are 0.44°C above the 1961-1990 normal period. From the mid-1998 till now, it began La Niña year, both of the above episodes affect the weather all around the world.

In this paper, we based on our observation data, find out some relationship between Anomaly and weather (precipitation and air temperature).

1. INTRODUCTION

In 1997, the warm episode of El Niño raised the global mean surface temperature for land and marine areas are 0.44 °C above the normal for the 1961-1990 period. On the other hand La Niña has had a profound influence on the atmosphere during the second half of 1998. Both of the above episodes have affected the weather all around the world. In 1997, Macao recorded 2716.6 mm of precipitation which ranked fifth highest in our historical record.

Table 1. Top Ten annual precipitation in Macao (E: El Niño year; L:La Niña year).

	Macao Annual Precipitation (1952-1998) Unit: mm									
Rank	1	2	3	4	5	6	7	8	9	10
Year	E1982	L1973	1983	L1975	E1997	E1972	1978	E1987	E1994	L1998
Precipit-ation	3041.4	2889	2883	2827	2716.6	2645.4	2547	2487.2	2474.1	2411

El Niño is an important phenomenon in the ocean involving a large area of sea surface temperature anomalies. During an El Niño period, the sea surface temperature over the eastern Pacific is abnormally warm along the coastlines of Peru and Ecuador. The Southern Oscillation is a large scale low frequency oscillation in the atmospheric circulation, i.e. there is an inverse pressure relationship between tropical Pacific and tropical Indian Ocean. Both phenomena are part of the large scale interaction between the ocean and the atmosphere in the tropical Pacific Ocean (Bjerknes). It is a cyclic oscillation, i.e. an irregular oscillation due to coupling with the change of atmospheric circulation over the ocean and the globe. On the contrary, La Niña is characterized by abnormally cold sea surface temperature in the Eastern Equatorial Pacific. El Niño or La Niña is generally defined by the calculation of the five months running average of SST in Nino3, specifically if it is 0.5°C higher or lower than climate average standard respectively.

Many studies indicate that its influence affects not only the Torrid Zone, but also the high latitude region. The influence of El Niño differs according to the season. Research indicates that the South Asian monsoon would be weakened if El Niño occurred in the Northern Hemisphere summer. For Eastern Asia, if it reached maturity before the summer, the Eastern

Asian monsoon would be enhanced. The activity of the Eastern Asian monsoon is a main reason for flood and drought conditions in this area. If El Niño occurs in the winter, the winter monsoon would be weakened and that winter would be warmer than normal. The objective of this research is to examine the validity of these observations for Macao using data collected by Macao Meteorological and Geophysical Service.

2. DATA SOURCE

The data used is the sea level pressure difference (SLPD, $P_{stand\ Tahiti}$ - $P_{stand\ Darwin}$) and SST($NINO_{1+2}$;$NINO_3$;$NINO_4$ and $NINO_{3.4}$) [1952-1998] from the Climate Prediction Center of the National Oceanic and Atmospheric Administration and NCEP reanalaysis data (1982-1994) and the observation data (precipitation and temperature) from Macao Meteorological and Geophysical Services.

3. STATISTICAL ANALYSIS OF THE IMPACT OF EL NIÑO AND LA NIÑA ON THE PRECIPITATION AND TEMPERATURE IN MACAO

(a) Monthly correlation between Sea Level Pressure Difference (SLPD) and Sea Surface Temperature (SST) on the precipitation and temperature in Macao:

Averaging the annual precipitation of 47 years in Macao indicates that, in abnormal years (including El Niño and La Niña), the annual precipitation is higher than in normal years (Fig. 1). In terms of annual total precipitation, eight of the ten wettest years were abnormal years (Table 1). This indicates a potential relationship between Macao precipitation and ENSO phenomena.

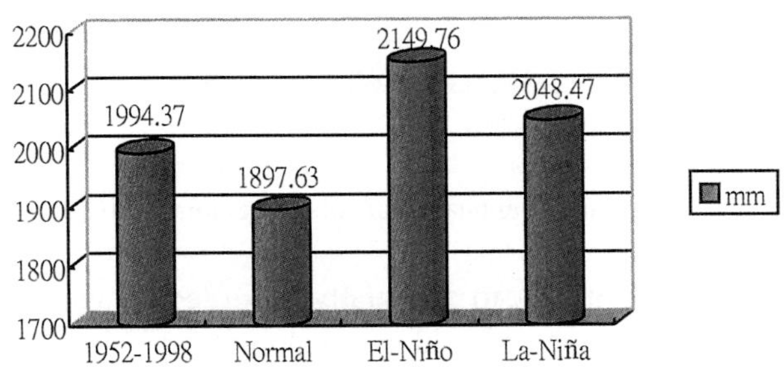

Fig. 1. Average of annual total precipitation of 1952-1998.

Depending on the season, the weather in Macao is markedly different. Two rainy seasons occur from April to October (from April to June, rain is caused by the quasi-stationary front and southwestly summer monsoon; from July to October, rain is mainly caused by thunderstorms and typhoons). Precipitation is low from November to February because of outflow from the intense cold air mass from Siberia. Fig. 2 shows monthly correlation between SLPD, SST ($NINO_{1+2}$;$NINO_3$;$NINO_4$;$NINO_{3.4}$) and mean precipitation. The result shows that the correlation for January, February and March is significant especially for March. The correlations are 0.495, 0.483, 0.537, 0.377 and 0.526 while for other months they are insignificant comparatively.

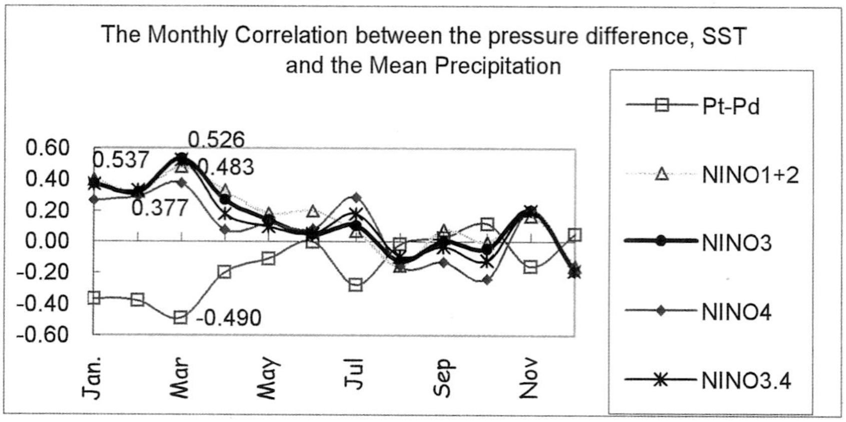

Fig. 2. Monthly correlation between the SLPD, SST and mean precipitation.(Pt: Pressure of Stand Tahiti, Pd: Pressure of Stand Darwin).

For temperature, data for the 47 year period shows that the temperature difference (Fig. 3), is insignificant since the abnormal period does not affect the whole year.

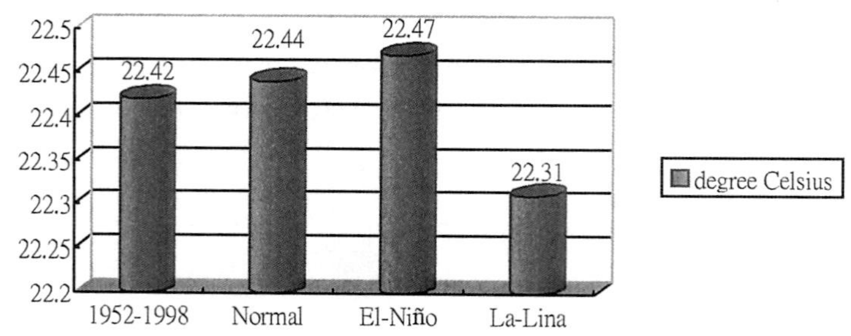

Fig. 3. Average of annual mean air temperature of 1952-1998.

The lowest temperature in Macao is usually recorded in January. Spring starts from March, by which time the influence of cold anticyclone from Siberia has weakened; summer starts from June and is hot. Fig. 4 shows the monthly correlation between SLPD, SST($NINO_{1+2}$;$NINO_3$;$NINO_4$;$NINO_{3.4}$) and mean air temperature. Three of the highest correlations are in November SLPD; $NINO_3$ and $NINO_{3.4}$, they are 0.35,0.29 and 0.32, while others were insignificant.

Summarizing the above findings, the correlation between SLPD, SST and precipitation is significant during January to March, and is fair with the temperature during November. Both of these periods coincide with the seasonal transition.

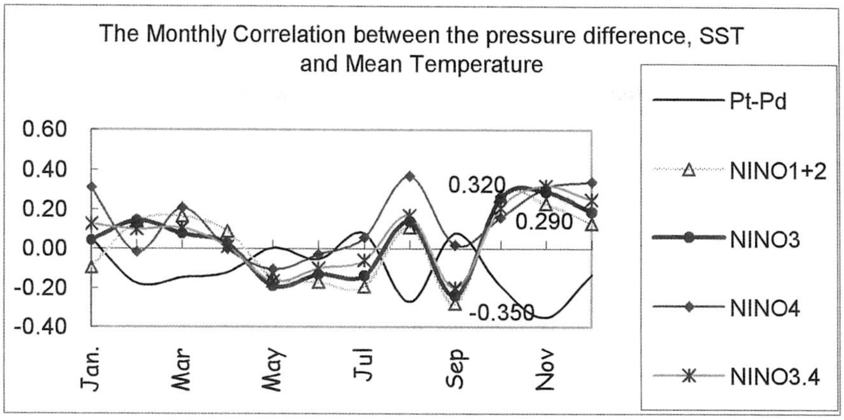

Fig. 4. Monthly correlation between the SLPD, SST and mean air temperature.

(b) Find out El Niño seasons by using Macao precipitation

To affect the atmospheric circulation, the warm episode must be strong enough and persist for a long period. To examine this further we selected Season as a unit for calculation. Winter (December to February), Spring (March to May), Summer (June to August), and Autumn (September to November) represent the four seasons in this context. Calculating the three months average of SLPD ($P_{stand\ Tahiti}$ - $P_{stand\ Darwin}$) , and SST in $NINO_{1+2}$, $NINO_3$, $NINO_4$ and $NINO_{3.4}$, we have five groups of data. After using the seasonal mean for calculating, it had already eliminate the monthly variability, it was not necessary to use the five months running average to eliminate the monthly variability. By using the five months running average, the two months which were before and after the season would have the weight of 1/5 respectively, it influenced the result. We consider that only a few seasons can be considered as abnormally seasons, so we select the seasons which is greater than 1.5σ (σ:Standard Derivation) as abnormally season, the others seasons are considered as normal seasons. The following Table 2 shows the result.

Table 2. El Niño season.

El Niño season					
	$P_{stand\ Tahiti}$ - $P_{stand\ Darwin}$	$NINO_{1+2}$	$NINO_3$	$NINO_4$	$NINO_{3.4}$
Winter	1983,1992,1998,	1983,1998	1983,1998		1983,1992, 1998
Spring	1987,1994,1997, 1998	1953,1957,1965, 1969,1972,1983, 1987,1992,1997, 1998	1953,1969, 1983,1987, 1992,1993, 1997,1998		1983,1987, 1992,1998
Summer	1965,1977,1982, 1987,1994,1997		1997	1997,1982,1987	1987,1997
Autumn	1982,1997		1997	1986,1987,1991, 1994,1997	1982,1997

We then divide the seasonal mean precipitation of Macao into El Niño season and non-El Niño season according to the result in Table 2. In order to understand whether the difference is significant or not, we then calculate the t-test on these two groups of data, selecting as a standard of reliability 95%, i.e. absolute value t = 1.645.

$$t = \frac{X^c - X^a}{S^2 * \mathrm{SQRT}(1/n + 1/m)}$$

where m: total number of non-El Niño year, n: total number of El Niño year ,

X^c :average precipitation value during non-El Niño year,

X^a :average precipitation value during El Niño year,

$$s^2 = \frac{\sum_{i-1}^{n} (x_i^c - x^c)^2 + \sum_{i-1}^{m} (x_i^a - x^a)^2}{n + m - 2}$$

Table 3. t-test result.

t	Winter	Spring	Summer	Autumn
Pt-Pd	2.48	1.81	2.09	0.13
$NINO_{1+2}$	3.42	3.06	0.00	0.00
$NINO_3$	3.42	1.88	2.11	0.52
$NINO_4$	0.41	0.00	0.00	0.84
$NINO_{3.4}$	0.00	0.00	2.13	1.74

From the result above (Table 3), we discover that the El Niño season defined by using SLPD and $NINO_3$ shows statistically significance for winter, spring and summer indicating that there were great differences in seasonal (Winter, Spring and Summer) mean precipitation between the El Niño season (defined in Table 2) and the non-El Niño season.

The calculated results presented in Fig. 5 and Fig. 6 verifies that the El Niño season defined above (Table 2) has a significant relationship with Macao's seasonal mean precipitation. We can see that using the method described above is reliable for the definition of an El Niño season, and that it has significant relationship with our precipitation, in Winter, Spring and Summer while the link is insignificant in Autumn.

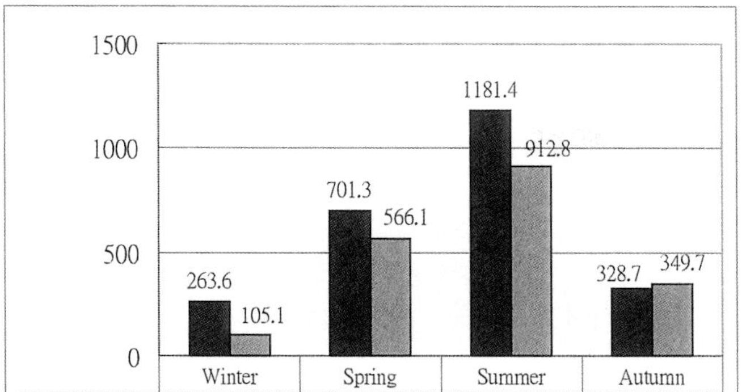

	Winter	Spring	Summer	Autumn
■ El Niño	263.6	701.3	1181.4	328.7
▨ Non-El Niño	105.1	566.1	912.8	349.7

Fig. 5. The comparison of seasonal mean precipitation between El Niño season and non-El Niño season which is defined by SLPD.

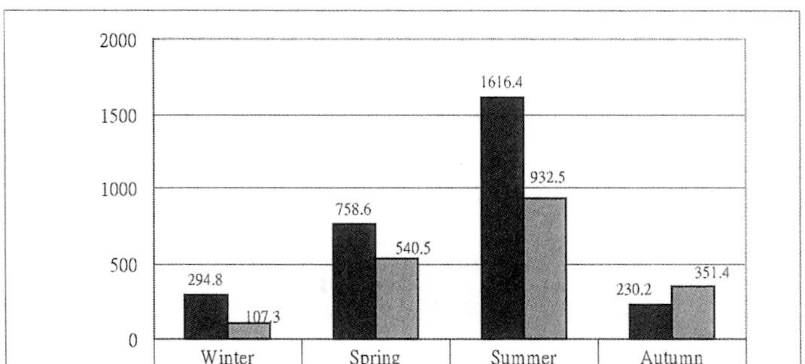

	Winter	Spring	Summer	Autumn
■ El Niño	294.8	758.6	1616.4	230.2
▨ Non-El Niño	107.3	540.5	932.5	351.4

Fig. 6. The comparison of seasonal mean precipitation between El Niño season and non-El Niño season which is defined by $NINO_3$.

(c) Monthly average value in 500 hPa geopotential height

Using the 1982-1994 NCEP reanalysis data, we calculate the monthly average 500hPa geopotential height between the El Niño season (defined by Table 2) and the non-El Niño season. The result is shown in Fig. 7 to Fig. 10. From December to February (Winter, Fig. 7), the area covered by the subtropical high (area above 5880) is obviously larger in the El Niño season and its location extends more northward and westward. Since the rainband over the northwest region of the subtropical high often affects the South China region, the precipitation can be increased in the winter during El Niño years. From March to May (Spring, Fig. 8), the situation is almost the same as in Winter, with the subtropical high being stronger in El Niño years. Therefore, the onset of summer monsoon in El Niño years is likely to be earlier and stronger which results in the increase of summer precipitation.

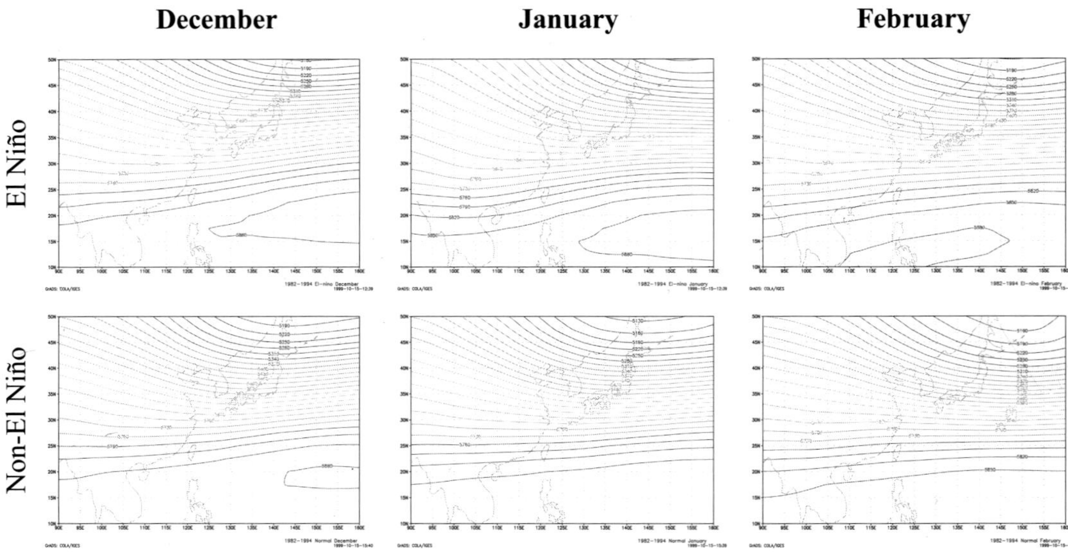

Fig. 7. WINTER December – February 500 hPa.

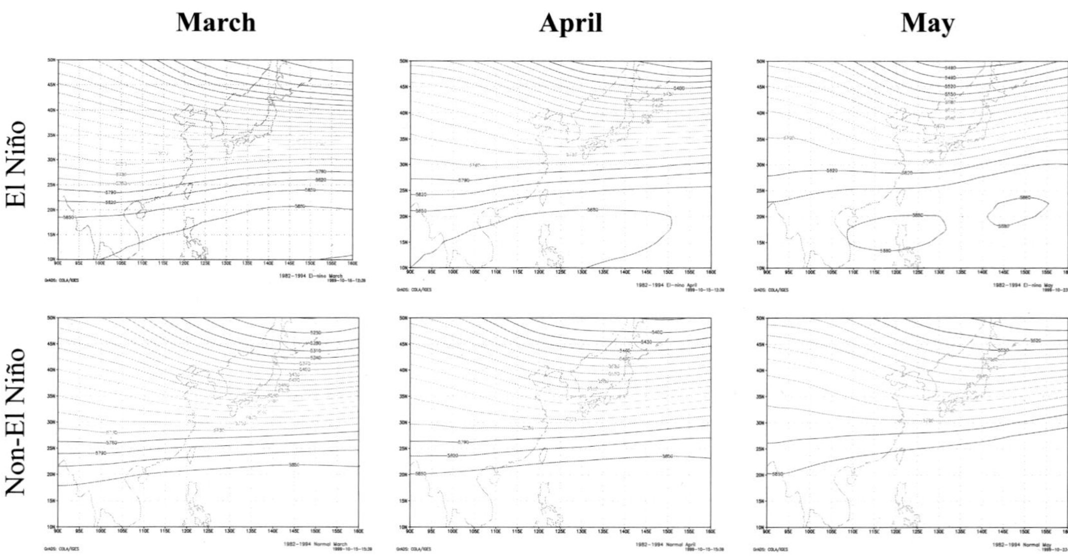

Fig. 8. SPRING March – May 500 hPa.

From June to August (Summer, Fig. 9), the trough in the Indo-China region is stronger during an El Niño year than in a non-El Niño year, even extending to the northern part of South China Sea thus resulting in an increase of precipitation in South China.

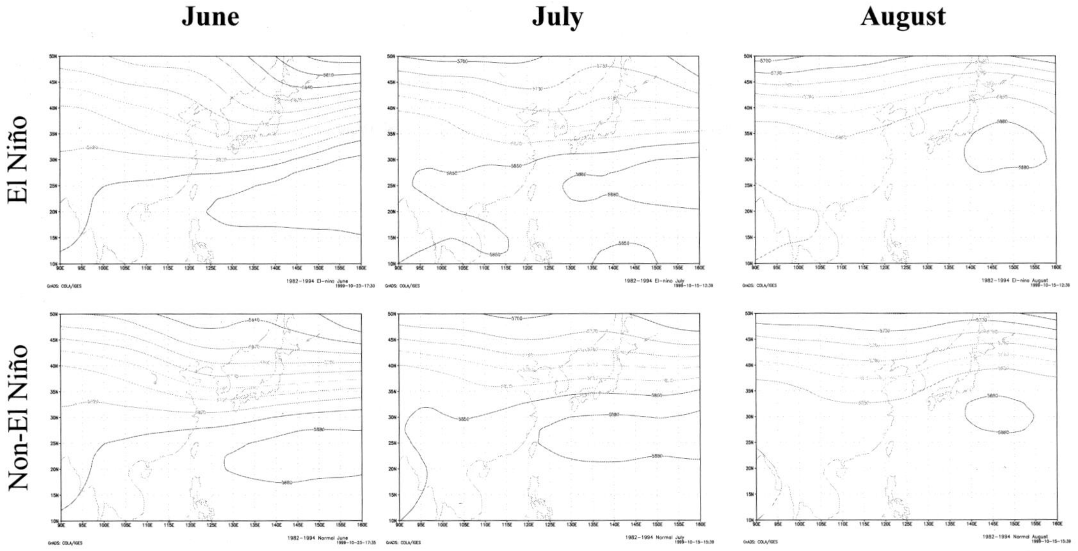

Fig. 9. SUMMER June – August 500 hPa.

From September to November (Autumn, Fig. 10), the synoptic situation is similar during El Niño and non-El Niño years. As the weather in South China is mainly affected by the mainland anticyclone, so it is mainly fine and with little precipitation. In fact, the precipitation basically relates to typhoon frequency.

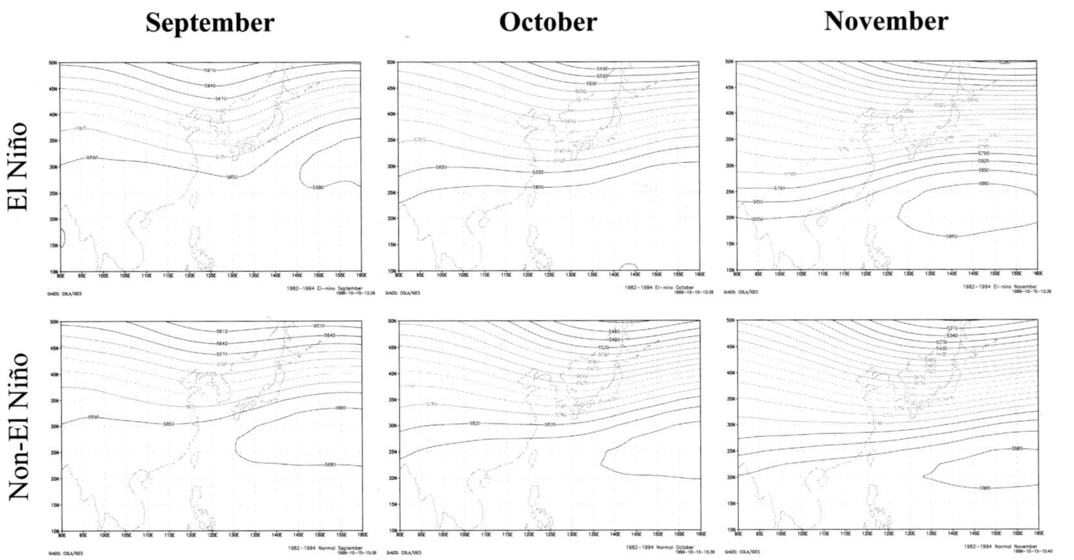

Fig. 10. AUTUMN September – November 500 hPa.

4. CONCLUSIONS

(a) From January to March, the correlations between SLPD, SST and Macao precipitation are quite good. For November, the correlations between SLPD, SST and Macao temperature is fair.

(b) El Niño has a stronger influence on Macao's precipitation than on its temperature.

(c) The analysis of seasonal mean precipitation corresponding to the El Niño seasons defined by intensity $\geq 1.5\sigma$, shows that both SLPD and $NINO_3$ provide excellent correlations. Except in Autumn, all other three seasons pass the t-test which indicates significant relationship between the seasonal mean precipitation of Macao and El Niño. In Autumn, the precipitation is less and mainly caused by tropical cyclones, so that there is no significant correlation with El Niño.

ACKNOWLEDGMENTS. We would like to thank Dr. Bill Kyle of the University of Hong Kong for reading the manuscript and helping in English editing.

REFERENCES

Ding Yihui, 1991: *Advanced Synoptic Meteorology*. China Meteorological Press, China, 717-751.

The 1997/1998 El Niño, 1998: *World Climate News* June, 1998 No.**13**, 5.

Walpole, Ronald E., and Raymond H. Myers: *Probability & Statistics for Engineers & Scientists*. Macmillan Publishing Co., Inc, New York, 574 pp.

Xiang Jingtian, Shi Jiuen, Du Jinguan, Zhou Qinfang, and Kong Nan, 1991: *Dynamic and Static Data Processing – Time Series and Statistical Analysis*. China Meteorological Press, China, 15-66.

Zhang Qin, Ding Yihui, and Zhou Qinfang, 1998: A Study about features of El Niño events during 1991-1994. *ACTA Meteorologica Sinica Vol.* **56** *No.* **5**. China Meteorological Press, China, 573-582.

RELATIONSHIP BETWEEN SSTA IN THE WESTERN PACIFIC AND THE JUNE PRECIPITATION IN THE MID ANDTHE LOWER CHANGJIANG VALLEYS*

Qian Yongfu

Department of Atmospheric Sciences , Nanjing University, Nanjing 210093

Wang Qianqian, Wang Yehong and Zhao Yuchun

Department of Atmospheric Sciences, Nanjing Institute of Meteorology, Nanjing 210044

ABSTRACT

The relationship between the SSTA in the western Pacific and the June precipitation in the mid and the lower Changjiang valleys(MLCJ) is discussed. A key ocean area is determined over the western Pacific (WP) in 134.5°E~147.5°E, 20.5°N~30.5°N. It is found that the period from the previous October to January in the same year is the WP SSTA leading period with the best positive correlation. The years with more precipitation (MPY) and the years with less precipitation (LPY) in June in the MLCJ are selected. Also, the years with higher SST (HTY) and the years with lower SST (LTY) in the WP area in the Oct. to Jan. period are chosen, too. The composed June anomaly patterns of precipitation in China are obtained for both the HTY and the LTY. It is seen that in HTY and LTY the distributions of June precipitation in China are almost in totally opposite phase. The composed June climatic anomaly patterns of meteorological fields are finally discussed in order to analyze the mechanism. It is found that the June climatic anomaly patterns of MPY and HTY are similar, and so are the patterns of LPY and LTY. Therefore, the SSTA in the WP area does have influence on the June precipitation anomalies in the MLCJ, and even in whole China.

Key words: regional climate; climate anomaly; June precipitation; key SSTA area

1. INTRODUCTION

The mid and the lower Changjiang valleys (MLCJ) are the regions with frequent meteorological disasters. Those disasters have resulted in a lot of economical losses. The main disasters in the MLCJ are the floods and the droughts, for example, the floods in 1954, 1983, 1991 and 1998 and the droughts in 1963, 1968, 1978 and 1985 are all very severe. What are the causes of the floods and the droughts in the MLCJ? Could we better predict those disastrous events and avoid great economical losses.

It is well known that the effects of El Nino events on the climate anomalies are remarkable. The El Nino events are usually characterized by the positive anomalies of sea surface temperature (SSTA) over the central and the eastern equatorial Pacific Ocean. Therefore, the effects of El Nino are found obviously and certainly in the tropical and the adjacent subtropical regions (see Chen 1977; Li et al. 1989; Fu 1987; Huang 1990; Yang et al. 1992). What are the effects of El Nino events on the climate in China? Are they the causes of floods and droughts in the MLCJ? And are they evident and certain? The answer from the statistical and the diagnostic studies in the past is not unique. The reason might be that the affecting factors of the climate anomalies in China are not only the SSTA in the Nino regions, but also the SSTA in the adjacent ocean areas, especially in the western or

* Sponsored by the Key Project of the National Natural Science Foundation: "Analyses and Mechanism Studies of Regional Climate change in China" under Grant No.49735170

the northwestern Pacific. Because in the El Nino years the distributive patterns of the SSTA in the adjacent ocean areas are not always the same (Wang and Qian 1995a, b). Due to the strong local interaction between the air and the ocean the climate is more easily influenced by the SSTA in the adjacent ocean areas than in the central and eastern Pacific. Moreover, as China is located in the evident monsoon area, the climate in China is strongly affected by the middle and the high latitude weather systems. Besides, the Tibetan Plateau is also of great importance. Therefore, the mechanisms of the climate anomalies in China are much more complex than in other places in the world.

It is also well known that the main rainfall processes in the MLCJ take place in June and anomalous Meiyu in MLCJ is the direct cause of floods and droughts. Moreover, there is good correlation between the anomalies of precipitation in the MLCJ and in the other regions. Therefore, in this paper, we shall only focus on the effects of SSTA in the adjacent ocean areas on the June precipitation anomalies in the MLCJ.

2. SELECTION OF THE KEY SSTA AREA IN THE CHINA'S ADJACENT OCEANS

In order to find the key SSTA area in the China's adjacent oceans of 40°E-150°E and 0°-60°N, the monthly precipitation data of 160 stations in China from 1951 to 1998 and the UKMO monthly SST data from 1951 to 1994 are used. The correlation coefficients between the June precipitation anomalies in the MLCJ and the monthly SSTA from January in the previous year to May of the same year are computed. The June monthly precipitation in the MLCJ is represented by the mean amount over 12 stations in that area including Nanjing, Hefei, Shanghai, Hangzhou, Anqing, Tunxi, Jiujiang, Hankou, Zhongxiang, Yueyang, Yichang and Changde. All those stations are located in a positive correlation area with correlation coefficients above 0.5 in the point-to-point correlation map with Nanjing station as the centre. Therefore, it is no doubt that they have consistent interannual variations of precipitation. It is found that the monthly correlation coefficients between the SSTA and the MLCJ June monthly precipitation anomalies exceeding the statistical significance level of 0.05 are found only in the previous October to February in the same year. Fig. 1 shows the distributions of individual monthly correlation coefficients orderly from October to February with the area of significance level above 0.05 shaded. It is seen from Fig. 1 that the maximum positive correlation areas in all months are located over the western Pacific with a little difference in different months, the maximum centres are almost always located near 140°E and 25°N of the Kuroshio region. The maximum correlation values are all higher than 0.35. In order to find the most important area with the highest correlation coefficient, for each month we select a rectangular area in which the confidence level is above 0.05 and then average the coefficients to get an area mean value. Such selected areas are not of same size in different months. The averaged correlation coefficients for individual months are given in Table 1.

From Table 1 we can find firstly that the highest correlation coefficient between the MLCJ June precipitation and the SST over the western pacific is 0.484 in the previous November. Secondly, the maximum area of the positive correlation is also in November with 13 longitudes and 10 latitudes. Therefore, this area may be taken as the key area in the adjacent oceans of China, and designated by the WP. In order to further check the representative of the month and the area, we compute the self-correlation coefficients of the SSTA in the WP of November with the leading and the lagging months. The result shows that the self-correlation is good and positive in two years and the self- correlation coefficients from previous August to present March are all above the significance level of 0.05. The SSTA persistence in the WP is at least more than half a year.

Table 1. Monthly averaged correlation coefficients between anomalies of June precipitation in MLCJ and SSTA in the WP.

Correl. Coefficients / SSTA month(area)	June Precipitation in the MLCJ
October (139.5E-142.5E,22.5N-25.5N)	0.428
November(134.5E-147.5E,20.5N-30.5N)	0.484
December(138.5E-145.5E,21.5N-24.5N)	0.408
January (133.5E-142.5E,19.5N-24.5N)	0.427
February (132.5E-141.5E,25.5N-31.5N)	0.364

Table 2. Correlation coefficients in different combinations of months between the June precipitation anomalies in MLCJ and SSTA in the WP area.

Month / Month	Oct.	Nov.	Dec.	Jan.	Feb.
Oct.	0.401	0.478	0.502	0.516	0.510
Nov.		0.484	0.476	0.479	0.465
Dec.			0.412	0.430	0.414
Jan.				0.399	0.381
Feb.					0.327

Note: in Table 2 the numeric in diagonal indicates coefficients in individual months and other numeric the coefficients of different month combinations, for example, the numeric 0.516 represents the coefficient of October to January 4 month combination.

By now we have still not known in what period the SSTA in the WP has the best correlation with the anomalies of the June precipitation in the MLCJ. We may find the best correlation from the correlation coefficients in individual months and in various combinations of months. Table 2 gives the correlation coefficients in different combinations of months. It is obviously seen from Table 2 that the SSTA in the WP has different correlation with the June precipitation in the MLCJ in different periods. The best correlation of individual months is in November as found in the above analysis, the best correlation of two, three and four month combinations is in October to November, October to December and October to January, respectively. The best correlation for all individual and month combinations is in October to January with a value of 0.516 that is much higher than the significance level of 0.001. Therefore, it is the best to use the SSTA in the period of October to January in the WP to predict the June precipitation anomalies in the MLCJ.

3. RELATIONSHIP BETWEEN SSTA IN THE WP AND JUNE PRECIPITATION ANOMALIES IN THE MLCJ AND THE SOUTHERN CHINA

Let us first look at the interannual variations of the monthly mean SSTA averaged from October to January in WP and the June precipitation anomalies over the MLCJ as shown in Fig. 2 that is normalized. It is seen from the figure that there is a good positive correlation between them. When the SST is higher than normal in WP the precipitation is usually more than normal in the MLCJ and vice versa. In order to analyze the common characteristics of the precipitation in the higher SST years as well as the lower SST years, we select them according to the SSTA curve in Fig. 2. The higher SST year is defined as that the positive SSTA is above 0.2°C (the variance is 0.26°C) and the lower SST year defined as that the negative SSTA is below -0.2°C. We call them "HTY" and "LTY", respectively. There are 12 HTY and 10 LTY. It is also seen that 4 of the 12 HTY happened in the early 1950's and 7 in the period from 1980's to nowadays, only 1 in 1970's. The LTY took place mainly in the late 1950's to the late 1970's (8/10).

From Fig. 2 we can also select the years with more precipitation (MPY) and the years with less precipitation (LPY). The MPY is defined as that the positive June precipitation anomaly is greater than 50 *mm* (the variance is 66.11 *mm*) and the LPY is defined as that the negative June precipitation anomaly is smaller than –50 *mm*. In such a way we find that MPY are 1954, 1955, 1956, 1964, 1971, 1975, 1983, 1984, 1986, and LPY are 1952, 1958, 1963, 1965, 1968, 1972, 1978, 1981, 1985. So, the number of MPY is equal to that of LPY, they are both 9 years. It is found again that the LPY appeared mainly in the late 1950's to the late 1970's (6/9). This is in correspondence with that for LTY.

It can be also seen from Fig. 2 that the HTY (LTY) do not match the MPY (LPY) one to one although, the period with more precipitation before the middle 1950's corresponds to the period of warmer SST. And the period of colder SST from 1957 to 1978 corresponds to the period of less precipitation in the MLCJ fairly. The MLCJ reenters a phase of more precipitation as a new warm SST period starts from 1980's.

The composed June precipitation anomaly distributions of the HTY and the LTY are shown in Fig. 3a, b, respectively. It is seen that the June precipitation anomalies in the HTY and the LTY are totally different. In HTY positive precipitation anomalies are located over the MLCJ and a large area north of 30°N except for the lower basin of the Yellow River, while in the south part of China, south of 30°N, there are negative anomalies. In LTY the case is almost completely reversed.

4. EFFECTS OF SSTA IN WP ON CIRCULATION AND DISCUSSIONS OF MECHANISM

The precipitation change is the final result of the changes of circulation. Analysis of circulation changes at different levels can reveal the mechanism of precipitation change. What are the characteristics of circulation anomalies resulted from the SSTA in the WP key ocean area? In order to discuss this problem, we compose the June anomaly fields of geopotential height at the 100 and 500 hPa levels and streamline at the 850 hPa level according to the HTY and the LTY, respectively.

Fig. 4a-f are the June anomaly distributions of the 100 hPa(a, b) and 500 hPa (c, d) level geopotential heights and the 850 hPa (e, f) level streamlines in the HTY (a, c, e) and the LTY (b, d, f), respectively. It is seen that at the 100 hPa level in the HTY (Fig. 4a) the positive anomalies are found in the low and the middle latitude belts from the east part of China westward to the Iranian Plateau with maximum values in between of Changjiang and Huanghe. There are another two positive centres, one is located to the northeast of Lake Baikal, and the other over the central and the eastern Pacific at the mid latitudes. The negative anomalies are located over the European continent, the north and northeast part of Japan and the northwestern Pacific. In the LTY (Fig. 4b), the case is almost reversed. At the 500 hPa, the patterns both in the HTY (Fig. 4c) and the LTY (Fig. 4d) are similar to those at the 100 hPa, correspondingly. The positive anomalies in the HTY are located over the Caspian Sea, the area between Lake Baikal and the Okhotsk Sea and the mid latitude eastern Pacific, a weak positive area is in the south of Changjiang. In the LTY, the situation is almost opposite. At the 850 hPa level, the anomalous streamlines indicate that in the HTY (Fig. 4e) there is a differential cyclonic circulation over the northwestern Pacific, Huanghe and Changjiang happen to locate in the west of the circulation. A differential anticyclone exists over the western Pacific. Along the MLCJ there is an evident shear line which results in a convergent zone and more precipitation over the MLCJ area. Compared with Fig. 4e, the anomaly streamlines in the LTY are nearly in an opposite phase.

The above phenomena indicate that the anomalous distributions of the geopotential height and the streamline are basically resulted from the anomalous surface heating which induces cyclonic anomalies at the low levels and anticyclonic anomalies at the high levels when heating is more. Moreover, the anomalous distributive patterns in the HTY and the LTY are opposite in deed. In order to check whether the circulation anomalies in the MPY and the LPY are similar to that of the HTY and the LTY, respectively, the composed circulation anomalies of the MPY and the LPY are also made (figures omitted) and compared to that of the HTY and the LTY. The answer is positive, especially at the 500 hPa and the 850 hPa levels. Therefore, the June precipitation anomalies in the MLCJ are at least determined by the SSTA values averaged in October to January in the WP area to a large extent. Therefore, we may consider the SSTA values and signs in the WP area as a predictive factor of the June precipitation anomalies in the MLCJ.

5. SSTA DISTRIBUTIONS OVER THE WORLD OCEANS IN HTY AND LTY OF THE WP

The above analyses point out that the SSTA in the key area of WP has important influences on the precipitation in the MLCJ and the southern China. In order to see whether there are other areas in the world oceans, in which the SSTA can greatly influence the June precipitation anomalies in the MLCJ as well, we compose the SSTA distributions of the world oceans in both the HTY and the LTY for individual months. The distributive patterns of the previous October, the present March and June are shown in Fig. 5. It is found that in the HTY (Fig. 5a-c), there is another positive SSTA area over the northwestern Atlantic (NWA) near 40°N and 40°W besides the WP area. The SSTA in the NWA persists for more than 8 months and does not change the value and area a lot. In the LTY (Fig. 5d-f), the SSTA in the NWA becomes negative, but the location of the area is kept the same as that in the HTY.

Xu et al. (1999) have already pointed out that the SSTA in the NWA has big impact on the June precipitation anomalies in the MLCJ. Therefore, the relationship between the SSTA in the WP and the NWA is not a happenstance. We can also see from the composed diagrams that either in the HTY or in the LTY the largest absolute SSTA are nearly located in the above two areas. The SSTA in the Nino regions are not the largest except for the previous Oct. in which the distribution of the SSTA is typical for the La Nina event. However, in June the pattern of SSTA changes to that of El Nino, while in the western and the northern Pacific as well as in the Indian Ocean the patterns of SSTA do not change much. This implies that the SSTA in the central and the eastern equatorial Pacific is not the dominant influencing factor for the June precipitation anomalies in the MLCJ and the south China. The pattern of SSTA outside the Nino region can be similar in the El Nino and the La Nina events. The SSTA in the WP and the NWA can be used as strong signals of the climate change to predict the June precipitation anomalies in the MLCJ. These results should be further proved by using more observational data and by sensitive numerical experiments in the future.

6. CONCLUSIONS

The above discussions may be concluded as follows:

(1) There may be some key regions in the adjacent oceans of China in the western Pacific where the SSTA have important influences on the precipitation in China. The SSTA in the Nino regions are not the only and the most important factor for the climate anomalies in China, especially in the MLCJ and the southern China. Therefore, besides the SSTA in the Nino areas we need to pay more attention to the SSTA distributive patterns in the adjacent oceans of China.

(2) The influence of the SSTA in WP comes to effect through the circulation changes at the atmospheric levels. The anomaly patterns of the geopotential height field at different levels in the warm and the cold SST years are totally different with opposite phases.

(3) The SSTA in the WP has good positive correlation with the SSTA in the NWA. Both the SSTA in the WP and the NWA can be used as strong signals of the climate change to predict the June precipitation anomalies in the MLCJ and even in whole China.

REFERENCES

Chen Liting, 1977: Influences of sea temperature anomalies over the equatorial eastern Pacific on tropical general circulation and precipitation in China in rainy season, *Sci. Atmos. Sinica*, **1**, 1-12 (in Chinese).

Fu Congbin, 1987: El Nino/Southern Oscillation and interannual climate change, *Sci. Atmos. Sinica*, **11**, 209-220 (in Chinese).

Huang Jiayou, 1990: Correlation between sea surface temperature in eastern equatorial Pacific and rain days over China in summer, *Adv. Atmos. Sci.*, **7**, 127-136.

Li Maichun et al., 1989: Relationship between monsoon precipitation in the east part of China and sea surface temperature over the equatorial eastern Pacific, *Sci. Atmos. Sinica*, **11**, 365-372 (in Chinese).

Wang Qianqian, and Qian Yongfu, 1995a: Numerical experiments of effects of sea temperature anomalies over the Pacific on precipitation in 1991, *J. Nanjing Institute of Meteorology*, **18(2)**, 200-205 (in Chinese).

Wang Qianqian, and Qian Yongfu, 1995b: Numerical experiments of mechanism of the floods in Changjing and Huaihe Basins in 1991, *Acta Meteo. Sinica*, **53(suppl.)**, 595 603 (in Chinese).

Xu Haiming, He Jinhai, and Yao Yonghon, 1999: Interannual variability of the Meiyu onset and its association with the atmospheric circulation in the previous winter and possible cause, *J. Nanjing Institute of Meteor.*, **22(2)**, 246-253 (in Chinese).

Yang Xiuqun, and Xie Qian, 1992: Correlation of the sea temperature over the mid and eastern equatorial Pacific and the sea ice in the arctic, *Tropical Meteorology*, **8**, 261-265 (in Chinese).

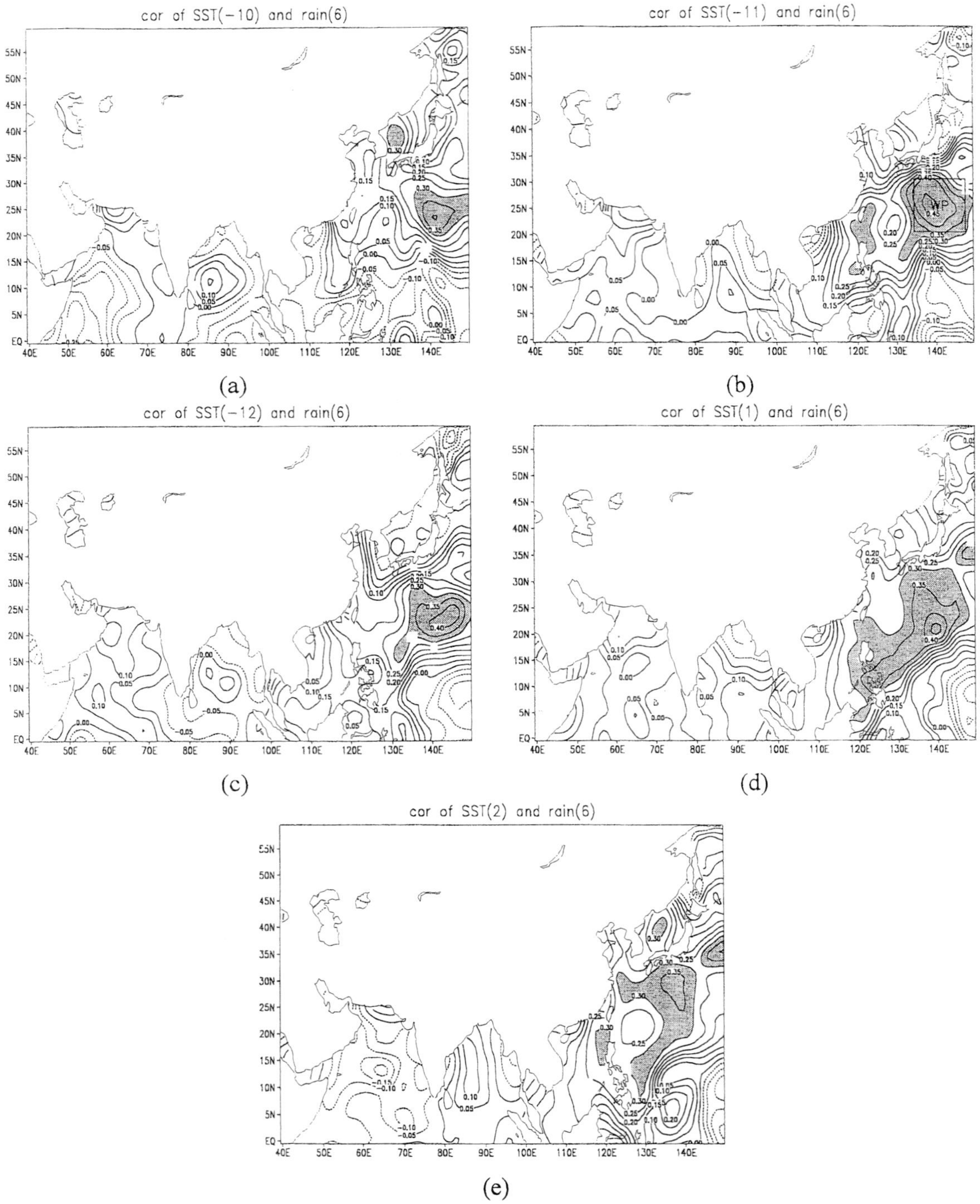

Fig.1 Distributions of correlation coefficients between the June precipitation anomalies in the MLCJ and the SSTA in the adjacent oceans of China in previous October(a), November(b), December(c), present January(d) and February(e) with the areas of coefficients above the significance level 0.5 shaded.

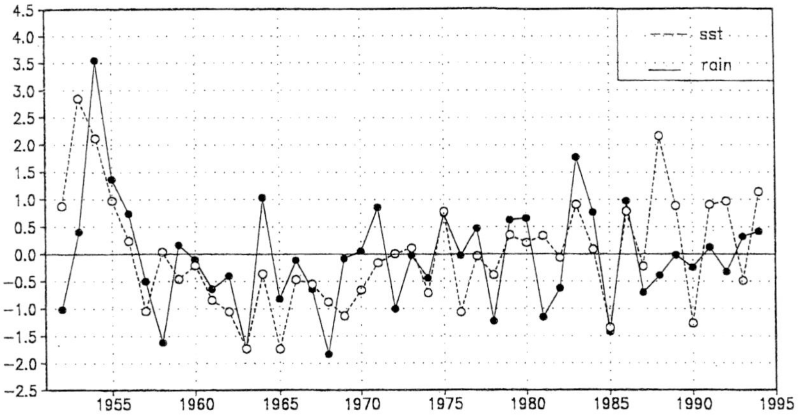

Fig.2 Interannual variations of monthly mean SSTA averaged from Oct.-Jan. in WP and June precipitation anomalies over the MLCJ

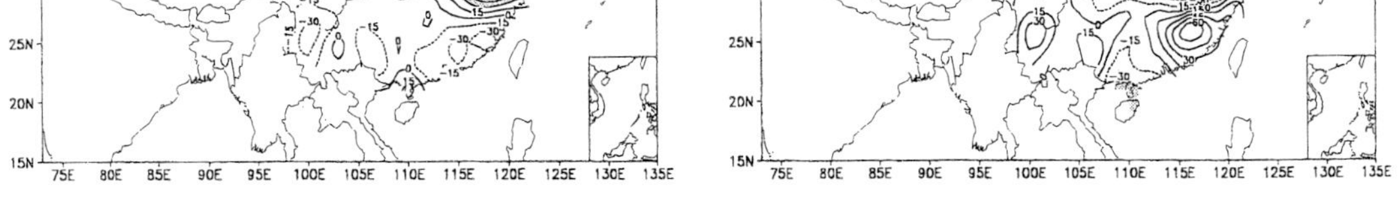

(a) (b)

Fig.3 The composed precipitation anomaly distributions of HTY(a) and LTY(b)

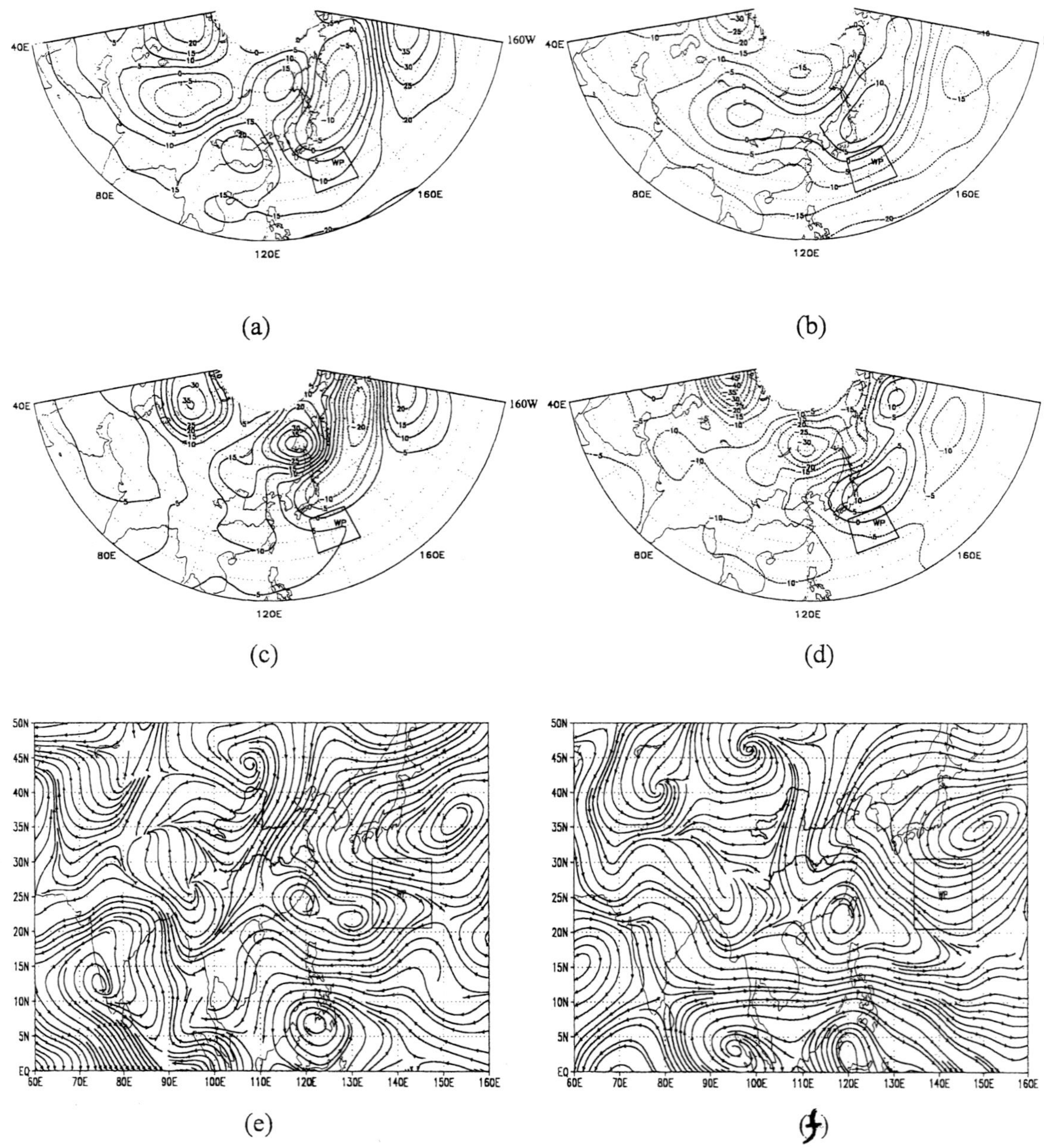

Fig.4 June anomaly distributions of the 100hPa(a, b) and 500hPa(c, d) level geopotential height and the 850hPa level streamlines(e,f) in the HTY(a,c,e) and the LTY(b,d,f)

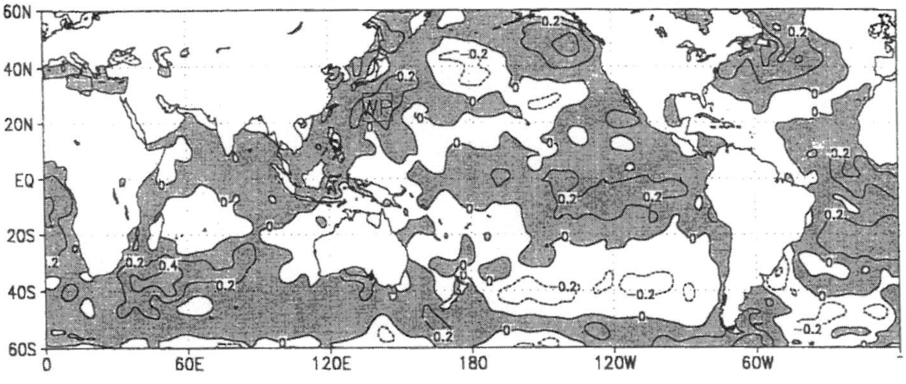

(a)

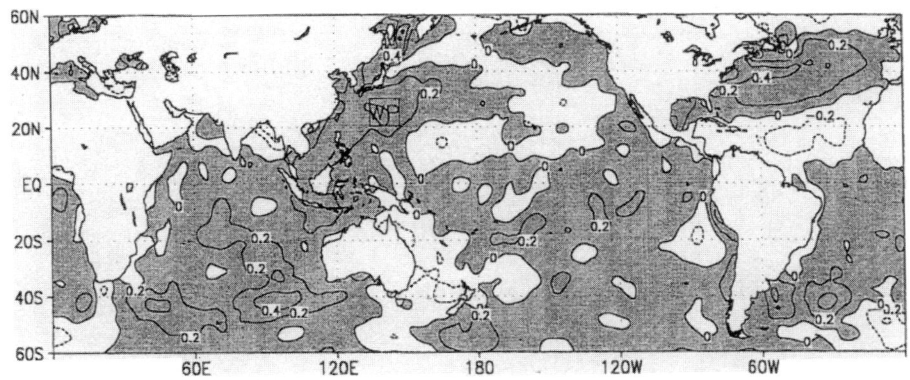

(b)

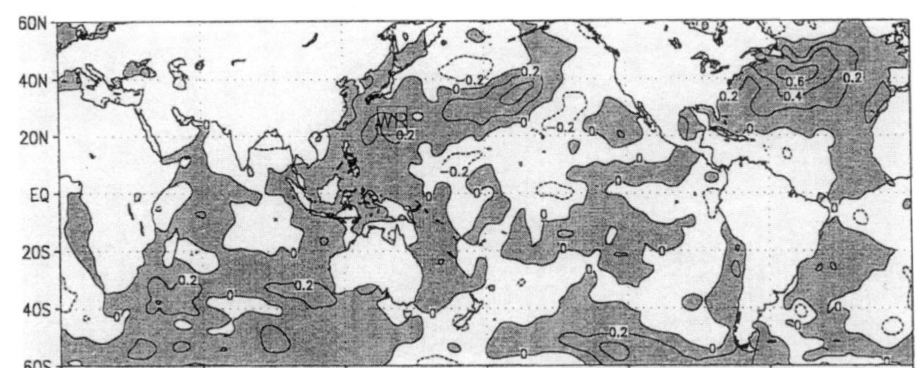

(c)

Fig.5 Composed SSTA in the world oceans in HTY(a-c) and LTY(d-f) in WP. Shaded areas for positive SSTA, (a,d) for previous Oct., (c, e) for Mar. and (d, f) for Jun.

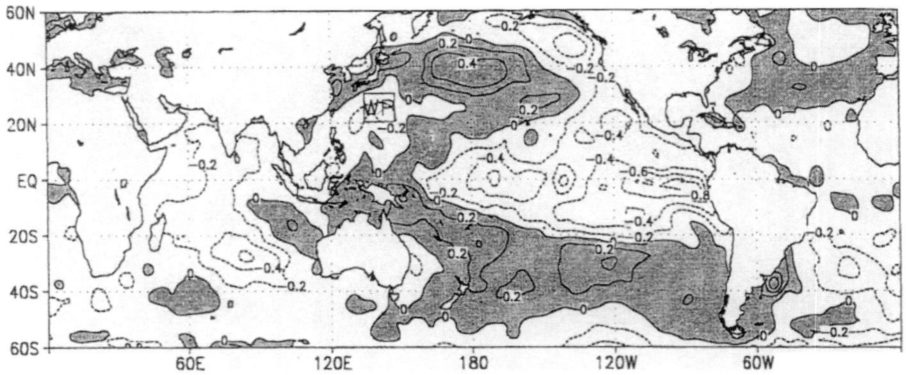

(d)

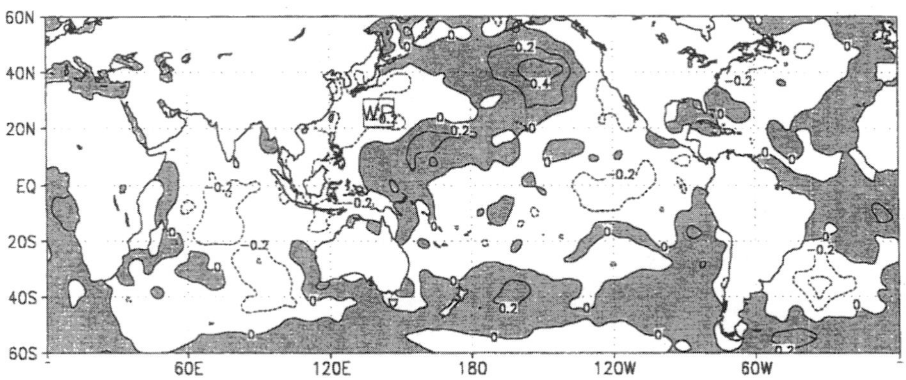

(e)

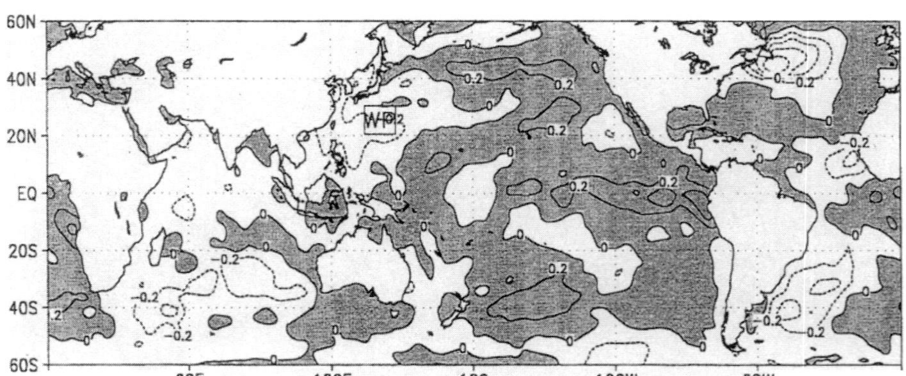

(f)

Fig. 5. (cont'd)

ANALYSIS ON CORRELATION AND CAUSES OF MONTHLY PRECIPITATION AND DROUGHTS/FLOODS ON BOTH SIDES OF THE STRAIT FROM 1951 TO 1998

CHEN Juying WANG Yuhong XIONG Minquan
Chinese Academy of Meteorological Sciences, Beijing 100081

ABSTRACT

Characteristics of climatological distributions of annual and monthly precipitations as well as the pattern of drought/flood variation in Taiwan over a 48-year period of 1951 - 1998 were statistically analyzed. It was found that the northeastern part had for many years a maximum in mean annual precipitation, while the mid-western and southwestern parts had a minimum. The mean annual precipitation over the southeastern coast of mainland was as much as that for southwestern Taiwan. Flood season over most parts of Taiwan occurred between April and June; but in the northeastern part, intense precipitation occurred between September and December. For the southeastern coast of mainland, intense precipitation usually occurred between April and June. Correlation of monthly precipitation and comparison of drought/flood years between Taiwan and the southeastern coast of mainland were made. They revealed a significant relationship regulating the drought/flood variation in the flood season over both sides of the Taiwan Strait and the corresponding changes in monthly precipitation. Causes of the observed phenomena were discussed.

Key words: Both sides of the strait, monthly precipitation, drought/flood variation, correlation cause research

1. DATA

Precipitation data from more than 19 meteorological stations in Taiwan were provided by Taiwan Meteorological Bureau. Precipitation data from the mainland were provided by CNCC.

2. DIVISION OF PRECIPITATION REGION

In view of the complex topography of Taiwan, regional division using agricultural climate as reference was adopted to conduct a thorough analysis on the characteristics of interannual variation of drought/flood on both sides of the Strait in the past 48 years (1951-1998). For the 25 available stations in Taiwan, we selected 19 stations for which the history of data was relatively long. They were then divided into eight regions: northeast (3 stations), northwest (4 stations), mid-west (2 stations), southwest (2 stations), south (2 stations), central (3 stations), and east (3 stations). The southeastern coast of mainland was divided into six regions: east of Guangdong (3 stations), south of Fujian (3 stations), north of Fujian and south of Zhejiang (3 stations), south of Jiangxi (3 stations), north of Jiangxi (3 stations), and the lower reaches of Yangtze River (10 stations).

3. ANALYSIS OF CLIMATIC DISTRIBUTION AND INTERANNUAL VARIATION OF PRECIPITATION ON BOTH SIDES OF THE STRAIT

(1) Climatic Distribution of Monthly Precipitations

As shown in Figs. 1 and 2, comparisons of the climatic distribution of annual precipitation were made between Taiwan and the southeastern coast of mainland. The annual precipitation for Taiwan as a whole was 2,566 mm (average of 19 stations). The annual total was 3,649 mm in northeast, 1,659 mm in mid-west and southwest, 1,324 mm in central, and 2,084 to 2,096 mm in the other three areas. The annual precipitation over the six regions of the southeastern coast of mainland ranged between 376 and 1,718 mm, close to that for the mid-west and southwest in Taiwan.

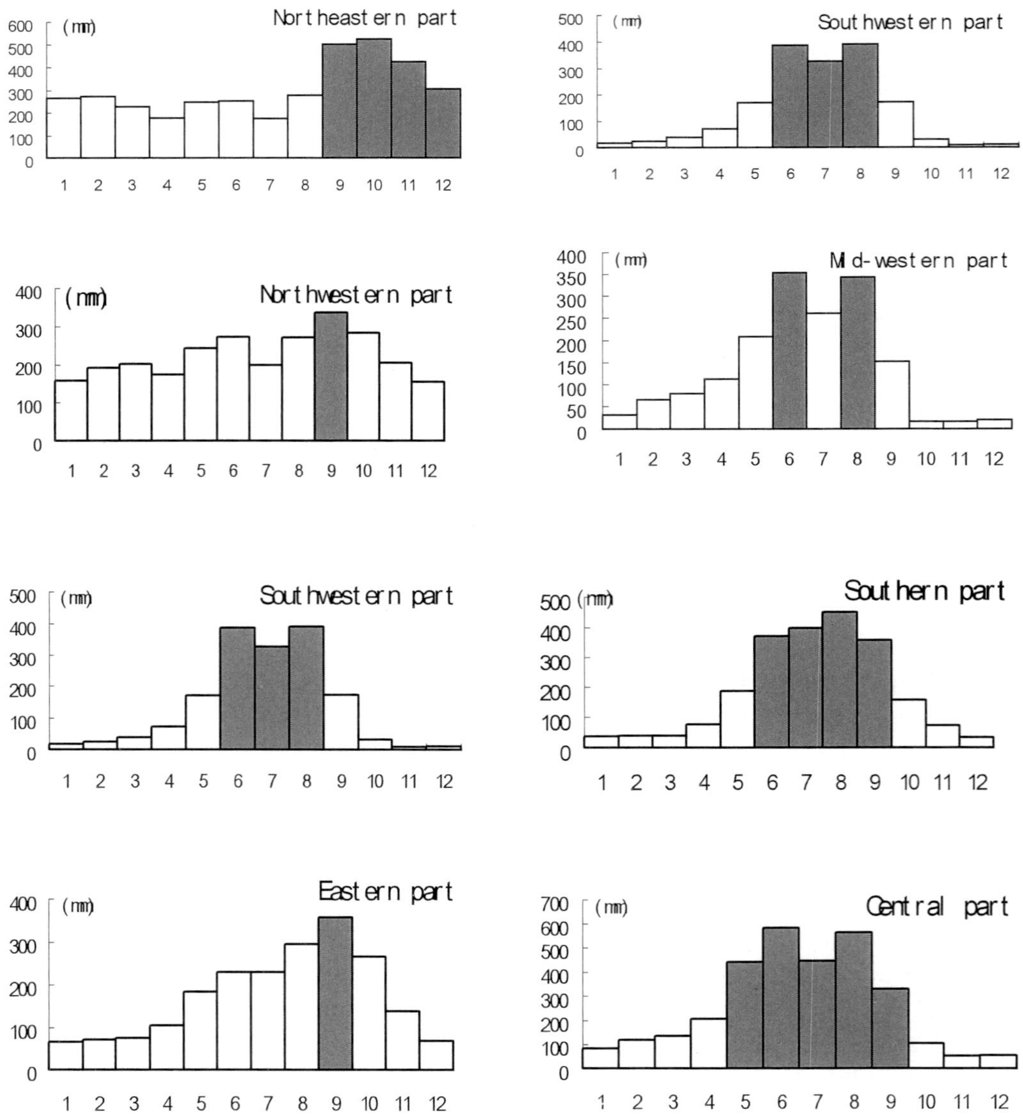

Fig. 1. Monthly precipitation over the eight regions of Taiwan (48-year average for 1951 - 1998).

Areas in Taiwan with monthly precipitation exceeding 300 mm included: northeast from autumn to winter (September to December), northwest and east in September, mid-west in June and August, southwest in summer (June to August), south from June to September, and

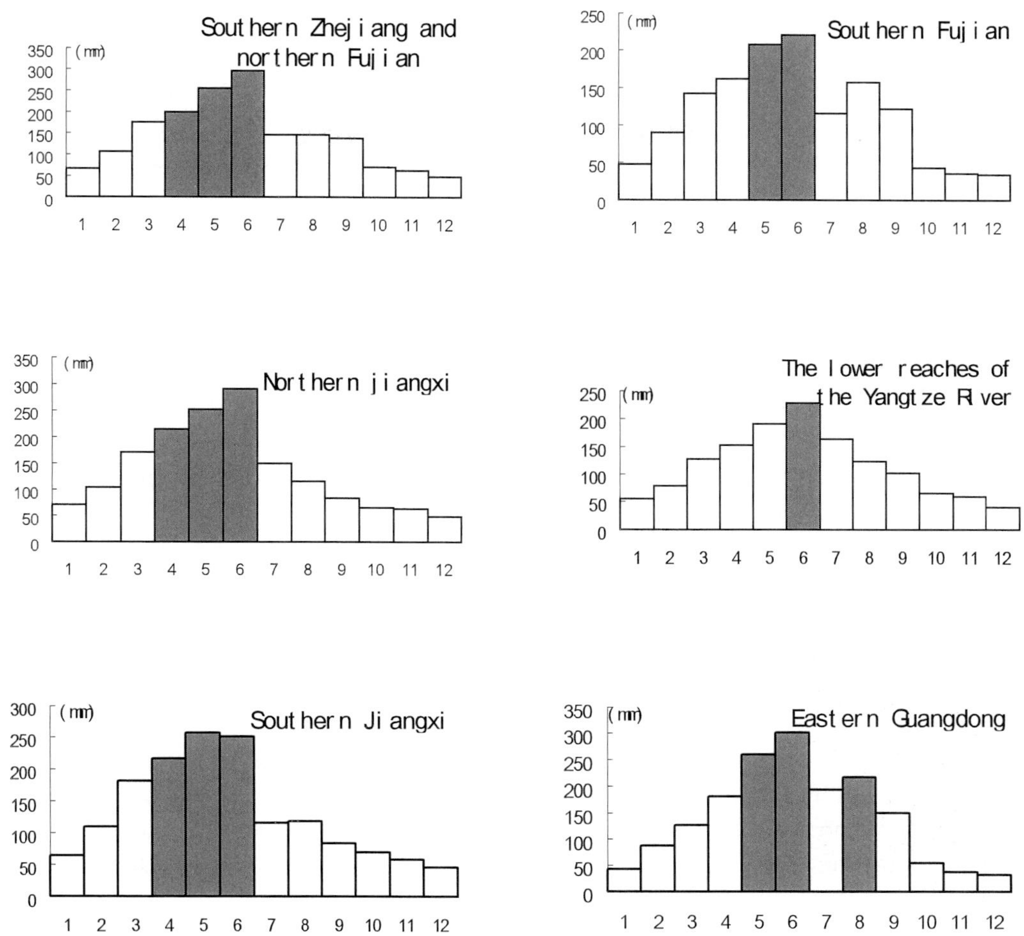

Fig. 2. Monthly precipitation over the six regions in the southeastern coast of mainland (48-year average for 1951 - 1998).

central from May to September. Monthly precipitations over the southeastern coast of mainland were generally less than 300 mm except for east of Guangdong where precipitation in June reached 301 mm. Areas with monthly precipitation exceeding 200 mm included: south of Zhejiang, north of Fujian, south of Jiangxi and north of Jiangxi from April to June; south of Fujian from May to June; east of Guangdong from May to June and August, and the lower reaches of Yangtze River in June.

As shown in Fig. 1, monthly precipitations over northeast and northwest of Taiwan were generally above 155 mm over the 48-year period, with the flood season in northeastern Taiwan coming in autumn and early winter. Monthly precipitations of all other regions were below 100 mm.

(2) Interannual Variations of Precipitation

The range of annual precipitation over various parts of Taiwan were: in northeast, a maximum of 5,178 mm in 1974 which was about 2.5 times of the minimum of 2,133 mm in 1963; in northwest, 3,761 mm (1969) and 1,760 mm (1963); in mid-west, 2,591 mm (1959) and 851 mm (1954); in southwest, 2,953 mm (1977) and 552 mm (1980); in south, 2,643 mm (1974) and 1,016 mm (1980); in central, 4,575 mm (1972) and 1,845 mm (1980). Thus the interannual variations of precipitation over Taiwan ranged from two to five times depending on locations. For the southeastern coast of mainland: in south of Zhejiang and north of Fujian, 2,311 mm (1975) and 1,172 mm (1971); in south of Fujian, 1,881 mm (1990) and 887 mm (1967); in east of Guangdong, 2,514 mm (1983) and 1,026 mm (1963); in the lower reaches of Yangtze River, 2,038 mm (1954) and 855 mm (1978); in south of Jiangxi, 2,130 mm (1961) and 1,082 mm (1963); in north of Jiangxi, 2,419 mm (1954) and 1,101 mm (1963). On the mainland side, the interannual precipitation variations were generally less than those for Taiwan.

In terms of monthly precipitations over Taiwan: in northeast, maximum variable argument ranged from 417 mm (May) to 1,894 mm (October) and minimum variable argument from 23 mm (January) to 99 mm (July); in northwest, maximum from 315 mm (January) to 1,099 mm (October) and minimum from 2 mm (April) to 57 mm (October); in mid-west, maximum from 74 mm (November) to 988 mm (August) and minimum from 0 mm (April) to 91 mm (August); in southwest, maximum from 51 mm (December) to 1,443 mm (August) and minimum from 0 mm (from October to May) to 41 mm (June); in south, maximum from 124 mm (March) to 1,042 mm (September) and minimum from 0 mm (March) to 53 mm (September); in east, maximum from 171 mm (March) to 1,388 mm (October) and minimum from 13 mm (August) to 47 mm (September); in central, maximum from 151 mm (November) to 1,470 mm (June) and minimum from 1 mm (November) to 141 mm (August).

4. STATISTICAL TEST OF CORRELATION SIGNIFICANCE BETWEEN MONTHLY PRECIPITATIONS ON BOTH SIDES OF THE STRAIT

The correlation significance of monthly precipitation between the eight regions of Taiwan and the six regions of southeastern coast of mainland was statistically analyzed and tested. As shown by the results in Table 2, month-to-month correlation was very significant between Taiwan and south of Fujian and east of Guangdong, with confidence level of 0.001 in the months of February to April, October and December. Correlation coefficient was as high as 0.9 in February between mid-west of Taiwan and east of Guangdong (Fig. 3). In May and June, correlation between northeast, northwest, mid-west, central of Taiwan and eastern Guangdong became more obvious. In July and August, there was a significant correlation between all parts of Taiwan and south of Fujian. Correlation coefficients of precipitation in the summer months of June-July-August (JJA) ranged from 0.34 to 0.57 with confidence levels of 0.001 to 0.01. The maximum correlation coefficients were: 0.59 for January between south of Fujian and central of Taiwan, 0.90 for February between east of Guangdong and mid-west of Taiwan, 0.74 for March between southern Fujian and northwest of Taiwan, 0.67 for April between east of Guangdong and northwest of Taiwan, 0.57 for May between east of Guangdong and northwest of Taiwan, 0.62 for June between east of Guangdong and mid-west of Taiwan, 0.51 for September between southern Zhejiang or northern Fujian and southwest of Taiwan, 0.59 for October between southern Fujian and central of Taiwan, 0.50 for November between east of Guangdong and mid-west of Taiwan, 0.61 for December between southern Zhejiang or northern Fujian and east of Taiwan, 0.57 for summer (JJA) between southern Fujian and the whole of Taiwan, and 0.51 from January to December between southern Fujian and mid-west of Taiwan.

Table 1. Correlation coefficients of Taiwan precipitations with southern Fujian and eastern Guangdong.

		1	2	3	4	5	6	7	8	9	10	11	12	6-8	6-9	1-12
Southern Fujian	THE WHOLE AREA	0.46	0.78	0.71	0.56	0.05	0.33	0.61	0.60	0.47	0.52	0.23	0.45	0.57	0.48	0.38
	Northeastern part	0.30	0.68	0.53	0.44	0.37	0.26	0.37	0.47	0.25	0.42	0.21	0.30	0.38	0.19	0.11
	Northwestern part	0.49	0.81	0.74	0.55	0.30	0.29	0.37	0.47	0.27	0.41	0.21	0.40	0.37	0.30	0.29
	Mid-western part	0.48	0.85	0.67	0.58	-0.06	0.47	0.47	0.62	0.37	0.55	0.46	0.27	0.56	0.52	0.51
	Southwestern part	0.20	0.80	0.50	0.47	-0.17	0.28	0.64	0.35	0.46	0.55	0.25	0.37	0.34	0.28	0.39
	Southern part	0.03	0.30	0.26	0.21	-0.17	0.07	0.57	0.28	0.25	0.55	0.06	0.48	0.39	0.28	0.27
	Eastern part	-0.02	0.53	0.23	0.25	-0.07	0.05	0.42	0.31	0.45	0.52	0.24	0.55	0.47	0.33	0.17
	Central part	0.59	0.70	0.60	0.55	-0.06	0.34	0.52	0.61	0.46	0.59	0.17	0.40	0.57	0.53	0.43
Eastern Guang-dong	THE WHOLE AREA	0.43	0.86	0.69	0.63	0.41	0.50	0.36	0.49	-0.01	0.44	0.25	0.38	0.49	0.28	0.25
	Northeastern part	0.27	0.76	0.43	0.56	0.41	0.41	0.10	0.34	-0.14	0.38	0.19	0.22	0.31	0.06	0.02
	Northwestern part	0.47	0.86	0.60	0.67	0.57	0.53	0.13	0.36	-0.13	0.33	0.22	0.29	0.38	0.22	0.26
	Mid-western part	0.47	0.90	0.72	0.62	0.37	0.62	0.43	0.37	0.13	0.44	0.50	0.28	0.52	0.41	0.46
	Southwestern part	0.32	0.83	0.62	0.54	0.31	0.32	0.45	0.39	0.04	0.40	0.32	0.44	0.38	0.25	0.36
	Southern part	0.05	0.33	0.36	0.19	0.08	0.15	0.31	0.44	0.05	0.47	0.18	0.48	0.29	0.19	0.16
	Eastern part	-0.01	0.59	0.38	0.24	-0.07	-0.01	0.22	0.44	0.07	0.44	0.25	0.59	0.29	0.05	-0.03
	Central part	0.55	0.81	0.66	0.56	0.36	0.52	0.39	0.31	0.03	0.54	0.25	0.40	0.48	0.29	0.31

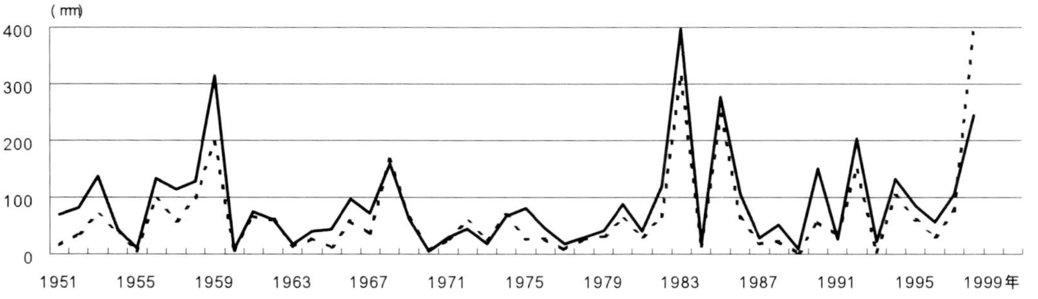

Fig. 3. Comparison curve showing annual changing precipitation in February between eastern Guangdong (solid curve) and mid-western Taiwan (dotted curve) from 1951 to 1998.

5. COMPARISON AND ANALYSIS OF INTERANNUAL DROUGHT/FLOOD VARIATIONS ON BOTH SIDES OF THE STRAIT

Wet (and dry) events were defined to be positive (negative) precipitation anomalies in excess of 30%. Precipitations in the flood season (June to September) over the whole of Taiwan (average of 19 stations) during the 48-year period were plotted by the lower curve in Fig. 4. Corresponding annual totals were also plotted (upper curve) for comparison.

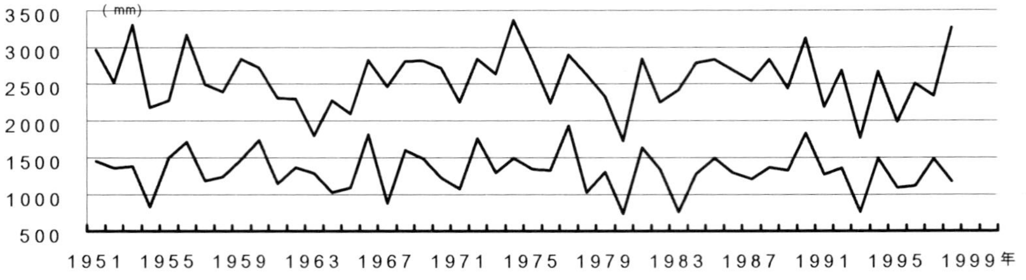

Fig. 4. Total precipitation from June to September (the lower curve) and from January to December (the upper curve) throughout Taiwan (average of 19 stations from 1951 to 1998).

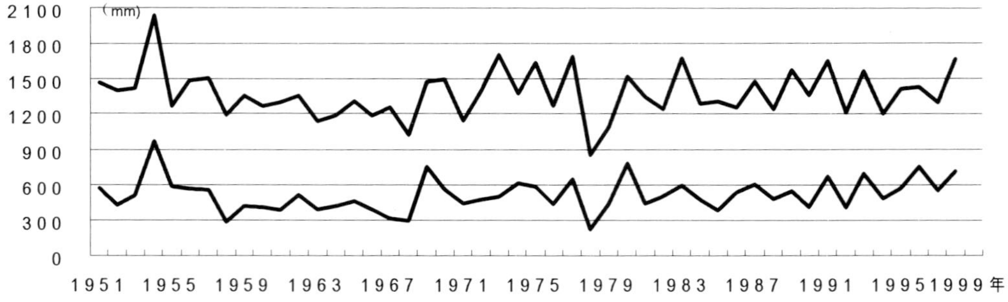

Fig. 5. Total precipitation from June to August (the lower curve) and from January to December (the upper curve) throughout the lower reaches of Yangtze River (average of 10 stations from 1951 to 1998).

The trends of both curves were generally consistent with each other. The exceptions were the wet years of 1969, 1973, 1974 and 1998 when heavy rain occurred in October outside the flood season. October precipitation anomalies in these years exceeded 100% with absolute values between 504 and 878 mm. Three out of these four years had wet episodes over the middle-lower reaches of Yangtze River, meaning a 75% probability that heavy precipitation would also occur in summer on the other side of the Strait over the middle-lower basin of Yangtze River.

The remaining six wet years and months in which flood occurred during the flood season were: 1956 (September), 1960 (August), 1966 (June), 1972 (June and August), 1977 (June and July), 1990 (June, August and September). Total precipitations (average of 19 stations) in the flood season (i.e. June to September) were within the range of 1,711 to 1,930 mm. Among these six wet years, there were five dry years (except 1977) over the Jianghui region and south of the Yangtze River. Therefore, there was a 83% probability that below-normal precipitation would occur over the middle-lower reaches of Yangtze River when heavy rain occurred during the flood season of Taiwan.

Table 2. Precipitation anomalies (△R/months), in %, for the seven wet years in the lower reaches of Yangtze River (LRYR, average of 10 stations) and the corresponding precipitation anomalies in Taiwan (TW, average of 19 stations).

	1954	1969	1980	1991	1993	1996	1998
△R/JJA (LRYR)	91	49	55	33	38	50	42
△R/JJA (TW)	-35	-16	-55	-16	-41	-10	-10
△R/JJAS (TW)	-37	12	-44	-4	-43	-15	-11
△R/July (TW)	-54	-35	-60	-4	-65	64	-57
△R/August (TW)	-24	-35	-28	-61	-60	-8	-26

For the flood season (JJAS) of Taiwan, there were five dry episodes in the years of 1954, 1967, 1980, 1983 and 1993. Apart from 1967, wet anomalies occurred in the middle-lower reaches of Yangtze River. On the other hand, as shown in Figs. 4 and 5 as well as Table 2, for the seven wet years (1954, 1969, 1980, 1991, 1993, 1996, 1998) in the lower reaches of Yangtze River, dry anomalies dominated in most parts of Taiwan during the flood season: seven out of seven had dry anomalies in June to August between 10% and 55%; six out of seven had dry anomalies in June to September between 4% and 44% and dry anomalies in August between 24 and 61%, five out of seven had dry anomalies in July between 35% and 65%. For June and September, it was mainly dry among the seven years.

For the flood season in Taiwan, the wet events (1956, 1960, 1966, 1972, 1977, 1990) never occurred in consecutive years. The same applied to the dry years (1954, 1967, 1980, 1983, 1993). The return periods for the wet and dry years were 3 to 6 years and 10 to 13 years respectively.

6. ANALYSIS OF FLOOD/DROUGHT DISTRIBUTION CHARACTERISTICS ON BOTH SIDES OF THE STRAIT

Our studies showed that for the wet years in Taiwan, droughts tended to occur over the areas south of Yangtze River, while deficient rainfall and droughts in the flood season of Taiwan were often associated with floods in the middle-lower reaches of Yangtze River. The main reason for such contrasts was that for the years with an active flood season in Taiwan, the Western Pacific subtropical high (WPSH) extended only to the longitudes of 120°E to 133°E with the ridge axis located further north at the latitudes of 25°N to 32°N. Under such conditions, moisture, in short supply over the southern part of mainland, was directed instead towards the Taiwan areas. On the other hand, for the wet years over the middle-lower reaches of Yangtze River, the WPSH was stronger and extended further to the west, reaching the longitudes of 98°E to 118°E and the ridge axis was also located more to the south at the latitudes of 17°N to 25°N. With this set up, Taiwan was under the dominance of the WPSH and as such was more likely to be hit by droughts.

7. SUMMARY

Based on the statistical analyses and studies of monthly precipitation patterns on both sides of the Strait over the 48-year period of 1951 to 1998, some significant results were obtained:

1. While the flood season in Taiwan was overall from June to September, there were also regional variations, e.g. May in central and from September to December in northeast. Over the southeastern coast of mainland and to the south of Yangtze River, the flood season was from April to June.
2. Precipitation distribution patterns in Taiwan showed more precipitation in northeast and less precipitation in southwest. The annual precipitation over the southeastern coast of mainland was close to that for southwestern Taiwan.
3. Precipitation over Taiwan was strongly linked to the rainfall over southern Fujian and eastern Guangdong in the same month; for example, the correlation coefficient between mid-western Taiwan and eastern Guangdong in February was as high as 0.90 over the 48-year period.
4. When Taiwan had an active flood season (JJAS), precipitation was likely to be below normal over the valleys of Yangtze River and Huai River. On the contrary, flood or heavy rain was highly probable over the valleys of Yangtze River when the flood season in Taiwan was relatively dry.
5. For most of the years when the basin of Yangtze River and the northern part of the areas south of Yangtze River were dominated by summer floods (JJA), the flood season in Taiwan tended to be relatively dry, especially in July and August.
6. The climatic trends in the spatial characteristics of precipitation distribution and the interannual variations of droughts/floods on both sides of the Strait were largely determined by the occurrence of typhoons and changes in the WPSH positions.

ACKNOWLEDGMENTS. We greatly appreciate the kindness of Mr. Xie Xinliang, Mr. Lin Mingsheng and Mr. Chen Zhenggai in providing the Taiwan data. We would also like to thank CNCC for their support.

REFERENCES

Guo Wen-Shuo, and Yang Zhi-Yuan, 1980: Division of agricultural climate regions in Taiwan, NSC−69b−0409−24, P55−107.

Chen Ju-Ying, 1991: The analysis of floods/droughts and study of long-range forecasting in China, Chinese Agriculture Press, Feb. 1991.

PACIFIC CLIMATE INFORMATION AND PREDICTION SYSTEM (PCIPS) – A NEW PROJECT IN THE U.S. CLIMATE PREDICTION CENTER

Yuxiang He and Anthony G. Barnston

Climate Prediction Center, National Centers for Environmental Predication

National Oceanic and Atmospheric Administration, USA

ABSTRACT

The Pacific Climate Information and Prediction System (PCIPS) is a U.S. Government Internet facility that provides climate data as well as climate forecasts for the tropical Pacific region. The Pacific region is strongly influenced by both the beneficial and detrimental rainfall impacts of the El Nino/Southern Oscillation (ENSO). This report highlights some of the prominent features of this information facility, and illustrates both the level of skill of the climate forecasts and the implied utility of information to the inhabitants of Pacific island nations.

1. INTRODUCTION

Research has shown that short-term global climate fluctuations, such as those related to the El Nino-Southern Oscillation (ENSO) phenomenon, play a dominant role in the climate variability in the tropical Pacific islands. Because the demand for climate information in the tropical Pacific islands has risen in the last decade, a website (http://www.cpc.ncep.noaa.gov/pacdir), called the Pacific Climate Information and Prediction System (PCIPS), has been established at the Climate Prediction Center (CPC) of the U.S. National Centers for Environmental Prediction (NCEP) to serve the tropical Pacific region.

PCIPS has seven components: (1) **Rainfall Prediction**, which includes the tropical Pacific islands rainfall prediction Bulletin, official rainfall forecast for Hawaii, U.S.-affiliated and non-U.S. affiliated Pacific islands using both a statistical CCA and experimental rainfall forecasts of the NCEP dynamical coupled model; (2) **ENSO Effects**; (3) **Climate Information**; (4) **Sea Level Analysis and Prediction**; (5) **Online Data**; (6) **Paper and Atlas;** and (7) **Useful links**. Here we discuss the first two components.

2. RAINFALL PREDICTION

(a) Tropical Pacific Islands Rainfall Prediction by NCEP CCA Model

Canonical correlation analysis (CCA) is a statistical method that identifies linear relationships between multi-component predictors and multi-component predictants. CCA is now being used commonly in the atmospheric sciences (e.g. Barnett and Preisendorfer 1987; Barnston and Ropelewski 1992; Barnston and He 1996). Here, CCA is used to predict 3-month precipitation anomalies in the Pacific Islands out to a year in advance, as described in He and Barnston (1996). Because rainfall in the tropical Pacific is strongly related to ENSO (Ropelewski and Halpert 1987,

1996), and ENSO has some predictability, it is worthwhile to establish a real-time prediction system to benefit the commercial interests in the Pacific Islands.

The forecasts are provided on a monthly basis on the Internet. The predictor fields used for the forecasts include quasi-global sea surface temperature (SST), Northern Hemisphere 700 mb geopotential height, and the predictand precipitation itself at an earlier time. CCA sensitivity experiments indicate that the SST field is the most valuable predictor field, with 700-mb heights and prior precipitation somewhat helpful. Further details about the skills and the predictors are provided in He and Barnston (1996). The predictors are configured as four consecutive 3-month periods prior to the time of the forecast, followed by a variable lead time, and then a single 3-month predictand period. The predictand includes 3-month total rainfall at 61 tropical Pacific Island stations. The rainfall data and climatology for 66 stations has been described in a recent atlas (He et al. 1998a). Figure 2 in Barnston 1994a illustrates the CCA procedure in greater detail.

The expected skill of the forecasts was estimated using a 1-year-out cross-validation scheme (He and Barnston 1996). During periods with significant ENSO signals, average skills are up to 60 percent higher. This would be expected in view of the increase in the signal-to-noise ratio in the rainfalls when ENSO-neutral periods are removed (Ropelewski and Halpert 1987, 1996). The climate variability across the tropical Pacific is seen to be governed largely but not entirely by the ENSO cycle. During the northern winter of mature warm (cold) SO events, there is a strong tendency toward negative (positive) precipitation anonymous across the northern tier of islands from Hawaii to the western Pacific, as well as at near-equatorial stations in the western Pacific and the southern tier of stations. Conversely, positive (negative) anomalies tend to occur near and slightly south of the equator in the central and eastern pacific, in a band that widens as one moves farther east. The skill of Jan-Feb-Mar forecasts made at the end of November (1-month lead) averaged over both the northern and southern Tropics is 0.34. Moderate skills (correlation >0.5) are found for certain stations in the northern Tropical at lead times of 3 months or less in late northern winter, especially in the western Pacific. The CCA modes (He and Barnston 1996) clearly show ENSO as the dominant influence on tropical Pacific climate, especially during the boreal winter and spring seasons (and during fall along the immediate equator near and somewhat east of the dateline).

Figure 1 shows forecasts for Jan-Feb--Mar 2000 (3.5 month lead), run in mid-September of 1999, of the standardized precipitation anomaly (X100) (i.e., the anomaly divided by the interannual standard deviation) for 61 Pacific Island stations using data through August 1999. The expected skill for these forecasts, based on cross-validation, is shown by the size of the numerals (as opposed to their value, which is the forecast itself): Large numerals indicate higher skill than small-sized numerals (see the figure caption).

The time series of the CCA forecasts for Hawaii are shown in Fig .2, for 1 to 13 seasons lead time (solid bars), along with their expected skills (lines). The horizontal axis reflects the lead time, whose corresponding actual target period for this forecast is indicated in the legend along the top of the figure (e.g. 1=Oct-Nov-Dec 1999). The same ordinate scale is used for both forecasts and skills (standardized anomaly and temporal correlation coefficient, respectively). Sometimes skill may increase as the lead is increased because a more forecastable target season has been reached. The forecasts and their skills differ not only due to their differing ENSO-responsiveness caused by general location differences the Pacific basin, but also due to differences in orientation with respect to the local orography (if any).

(b) Tropical Pacific Islands Rainfall Prediction by NCEP Coupled Model

The NCEP comprehensive coupled ocean-atmosphere model was developed for long-lead climate forecasts (Ji et al. 1994a,b). The atmospheric general circulation model (AGCM) forecast is the ensemble mean of 18 individual integrations, each using a different atmospheric initial condition with the same forecasted SST boundary conditions. The skill of the model's mid-latitude atmospheric response to the ENSO state, specifically over the PNA region, is found to be moderate during the northern winter during warm and cold ENSO episodes (Livezey et al. 1996). Skill during non-ENSO years is relatively poor. This ENSO-dependency of skill is plausible, given signal-to-noise ratio considerations in the extratropics with respect to ENSO-related anomalous tropical Pacific SST boundary forcing (Kumar and Hoerling 1998).

Here, we present GCM forecasts for 3-month mean tropical Pacific precipitation out to 8 months of lead time, where lead time refers to the time between the forecast time and the middle month of the 3-month target period. The final skill is presently approximately comparable to that of the statistical CCA forecasts, also produced at CPC (Barnston et al. 1994b and 1999); however, in specific circumstances the skills of the two forecasting approaches may differ substantially. In the regions that experience strong ENSO impacts, boreal winter skills are locally moderately high during ENSO episodes at lead times of up to 6 months. The coupled model forecasts for tropical Pacific precipitation are shown in Fig. 3.

3. Effects of ENSO

The influence of ENSO episodes on Pacific Basin precipitation can be described using composite analysis, in which responses to warm ENSO episodes are considered separately from responses to cold episodes. The degree of realism of the assumption of linearity in the ENSO-rainfall relationship, which is needed in the overall interpretation of full sample ENSO-rainfall correlations, is evaluated in composite analysis. If rainfall anomaly composites are fairly equal-but-opposite for warm versus cold ENSO conditions, approximate linearity is confirmed. The composite analysis is conducted as follows: Three-month running total precipitation is composited (averaged) over the cases assigned to a particular ENSO status (warm or cold episode), for 3-year periods centered on the boreal winter of the mature phase of the episode, and compared with the climatological average rainfall. The set of winters defined as warm or cold episodes, shown in Table 1, is based on an examination of the Tahiti-minus Darwin sea level pressure-based Southern Oscillation Index (SOI) and the tropical Pacific SST leading up to and including the boreal winter season in question. While the resulting set of winters is in general agreement with the sets as defined earlier by Van Loon and Madden (1981), Rasmusson and Carpenter (1983) and Ropelewski and Jones (1987), a few differences exist.

As an example, Fig. 4 shows ENSO composite rainfall results for Kahului, Hawaii. The dotted line shows the climatological mean rainfall for all years within 1955-96, the solid line the mean for the composited 12 warm ENSO episode years (57/58, 63/64, 65/66, 68/69, 69/70, 72/73, 77/78, 82/83, 86/87, 87/88, 91/92, 94/95), and the dashed line that for the 8 cold episode years (55/56, 64/65, 70/71, 73/74, 75/76, 84/85, 88/89, 95/96). Vertical lines denote calendar year changes. Each plot shows composites for the period beginning about 5 seasons before the boreal winter of a mature episode and ending about 5 seasons after that winter. The notation for relative year identity (0 and

+1) in Fig. 4 is as defined in Ropelewski and Halpert (1987). The mature episode winter (year 0 to +1) is located near the midpoint of the abscissa range. For example, the El Nino winter of 1997-98 would fit into the middle of this range. The differences between the composited rainfall totals for the samples representing the warm or cold phases of ENSO versus the totals of the remaining years (neutral-plus-oppositely phased years) were statistically tested with the Student's t-test. It should be noted that the statistical assumptions underlying the t-test (e.g. Gaussian distributions) might not be sufficiently satisfied. Nonetheless, we use it as a rough guide for indicating significant mean differences in rainfalls as a function of ENSO category. Those seasons whose warm or cold episode years' mean passed a 2-sided test with a significance level of 0.05 or better are indicated with a hollow or solid square, respectively, along the solid or dashed curve in Fig. 4. Near the time of the mature episode boreal winter at Kahului, warm episodes are significantly associated with deficient precipitation in Dec-Jan-Feb, Jan-Feb-Mar, Feb-Mar-Apr and Mar-Apr-May. Cold episodes associate significantly with enhanced rainfall only during Feb-Mar-Apr and Mar-Apr-May. More detailed information (e.g., the seasonal variation of the tropical rainfall with ENSO) can be found in the He et al. 1998a.

In looking at the time of mature episode boreal winter over 66 tropical Pacific stations (not shown), the general relationship between ENSO and rainfall described in Ropelewski and Halpert (1987) is found: At the off-equator stations (such as many of the U.S.-affiliated stations) warm ENSO is associated with suppressed rainfall, while at the near-equatorial stations from the date line eastward to the South American coast, where the SST anomaly is positive, rainfall is enhanced. Enhanced rainfall with El Nino is particularly dramatic at Kiribati stations (Banaba, Butaritari, Tarawa, Beru, Arorae, Fanning and Christmas), at Nauru, and at two of four Tuvalu stations (Nui and Atafu). All of these effects tend to occur in reverse for cold ENSO episodes. Because equatorial Fanning and Christmas islands are surrounded by a somewhat cool ocean most of the year, they receive fairly light climatological rainfall, but with very large positive deviations occurring during El Nino episodes. Many of the more off-equator locations, especially west of 170W, experience drought with El Nino.

A biennial oscillation is suggested in Fig. 4. Recent ENSO studies have documented such a biennial component superposed on the well-accepted irregular 4- to 5-year component of ENSO variability (Meehl 1990; Rasmusson et al. 1990; Barnett 1991; Ropelewski et al. 1992).

The influence of ENSO on Pacific Basin precipitation is described in a general way by spatial maps of the linear correlation between an index of ENSO and the rainfall at 66 stations. ENSO-rainfall correlations can be examined both contemporaneously and at lag (in which the ENSO index occurs before the rainfall, and is thus viewed as a predictor). Fig. 5 shows the spatial distribution of correlation between a standardized Southern Oscillation Index (SOI) and rainfall, where SOI leads rainfall by 6 months. To simplify interpretation, the sign of the SOI is reversed so that positive SOI is associated with El Nino. The correlation maps are shown for four target (predictand) seasons: Jan-Feb-Mar, Apr-May-Jun, Jul-Aug-Sep and Oct-Nov-Dec. Light shading denotes statistically significant negative correlations at the 0.05 significance level, and dark shading shows significant positive correlations. The maps show that the SOI offers some meaningful hints about rainfall anomalies to occur 6 months later, especially for northern winter rainfall. The Hawaiian result at 2 season lag was also obtained by Chu and He (1994), and implies that useful precipitation forecasts can be made for Jan-Feb-Mar as early as early autumn, providing several months for impact mitigation efforts by water managers in affected regions. Significant relationships at 6 months lead also exist in parts of Fiji and southern Tonga, and form a familiar horseshoe-shaped pattern of off-equator El Nino-related dryness surrounding the equatorial wet zone, with northern and southern dry regions nearly meeting in the western equatorial Pacific. Moderately strong correlations for Jan-Feb-Mar at 6 months lag appear close to the equator both east and slightly west of the date line, at most of the Kiribati stations. The general geographical extent of the predictive potential shown here is qualitatively similar to that described by Ropelewski and Halpert (1987, 1996), the CCA studies of Barnston and He (1996) for Hawaii, and He and Barnston (1996) for the tropical Pacific islands in general. When an ENSO phase has developed by boreal mid-summer, that phase tends to persist through the remainder of the calendar year (Barnston and Ropelewski 1992). This causes the lagged correlation relationships with Jan-Feb-Mar rainfall to be somewhat more similar to the Jan-Feb-Mar simultaneous relationships than is the case for other target seasons. By contrast, the "spring barrier" in the continuity of the ENSO state causes the ENSO-rainfall relationships to weaken more quickly for boreal summer and fall target periods when lag time is introduced (not shown). Some ENSO phase-specific rainfall impacts are distinguishable in Fig. 4 that are not visible in the overall correlation results of Fig. 5. Recently, attention has been given to the "spring barrier" in order to improve the forecast skill (e.g. Chen et al. 1995).

In general, ENSO effects during periods other than the boreal late summer through fall, winter and spring of the mature phase, are not very noteworthy. However, in some cases an apparent ENSO effect can be noted in the boreal winters a year before or a year after the mature episode boreal winter. At Kahului, for example, there is a significant tendency toward a wet winter the year following the mature warm episode, which itself tends to be dry. Adjacent winter responses may be associated with the episode that peaks a year beforehand or a year afterward (e.g. positive temperature anomalies in Hawaii are seen to occur the boreal winter one year after a mature El Nino as much as during the El Nino boreal winter itself; Barnston and He 1996). However, they may also be explained in part by adjacent year mature episodes in their own right.

4. Future plans

The utility of seasonal precipitation prediction and climate information to populated tropical Pacific islands is clear, given their water-dependent economies. The strong ENSO events of 1997-98-99 (one warm, one cold) give us an opportunity to update our analysis of the oceanic and atmospheric impacts, to confirm relationships documented using 1955-1996 data. In the near future we intend to: (1) Develop a section on the CPC web site devoted to the 1997-98-99 ENSO cycle impacts on the tropical Pacific climate anomalies; (2) Develop a more reliable system through which rainfall data observations throughout the Pacific basin can be archived in real-time; (3) Forecasts of the NCEP coupled model and Scripps/Max Planck Institute ECHAM model will be skill-evaluated and optimally combined with one another and with the CCA forecasts to produce a consolidated forecast whose skill exceeds that of any tool alone; and (4) Develop innovative, presently nonexistent empirical and dynamical predictive tools to serve the tropical Pacific region. Such tools would, hopefully, include signals in addition to ENSO, such as lower frequency (e.g. interdecadal) variability. Recently, CPC has developed a new outlook, Probability of Exceedance (POE) outlook, which gives the detailed quantity information and render the CPC seasonal forecast product more universally useful (He et al, 1998b, Barnston et al, 2000). We will provide the same kind of outlook to the tropical Pacific region.

References

Barnett, T.P. and R. Preisendorfer, 1987: Origins and levels of monthly and seasonal forecast skill for United States surface air temperatures determined by canonical correlation analysis. Mon. Wea. Rev., 115, 1825-1850.

----, M. Latif, E. Kirk, and E. Roekner, 1991: On ENSO physics. J. Climate, 4, 487-515.

Barnston, and C.F. Ropelewski, 1992: Prediction of ENSO episodes using canonical correlation analysis. J. Climate, 5, 1316-1345.

----, 1994a: Linear Statistical short-term climate predictive skill in the Northern Hemisphere. J. Climate, 6, 1545-1566.

---- and Coauthors, 1994b: Long-lead seasonal forecasts-Where do we stand? *Bull. Amer. Meteor. Soc.*, **75,** 2097-2114.

----, and Y. He, 1996: Skill of CCA forecasts of 3-month mean surface climate in Hawaii and Alaska. J. Climate, 9, 2579-2605.

----, M.H. Glantz, and Y. He, 1999: predictive skill of statistical and dynamical climate models in SST forecasts during the 1997-98 El Niño episode and the 1998 La Niña onset. Bull. Amer. Meteor. Soc., 80, 217-243.

----, Y. He, and D Unger, 2000: A Forecast Product that Maximizes Utility for State-of the-Art Seasonal Climate Prediction. J. Climate, in press.

Chen, D., S.E .Zebiak, A.J. Busalacci, and M.A. Cane, 1995: An improved procedure for El Niño Forecasting: Implications for predictability. *Science,* **269**, 1699-1702.

Chu, P.S. and Y. He, 1994: Long-range prediction of Hawaiian winter rainfall using canonical correlation analysis. Int. J. Climatol., 14, 659-669.

He, Y. and A.G. Barnston, 1996: Long-lead forecasts of seasonal precipitation in the tropical Pacific islands Using CCA. J. Climate, 9, 2020-2035.

He, Y and A.G. Barnston and A.C. Hilton, 1998a: NCEP/Climate Prediction Center Atlas No. 5: A precipitation climatology for stations in the tropical Pacific basin; effects of ENSO. U.S. Dept. of Commerce, NOAA, 280pp.

He, Y. A.G. Barnston and D. Unger, 1998b: An improved Seasonal Forecast Product from CPC. Proceeding of the 23rd Annual Climate Diagnostics and Prediction Workshop. 112-114, Miami, Florida.

Ji, M., A. Kumar and A. Leetmaa, 1994a: A multi-season climate forecast system at the National Meteorological Center. *Bull. Am. Meteor. Soc.*, **75**, 569-577.

----, ----, ----, 1994b: An experimental coupled forecast system at the National Meteorological Center: Some early results. *Tellus*, **46**A, 398-418.

Kumar, A. and M.P. Hoerling, 1998: Annual cycle of Pacific/North American seasonal predictability associated with different phases of ENSO. *J. Climate*, **11**, in press.

Livezey, R.E., M. Masutani and M. Ji, 1996: SST-forced seasonal simulation and prediction skill for versions of the NCEP/MRF model. *Bull. Am. Meteor. Soc.*, **77**, 507-517.

Meehl, G. A., 1990: Seasonal cycle forcing of El Niño-Southern Oscillation in a global, couple ocean-atmosphere GCM. J. Climate, 3, 72-98.

Ropelewski, C.F. and M.S. Halpert, 1987a: Global and regional scale precipitation patterns associated with the El Niño/Southern Oscillation. Mon. Wea. Rev., 115, 1606-1626.

----, and P.D. Jones, 1987b: An extension of the Tahiti-Darwin Southern Oscillation Index. Mon. Wea. Rev, 115, 2161-2165.

----, C.F., M.S. Halpert, and X. Wang, 1992: Observed tropospheric biennial variability in the global tropics. J. Climate, 5, 594-614.
----, and M.S. Halpert, 1996: Quantifying Southern Oscillation-precipitation relationships. J. Climate, 9, 1043-1059.
Rusmusson, E. M., and T. H. carpenter, 1983: the relationship between eastern equatorial Pacific sea surface temperatures and rainfall over India and SriLanka. Mon. Wea. Rev., 111, 517-528.
----, X. Wang, and C.F. Ropelewski, 1990: The biennial component of ENSO variability. J. Mar. Sys., 1, 71-96.
Van Loon, H. and R.A. Madden, 1981: The southern Oscillation. Part I: Global associations with pressure and temperature in Northern winter. Mon. Wea. Rev, 109, 1150-1168

Table 1. Year-pairs whose connecting boreal winters are defined as warm or cold episode ENSO cases for this study, determined by the SOI and the Niño 3.4 SST in accordance with the procedure described in the appendix. The estimated relative strength of each episode is indicated by asterisks, where one asterisk denotes a moderately strong episode, two asterisks a strong episode, and no asterisk an episode of standard strength.

Warm episodes	Cold Episodes
1957/58 *	1955/56 *
1963/64	1964/65
1965/66 *	1970/71 *
1968/69	1973/74 **
1969/70	1975/76 **
1972/73 **	1984/85
1977/78	1988/89 **
1982/83 **	1995/96
1986/87 *	
1987/88 *	
1991/92 **	
1994/95	

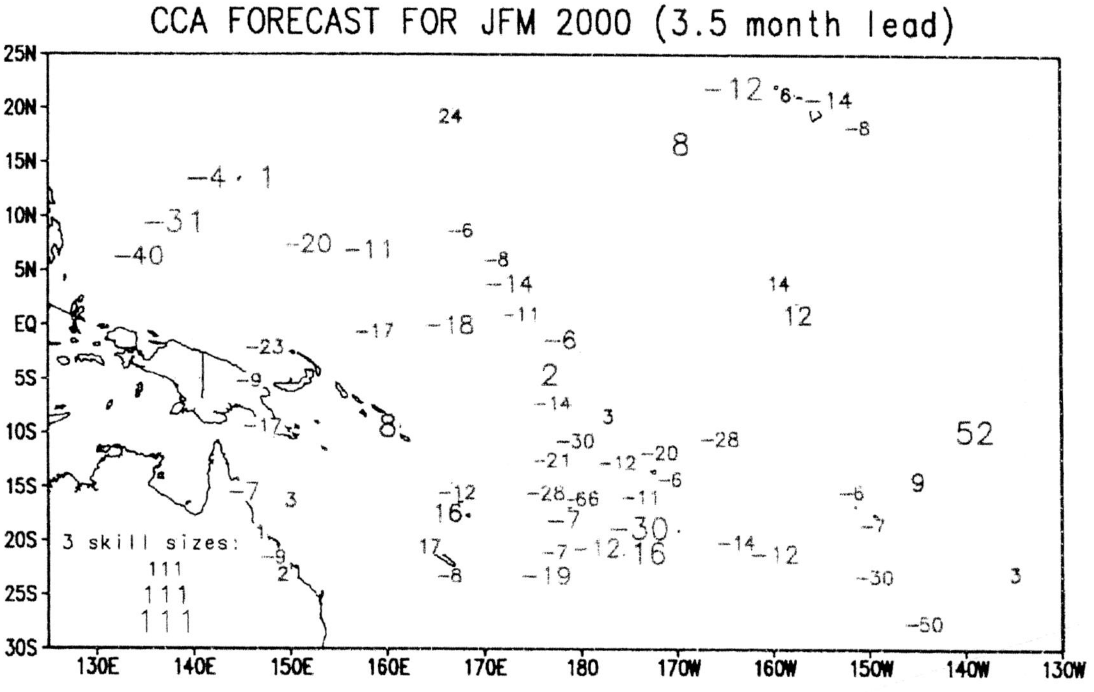

Fig. 1. CCA-derived precipitation standardized anomaly forecast (X100) for 61 tropical Pacific Island stations for Jan-Feb-Mar 2000. The cross-validated skill expected for the forecasts is indicated by the *size* of the numerals (not their *value*, which shows the forecast itself). Small numeral size indicates correlation skill of less than 0.30, considered unusable; medium size is used for 0.30<skill<0.45 which is modest but usable; large size indicates skill>0.45, considered a relatively more reliable forecast.

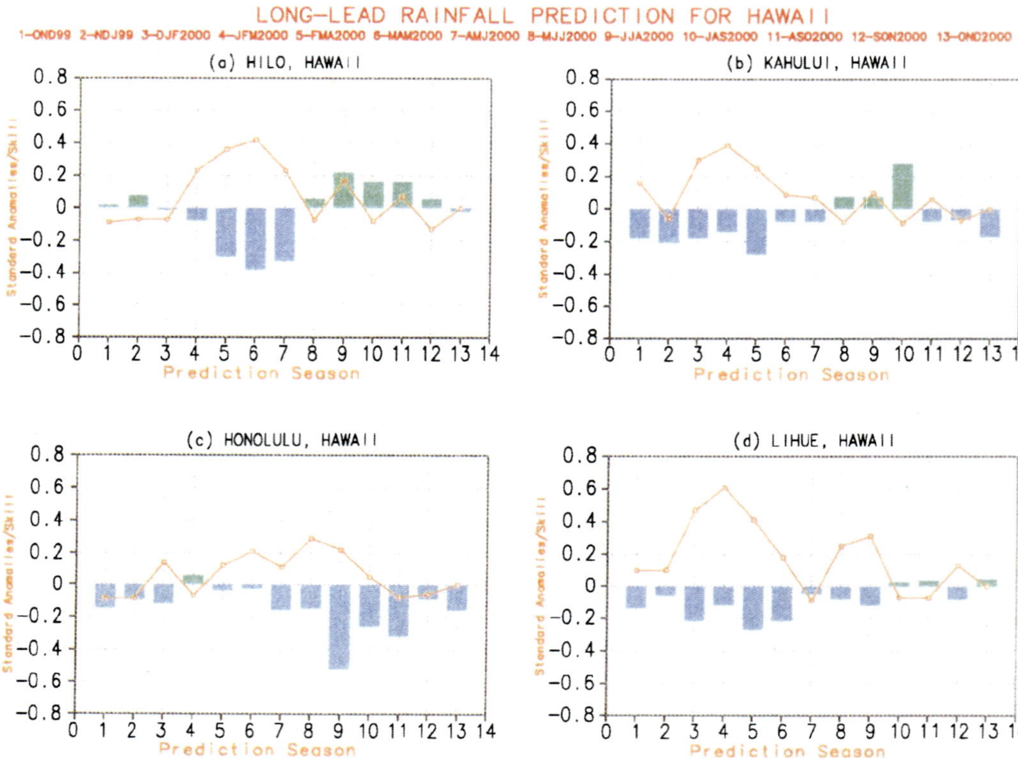

Fig. 2 Time series of CCA-based long-lead precipitation anomaly forecasts, and their expected skills, out to one year into the future for Hawaii. The bars indicate the forecast values (as standardized anomalies) and the lines indicate the associated skills (as correlation coefficients). Both forecasts and skills use the same ordinate scale. The Target season is indicated on the abscissa, ranging from 1 (Oct-Nov-Dec 1999) through 13 (Oct-Nov-Dec 20000); see the legend at top.

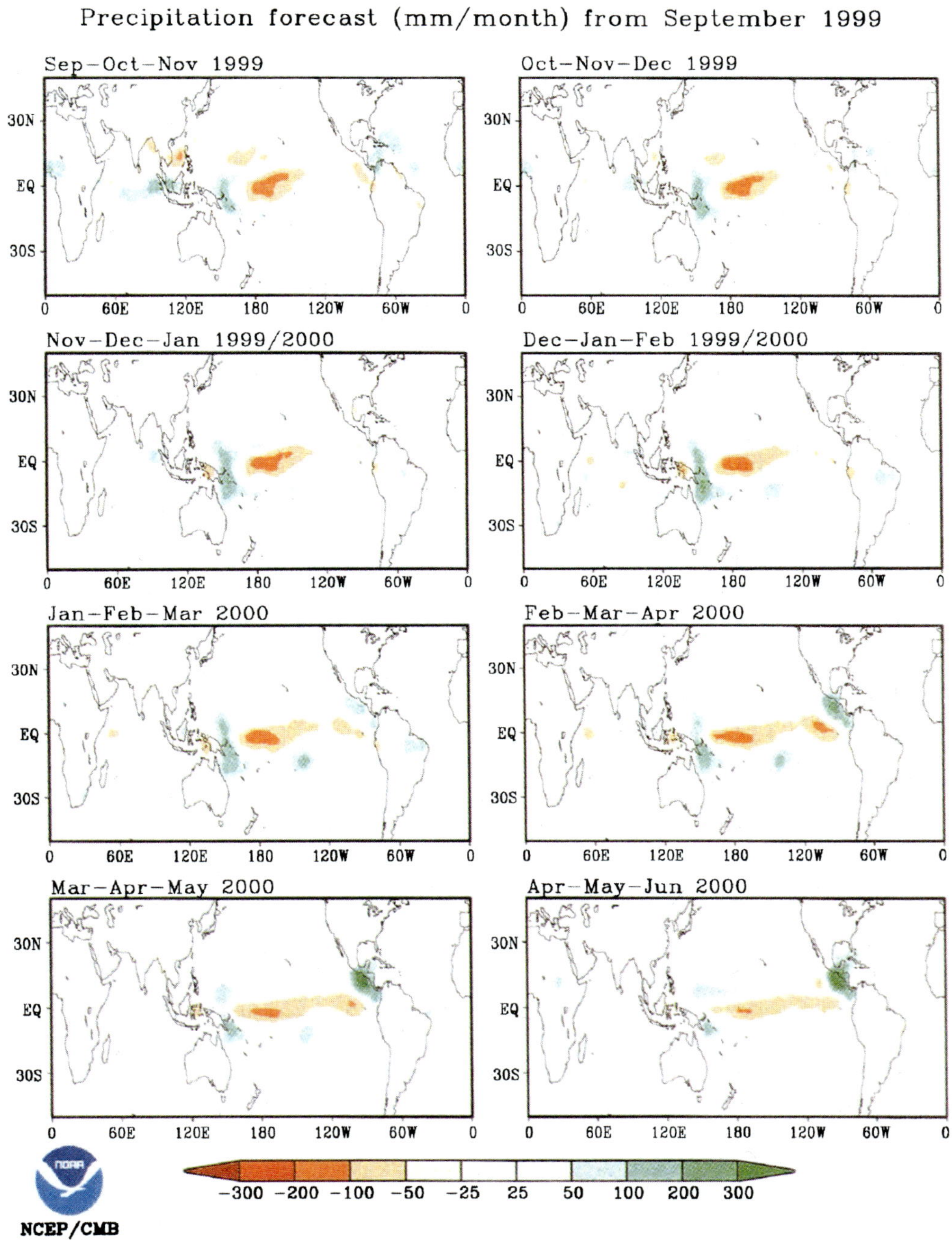

Fig.3 Tropical precipitation anomaly forecast of the two-tiered NCEP model for the coming 8 overlapping 3-month periods. The mean of an 18-member ensemble of AGCM runs is shown.

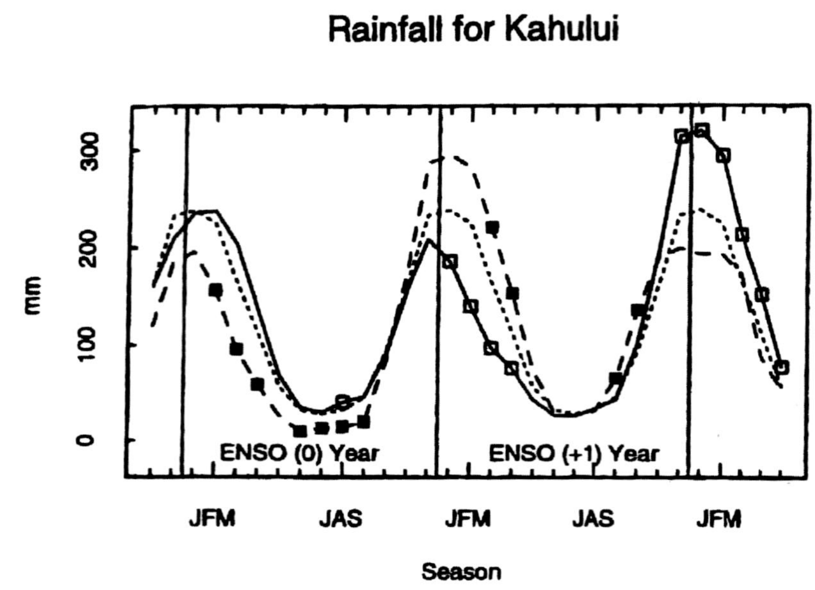

Fig. 4 Composite rainfall amounts for Kahului, Hawaii, by ENSO status. The dotted line denotes the climatological mean rainfall for all years in the 1955-96 period, the solid line the mean for the12 composited warm ENSO episode years, and the dashed line the mean for the 8 cold episode years. Notation for relative year identity (0 and +1) is as defined as in Ropelewski and Halpert (1987). Differences between the composited rainfall for a stratified sample versus that of the remaining years passing a 2-tailed significance test at the 0.05 level are indicated by a hollow along the line.

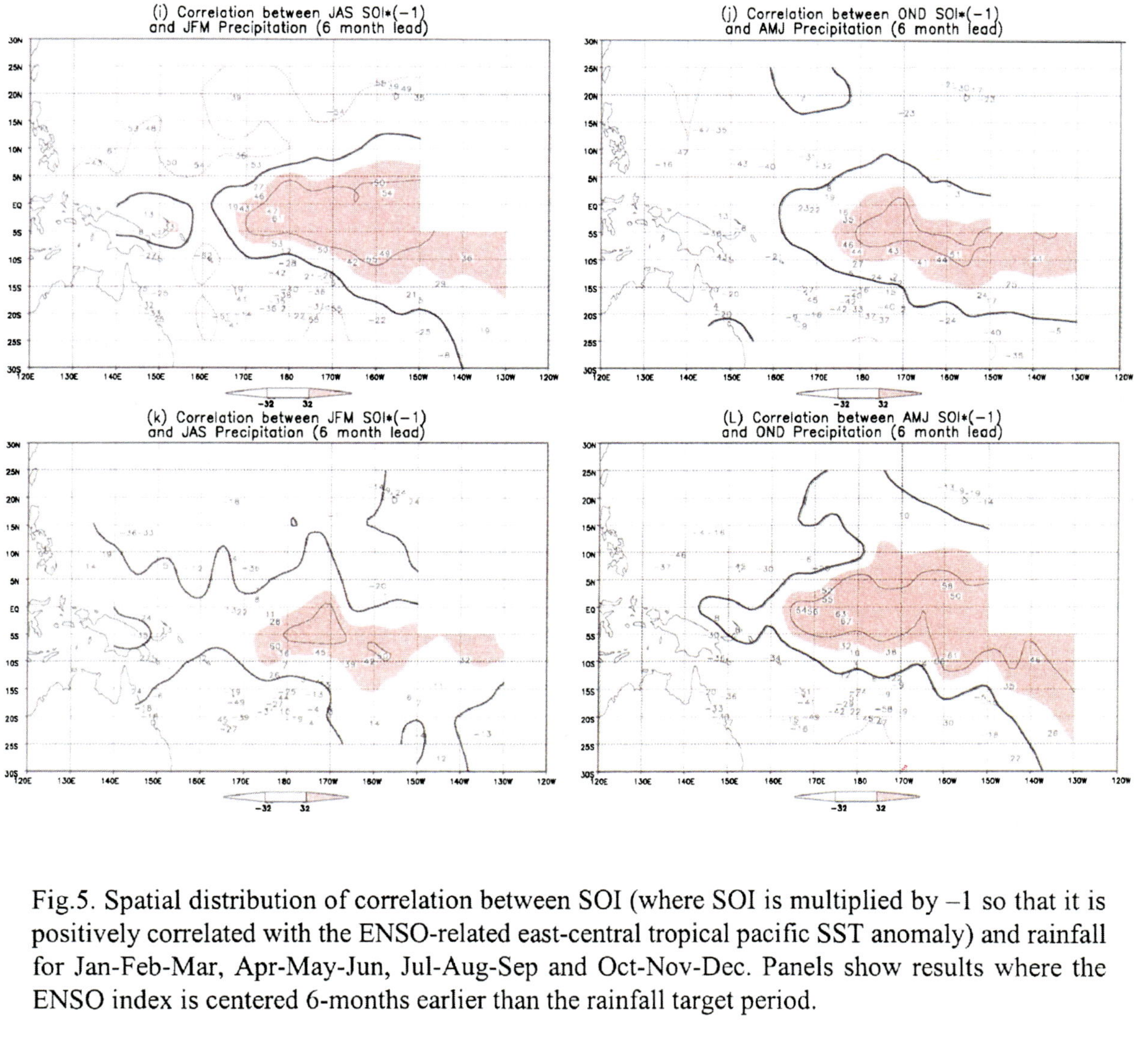

Fig.5. Spatial distribution of correlation between SOI (where SOI is multiplied by –1 so that it is positively correlated with the ENSO-related east-central tropical pacific SST anomaly) and rainfall for Jan-Feb-Mar, Apr-May-Jun, Jul-Aug-Sep and Oct-Nov-Dec. Panels show results where the ENSO index is centered 6-months earlier than the rainfall target period.

~ Theoretical Studies ~

TROPICAL TROPOSPHERIC BIENNIAL OSCILLATION AND ENSO

C.-P. Chang and Tim Li[1]
Department of Meteorology, Naval Postgraduate School, Monterey, CA 93943

ABSTRACT

The tropospheric biennial oscillation (TBO) and El Nino-Southern Oscillation (ENSO) are interannual variations that affect the Asian-Australian monsoon region and the tropical Pacific. Some investigators view the two oscillations are actually one system, with TBO a "weaker sister" of ENSO. Others view the two as separate modes, with TBO primarily driven by the monsoon and ENSO primarily driven by atmosphere-coupling of the tropical eastern Pacific. In this paper the Chang-Li theory, which belongs to the latter category, is used as the basis to study the interactions of the two oscillations. It is shown that the nonlinear interaction of the two modes may produce a variety of solutions that may explain the complexity of the observed monsoon-ENSO relationship.

1. INTRODUCTION

Based on previous strong El Nino events, prediction of the effects of the 1997-98 El Nino on the precipitation anomalies in North America and the equatorial maritime continent were quite successful. However, the prediction of large deficits in rainfall over Australia and India did not materialize. This is indicative of important processes of different time scales that interfere with the effects of ENSO in these regions. We propose that the interactions between TBO and ENSO modes may determine the main interannual variations in the Asian-Pacific sector.

It is now well known that the El Nino and La Nina events result from active interactions between the ocean and atmosphere in the tropical Pacific. The anomalous sea-surface temperature (SST) conditions in the Pacific has be predicted with some skill up to several seasons in advance, using either a relatively simple model such as Cane and Zebiak (1985) or complex coupled ocean-atmosphere general circulation models. There are growing observational evidences that anomalous SST conditions in the equatorial Pacific can influence global weather and climate patterns. However, the predictions of these anomalous patterns using either a statistical or a dynamic method are still difficult. In particular, although observational records indicate that rainfall anomalies over Australia and Asia are somewhat correlated to the SST anomalies (SSTA) in the eastern Pacific, from time to time such a correlation is rather vague.

In addition to the external influence from ENSO which generally has a time scale of 3-7 years, the Asian-Australia monsoon system also experiences its own interannual variability, with a dominant time scale of 2-3 years (Meehl 1987, 1993, 1994, Yasunari 1990). The latter is referred to as the tropospheric biennial oscillation (TBO). Meehl (1997) showed that it is the result of coupling between atmosphere, ocean, and land. Chang and Li (2000) proposed that it involves active interactions between the northern summer and winter monsoon and between the anomalous monsoon heating and the planetary-scale east-west circulation. Whereas ENSO involves ocean-atmosphere interactions mainly in the Pacific, TBO covers an

---

[1] Present Affiliation: IPRC, School of Ocean and Earth Science and Technology, University of Hawaii, Honolulu, HI 96822

even wider spatial scale, including the Asian-Australian monsoon sector and the entire tropical Indian and Pacific oceans.

Observations have established that the signals of the TBO are found not only in the Indian-Australia rainfall records, but also in the tropospheric circulation, SST, and upper-ocean thermal fields (e. g, Yasunari 1991, Ropelewski et al. 1992, Lau and Yang 1996). One of the most remarkable features of TBO is its characteristic seasonal progression and dynamically coherent spatial structure. The conditions of a strong (or weak) monsoon over India and Southeast Asia in the northern summer often continues in the succeeding autumn and winter over the maritime continent and Australia monsoon region so that a strong (weak) Asian monsoon is often followed by a strong (weak) Australian monsoon. Using a simple five-box model that considers SST-monsoon, evaporation-wind, monsoon-Walker circulation, and wind-thermocline feedback, Chang and Li (2000) illustrated that these processes can give rise to the seasonal evolution of TBO. Their theory explains why TBO can maintain the same phase from northern summer to northern winter and why a reversed phase of TBO can last three locally inactive seasons to affect the next year's monsoon.

Without the interference from and interaction with motions of other scale, Chang and Li (2000) obtained a pure, regular biennial oscillation. But in reality, TBO exhibits a rich, irregular spectrum. Such irregularity may, in large part, result from nonlinear interaction with ENSO. Since TBO is an inherent mode of the coupled ocean-atmosphere system in which monsoon is a major part, the prediction of anomalous rainfall over the Asian-Australian monsoon region based on ENSO forecast must consider the combined effects of TBO and ENSO.

The purpose of this study is to investigate the role of the nonlinear interaction between TBO and ENSO in the monsoon rainfall anomalies and the effect of ENSO on the irregularity of TBO. We will begin by summarizing Chang and Li's (2000) model for TBO, and then incorporate a simple ENSO mode based on Li (1997) in the model to study the interaction of the two modes in a unified dynamic framework.

2. THE CHANG-LI TBO MODEL

In the TBO model proposed by Chang and Li (2000) consists of five boxes representing, respectively, the South Asian and Australian monsoon regions and the equatorial Indian, and western and eastern Pacific oceans. The five regions interact among themselves through the SST-monsoon, evaporation-wind, monsoon-Walker circulation, and wind stress-ocean thermocline feedback.

The change of SSTA in the equatorial Indian Ocean (T_I), and the western (T_W) and eastern Pacific (T_E), is primarily determined by horizontal and vertical temperature advection and surface latent heat fluxes. The vertical temperature advection includes effects of both anomalous upwelling and thermocline-depth related subsurface temperature changes. Anomalous heating over the monsoon region excites both planetary-scale east-west circulation and local cyclonic flows near the surface, both of which further change the SST through the dynamic and thermodynamic processes. These SST equations may be found in Chang and Li (2000).

In the following equations we shall define δ_I and δ_A as switch-on coefficients for the Indian and Australian monsoons, respectively. They assume value of unity during the respective monsoon season and zero otherwise. On the interannual time scale, surface zonal winds over the Indian Ocean consist of two major parts, as illustrated in the following equation:

$$U_I = c_1'Q_I + c_2T_W = \delta_I c_1 T_I + c_2 T_W . \qquad (1)$$

Here the first term on the right hand side gives the direct response to anomalous monsoon heating and the second part reflects the change of intensity of a Walker cell over the Indian Ocean due to the SST change in the maritime continent. Eq. (1) is schematically illustrated in Fig. 1a.

The surface wind in the western Pacific, on the other hand, is determined by the anomalous heating over Australia and by the Indian monsoon which, by altering the planetary-scale east-west circulation, changes the surface wind in both the western and central equatorial Pacific (Barnett et al., 1989; Meehl, 1997):

$$U_W = c_3'Q_A - c_4'Q_I = \delta_A c_3 T_W - \delta_I c_4 T_I \qquad (2)$$

This equation is shown schematically in Fig. 1b.

In addition to the influence by anomalous South Asian and Australia monsoon heating, the central equatorial Pacific wind is also affected by zonal SST gradients along the equator (Lindzen and Nigam 1987) so that the total anomalous zonal wind in the central Pacific is written as:

$$U_C = -c_5'Q_A - c_6'Q_I + \frac{A}{\varepsilon L_{EW}}(T_E - T_W)$$

$$= -\delta_A c_5 T_W - \delta_I c_6 T_I + \frac{A}{\varepsilon L_{EW}}(T_E - T_W), \qquad (3)$$

where A is a SST-gradient momentum forcing coefficient (Wang and Li 1993), and ε is an atmospheric Rayleigh friction coefficient. The interactive coefficients c1, c2 .... c6 are determined from a scale analysis [see Chang and Li (2000) for details]. Eq. (3) is represented schematically in Fig. 1c.

The surface wind in turn drives the ocean surface current and induces ocean vertical overturning. A simplified Cane-Zebiak (Zebiak and Cane 1987) model is used to calculate the ocean surface currents and Ekman pumping velocity along the equatorial Indian and Pacific Oceans:

$$u_I = \frac{\alpha U_I}{\rho h r}, \qquad (4)$$

$$w_I = -\frac{(H-h)\beta\alpha U_I}{\rho H r^2}, \qquad (5)$$

$$u_C = \frac{\alpha U_C}{\rho h r}, \qquad (6)$$

$$w_W = -\frac{(H-h)\beta\alpha U_W}{\rho H r^2} + \frac{2h(H-h)u_C}{L_{EW}}, \tag{7}$$

$$w_E = -\frac{(H-h)\beta\alpha U_C\sigma}{\rho H r^2} - \frac{2h(H-h)u_C}{L_{EW}} \tag{8}$$

Here β denotes the planetary vorticity gradient; H the mean depth of the ocean thermocline; and r a friction coefficient in the oceanic Ekman layer. The vertical velocities in (7) and (8) consist of both meridional and zonal convergence components. The meridional component of the Ekman convergence in the eastern Pacific is estimated with a ratio of σ from its counterpart in the central Pacific. The interactive coefficients c_1, $c_2 ... c_6$ are determined from a scale analysis of the relevant dynamic and thermodynamic balances.

After integrating the model in time, a biennial oscillation emerges in a reasonable parameter regime, with model SST and wind variations resembling many aspects of the observed TBO (Fig. 2). A strong (weak) South Asian monsoon is always followed by a strong (weak) Australian monsoon.

The physical mechanism that causes the TBO in the model can be summarized as follows. Since the starting point of an ideal TBO cycle is arbitrary, we will start from a warm SSTA condition in the Indian Ocean prior to the South Asian summer monsoon season. To aid our discussion, a schematic diagram illustrating how the interactions between the monsoon, ocean, and atmosphere promote a biennial oscillation in the model is given in Fig. 3. In this figure the main cause-effect for the strong monsoon phase are given in the left hand side, with the sequence of development driven by the northern summer events indicated by red solid thin and double arrows, while those driven by the northern winter events by black dashed double arrows. The chain of events in the weak monsoon phase is only symbolically sketched on the right hand side since the details are exactly mirror images of the left hand side. The two phases are separated by a ribbon. Fig. 4 graphically summarizes all the important interactive processes shown in Fig. 1a-c, and is presented as a companion diagram to Fig. 3. A warm SSTA condition in the equatorial Indian ocean increases the local lower tropospheric moisture through surface evaporation. As the South Asian summer monsoon develops, southerly winds to the south of the Indian subcontinent bring the excess water vapor into South Asia, which intensifies the convective rainfall and leads to a strong monsoon. The convective heating associated with the stronger monsoon on one hand induces a westerly wind anomaly over the Indian Ocean due to the enhanced low-level vorticity (indicated by a thin arrow), and on the other hand intensifies a planetary-scale east-west circulation leading to anomalous easterlies over the western and central Pacific (indicated by the solid double arrow). The westerly anomaly over the Indian Ocean decreases the local SST, primarily through the evaporation-wind feedback. This cooling will help the development of a weak phase of the South Asian summer monsoon if the cold SST can persist for nearly one year.

The anomalous easterlies associated with the intensified east-west circulation affect the western Pacific SST in two ways. Over the western Pacific the seasonal mean is westerly, therefore the mean wind is reduced resulting in a reduced evaporation, leading to a warming effect. Meanwhile, the central Pacific easterly anomalies deepen the ocean thermocline in the western Pacific, so that the SST also increases due to anomalous vertical temperature advection. (The easterly anomaly also tends to cool the local SST through zonal advection and Ekman-induced upwelling, but both effects are small – the latter due to the weak mean vertical stratification in the western Pacific.) As a result, a warm SST persists in the western Pacific through the northern fall, which eventually leads to a stronger Australia monsoon in the northern winter and the beginning of the winter-driven event sequence indicated by the dashed double arrows.

Whenever there is warming in the western Pacific, it strengthens the western branch of the Walker cell and thus a surface westerly anomaly over the Indian Ocean. Therefore, the westerly anomalies over the Indian Ocean are reinforced during the northern fall (the solid double arrow). This helps the cold SST to persist throughout the succeeding seasons. Meanwhile, the easterly anomaly in the central Pacific, through anomalous upwelling and shoaling of the thermocline, also cools the SST in the eastern Pacific. The east-west SST gradient further intensifies the anomalous easterly over the central Pacific, reinforcing the tilting of the thermocline and the warming of the western Pacific SST.

The warm SSTA in the western Pacific are weakened in late northern fall by anomalous cold advection from the eastern Pacific. However, they are reinvigorated in the northern winter when the Australian monsoon becomes stronger. This is accomplished through two processes: the western Pacific westerly anomaly that weakens the seasonal mean easterlies resulting in a reduction of evaporation, and the central Pacific easterly anomaly (associated with the strengthened east-west circulation) through the change in the tilting of the thermocline. Thus, the winter driven events (indicated by the dashed double arrows) become a third mechanism to keep the Indian Ocean's surface wind anomaly westerly and SST cold, leading to a weaker Asian monsoon in the following summer. (The warm western Pacific SST further intensifies the eastern Walker cell and helps to keep the eastern Pacific cold.) This interactive system explains why the Indian Ocean SSTA, can be kept from rapid weakening for several seasons.

The reversed cycle associated with a weak Asian monsoon is sketched to the upper-right of the ribbon in Fig 3. The thin arrows that form a closed loop surrounding this ribbon represent the local wind-evaporation negative discussed by Meehl (1987, 1997) for the TBO of the South Asian monsoon. In order to explain the coherent temporal and spatial structure, the interactions between the monsoons, east-west circulation, and ocean thermocline variations, as represented by the double arrows (both solid and dashed) in the strong monsoon phase in Fig. 3, are required. These processes are responsible for an Australian monsoon whose TBO phase follows the South Asian monsoon, and for a sustainable SST anomaly in the Indian Ocean to cause a phase reversal of the TBO cycle one year later.

The proposed TBO theory is based on processes within the tropics, thus it excludes the possible feedback through tropical-midlatitude interactions. However, such interactions will not change the phase of the simulated oscillation, because their effects would give rise to the same sign of response in the biennial cycle as that of the equatorial SST produced by the present model.

3. INCORPORATING THE STATIONARY SST ENSO MODE

When ENSO is included in the model, the wind anomalies over the Indian Ocean and the western and central Pacific consist of two parts. One is related to the internal TBO mode that is associated with the east-west circulation due to anomalous monsoon heating and SST gradients as described in the preceding section. The other part describes the effect of ENSO on TBO. Thus, we have,

$$U_I = \delta_I c_1(T_I - \delta_{ET} T_E) + c_2 T_W, \qquad (9)$$

$$U_W = \delta_A c_3(T_W - \delta_{ET} T_E) - \delta_I c_4(T_I - \delta_{ET} T_E), \qquad (10)$$

$$U_C = -\delta_A c_5(T_W - \delta_{ET} T_E) - \delta_I c_6(T_I - \delta_{ET} T_E) + c_7[\delta_{ET} T_E - (1+\psi)T_W], \qquad (11)$$

where δ_{ET} is a switch-on coefficient representing the strength of ENSO influence on TBO. The interactive coefficient c7 is defined in (3).

In the TBO model, we intentionally filtered out the ENSO mode by considering only the zonally asymmetric thermocline depth variation that is in Sverdrup balance with the wind or SST anomaly. However, on the ENSO time scale the zonal mean thermocline depth variation is clearly not in equilibrium with the wind or SST anomaly. Li (1997) pointed out that while the asymmetric variation influences the coupled instability through a positive feedback between the ocean and atmosphere, the zonally symmetric variation is associated with a negative feedback and is essential for the phase transition of ENSO. Following Li's stationary SST mode ENSO model, with <h> denoting the zonal mean thermocline depth anomaly, the eastern Pacific SSTA equation may be written,

$$\frac{dT_E}{dt} = \Psi U_C^{(ENSO)} - (\lambda V_0 \kappa + \frac{\overline{w}_E}{h})T_E + \frac{\overline{w}_E}{h}\gamma < h >, \qquad (12)$$

$$\frac{d<h>}{dt} = -\Lambda T_E - \varepsilon_h < h >, \qquad (13)$$

where $\Psi = \frac{\overline{T}_E^{(z)}(H-h)\beta\alpha\sigma}{\rho H r^2} + \frac{2h(H-h)\alpha \overline{T}_E^{(z)}}{\rho h r L_{EW}} + \frac{\overline{w}_E \gamma L_{EW} \alpha}{2h\rho g' H}$, $\Lambda = 1.3 \times 10^{-7} ms^{-1}K^{-1}$ and $\varepsilon_h = (2year)^{-1}$ are constants.

The TBO mode influences the ENSO mode through the central Pacific zonal wind that is a function of both eastern and western equatorial Pacific SSTs:

$$U_C^{(ENSO)} = c_7(T_E - \delta_{TE} T_W), \qquad (14)$$

where δ_{TE} is a switch-on coefficient describing the influence of TBO on ENSO.

Equations (9) to (14), together with the SSTA equation for T_I and T_W, form a simple dynamic framework for ENSO-TBO interactions. The two modes interact in such a way that a strong monsoon due to TBO would produce anomalous easterly in the central Pacific that further leads to an anomalous cooling tendency for the eastern Pacific SST, as represented by the δ_{TE} term. The anomalous warming in the eastern Pacific associated with ENSO, on the other hand, has both direct and indirect effects on the monsoon rainfall anomalies. Due to the associated Walker cells, the SST anomaly in the Indian Ocean in general tends to be in-phase with the eastern Pacific but out-of-phase with the western Pacific. This indirect effect causes the eastern Pacific El Nino condition to lead to a warm SST anomaly over the Indian Ocean and thus a stronger Indian monsoon. Meanwhile, a large-scale descent over the Asian-Australian monsoon region due to the anomalous vertical overturning associated with El Nino directly produces a deficient monsoon rainfall anomaly. The combination of these two effects would determine the net effect of ENSO on TBO-monsoon rainfall anomalies.

4. NONLINEAR TBO-ENSO INTERACTION

When $\delta_{ET} = \delta_{TE} = 0$, this system has two independent solutions, with one being the TBO mode and the other the ENSO mode. The former involves the dynamic feedback between the Asian-Australia monsoon heating and the planetary-scale east-west circulation, and the latter involves interactions between the zonal mean (and zonally asymmetric) thermocline depth variation and SST anomaly. Fig. 5 shows the model solutions for the two modes. The TBO mode (top panel) is characterized by regular biennial oscillations in India and Australia rainfall anomalies. A strong Australian monsoon follows a strong Indian monsoon, as observed. The ENSO mode (bottom panel) is characterized by SST variations in the eastern equatorial Pacific, which has an oscillation period of 3 to 4 years and an amplitude of 4°C.

The top panel of Fig. 6 illustrates the net effect of the model ENSO on TBO ($\delta_{ET} = 1, \delta_{TE} = 0$). Note that the regular biennial oscillations of anomalous rainfall over both the Asian and Australia monsoon regions have been perturbed. A weak rainfall anomaly is not always followed by a strong one. The amplitude of the rainfall anomaly varies with time. A strong warm event in the eastern Pacific associated with El Nino does not always coincide with a drought condition in Australia (e.g., compare rainfall anomaly in Australia at year 9 in Fig. 6 with that in Fig. 5). When $\delta_{ET} = 0, \delta_{TE} = 1$, ENSO is also interfered by TBO as shown in the bottom panel of Fig. 6. The eastern Pacific SST oscillation becomes more irregular, owing to the change of the central Pacific wind associated with TBO.

The results presented above illustrate one-way interactions, either TBO affecting ENSO or ENSO affecting TBO. The nonlinear two-way interactions between the two modes exhibit even richer characteristics. Fig. 7 is a scatter diagram in the phase space of eastern Pacific SST anomaly versus the Australian rainfall anomaly in northern winter, obtained from a 1000-year integration of the model with $\delta_{ET} = \delta_{TE} = 0.1$. In most cases there is no sequential relationship between adjacent solutions in this diagram. There are at least three types of limit cycles in the phase space. The first is a narrow cross-centre type, diagonally oriented from top-left to bottom-right, that reflects the conventional ENSO theory: A positive (negative) SST anomaly in the eastern Pacific leads to a deficient (excess) rainfall in northern Australia. This narrow cycle is accompanied by a round cycle around the centre (0,0). This limit cycle is related to the regular TBO mode, since it does not depend on SST anomaly in the eastern Pacific. A maximum (minimum) Australian anomaly occurs near $T_E = 0$. The third type is like a bow with its two ends bounded by the previous two limit cycles. It may result from the nonlinear interaction between the TBO and ENSO modes. Overall, this phase solution indicates that the anomalous rainfall in Australia depends on the combination and competition of both the ENSO and TBO modes, and that a strong El Nino may cause either a deficient rainfall (greater than a standard deviation) or small positive or negative rainfall, but it is unlikely to reach an opposite extreme. The 1997-98 El Nino event may be described by the circle area in the phase space.

The results presented here demonstrate that it may be possible to predict certain important aspects of anomalous rainfall over the Asian-Australian monsoon regions, but only when both ENSO and TBO modes are included in a dynamic coupled ocean-atmosphere model. Many previous models, either statistical or dynamic ones, failed because they only considered the ENSO effect.

ACKNOWLEDGMENTS. This work is supported by National Science Foundation, Division of Atmospheric Sciences, under Grant ATM9613746.

REFERENCES

Barnett, T.P., L. Dumenil, U. Schlese, E. Roeckner, and M. Latif, 1989: The effect of Eurasian snow cover on regional and global climate variations. *J. Atmos. Sci.*, **46**, 661-685.

Chang, C.-P., and T. Li, 2000: A Theory for the tropical tropospheric biennial oscillation. *J. Atmos. Sci.*, **57**, 2209-2224.

Lau, K. M., and S. Yang, 1996: The Asian monsoon and predictability of the tropical ocean-atmosphere system. *Q. J. Roy. Meteorol. Soc.*, **122**, 945-957.

Li, T., 1997: Phase transition of El Nino-southern Oscillation: A stationary SST mode. *J. Atmos. Sci.*, **54,** 2872-2887.

Lindzen, R. S., and S. Nigam, 1987: On the role of sea surface temperature gradients in forcing low level winds and convergence in the tropics. *J. Atmos. Sci.*, **44**, 2240-2458.

Meehl, G.A., 1987: The annual cycle and interannual variability in the tropical Pacific and Indian Ocean region. *Mon. Wea. Rev.*, **115**, 27-50.

______, 1994: Influence of the land surface in the Asian summer monsoon: external conditions versus internal feedback. *J. Climate*, **7**, 1033-1049.

______, 1997: The South Asian monsoon and the tropospheric biennial oscillation (TBO). *J. Climate*, **10**. 1921-1943.

Mooley, D. A., and B. Parthasarathy, 1984: Fluctuations in all-India summer monsoon rainfall during 1871-1978. *Climatic Change*, **6**, 287-301.

Ropelewskii, C. F., M. S. Halpert, and X. Wang, 1992: Observed tropospheric biennial variability and its relationship to the Southern Oscillation. *J. Climate*, **5**, 594-614.

Wang, B., and T. Li, 1993: A simple tropical atmospheric model of relevance to short-term climate variations. *J. Atmos. Sci.*, **50**, 260-284.

Yasunari, T., 1990: Impact of Indian monsoon on coupled atmosphere ocean system in tropical Pacific. *Meteor. Atmos. Phys.*, **44**, 29-41.

Zebiak, S.E., and M.A. Cane, 1987: A model ENSO. *Mon. Wea. Rev.*, **115**, 2262-2278.

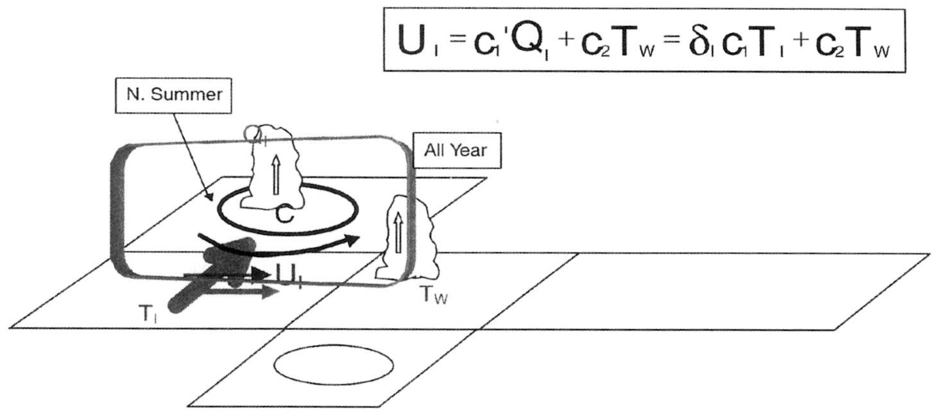

Fig. 1a. Schematic diagram showing the relationship between U_I and T_I and T_W in the TBO model. The Indian Ocean SST determines the anomalous moisture convergence (green thick arrow) into South Asia, which in turn determines the convective heating (Q_I) and cyclonic vorticity (*C*) of the northern summer monsoon, therefore relating T_I and U_I during northern summer only. The western equatorial Pacific SST determines the strength of the western Walker cell. This cell operates in all four seasons and causes T_I to influence U_I continuously throughout the year. (from Chang and Li 2000).

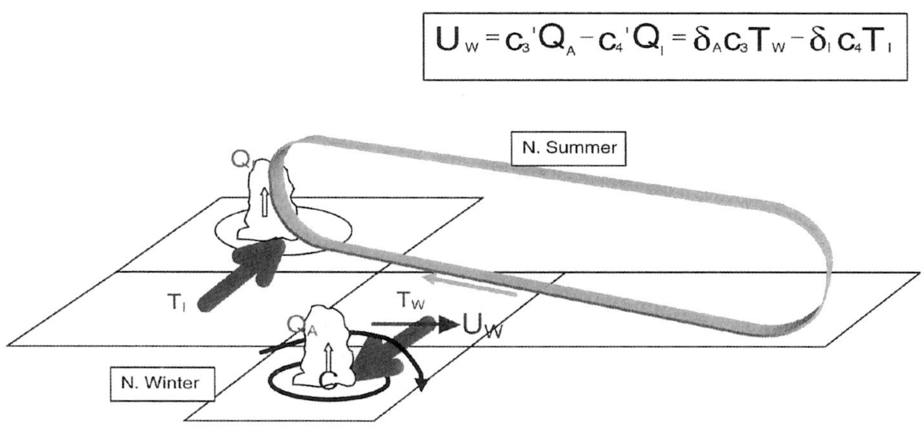

Fig. 1b. Schematic diagram showing the relationship between U_W and T_I and T_W in the TBO model. The western equatorial Pacific SST determines anomalous moisture convergence (green thick arrow) into Australia, which in turn determines the convective heating (Q_A) and cyclonic vorticity (*C*) of the southern summer monsoon, therefore relating T_W and U_W during northern winter only. The strength of the east-west circulation depends on the convective heating of the South Asian monsoon, thus it links the Indian Ocean SST with U_W during northern summer. (From Chang and Li 2000).

$$U_C = -c_5'Q_A - c_6'Q_I + \frac{A}{\varepsilon L_{EW}}(T_E - T_W)$$
$$= -\delta_A c_5 T_W - \delta_I c_6 T_I + \frac{A}{\varepsilon L_{EW}}(T_E - T_W)$$

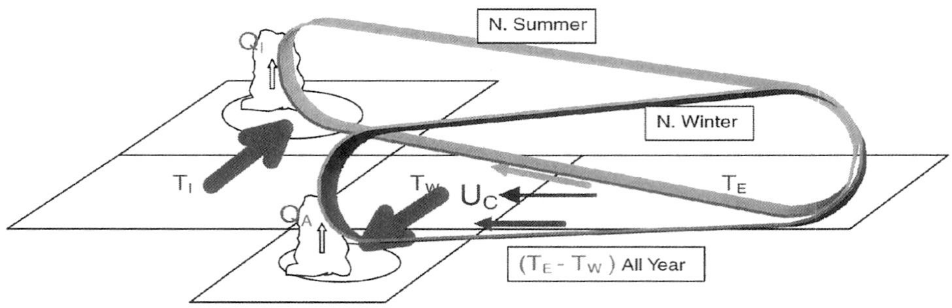

Fig. 1c. Schematic diagram showing the relationship between U_C and T_I , T_W and T_E in the TBO model. Through the strength of the east-west circulations, U_C is affected by the Indian Ocean SST during northern summer and the western equatorial Pacific SST during northern winter. In addition, the SST gradient between the eastern and western equatorial Pacific influences U_C throughout the year. (From Chang and Li 2000).

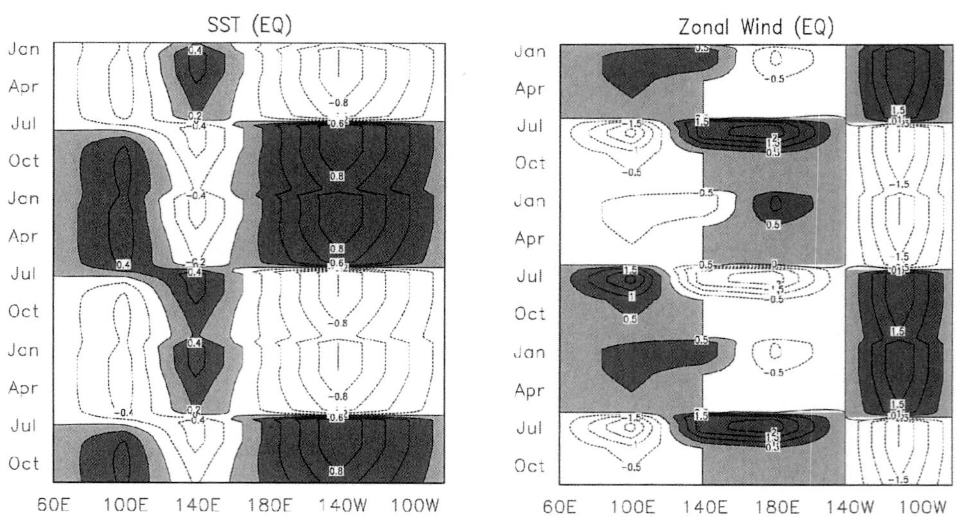

Fig. 2. Time evolution of the model SST (K, Intervals 0.2) and zonal wind (m s^{-1}, Intervals 0.5) anomalies, positive areas shaded. The eastern Pacific zonal wind is not a model variable. The wind anomalies east of 140°W are computed from zonal SST gradient anomalies due to T_E. (From Chang and Li 2000).

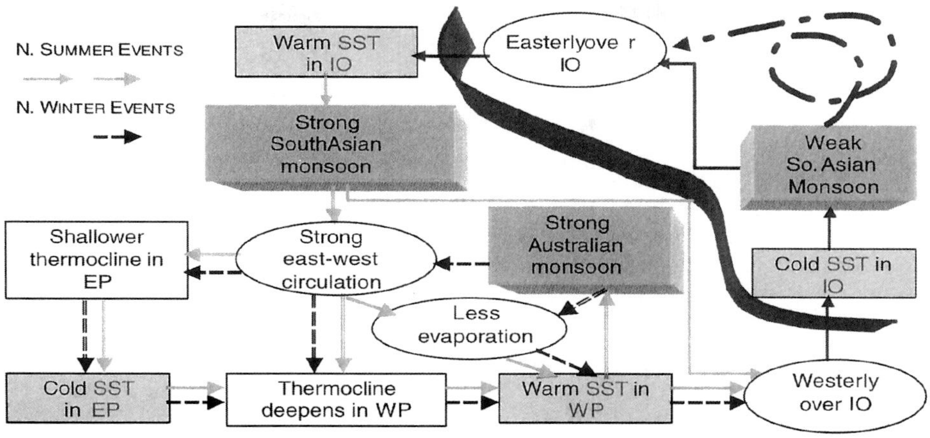

Fig. 3. Schematic diagram indicating the interactive processes leading to the TBO. Land and oceans regions are shaded. Non-shaded boxes indicate atmospheric (oval) and oceanic (rectangular) processes. The strong monsoon phase starts with warm Indian Ocean SST leading to a strong South Asian summer monsoon. In addition to a surface westerly anomaly during summer (thin arrow) that cools the Indian Ocean SST, complex interactive processes involving the tropical western Pacific and the Australian monsoon are required to reinvigorate this westerly anomaly until next summer. The processes also lead to a strong winter monsoon that follows the strong summer monsoon. The reversed phase is sketched in the upper right side. (From Chang and Li 2000).

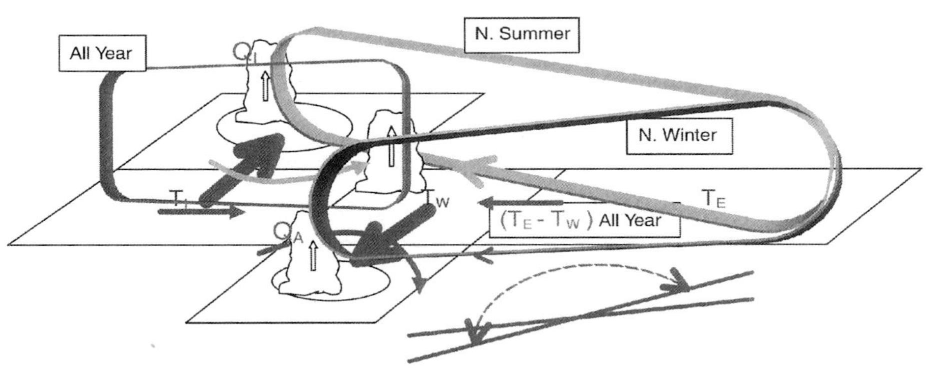

Fig. 4. Schematic diagram summarizing the important interactions. The primary SST change mechanism is through wind-evaporation for the Indian Ocean, thermocline tilting for the eastern equatorial Pacific, and both for the western equatorial Pacific. Notice the key roles played by the western equatorial Pacific both in allowing the northern summer monsoon to influence the northern winter monsoon, and in the maintenance of the Indian Ocean surface wind anomalies until the next monsoon by linking the feedback of the Australian monsoon to the Indian Ocean. (From Chang and Li 2000).

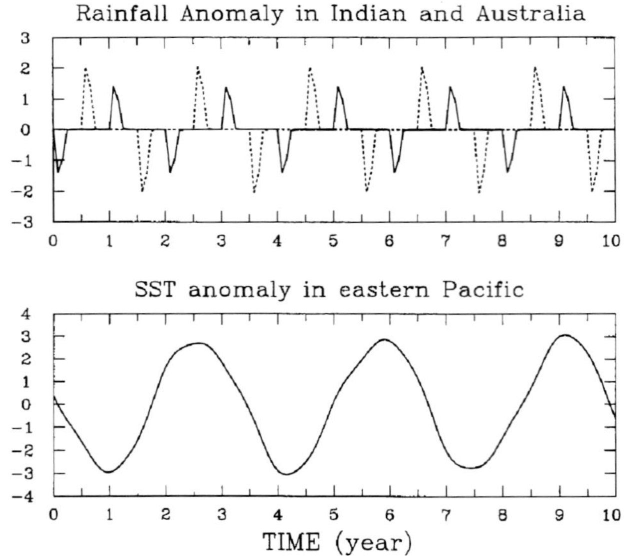

Fig. 5. Time series of anomalous rainfall rate (mm/day) over India (dotted line) and Australia (solid line) and anomalous SST (°C) in the eastern Pacific (bottom panel) in the case of no TBO-ENSO interactions ($\delta_{ET} = \delta_{TE} = 0$).

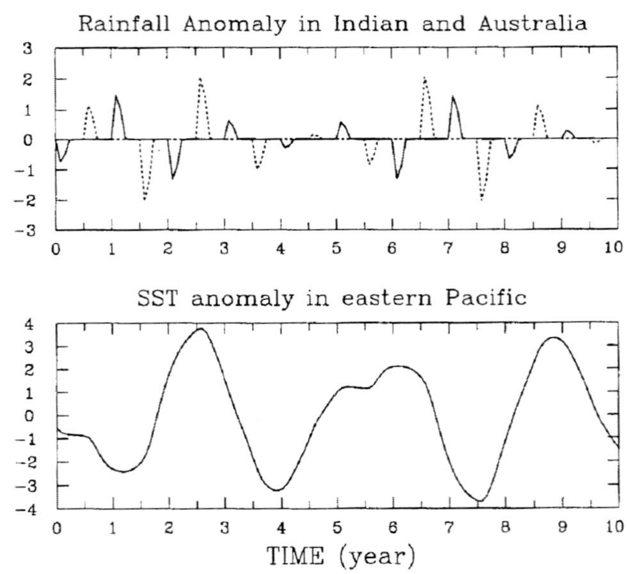

Fig. 6. Time series of anomalous rainfall rate (mm/day) over India (dotted line) when only ENSO influences TBO, and Australia (solid line) and anomalous SST (°C) in the eastern Pacific (bottom panel) when only TBO influences ENSO.

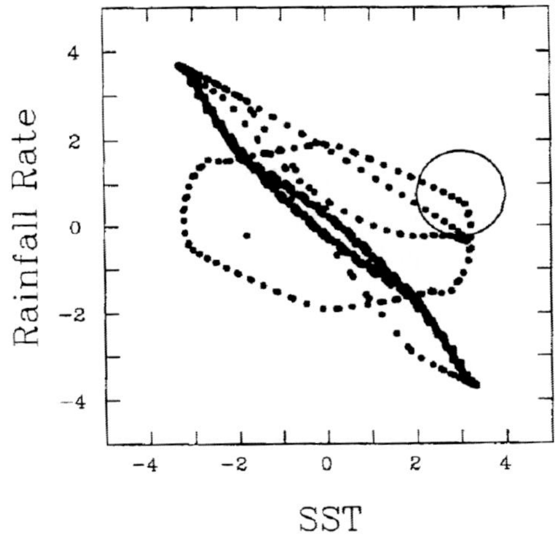

Fig. 7. Scatter diagram of December Australian rainfall anomaly (mm/day) versus eastern Pacific SST anomaly (°C), obtained from a 1000-year model integration under nonlinear two-way interactions ($\delta_{ET} = 0.1$, $\delta_{TE} = 0.1$). A circle denotes a possible phase regime for the 97-98 El Nino.

NONLINEAR EQUATORIAL KELVIN WAVES AND CISK

Han-Ru Cho
Department of Atmospheric Science, National Central University, Chungli

1. INTRODUCTION

The Kelvin wave-CISK has been proposed as the cause for intra-seasonal oscillations in the equatorial troposphere (Chang and Lim 1968; Lau and Peng 1987; Cho et al., 1994).

Since I will be discussing CISK, I need to introduce the cumulus parameterization scheme that we will be using. The cumulus parameterization scheme I will be using is based on low level convergence represented approximately by the mean vertical motion at the cloud base level:

$$Q = -\varepsilon\sigma\omega^* k(p) \qquad (1)$$

where Q is the cumulus heating rate, σ is the vertical stability parameter and k(p) gives the normalized vertical distribution function. ε here can be regarded as a proportional constant. It has been recognized for some time now that this parameterization is a nonlinear function of ω^*. Since there should be no cumulus heating when the vertical motion is downward, $\varepsilon=0$. On the other hand, if we have an upward vertical motion field $\varepsilon > 0$. Furthermore, it is difficult to linearize this kind of nonlinear dependence.

2. GOVERNING EQUATIONS FOR NONLINEAR KELVIN WAVES

The nonlinear equations for equatorial disturbances written on the equatorial β-plane are

$$\frac{\partial u}{\partial t} + u\frac{\partial u}{\partial x} + v\frac{\partial u}{\partial y} + \omega\frac{\partial u}{\partial p} + \frac{\partial \Phi}{\partial x} - \beta y v = 0 \qquad (2a)$$

$$\frac{\partial v}{\partial t} + u\frac{\partial v}{\partial x} + v\frac{\partial v}{\partial y} + \omega\frac{\partial v}{\partial p} + \frac{\partial \Phi}{\partial y} + \beta y u = 0 \qquad (2b)$$

$$\left(\frac{\partial}{\partial t} + u\frac{\partial}{\partial x} + v\frac{\partial}{\partial y}\right)\left(\frac{\partial \Phi}{\partial p}\right) + \sigma\omega = -\frac{\alpha}{C_p}\frac{ds}{dt} \qquad (2c)$$

$$\frac{\partial u}{\partial x} + \frac{\partial v}{\partial y} + \frac{\partial \omega}{\partial p} = 0 \qquad (2d)$$

$$\frac{\partial \Phi}{\partial p} = -\alpha = -\frac{RT}{p} \qquad (2e)$$

The right hand side of (2c) represents heating by cumulus convection where s is the entropy. σ in this equation is the vertical stability parameter which is assumed to be a constant. Equation (2d) is the mass continuity equation and (2e) is the hydrostatic relation. Since in equatorial Kelvin waves the y-component of the velocity is zero, the y-component of the equation of motion becomes a diagnostic relation describing geostrophic balance in that direction. The governing equations for Equatorial Kelvin waves become then

$$\frac{\partial u}{\partial t}+u\frac{\partial u}{\partial x}+\omega\frac{\partial u}{\partial p}+\frac{\partial \Phi}{\partial x}=0 \tag{3a}$$

$$\frac{\partial}{\partial t}\left(\frac{\partial \Phi}{\partial p}\right)+u\frac{\partial}{\partial x}\left(\frac{\partial \Phi}{\partial p}\right)+\sigma\omega=Q \tag{3b}$$

Here we have kept only the two prognostic equations. The three diagnostic relationships are the geostrophic balance condition in the y-direction

$$\frac{\partial \Phi}{\partial y}+\beta y u=0 \tag{3c}$$

together with the mass continuity equation (2d) and the hydrostatic relation (2e). The term of vertical advection of zonal momentum in (3a) is often ignored. This can be justified by assuming U ~ 10 m/s, $\omega \sim 5 \times 10^{-5}$ mb/s, a horizontal scale L ~ 10^7 m, and a vertical scale 10^3 mb. This gives

$$u\frac{\partial u}{\partial x}\sim 10^{-5}$$

$$\omega\frac{\partial u}{\partial p}\sim 5\times 10^{-7}$$

We shall adopt (1) to parameterize cumulus convection. We assume that k(p) is a normalized vertical distribution of cumulus heating:

$$\frac{1}{\Delta p}\int_{P_t}^{P_s} k(p)\,dp=1$$

Where P_t and P_s denote the pressures at the tropopause and at the surface. $\Delta p = P_s - P_t$. At the present time we keep the function f as general as possible. As pointed out in Lim et al. (1990), the values of ε should depend on the values of ω*. For example ω* < 0, then we should have convection and therefore $\varepsilon > 0$. On the other hand, if $\omega^* > 0$, then convection should be suppressed and we have $\varepsilon = 0$, i.e.,

$$\varepsilon = \begin{cases} \varepsilon_0 = const. > 0 & if\ \omega^* < 0 \\ 0 & if\ \omega^* > 0 \end{cases}$$

As pointed out by Lim et al. (1990) this makes ε a nonlinear function of ω^*, Furthermore, this nonlinear dependence doesn't depend on the amplitude of ω^* and hence cannot be linearized.

3. Small amplitude approximation

If one assumes the amplitudes of the perturbations from a resting atmosphere is very small, then the second terms on the left hand sides of (3a) and (3b), as well as the third term on the left side of (3a), can be neglected. The governing equations in this case becomes

$$\frac{\partial u}{\partial t} + \frac{\partial \Phi}{\partial x} = 0$$

$$\frac{\partial}{\partial t}\left(\frac{\partial \Phi}{\partial p}\right) + \sigma\omega = Q$$

$$\frac{\partial u}{\partial x} + \frac{\partial \omega}{\partial p} = 0 \tag{4}$$

$$\frac{\partial \Phi}{\partial y} + \beta y u = 0$$

$$\frac{\partial \Phi}{\partial p} = -\alpha = \frac{RT}{p}$$

Note that the thermodynamic equation is still a nonlinear equation because the term Q. All the left side terms have been linearized, however. Further we can assume the separability of the three spatial coordinate axes as follows:

$$u = u_x(x,t)\, u_y(y)\, u_p(p) \qquad \Phi = \Phi_x(x,t)\, \Phi_y(y)\, \Phi_p(p)$$
$$\omega = \omega_x(x,t)\, \omega_y(y)\, \omega_p(p)$$

From the third and the fourth equations in (4) we can assume that $u_p = \Phi_p = \omega_p'$, $u_y = \omega_y$, and $\partial u_x/\partial x + \omega_x = 0$, where the superscript ' denotes in this section the vertical derivative. If one assumes that $u_p = \Phi_p = g(p)$ and $\omega_p = h(p)$. then the second equation in (4) implies that

$$\frac{\partial \Phi_x}{\partial t} h''(p) + \frac{\partial u_x}{\partial x} \left(\varepsilon h(p^*)\, k(p) - \sigma h(p) \right) = 0$$

Divide the above equation by h"(p) at height levels where h"(p) is not zero, it is obvious then

$$\frac{-h(p)\left(\sigma - \varepsilon h(p^*) \dfrac{k(p)}{h(p)} \right)}{h''(p)} = constant = b^2 \tag{5a}$$

since Φ_x and u_x are functions of x and t only. Note that b^2 defined in the equation is a real constant in each of the areas of cloud region or the cloud-free region, and its value is proportional to the difference between the adiabatic cooling term and the cumulus heating term in the thermodynamic equation. Because of the requirement that $\omega = 0$ at $p = p_s$ and p_t, the ratio -h(p)/h"(p) is usually a positive quantity. Therefore $\varepsilon = 0$ and $b^2 > 0$ in a cloud-free region. In the cloud region, $b^2 \geq 0$ when the adiabatic cooling effect is larger than or equal to the cumulus heating term at all height levels. Furthermore, as we shall see later in this section that $\pm b$ is the propagation speed of the hyperbolic waves governed by the equations. If in the cloud region the cumulus heating is larger than adiabatic cooling at all levels, $b^2 < 0$, or b is a pure imaginary number in the cloud region. From (5a) a differential equation can be derived for h:

$$h''(p) + \frac{\sigma}{b^2} h(p) = \frac{\varepsilon}{b^2} h(p^*) k(p) \tag{5b}$$

The solution must satisfy the requirement that at the cloud base level, h(p) must be equal to $h(p^*)$ on the right hand of the equation. This condition can be used to determine the value of b^2. This equation can be solved formally in terms of a Green's functions and b^2 is determined from a integral relationship. In previous linear studies people are looking for imaginary or copmplex solutions of the equation for b^2. In this paper we merely point out that b^2 is physically realistic only when it is a real number. Further, b^2 is larger in the cloud-free region than in the cloudy region.

From the second equation of (4):

$$\frac{\partial \Phi_x}{\partial t} + b^2 \frac{\partial u_x}{\partial x} = 0 \tag{6}$$

When this result is coupled with the first equation we obtain

$$\begin{aligned} \frac{\partial u}{\partial t} + \frac{\partial \Phi}{\partial x} &= 0 \\ \frac{\partial \Phi}{\partial t} + b^2 \frac{\partial u}{\partial x} &= 0 \end{aligned} \tag{7}$$

We have dropped the subscript x in these equations, and this convention will be adopted throughout this small amplitude analysis.

If we let

$$\vec{\psi} = \begin{pmatrix} u \\ \Phi \end{pmatrix}$$

Then (7) can be written as:

$$\frac{\partial \vec{\psi}}{\partial t} + A \frac{\partial \vec{\psi}}{\partial x} = 0 \tag{8}$$

where A is the matrix

$$A = \begin{pmatrix} 0 & 1 \\ b^2 & 0 \end{pmatrix}$$

The eigenvalues of A are

$$\lambda_1 = +b \qquad \lambda_2 = -b \tag{9a}$$

These are the propagating speed c of the hyperbolic waves if b^2 is positive. The eigenvectors are

$$\vec{r}_1 = \begin{pmatrix} 1 \\ b \end{pmatrix} \qquad \vec{r}_2 = \begin{pmatrix} 1 \\ -b \end{pmatrix} \tag{9b}$$

Note that because A is not symmetric, the eigenvectors are not mutually perpendicular. This can be resolved in two ways. First by changing the scales for time and length so that the unit for velocity becomes b. In such a set of scales, we have

$$b^2 \frac{\partial u}{\partial t} + b^2 \frac{\partial \Phi}{\partial x} = 0$$

$$b^3 \frac{\partial \Phi}{\partial t} + b^3 \frac{\partial u}{\partial x} = 0$$

We have assumed that b $\neq$ 0. The equation can be wriiten as

$$\frac{\partial \vec{\psi}}{\partial t} + B \frac{\partial \vec{\psi}}{\partial x} = 0$$

where $\vec{\psi}$ is the same as defined earlier. B is given by

$$B = \begin{pmatrix} 0 & 1 \\ 1 & 0 \end{pmatrix}$$

Note that B is a symmetric matrix.

The second approach is based on the fact that Kelvin waves are allowed to propagates only eastward (to be explained later in this section). Since only one mode is allowed to propagate, the governing equation system is reduced to a single first order equation. Therefore unless stated otherwise we shall use the old units.

The properties of the governing equations clearly depends on the sign of b^2. If $b^2 > 0$, (i.e., when the adiabatic cooling rate is larger than the cumulus heating rate) then the governing equations are a hyperbolic system. If $b^2 < 0$, then the eigenvalues is purely imaginary, hence is not allowed for the reason that the solutions is not equatorially confined (to be seen later in the section).

From (9b) we can see that

$$\Phi_{1,2} = \pm\, b\, u_{1,2}$$

This can be substituted into (3c) to obtain the y-structure of the eigenvectors:

$$\Phi_i = \Phi_0 \exp\left[-\frac{\beta}{2c} y^2 \right]$$

where

$$c = \pm\, b \qquad \textit{for } i = 1,\ 2 \tag{10}$$

Therefore, an equatorially trapped mode is possible only when $b^2 > 0$ and only for the first eigenmode. Since we are dealing with equatorial Kelvin waves, we will assume the case of $b^2 > 0$ in the remaining of this paper.

The standard method of solving hyperbolic sets of equations is the method of characteristics. Let the initial value of the solution $\vec{\psi}(x, t = 0) = \vec{\psi}_0(x)$. One can make the expansion if $\vec{r}_i$ for i = 1, 2 are mutually perpendicular:

$$\vec{\psi}_0 = \sum_{i=1}^{2} v_i(x, 0)\, \vec{r}_i$$

and

$$\psi(x, t) = \sum_{i=1}^{2} v_i(x, t)\, \vec{r}_i = \sum_{i=1}^{2} v_i(x - \lambda_i t, 0)\, \vec{r}_i$$

if the system is hyperbolic. The lines in the (x, t) plane given by

$$\frac{dx}{dt} = \lambda_i$$

are called the characteristics of the hyperbolic system of equations. We shall keep only the first eigenmode in the solution to represent equatorially trapped waves. Or in other words, we shall keep only

$$\frac{\partial v_1}{\partial t} + \lambda_1 \frac{\partial v_1}{\partial x} = 0 \tag{11}$$

Since

$$r_1 = \begin{pmatrix} 1 \\ b \end{pmatrix}$$

it is obvious that $v_1 = u$ and $\Phi_1 = bu$. We can also write (11) into a flux form by defining

$$f(u) = bu = \Phi_1 \tag{12}$$

(11) then becomes

$$\frac{\partial u}{\partial t} + \frac{\partial f(u)}{\partial x} = 0 \tag{13}$$

Since we consider only one eigenmode, the system reduces to a first-order conservation law.

If there are abrupt change in ε, say at the boundary between clear area and cloudy region where $\varepsilon = 0$ in the clear area and ε has a finite value in the convective region, then there is also an abrupt change in the propagation speed of the wave c. Therefore, weak solutions should be considered for the present system of equations (See, for example, LeVeque, 1992). We will consider two different parts of the of the cloud region: namely the trailing edge and the leading edge where there are discontinuities in c. Let us first consider in detail only the trailing edge in this part of the paper.

a. Shock wave

Keep in mind that the vertical profiles, say h(p) for ω_p, is different to the left of a cloud region from the cloud region itself if the cloud region is propagating eastward, in addition to the difference between the values of ε. In this case the air on the left side of the trailing edge of a cloud region should be clear, and $\lambda = \lambda_l = + b_l$ where b_l corresponds to $\varepsilon = 0$ and the vertical profile h(p) in the clear area. On the right side, $\lambda = \lambda_r = + b_r$ where b_r corresponds to $\varepsilon > 0$ and the vertical profile h(p) in the cloud area. It is not difficult to deduce that $\lambda_l > \lambda_r$ (See Appendix A.). Since the wave on the left hand side will catch up with the waves on the right hand side of the boundary region, a discontinuity (or a shock) will form. The propagation speed s of the discontinuity is given by the Rankine-Hugoniot jump condition (see, for example, LeVeque, 1992):

$$s = \frac{f(u_l) - f(u_r)}{u_l - u_r} = \frac{[f]}{[u]} \tag{14}$$

where [*] is used to denote the jump of * across the discontinuity. This jump condition is merely a restatement of the conservation law as given in (13) when applied to a discontinuity, namely any local change in u must be balanced by the convergence or divergence of the flux f. This is illustrated schematically in Figure 1 where a jump in u is shown across the discontinuity. Due to the propagation of the discontinuity, there is a jump in the content of u. This jump, if u is conserved, has to be supplied by a discontinuity in the flux of u, namely [f].

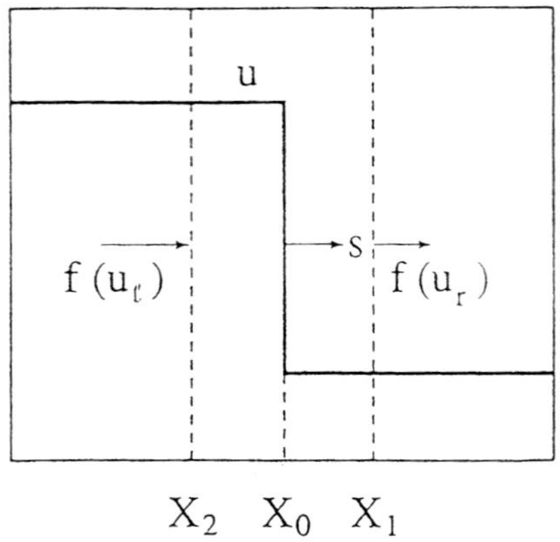

FIGURE 1: Jump condition.

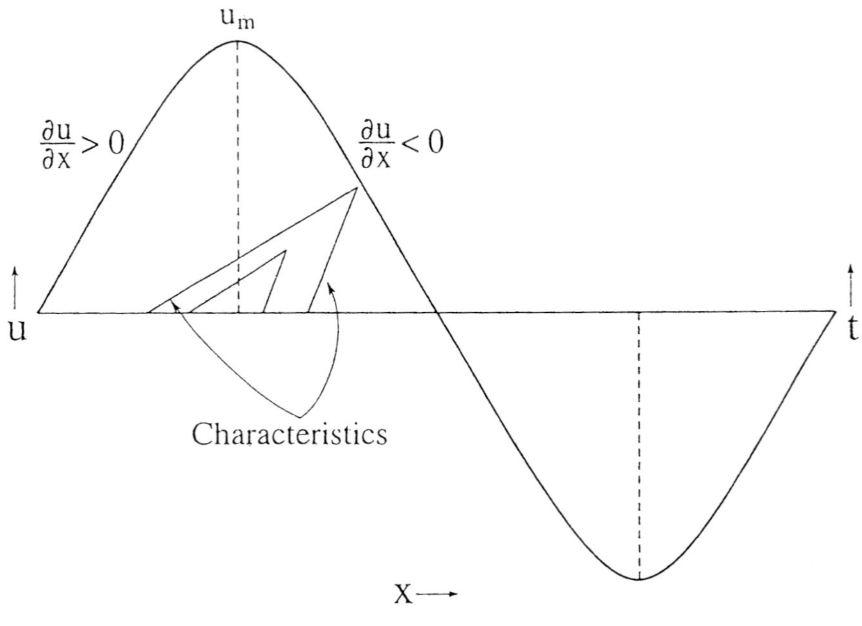

Figure 2: The cloudy area in a disturbance.

At the trailing edge of a cloud region, we must have $\partial u/\partial x < 0$ to the right and $\partial u/\partial x > 0$ to the left (as illustrated schematically in Figure 2). If u is continuous before the formation of the shock, at the trailing edge $\partial u/\partial x = 0$ and u has a local maximum u_m. Consider a time $t = 0$ when the discontinuity is just forming at the trailing edge. At this time $u_l \to u_r \to u_m$, therefore $s = (b_l - b_r)u_m/(u_l - u_r)$. Since there is a finite jump in ε, we have $b_l > b_r$. The jump condition leads to a physically unacceptable situation where $s = \infty$. Therefore the discontinuity in ε seems to be too harsh a condition for the cumulus parameterization scheme at $t = 0$ if the initial u is continuous.

A simple remedy to this situation may be to assume, for example, that the proportional constant in the parameterization (epsilon), instead of being a constant, is itself proportional to $\partial u/\partial x$ when there is convergence and zero when we have a divergent flow. The only drawback of this procedure is that CISK is then a truly nonlinear problem.

b. rarefaction wave

The situation is entirely different at the leading edge of a cloud region, because at the leading edge, the propagation speed on the right side is larger than the propagation speed on the left side:

$$\lambda_l < \lambda_r$$

At such a boundary, it has been shown that the solution takes the form of a rarefaction wave (Courant and Friedrichs, 1948; LeVeque, 1992):

$$\vec{\psi} = \vec{w}(x/t) \qquad for\ \lambda_l \le x/t \le \lambda_r$$

However, because of the linearization we have made, this is not a genuinely nonlinear problem (Lax, 1957). The magnitude of u cannot be determined in the region of rarefaction wave.

One thing surprising about these small amplitude results is that there is no instability or growth of the wave as it propagates. Since u is conserved along characteristics, there is no exponential amplification as predicted according to the linear CISK theory. This is independent of heating profile f(p). In a linear CISK analysis on equatorial waves, Cho and Pendlebury (1997) found the sensitivity of the results on the vertical profile of cumulus heating. We note that from (5a) b^2 is always a real quantity, either positive or negative. When b^2 is negative, we do not have a equatorially trapped wave. In the linear CISK analysis, b is complex number if Kelvin waves are unstable. We suggest that this is not a physically reasonable result.

References

Chang, C. P. and H. Lim, 1988: Kelvin wave-CISK: A possible mechanism for the 30-50 day oscillations. *J. Atmos. Sci.* **45**, 1709-1720.

Cho, H. R., K, Fraedrich, and J. T. Wang, 1994: Cloud clusters, Kelvin Wave-CISK, and the Maden-Julian Oscillations in the equatorial troposphere. *J. Atmos. Sci.*, **51**, 68-76.

Cho, H. R., and D. Pendlebury, 1997: Wave-CISK of equatorial waves and the vertical distribution of cumulus heating. *J. Atmos. Sci.* 54, 2429-2440.

Courant, R., and K. O. Friedrichs, 1948: Supersonic flows and shock waves. Springer, 464pp

Emanuel, K.A., 1987: An air-sea interaction model of intraseasonal oscillations in the tropics. *J. Atmos. Sci.*, **44**, 2324-2340.

Lau, K.M., and L. Peng, 1987: Origin of low frequency (intraseasonal) oscillations in the tropical atmosphere. Part 1: Basic Theory. *J. Atmos. Sci.*, **44**, 950-972.

Lau, K.M., L. Peng, C. H. Sui, and T. Nakazawa, 1989: Dynamics of super clusters, wind burst, 30-60 day oscillations and NESO: A unified view. *J. Meteor. Soc. Japan*, **57**, 205-219.

Lax, P.D., 1957: Hyperbolic systems of conservation laws, II., *Comm. Pure. Appl. Math.*, **10**, 537-566.

LeVeque R.J., 1992: Numerical methods for conservation laws. Birkhauser, Basel, 214pp.

Lim, H., T. K. Lim, and C. P. Chang, 1990: Reexamination of wave-CISK theory: Existence and properties of nonlinear wave-CISK mode. *J. Atmos. Sci.*, **47**, 3078-3091.

Madden, R., and P. R. Julian, 1971: Detection of a 40-50 day oscillation in the zonal wind in the tropical Pacific., *J. Atmos. Sci.*, **28**, 702-708.

Neelin, J. D., I. M. Held, K. H. Cook, 1987: Evaporation-wind feedback and low frequency variability in the tropical atmosphere. *J. Atmos. Sci.*, **44**, 2341-2348.

Sui, C. H., and K. M. Lau, 1989: Origin of low-frequency (intraseasonal) oscillations in the tropical atmosphere. Part II: Effect of an improved treatment of moist processes. *J. Atmos. Sci.*, **46**, 37-56.

Wang, B., and H. Rui, 1990: Dynamics of the coupled moist Kelvin-Rossby wave on an equatorial beta-plane. *J. Atmos. Sci.*, **47**, 397-413.

USING THE EQUIVALENT GEOPOTENTIAL AND GEO-STREAMFUNCTION TO REPRESENT THE PRESSURE GRADIENT FORCE IN TERRAIN-FOLLOWING COORDINATES OVER MOUNTAINOUS AREA

Qiu-shi Chen
Byrd Polar Research Center, The Ohio State University, Columbus, OH 43210

1. INTRODUCTION

Mountains have important influence on large- and mesoscale atmospheric motion, especially over East Asia due to the highest Qinghai-Xizang (Q-X) Plateau (or the Tibetan Plateau). As has been understood for decades, the horizontal pressure gradient force (PGF) in terrain-following (including pressure-based (σ) and height-based) coordinates is a small difference between two large terms over steep slopes and its computational errors are very large. A number of authors, for example, Corby et al. (1972), Gary (1973), Sunqvist (1976), Nakamura (1978) and Mesinger et al. (1988), examined how these errors can be reduced based on the method of computation. The same problem arises in the nonhydrostatic Penn State/NCAR Mesoscale Model (MM5) formulated in terms of perturbations relative to a hydrostatic basic state. Recently, Colle et al. (1999) verified the 36- and 12-km resolution MM5 precipitation forecasts from 9 December 1996 through 30 April 1997 and NCEP's 10-km resolution Eta Model (Eta-10) forecasts from 7 January 1997 through 30 April 1997 across Pacific Northwest of US. It is found that the 12-km MM5 tends to generate too much precipitation along the steep windward slopes and not enough precipitation in the lee of major barriers. In the latter regions, the 36-km grid may actually be providing a more skillful forecast. The Eta-10 over-predicts precipitation along the windward slopes even more than the 12-km MM5.

The over-prediction of precipitation along the windward slopes may be attributed to several problems. One of the problems, pointed out by Colle et al. (1999) based on the citation of a personal communication from Z. Janjic (1998), is that the Eta-10 produces excessive flow blocking upwind of major barriers, which can result in too much upward motion and precipitation well upwind of the orographic crest. To a lesser degree than the Eta-10, the MM5 also had over-prediction problems along the windward slopes and under-prediction in the lee of barriers. This suggests that the MM5 may be over-predicting upward motion on the windward slopes and the leeside subsidence. Dempsay and Davis (1998) gave an error analysis of the MM5's PGF schemes, and found that the standard PGF schemes in MM5 can produce significant velocity errors above steep terrain. Some alternative schemes can reduce, but cannot eliminate, these errors.

Mesinger (1984) developed the step-mountain approach of the Eta-coordinate system which is a good method to calculate the pressure gradient in a region of complex terrain. This method has also been further developed by Mesinger et al. (1988), Janjic (1990, 1994), and is now used in the NCEP Eta-model. However, the piecewise constant representation of the terrain is only first-order accurate in mathematics, whereas the continuous representation of the terrain in the terrain-following coordinates is at least second order accurate (depending on the chosen method of discretization). Furthermore, physical parameterizations in the planetary boundary layer are straightforward with terrain-following coordinates. At present, many global and limited-area models still use terrain-following coordinates. In this paper, we will develop a new method to compute the horizontal PGF and to simulate effects of mountains on atmospheric motions instead of the ones used in the MM5 and Eta model. This method uses terrain-following coordinates but can eliminate large errors in computing the PGF over steep slopes of mountains.

The horizontal wind can be separated into its irrotational and rotational parts and expressed by

$$\vec{V} = -\nabla\chi - \vec{k}\times\nabla\psi \tag{1.1}$$

where χ and ψ denote velocity potential and streamfunction, respectively. The horizontal PGF $\vec{G}$ in σ-coordinates in a hydrostatic model is expressed by

$$\vec{G} = -\nabla\phi(x,y,\sigma,t) - RT\nabla lnp_*(x,y,t) \tag{1.2}$$

where $\phi(x,y,\sigma,t)$ and $p_*(x,y,t)$ are the geopotential and surface pressure, respectively. Here $\vec{G}$ is also an horizontal vector, and it can also be separated into its irrotational and rotational components and expressed by

$$\vec{G} = -\nabla\phi_e - \vec{k}\times\nabla\eta \tag{1.3}$$

where $\phi_e(x,y,\sigma,t)$ is referred to as an equivalent geopotential, and $\eta(x,y,\sigma,t)$ is called a geo-streamfunction (Chen and Bromwich 1999, hereafter referred to as CB99).

Because the horizontal PGF $-\nabla\phi(x,y,p,t)$ in p-coordinates is also irrotational, the equivalent geopotential $\phi_e(x,y,\sigma,t)$ can be used in σ-coordinates in the same way as $\phi(x,y,p,t)$ is used in p-coordinates. The equivalent geopotential ϕ_e can be used in synoptic analysis on a constant σ surface, and several examples of the equivalent geopotential analysis at σ=0.995 are shown in CB99. Artificial anomalous systems over the Tibetan Plateau and Greenland caused by pressure reduction to sea level are not present and weather systems are shown clearly in high mountain regions. The geostrophic relation between the equivalent geopotential and wind is approximately satisfied for synoptic scale motions on the constant σ surface. The equivalent geopotential ϕ_e has been used to develop a generalized ω-equation in σ-coordinates, and the overpredicted precipitation caused by the steep slopes of the Greenland Ice Sheet from the NCEP/NCAR reanalysis has been successfully improved by this method (Chen et al. 1997b; Bromwich et al., 1999).

Because the horizontal PGF $-\nabla\phi(x,y,p,t)$ is irrotational, the rotational part, corresponding the second term on the right hand side of (1.3), is not included in $-\nabla\phi(x,y,p,t)$ in p-coordinates. The corresponding rotational part in p-coordinates can only be computed through the lower boundary condition at $p = p_*(x,y,t)$ or $z = H_*(x,y)$ and expressed implicitly.

Expression (1.3) has two important advantages: The first is that the irrotational part of (1.3) can be computed accurately. This irrotational part corresponds to $-\nabla\phi(x,y,p,t)$ in p-coordinates. It is shown in Table 1 of CB99 that the absolute value of the irrotational part is much larger than that of the rotational part in (1.3) over the Greenland region. Thus, the small difference between two large terms over steep slopes occurred in (1.2) is eliminated automatically by using expression (1.3).

The second advantage is that the rotational part can also be computed accurately because it is expressed explicitly and separately in (1.3). Due to smallness of the rotational part, if this part is computed by (1.2), its relative computational errors must be even larger than those of the irrotational part. Thus, the rotational part cannot be computed accurately by (1.2). This rotational part is also difficult to be computed accurately in p-coordinates because it cannot be expressed explicitly. Although the rotational part is small, its distribution is closely related to position of mountains and to exchange of momentum or vorticity between the mountain and atmosphere. The accurate computation of the rotational part is also important in determining the orographic effects.

2. Separation of a horizontal vector into its nondivergent and irrotational parts in a limited region and the external wind boundary value for limited-area models

(a) The inner and harmonic parts of the streamfunction and velocity potential and the internal and external wind in a limited region

In coordinates of a conformal map projection, let U, V, Ω, D be

$$U = u/m, \qquad V = v/m, \qquad \Omega = \zeta/(m^2), \qquad D = \delta/(m^2) \tag{2.1}$$

where u and v are the horizontal components of the wind, ζ the vertical component of the vorticity, δ the horizontal divergence, and m the map scale factor. The wind components are expressed by:

$$U = -\frac{\partial\psi}{\partial y} + \frac{\partial\chi}{\partial x}, \qquad V = \frac{\partial\psi}{\partial x} + \frac{\partial\chi}{\partial y} \tag{2.2}$$

where the streamfunction ψ and velocity potential χ satisfy the following Poisson equations

$$\nabla^2\psi = \Omega, \qquad \nabla^2\chi = D \tag{2.3}$$

within the limited region, and at the boundary they satisfy the coupled conditions

$$\vec{s}\cdot\vec{V} = \frac{\partial\psi}{\partial n} + \frac{\partial\chi}{\partial s} = V_t, \qquad \vec{n}\cdot\vec{V} = -\frac{\partial\psi}{\partial s} + \frac{\partial\chi}{\partial n} = V_n. \tag{2.4}$$

Here $\vec{s}$ and $\vec{n}$ are tangential and normal unit vectors, respectively, and s and n are distances along and normal to the boundary.

Lynch (1989) and Chen and Kuo (1992) found that partitioning the wind in a limited region into nondivergent and irrotational components is not unique. The nonuniqueness is closely related to a portion of the wind being both nondivergent and irrotational in a limited region. A limited region R is only a portion of the global area. The region over the globe outside region R is referred to as an external region. From a global perspective, the streamfunction and velocity potential, as well as the horizontal wind, in region R depend upon the vorticity and divergence not only within region R but also in the external region. The portion of the wind field that is both nondivergent and irrotational within region R is referred to as an external wind. Due to the properties of nondivergence and irrotation, no matter how the vorticity and divergence vary within the region R, the external wind does not change. Thus, the external wind has no relation to the vorticity and divergence within the region R. However, from a global view, the external wind must depend only on the vorticity and divergence in the external region, which is why it is called the "external" wind. For the global area, a both nondivergent and irrotational wind always zero, thus, the external wind must vanish, and also there is no external region.

In order to avoid the nonuniqueness, Chen and Kuo (1992) separated the streamfunction and velocity potential in a limited region into the inner and harmonic parts as

$$\psi = \psi_i + \psi_h, \qquad \chi = \chi_i + \chi_h \tag{2.5}$$

where the inner parts, ψ_i and χ_i, satisfy the Poisson equations

$$\nabla^2 \psi_i = \Omega, \qquad \nabla^2 \chi_i = D \tag{2.6}$$

with zero Dirichlet boundary value. The wind computed from the inner part as

$$U_I = -\frac{\partial \psi_i}{\partial y} + \frac{\partial \chi_i}{\partial x}, \qquad V_I = \frac{\partial \psi_i}{\partial x} + \frac{\partial \chi_i}{\partial y}. \tag{2.7}$$

is called an internal wind. Because the solution of a Poisson equation with a Dirichlet boundary value is unique, these inner parts and the internal wind are uniquely determined by the vorticity and divergence within the limited region R. No matter how the vorticity and divergence vary in the external region (outside the region R), the inner parts of the streamfunction and velocity potential, as well as the internal wind, do not vary.

The difference between the wind and internal wind can be expressed by

$$U_E = U - U_I, \qquad V_E = V - V_I. \tag{2.8}$$

According to (2.2) and (2.7), the relationship between $\vec{V}_E$ and the harmonic parts of the streamfunction and velocity potential is expressed by

$$U_E = -\frac{\partial \psi_h}{\partial y} + \frac{\partial \chi_h}{\partial x}, \qquad V_E = \frac{\partial \psi_h}{\partial x} + \frac{\partial \chi_h}{\partial y}. \tag{2.9}$$

From (2.3) and (2.6), the harmonic parts satisfy Laplace equations

$$\nabla^2 \psi_h = 0, \qquad \nabla^2 \chi_h = 0 \tag{2.10}$$

with the coupled boundary conditions

$$U_E\,|_\Sigma = \left(-\frac{\partial \psi_h}{\partial y} + \frac{\partial \chi_h}{\partial x}\right)\Big|_\Sigma, \qquad V_E\,|_\Sigma = \left(\frac{\partial \psi_h}{\partial x} + \frac{\partial \chi_h}{\partial y}\right)\Big|_\Sigma . \tag{2.11}$$

The wind $\vec{V}_E$ that satisfies (2.9), (2.10) and (2.11) must be both nondivergent and irrotational within the limited region R, thus, it must be the external wind. By using the above method, the horizontal wind can be separated into the internal and external winds in a limited region.

(b) Separation of a horizontal vector in a limited region into three components

The internal wind can be further separated into its internal nondivergent and irrotational components based on (2.7). Thus, the horizontal wind in a limited area is expressed by a sum of three parts as

$$\vec{V} = \vec{k} \times \nabla \psi_i + \nabla \chi_i + \vec{V}_E \tag{2.12}$$

where $\vec{k} \times \nabla \psi_i$ is the internal nondivergent component and $\nabla \chi_i$ is the internal irrotational component. Using the method (2.12), any horizontal vector in a limited region can be separated into its internal nondivergent, internal irrotational and external components.

(c) The external wind lateral boundary value for limited-area models (LAMs)

The separation of the wind into the internal and external components represents a separation of the wind into two parts so that each part is related to the vorticity and divergence only in one region. The internal wind is related to the vorticity and divergence only within the limited region, while the external wind depends on those only in the external region. This is a principle of the wind separation in a limited region, and it can be used to solve the lateral boundary condition (LBC) of a LAM.

The one-way nesting is fundamental for the LBC of a LAM. If a LAM is formulated in terms of the vorticity and divergence equations and is nested in a global model, only the vorticity and divergence within the region at each time step can be predicted by the LAM. This means that only the internal wind can be predicted from the LAM, while the external wind must be derived from the prediction of the global model. The total wind in the limited region including its boundary at each time step is the sum of the internal and external winds. If at the boundary only the external wind is given by the global model and the internal wind is determined by the LAM itself in an one-way nesting method, this method is referred to as an external wind lateral boundary value (Chen et al. 1997a). In this boundary method, the wind at the boundary of the LAM is not only given by the global model but also determined by the LAM itself.

Because the components of the external wind is a harmonic function, in the external wind boundary method, the boundary values are not only given at the closed boundary line of the limited region, but also given by the harmonic functions defined in the whole limited domain. The derivatives of the harmonic functions up to second order are continuous within the region up to the boundary. The procedure of adding harmonic functions to the inner parts represents the effect of the lateral boundary values from the global model on the prediction of the LAM. This procedure cannot cause any discontinuity of the variables near the boundary, thus, the external wind boundary method does not need any artificial boundary treatment.

In the one-way method of the current MM5 or many other LAMs, lateral boundary winds are prescribed based on the output from a previous integration of a global model, while the wind within the region is predicted based on a fine-mesh LAM. In this one-way method, the wind at the boundary of the LAM is determined by the global model only. In this boundary method, in order to obtain a continuous wind solution, some artificial boundary techniques, such as a boundary relaxation (Davies 1976, 1983; Davies and Turner 1977) or a sponge zone (Perkey and Kreitzberg 1976; Kar and Turco 1995), are necessary. However, these artificial techniques do not exist in the real atmosphere and produce errors in the solution near the boundary. This lateral boundary method needs to use a buffer zone to remove the lateral boundaries in each direction to a distance that the LBC errors cannot penetrate to the area of meteorological interest during a forecast, and it results in serious limitations to the utility of LAMs (Warner et al. 1997).

The external wind boundary values have been successfully tested in the harmonic-Fourier (H-F) spectral LAM (Chen et al. 1997a). In one example, the cross-boundary flow in upwind direction is predicted results show that the cyclone gradually and smoothly moves into the region across the western boundary. The external wind boundary method is capable of transmitting motion systems smoothly into and out of the limited region. All the fields near the boundary are very smooth and there is no discontinuity near the boundary. Another example studied by Chen et al. (1997a) shows that the external wind boundary method can also be used at the boundary where high mountains are intersected. The limited-area model can run very smoothly near the boundary when complex topography is present.

3. The equivalent geopotential and geo-streamfunction in σ-coordinates

(a) Derivation of the equivalent geopotential and geo-streamfunction in a limited region

If the horizontal hydrostatic pressure gradient force in σ-coordinates (1.2) is separated into the rotational and irrotational parts as in (1.3), the equivalent geopotential ϕ_e and the geo-streamfunction η satisfy the Poisson equations,

$$\nabla^2\phi_e = \nabla^2\phi + \frac{\partial}{\partial x}\left(RT\frac{\partial lnp_*}{\partial x}\right) + \frac{\partial}{\partial y}\left(RT\frac{\partial lnp_*}{\partial y}\right), \qquad \nabla^2\eta = R\left(\frac{\partial T}{\partial x}\frac{\partial lnp_*}{\partial y} - \frac{\partial T}{\partial y}\frac{\partial lnp_*}{\partial x}\right) \tag{3.1}$$

Over the globe, the Poisson equations (3.1) are easily solved without lateral boundary conditions. If equations (3.1) are solved in a limited region, ϕ_e and η can be separated into their inner and harmonic parts, and their inner parts satisfy

$$\nabla^2\phi_{ei} = \nabla^2\phi_i + \frac{\partial}{\partial x}\left(RT\frac{\partial lnp_*}{\partial x}\right) + \frac{\partial}{\partial y}\left(RT\frac{\partial lnp_*}{\partial y}\right), \qquad \nabla^2\eta_i = R\left(\frac{\partial T}{\partial x}\frac{\partial lnp_*}{\partial y} - \frac{\partial T}{\partial y}\frac{\partial lnp_*}{\partial x}\right) \tag{3.2}$$

with zero Dirichlet boundary value. In order to give a clear picture of the equivalent geopotential, we will compare it with the generalized geopotential for synoptic scale motions.

(b) The differences between the equivalent geopotential and generalized geopotential

Many global models and the nonlinear normal mode initialization (NNMI) for global and limited-area models (Errico and Bates 1988; Takano et al. 1990) in σ-coordinates use the generalized geopotential. The generalized geopotential is often expressed by

$$\Phi = \phi + RT_0(\sigma)\ln p_* \tag{3.3}$$

where ϕ is the geopotential height in σ-coordinates, and $T_0(\sigma)$ is the mean temperature averaged on constant σ-surfaces and depends on σ only. Errico and Bates (1988) called Φ a pseudo-geopotential.

The inner part of the generalized geopotential for σ=0.510 at 0000 UTC 2 January 1985 over a limited region of North America is shown in Fig. 1(a). Figures 1(b) and (c) are the inner parts of the streamfunction and equivalent geopotential for σ=0.510 at 0000 UTC 2 January 1985, respectively. The geostrophic approximation for rotational wind in σ-coordinates (CB99) is expressed by $f_0\psi_i = \phi_{ei}$. Comparing Figs. 1(b) and (c), these two figures are similar. Thus, the geostrophic relation is approximately satisfied. The difference between the generalized and equivalent geopotential (Fig. 1(a) - Fig. 1(c)) is shown in Fig. 1(d). A negative center is over the Rocky Mountain region and it means that the generalized geopotential is smaller than the equivalent geopotential in that area. Because the equivalent geopotential has good geostrophic approximation, the generalized

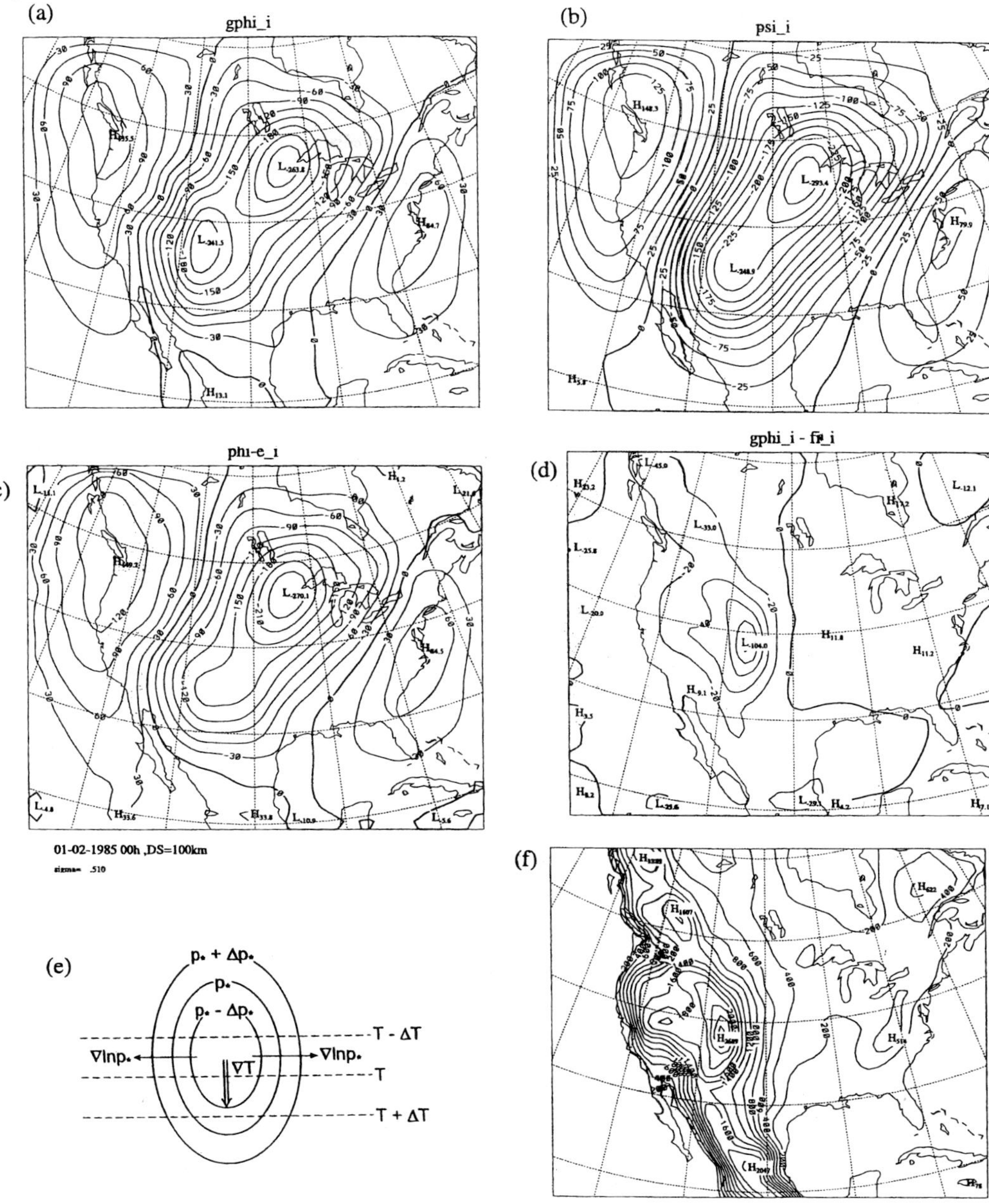

Fig.1 (a) The inner part of the generalized geopotential at 0000 UTC 2 January 1985 at σ=0.510 (in $10\,m^2s^{-2}$ and contour interval: $30\times 10\,m^2s^{-2}$); (b) Same as (a) except for the streamfunction (in $10^6\,m^2s^{-1}$ and contour interval: $25\times 10^5\,m^2s^{-1}$); (c) Same as (a) except for the equivalent geopotential; (d) Same as (a) except for the difference between the generalized geopotential and equivalent geopotential (in $10\,m^2s^{-2}$ and contour interval: $20\times 10\,m^2s^{-2}$); (e) Schematic diagram of a mountain with an idealized elliptical shape and a temperature field which decreases with latitude; (f) The topography of part of North America (contour interval of 200 m).

geopotential must have large deviations from the geostrophic approximation over such high mountain regions.

The generalized geopotential is often used for initializations of LAMs (Errico and Bates 1988; Takano et al. 1990; Sashegyi and Madala 1993) and global models in σ-coordinates. The balance conditions used by Temperton (1988) for initialization is based on

$$\partial D/\partial t = 0, \qquad \partial(\nabla^2\Phi - f_0\Omega)/\partial t = 0 \tag{3.4}$$

where Φ is the generalized geopotential. The second balance condition of (3.4) means that the tendency of the ageostrophic vorticity vanishes (Chen et al 1996). The large deviations from the geostrophic approximation in the generalized geopotential over high mountain regions must affect the initialization and prediction in these areas if the generalized geopotential is used in (3.4). Thus, it is necessary to use the equivalent geopotential instead of the generalized geopotential over high mountain regions.

(c) The geo-streamfunction and pressure drag on vorticity

The quantity $-RT\nabla lnp_*$ in the second term on the right hand side of (1.2) is referred to as a surface pressure stress gradient force. According (5.4) and (5.6′) of CB99, the vorticity equation is expressed by

$$\partial\Omega/\partial t = -f_0 D - \nabla^2\eta + \Omega_{adv} \tag{3.5}$$

where Ω_{adv} is the generalized vorticity advection and expressed by

$$\Omega_{adv} = -\left(\frac{u}{m}\frac{\partial(f'+\zeta)}{\partial x} + \frac{v}{m}\frac{\partial(f'+\zeta)}{\partial y}\right) - (f'+\zeta)\frac{\delta}{m^2} + \frac{\partial}{\partial x}\left(\frac{P_y}{m} - \dot{\sigma}\frac{\partial v/m}{\partial\sigma}\right) - \frac{\partial}{\partial y}\left(\frac{P_x}{m} - \dot{\sigma}\frac{\partial u/m}{\partial\sigma}\right) \tag{3.6}$$

where P_x and P_y are momentum changes produced by the vertical fluxes due to turbulence, and $f = f_0 + f'$. Here f_0 is the averaged value of the Coriolis parameter over the limited region and f' is its deviation.

The quantity $-\nabla^2\eta$ is the vorticity production rate caused by the surface pressure stress, and it represents the exchange of vorticity between the mountains and atmosphere. In the region where $-\nabla^2\eta < 0$, vorticity decreases in the atmosphere, and vorticity in the atmosphere transfers to the surface of mountains through the surface pressure, and vice versa. Thus, $-\nabla^2\eta$ is referred to as a surface pressure drag on vorticity, and it acts as a pressure torque.

There are three kinds of mountain drag. One is due purely to a viscous surface stress at the interface, and it may be referred to as a frictional drag. The second is due to horizontal variation of the surface pressure stress in mountain regions, and it is referred to as a pressure drag. The third is based on the concept that orographically generated gravity waves may exert a drag on large scale flow, and it is called a gravity wave drag (Sawyer 1959; Lilly 1972; Palmer et al. 1986; McFarelane 1987; Durran 1995).

Studies of exchange processes between the atmosphere and mountains caused by the surface pressure stress, which appear in the literature, primarily concern the angular momentum balance of the zonal circulation (White 1949; Widger 1949; Holton 1992, p. 333). Recently, the pressure stress effects have also been studied locally. Davies and Phillips (1985) found pressure stresses along Alps to be a major sink of atmospheric angular momentum, whereas Hafner and Smith (1985) found that pressure drag may be up to 20 times larger than surface frictional drag during shorter period synoptic events like lee cyclogenesis and foehns. Czarnetzki and Johnson (1996) diagnosed the role of momentum transfer by pressure stresses during the life circle of a cyclone in the lee of the Rocky Mountains based on Bjerknes's tendency equation of absolute dynamic circulation. They found that the mountain-induced pressure torque is negative in the lower layers of the cyclone vortex, which indicates the transfer of absolute dynamic circulation from cyclone to the mountain. Czarnetzki (1997) also found that the pressure stress torque is generally much larger in magnitude than the frictional torque associated with lee cyclones. However, the method used by Hafner and Smith (1985) and Czarnetzki and Johnson (1996) can only be applied to diagnostic studies. How the local surface pressure stress torque can be computed exactly at each time step in a numerical model is unsolved. This problem can be resolved by using the geo-streamfunction.

(d) pressure drag solenoids

The value of $\nabla^2\eta_i$ can be estimated from the distribution of temperature and slopes of mountains. For this purpose, equation (3.2) can be rewritten as

$$\nabla^2\eta_i = \vec{k}\cdot(R\nabla T\times\nabla lnp_*) \tag{3.7}$$

Near high mountains, the horizontal variation of the surface pressure $p_*(x,y,t)$ caused by topography is much larger than that caused by synoptic disturbances. Thus, the contours of the height of the earth's surface, $H_*(x,y)$, is roughly parallel to the contours of p_* or $\ln p_*$. A low value center of p_* is located in the region where $H_*(x,y)$ is high.

The temperature $T(x,y,\sigma,t)$ in σ-coordinates can be separated into two parts as

$$T(x,y,\sigma,t) = T_{or}(x,y,\sigma,t) + T_{SL}(x,y,\sigma,t) \tag{3.8}$$

where $T_{SL}(x,y,\sigma,t)$ is the temperature in σ-coordinates under condition that there is no orography, i.e., the elevation of the earth's surface $H_*(x,y)=0$ and the surface pressure $p_* = p_{SL}(x,y,t)$, where $p_{SL}(x,y,t)$ is the sea level pressure. Thus, temperature T_{SL} is equivalent to the temperature in p-coordinates. The temperature $T_{or}(x,y,\sigma,t)$ is produced by orography only. Because if there is no orography, i.e., $H_*=0$ and $p_*=p_{SL}(x,y,t)$, then $T(x,y,\sigma,t)=T_{SL}$. From (3.8), we have $T_{or}=T(x,y,\sigma,t)-T_{SL}=0$.

The temperature T_{or} caused by orography is due to the temperature decrease with height, thus, the isotherms of T_{or} are almost parallel to the contours of $p_*(x,y,t)$ over a high mountain region. The cold center of T_{or} is located in the region where p_* is low (or the elevation is high). Thus, $\vec{k}\cdot(\nabla T\times\nabla lnp_*) = \vec{k}\cdot(\nabla(T_{SL}+T_{or})\times\nabla lnp_*) \cong \vec{k}\cdot(\nabla T_{SL}\times\nabla lnp_*)$, which means that the value of $\vec{k}\cdot(\nabla T\times\nabla lnp_*)$ can be estimated from the distribution of p_* and T_{SL}, where T_{SL} is similar to the temperature in p-coordinates.

If a mountain has an idealized elliptical shape with its long axis along the north-south direction and the temperature T_{SL} decreases from south to north, the contours of p_* and T_{SL} are shown in Fig.

1(e). In this case, the quantity $\vec{k} \cdot (\nabla T_{SL} \times \nabla lnp_*)$ is positive in the east slope of the mountain and negative in the west slope. If this distribution of $k \cdot (\nabla T_{SL} \times \nabla lnp_*)$ is solved from the Poisson equation (3.7), a negative region of the geo-streamfunction η will be located over the east side of the mountain and a positive region over the west side.

In Fig. 1(e), lines of constant T_{SL} cut across lines of constant $\ln p_*$ on constant σ-surfaces to form parallelograms which may be called pressure drag solenoids. The larger the number of these solenoids per unit area, the larger is the absolute value of $\vec{k} \cdot (\nabla T_{SL} \times \nabla lnp_*)$. Thus, the value of $\nabla^2\eta_i$ is closely related to the slope of the mountain and strength of synoptic fronts.

Figure 1(f) is the topography over a limited region of North America. Figures 2(a) and (b) are the observed sea level pressure (SLP) at 0000 UTC 10 and 11 March 1988, respectively. It is seen from Figs. 2(a) and (b) that a cyclone in the lee of Rockies develops in 24 hours, and its central pressure reduces from 999 hPa to 991 hPa.

The distributions of the inner part of the geo-streamfunction at 0000 UTC 10 and 11 March 1988 are shown in Figs. 2 (c) and (d), respectively. They are derived from (3.2) with zero Dirichlet boundary value. Figs. 2(c) and (d) show that $-\nabla^2\eta$ is negative on the east side of the Rocky Mountains and is positive on the west side, which is in some degree similar to that produced by the idealized case Fig. 1(e). It is also seen from Figs. 2(c) and (d) that the absolute value of the negative center of the geo-streamfunction develops in the lee of the Rockies during growth of the lee cyclone. Thus, transfer of vorticity from the atmosphere to the surface of the Rockies also grows in the region where the lee cyclone develops.

4. The equivalent geopotential and geo-streamfunction related to lee cyclone development

As pointed out recently by Smith et al. (1997): "One of the key problems with the current operational models is the inability to predict the timing and strength of lee cyclogenesis when synoptic-scale troughs cross the Rockies. This has been identified by NCEP as a common problem for the NGM, the Eta, and the global model during winter. The 24-48-h forecasts consistently predict early and excessive cyclogenesis situated too far to the north." It is natural to infer that these forecast errors may be caused by inaccurate computation of vorticity change produced by mountains. An accurate method of computing the mountain-induced pressure stress torque and the horizontal PGF may have a significant influence on the prediction of the timing and strength of lee cyclogenesis.

Let a variable $\psi_{adv,i}$ satisfy the Poisson equation $\nabla^2\psi_{adv,i} = \Omega_{adv}$ with zero Dirichlet boundary value. The solution of (3.5) with zero boundary value is written as

$$\frac{\partial\psi_i}{\partial t} = -f_0\chi_i - \eta_i + \psi_{adv,i} \tag{4.1}$$

The tendency of streamfunction $\partial\psi_i/\partial t$ in the center of the lee cyclone produced by $-\eta_i$ alone increases from $1.928 \times 10^2\, m^2s^{-2}$ to $3.170 \times 10^2\, m^2s^{-2}$ in 24 hr. If these values are used to estimate the tendency of the equivalent geopotential based on the geostrophic approximation $\phi_{ei} \cong f_0\psi_i$ and $f_0 \cong 10^{-4}s^{-1}$, their corresponding values increase from 20.5 tc 33.6 $gpm/(3hr)$ in the centers of the lee cyclone. These positive tendency values show that transfer of vorticity from the atmosphere to

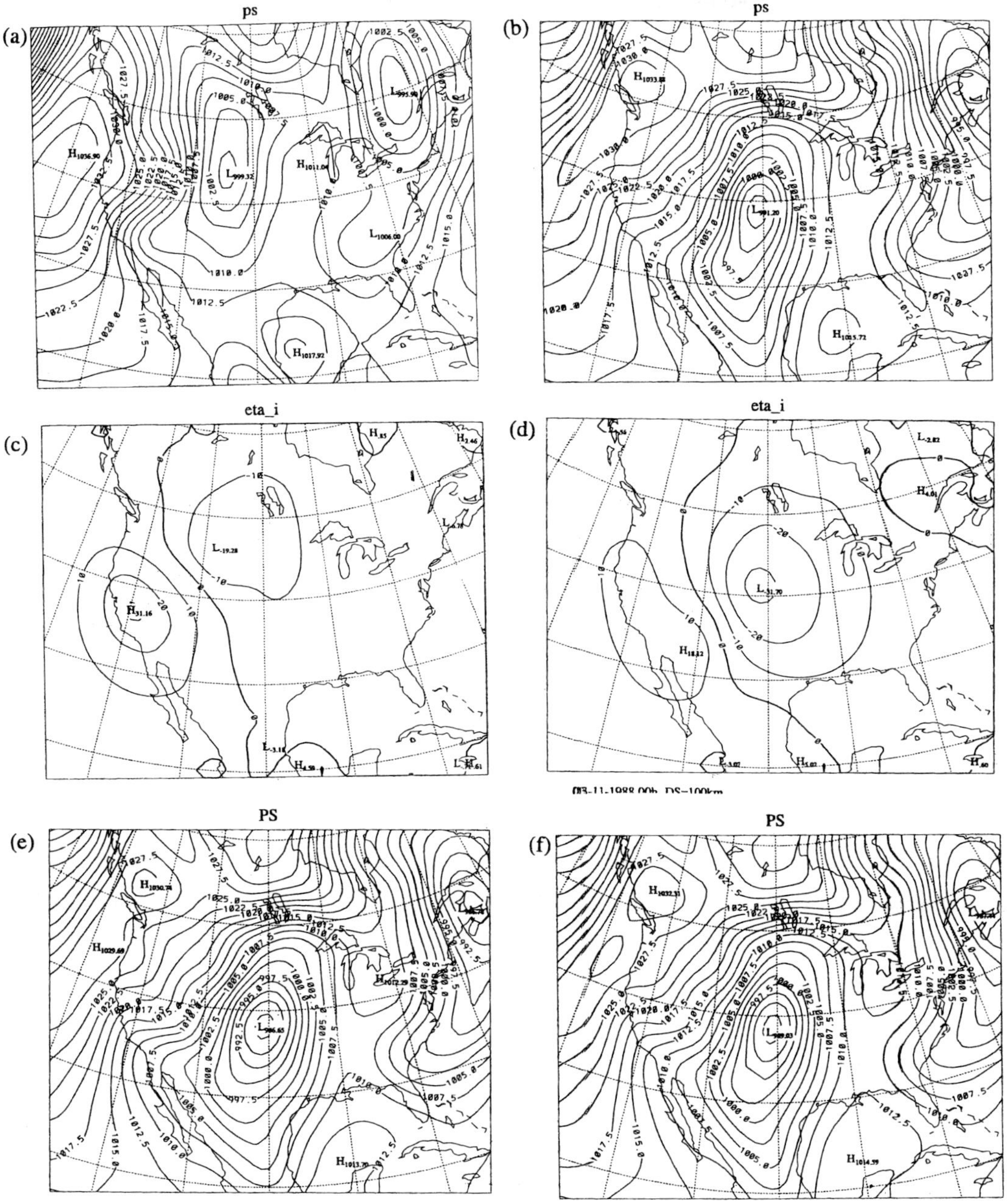

Fig.2 (a) The observed sea level pressure (SLP) at 0000 UTC 10 March 1988; (b) The observed SLP at 0000 UTC 11 March 1988; (c) The inner part of the geo-streamfunction at 0000 UTC 10 March 1988 at σ=0.995 (in 10 $m^2 s^{-2}$ and contour interval: $10 \times 10 m^2 s^{-2}$); (d) Same as (c) but at 0000 UTC 11 March 1988; (e) A 24 h forecast map of the SLP at 0000 UTC 11 March 1988 predicted by the model using the generalized geopotential without ϕ_e and η; (f) Same as (e) but predicted by the model using ϕ_e and η.

the Rockies may play a considerable role in timing and strength of lee cyclone development.

The H-F spectral model developed by Chen et al. (1997a) uses the generalized geopotential without ϕ_e and η. Recently, the equivalent geopotential and geo-streamfunction have been used in this H-F spectral LAM. In this new version, the inner parts of the streamfunction, velocity potential and equivalent geopotential, (ψ_i, χ_i and ϕ_{ei}) are the basic predicted variables. A preliminary test is encouraging. Figure 2(e) is a 24 h forecast map of the SLP at 0000 UTC 11 March 1988 predicted by the H-F spectral model using the generalized geopotential without ϕ_e and η. The horizontal grid-length is 100 km and the initial condition at 0000 UTC 10 March 1988 for the simulation is shown in Fig. 2(a). The observed SLP at 0000 UTC 11 March 1988 is shown in Fig. 2(b). Comparing Figs. 2(e) and (b), it can be seen that the predicted result in general is very good, but the central pressure decreases excessively. The central pressure in Fig. 2(e) is about 987 hPa. Figure 2(f) is a 24 h forecast map of the SLP at 0000 UTC 11 March 1988 predicted from the same initial value by the H-F spectral model with ϕ_e and η. The other parts of the model do not change. The prediction in Fig. 2(f) is better than that in Fig. 2(e), and the central pressure has improved to be about 989 hPa.

5. Conclusion

1. The horizontal PGF $\vec{G}$ in σ-coordinates can be separated into its irrotational and rotational components and expressed by $-\nabla\phi_e - \vec{k}\times\nabla\eta$, where ϕ_e is referred to as an equivalent geopotential, and η is called a geo-streamfunction. The equivalent geopotential $\phi_e(x,y,\sigma,t)$ can be used in σ-coordinates in the same way as $\phi(x,y,p,t)$ is used in p-coordinates, and it can be analyzed on constant σ surfaces for synoptic analyses. The equivalent geopotential and streamfunction on the constant σ surface approximately satisfy the geostrophic relation for the synoptic scale motion. The small difference between two large terms over steep slopes on the right hand of (1.2) is eliminated automatically by using expression (1.3).

2. The quantity $-\nabla^2\eta$ represents the exchange of vorticity between the mountains and atmosphere. In the region where $-\nabla^2\eta < 0$, the vorticity decreases in the atmosphere and it transfers to the surface of mountains through the surface pressure, and *vice versa*. The quantity $-\nabla^2\eta$ is referred to as a surface pressure drag on vorticity, and it acts as a pressure torque. Thus, the local surface pressure stress torque can be computed exactly at each time step in a numerical model by using the geo-streamfunction.

4. A computed example shows that a negative center of the geo-streamfunction develops in the lee of the Rockies during the growth of a lee cyclone, and the vorticity transfer from the atmosphere to the surface of the Rockies also grows in the region where the lee cyclone develops. In this example, the prediction of the strength of the lee cyclone is improved by using the equivalent geopotential and geo-streamfunction.

References

Bromwich, D. H., Q.-S. Chen, Y. Li and R. I. Cullather, 1999: Precipitation over Greenland and its relation to the North Atlantic Oscillation. *J. Geophys. Res.*, **104,** (D18), 22,103-22,115.

Chen, Q.-S., and Y.-H. Kuo, 1992: A harmonic-sine series expansion and its application to the partitioning and reconstruction problem in a limited area. *Mon. Wea. Rev.*, **120,** 91-112.

Chen, Q.-S., Y.-H. Kuo and D. H. Bromwich, 1996: A balanced ageostrophic initialization with a fixed external wind lateral boundary value for limited-area models. *J. Meteor. Soc. Japan.* **74**, 325-342.

Chen, Q.-S., L. Bai and D. H. Bromwich, 1997a: A harmonic-Fourier spectral limited-area model with an external wind lateral boundary condition. *Mon. Wea. Rev.*, **125**, 143-167.

Chen, Q.-S., D. H. Bromwich and L. Bai, 1997b: Precipitation over Greenland retrieved by a dynamic method and its relation to cyclonic activity. *J. Climate,* **10,** 839-870.

Chen, Q.-S., And D. H. Bromwich, 1999: An equivalent isobaric geopotential height and its application to synoptic analysis and to a generalized ω-equation in σ–coordinates. *Mon. Wea. Rev.*, **127**, 145-172.

Colle B. A., K. J. Westrick, and C. F. Mass, 1999: Evaluation of MM5 and Eta-10 precipitation forecasts over the Pacific Northwest during the cool season. *Weather and Forecasting,* **14,** 137-154.

Corby, G. A., A. Gilchrist and R. L. Newson, 1972: A general circulation model of the atmosphere suitable for long period integration. *Quart. J. Roy. Met. Soc.,* **98,** 809-832.

Czarnetzki, A. C., and D. R. Johnson, 1996: The role of terrain and pressure stresses in Rocky Mountain lee cyclones. *Mon. Wea. Rev.*, **124**, 553-570.

Czarnetzki, A. C., 1997: Regional mountain torque estimates over the Rocky Mountains in lee cyclones. *J. Atmos. Sci.,* **54**, 1986-1997.

Davies, H. C. 1976: A lateral boundary formulation for multi-level prediction models. *Quart. J. Roy. Met. Soc.,* **102,** 405-418.

Davies, H. C., 1983: Limitations of some common lateral boundary schemes used in regional NWP models. *Mon. Wea. Rev.,* **111,** 1002-1012.

Davies, H. C., and R. E. Turner, 1977: Updating prediction models by dynamic relaxation: An examination of the technique. *Quart. J. Roy. Met. Soc.,* **103**, 225-245.

Davies, H. C., and P. D. Phillips, 1985: Mountain drag along the Gotthard section during ALPEX. *J. Atmos. Sci.,* **42,** 2093-2109.

Dempsey, D. and C. Davis, 1998: Error analyses and test of pressure gradient force schemes in nonhydrostatic, mesoscale model. Preprints, *12th Conf. on Numerical Weather Prediction*, 11-16 January 1998, Phoenix, AZ, Amer. Meteor. Soc., 236-239.

Durran, D. R., 1995: Do breaking mountain waves decelerate the local mean flow? *J. Atmos. Sci.,* **52,** 4010-4032.

Errico, R. M., and G. T. Bates, 1988: Implicit normal-mode initialization of the PSU/NCAR mesoscale model. *NCAR Technical Note*, NCAR/TN-312+1A, 112 pp.

Gary, J. M., 1973: Estimate of truncation error in transformed coordinate, primitive equation atmospheric models. *J. Atmos. Sci.,* **30,** 223-233.

Grell, G. A., J. Dudhia, and D. R. Stauffer, 1994: A description of the fifth-generation Penn State/NCAR mesoscale model (MM5). *NCAR Technical Note*, NCAR/TN-398+STR, 122 pp.

Hafner, T. A., and R. B. Smith, 1985: Pressure drag on the European Alps in relation to synoptic events. *J. Atmos. Sci.,* **42,** 562-575.

Holton, J. R., 1992: *An introduction to dynamic meteorology.* Third edition, Academic Press, INC., 507pp.

Janjic, Z. I., 1990: The step mountain coordinate: physical package. *Mon. Wea. Rev.,* **118**, 1429-1443.

Janjic, Z. I., 1994: The step mountain eta coordinate model: Further development of the convection, viscus sublayer, and turbulence closure schemes. *Mon. Wea. Rev.,* **122**, 927-945.

Kar, S. K., and R. P. Turco, 1995: Formulation of a lateral sponge layer for Limited-area shallow water models and an extension for the vertically stratified case. *Mon. Wea. Rev.,* **123**, 1542-1559.

Lilly, D. K., 1972: Wave momentum flux. A GARP problem. *Bull. Amer. Met. Soc.,* **53,** 17-23.

Lynch, P., 1989: Partitioning the wind in a limited domain. *Mon. Wea. Rev.,* **117**, 1492-1500.

McFarlane, N. A., 1987: The effect of orographically excited gravity wave drag on the general circulation of the lower stratosphere and troposphere. *J. Atmos. Sci.,* **44,** 1775-1800.

Mesinger, F., 1984: A blocking technique for representation of mountains in atmospheric models. *Rev. Meteorol. Aeronautica.* **44**, 195-202.

Mesinger, F., Z. Janjic, S. Nickovic, D. Gavrilov, and D. G. Deaven, 1988: The step mountain coordinate: model description and performance for cases of Alpine lee cyclogenesis and for a case of an Appalachian redevelopment. *Mon. Wea. Rev.,* **116**, 1493-1518.

Nakamura, H., 1978: Dynamical effects of mountains on the general circulation of atmosphere. I. Development of finite-difference scheme suitable for incorporating mountains. *J. Meteor. Soc. Japan.* **56**, 317-340.

Palmer, T. N., G. J. Shuts, and R. Swinbank, 1986: Alleviation of a systematic westerly bias in general circulation and numerical weather prediction models through an orographic gravity wave drag parameterization. *Quart. J. Roy. Meteor. Soc.,* **112**, 1001-1039.

Perkey, D. J., and C. W. Kreitzberg, 1976: A time-dependent lateral boundary scheme for limited-area primitive equation models. *Mon. Wea. Rev.,* **104**, 744-755.

Sasheghi, K. D. and R. V. Madala, 1993: Application of vertical mode initialization to a limited-area model in flux form. *Mon. Wea. Rev.*, **121**, 207-220.

Sawyer, J. S., 1959: The introduction of the effects of topography into methods of numerical forecasting. *Quart. J. Roy. Meteor. Soc.*, **85**, 31-43.

Smith, R. B., J. Paegle, T. Clark, W. Cotton, D. Durran, G. Forbes, J. Marwitz, C. Mass, J. McGinley, H.-L. Pan and M. Ralph, 1997: Local and remote effects of mountains on weather: Research needs and opportunities. *Bull. Amer. Met. Soc.*, **78,** 877-892.

Sundqvist, P. J., 1976: On vertical interpolation and truncation in connection with use of sigma models. *Atmosphere,* **14,** 37-52.

Takano, I., H. Nakamura and Y. Tatsumi, 1990: Non-linear normal mode initialization for a spectral limited-area model. *J. Meteor. Soc. Japan.* **68**, 265-280.

Temperton, C., 1988: Implicit normal mode initialization. *Mon. Wea. Rev.*, **116**, 1013-1131.

Warner, T. T., R. A. Peterson, and R. E. Treadon, 1997: A tutorial on lateral boundary conditions as a basic and potentially serious limitation to regional numerical weather prediction. *Bull. Amer. Met. Soc.*, **78,** 2599-2617.

White, R. M., 1949: The role of mountains in the angular momentum balance of the atmosphere. *J. Meteor.*, **6,** 353-355.

Widger, W. K., 1949: A study of flow of angular momentum in the atmosphere. *J. Meteor.*, **6,** 291-299.

Abstract

The horizontal PGF $\vec{G}$ in σ-coordinates can be separated into its irrotational and rotational components and expressed by $-\nabla\phi_e - \vec{k}\times\nabla\eta$, where ϕ_e is referred to as an equivalent geopotential, and η is called a geo-streamfunction. The equivalent geopotential $\phi_e(x,y,\sigma,t)$ can be used in σ-coordinates in the same way as $\phi(x,y,p,t)$ is used in p-coordinates. The small difference between two large terms over steep slopes on the right hand of (1.2) is eliminated automatically by using expression (1.3). The quantity $-\nabla^2\eta$ is referred to as a surface pressure drag on vorticity, and it acts as a pressure torque. In the region where $-\nabla^2\eta < 0$, the vorticity decreases in the atmosphere and it transfers to the surface of mountains through the surface pressure stress, and *vice versa*. The surface pressure stress torque can be computed exactly at each time step in a numerical model by using the geo-streamfunction. A computed example shows that a negative center of the geo-streamfunction develops in the lee of the Rockies during the growth of a lee cyclone, and the vorticity transfer from the atmosphere to the surface of the Rockies also grows in the region where the lee cyclone develops. The prediction of the strength of the lee cyclone is improved by using the equivalent geopotential and geo-streamfunction.

~ Typhoon ~

A STATISTICAL ANALYSIS OF TYPHOONS AFFECTING TAIWAN IN 1897-1996

Shinn-Liang Shieh and Shih-Ting Wang
Central Weather Bureau, Taipei

ABSTRACT

The typhoon-track data over the Western North Pacific (WNP) were collected since 1897 when routine hourly observations were first available from the weather stations around Taiwan area. Among the 666 cases recorded in 1897-1996, 374 either approached or made landfalls on Taiwan. These storms are thus referred to as the "invading" typhoons in this study, and are also further categorized into 9 sub-groups mainly according to their moving paths and partly their origins. The averaged yearly total invading typhoons are 3.7, but the number can be as high as 10 or none at all for individual years. The yearly total of the invading typhoons seemed to exhibit a significant 30-year oscillation according to a power spectral analysis. We have also ranked the intensity of storms in terms of locally observed wind and precipitation data. From the statistics of the first 10 such most intensive invading typhoons in each local area, the occurrence of local maximum wind and rainfall are often strongly related to local topography.

The orographic effects in changing the storm motion when passing over the Central Mountain Range (CMR) were characterized into 15 conceptual models from the historical data set, which can then be readily applied to operational use. In analyzing the occurring frequency of the storms that locally reached the hazardous intensity for office-closing, we also found the great importance to local wind forecasts of the blocking effect by the CMR and the channel effect associated with the Taiwan Strait. As to the distribution of precipitation, it is again noted to be mostly phase-locked with the topography of Taiwan.

1. INTRODUCTION

The damage caused by the invading typhoons* is one of the most hazardous natural disasters in Taiwan. Therefore, to do typhoon-related studies is always the main emphasis at the Central Weather Bureau. However, being an island with a rugged and complicated topography, it is very difficult to predict even the track of approaching typhoons, not to mention the would-be local distributions of wind or precipitation associated with the storms. At current status, we still heavily rely on the statistical information for operational references.

The weather stations on the island were first set up by the Japanese in 1897. The hourly observation system, especially for approaching typhoons, was then implemented and the records were well kept to present. Since the data collection has reached the one-hundred-year milestone in 1996, we thus focus on reporting the results from a thorough analysis of those invading typhoons (either approaching or making landfalls on Taiwan) occurred in these past 100 years (i.e., 1897-1996).

2. DATA

The local observation data related to the invading typhoons in Taiwan were collected and analyzed since 1897. The total number of local weather stations during 1930-40's was nearly the same as present. It is noted that some half-an-hour observations or even

---

* Here typhoon is loosely defined to include tropical storm (wind speed $\geqq 18$ m s^{-1})

10-minute ones were available, especially during typhoon passage.

In addition, hourly typhoon track data are established as follows: a) All tracks on land are re-determined by re-analyzing all the available observation data (Wang 1992 and Shieh et al. 1998). b) Tracks over the neighbouring seas are obtained from several sources: For typhoons before 1949, best tracks are much relied on both Chin (1958) of the Hong-Kong Observatory and Hsu (1979) of the Central Weather Bureau. For typhoons in 1949 and after, best tracks are established (see Wang 1992 and Shieh et al., 1998) based on those issued by the Joint Typhoon Warning Center (JTWC) at Guam.

3. STATISTICAL ANALYSES

(a) Occurring frequency

There are totally 374 cases of typhoons affecting Taiwan during the period of 1897-1996. They can be generally categorized into 9 sub-groups in terms of their tracking patterns (Fig. 1). The first 5 types are all westward moving typhoons except that the centres of storms in the first and the fifth ones do not make landfalls on Taiwan but by-passing the island over neighboring seas either to the north (1st) or the south (5th). The storms belong to the 2nd, 3rd, and 4th do make landfalls on the northern, central, and southern parts of Taiwan, respectively. Both types 6 and 7 are characterized by northward re-curving storms passing Taiwan through either east or west sides of the CMR. Finally, the clusters number 8 and 9 are storms coming from the South China Sea areas, except that those of former group are originally coming from the WNP, not locally formed as those in the 9th group. The distribution of occurring frequency for each sub-group is also shown in Fig. 1. On the other hand, Fig. 2 illustrates the geographic distribution of occurring frequency for those typhoons that made landfalls on Taiwan. Obviously, most of the storms made landfalls on the eastern parts of the island, and only a few were from the west and the southwest.

The time series of yearly total invading typhoons are plotted in Fig. 3. There are 3.7 affecting storms per year, but can be as frequently as 10 (in 1959), or none at all (in 1941). The standard deviation (indicated by the dashed lines in Fig. 3) is around 1.7. There may exist a 30-year cycle as suggested from the power spectral analysis (not shown).

(b) Statistics in terms of wind intensity

Due to data limitation before 1949, we have selected the first 10 strongest storms (with local wind speed over 17.2 m/s) from the period of 1949-96 only. Their moving tracks are plotted in Fig. 4. All of them were westward moving typhoons that made landfalls on eastern Taiwan. These storms were sufficiently strong that they passed continuously over the mountains. As a matter of fact, they were all categorized as "strong" typhoons before making landfall, but quickly turning into "moderate" ones after landfall.

(c) Statistics in terms of precipitation

We have compiled the first 10 strongest storms in terms of total rainfall amount during their individual affecting period for the northern, eastern and western parts of Taiwan, respectively. It is found that the topographic effect is the most dominant factor in modulating the corresponding patterns of track for different areas. All the first 10 storms causing a great amount of rainfall over the eastern Taiwan (Fig. 5a) were associated with winter monsoon systems, except for Typhoon Nadine (1968), which was due to the interaction with the circulation of Typhoon Mary. These typhoons are mostly with tracks characterized as sub-groups 5 or 6 defined in the previous section. Similarly, seven of the ten storms that affected the northern Taiwan most (Fig. 5b) were also associated with the winter monsoon systems. For example, Typhoon Lynn that had resulted in a severe flood at Taipei in October 1987, is a typical case. On the other hand, the typhoons causing heavy rainfalls over the western Taiwan did not have the coupling effects (Fig. 5c). The notorious Typhoon Herb in 1996 is the leading example. In fact, it is Typhoon Herb that recorded the highest rainfall amount in the period examined.

The occurrence of strong local wind is generally associated with the direct impinging of a strong typhoon. However, the occurrence of local heavy rainfall does not necessarily correspond to a typhoon with strong wind intensity (see Fig. 4). For example, the centres of those typhoon that caused a lot of rainfall over the eastern and northern parts of the island were mostly located over the neighboring sea, and sometime may even be quite distant from Taiwan (cf. Fig. 5).

4. APPLICATIONS

(a) Orographic effects on typhoon motion

From analyses of hourly synoptic charts, topographically induced changes in typhoon motion can often be observed during the time when centres of storms moving across the mountain areas. Fig. 6 summarizes the 15 major types of tracking change caused by the CMR from the historical data set. It is seen from Fig. 6 that the typhoon track can sometimes be bent, become quasi-stationary, be replaced by the secondary centres, or gradually disappeared while trying to move over the CMR. Actually, the time-scale for a typhoon passing a mesoscale mountain like the CMR is only an order of a few hours, accordingly, only careful analyses with hourly data can made these results meaningful and fruitful. Since the pronounced deviation or distortion of a typhoon track as mentioned here certainly involves many problems, a detailed discussion on the subject is beyond the scope of this paper. However, further study to these problems may be addressed in the future. All these changes mentioned will directly affect the accuracy of our forecasts. Those conceptual models can readily be used as an analog forecast for reference.

(b) Local wind distribution

According to the 211 cases of invaded typhoons in 1949-1996, we have shown in Fig. 7 the distribution of occurring frequency of storms that reached the local office-closing intensity (i.e., a mean wind over 7 grade in Beaufort scale, or a gusting wind over 11 grade). From Fig. 7, it is apparent that the orographic effect is a major consideration to an accurate local wind forecast. For example, the occurring frequency is relatively low (less than 10%) at Taichung due to the blocking of CMR, while at Wuchi, another station only 30 km away from Taichung, the occurring frequency jumps to 73% primarily due to the channel effect in the Taiwan Strait.

(c) Local rainfall distribution

As to the rainfall associated with an invaded typhoon, it can be influenced by a much more complicated local effect. Fortunately, we have a dense network (about 900 station in Taiwan) in recording the precipitation data, so that the local distribution of rainfall corresponding to a particular typhoon can be made available for further analysis and applications. Fig. 8 shows the distribution of total rainfall amounts brought with Typhoon

Herb in 1996 that caused the most devastating damages over the northern and central mountain regions in Taiwan history. It had set a record-high daily rainfall at the Alishan station at 1094 mm in July 31, 1996. It is also Typhoon Herb that has poured down the most total precipitation over the Taiwan area since 1897. Although the total rainfall set at Alishan was 1987 mm, the total amount over the champaign areas of eastern Taiwan on the other hand did not even accumulate over 50 mm, according to Fig. 8. From the tracking information and associated streamlines (see the lower-right corner of Fig. 8), there is no surprise why the rainfall distribution is so deviated. The prevailing westerly wind blew into the western (upslope) mountain region thus causing torrential rainfall, while the eastern part of the island was on the lee-side so that there was not much rainfall. In fact, typhoons with similar tracks passing northern Taiwan will cause maximum rainfalls over the same western mountainous areas.

From the discussions above, we have noted that the changes in typhoon motion and local distributions of wind and precipitation associated with the invading typhoons are closely related to Taiwan's topography. Those storms with similar track, intensity and size will bear some resemblance in their associated distribution of local wind and rainfall. This explains why the historical statistics are applicable to practical purpose.

5. CONCLUDING REMARKS

It is the first try to compile a one-hundred-year data set. The invasion of typhoons is the most severe natural disaster in Taiwan that causes an averaged property loss of about 1.3 billion NT dollars per year. It is thus hoped that the results from this study can be beneficial to operational applications in mitigating the damages that would be caused, and to future typhoon studies, especially those related to the interaction of typhoons with the complex topography in Taiwan.

ACKNOWLEDGMENTS. The authors wish to thank Drs. Jen-Cheng Chang, Ben Jong-Dao Jou and Li Peng for their valuable discussion and fruitful suggestion. The authors also wish to thank Hsiao-Lin Chung, Jin-Liarn Lee and Chang Yu for their assistance in the preparation of this paper.

REFERENCES

Chin, P. C., 1958: *Tropical cyclones in the Western Pacific and China Sea Area from 1884 to1953.* Hong Kong Observatory TM7, 93pp.

Hsu, C.-H., 1979: *Tropical cyclone tracks in the Western Pacific from 1892 to 1977.* Technical report of the NSC, 573pp.

JTWC, 1997: *Digital Database for the Tropical cyclone tracks in the Western North Pacific from 1947 to 1996.*

Shinn, S.-L., S.-T., Wang, M.-D. Cheng, and T.-C. Yeh, 1998: *Tropical cyclone tracks over Taiwan and its vicinity from 1897 to 1996.* Technical report of the Central Weather Bureau, Taipei, 497pp.

Wang, S.-T., 1992: *An integrated study of the impact of the orography in Taiwan on the movement, intensity, structure, wind and rainfall of invading typhoons.* Research report No. 80-73, NSC 80-0414-P052-02B, 285pp.

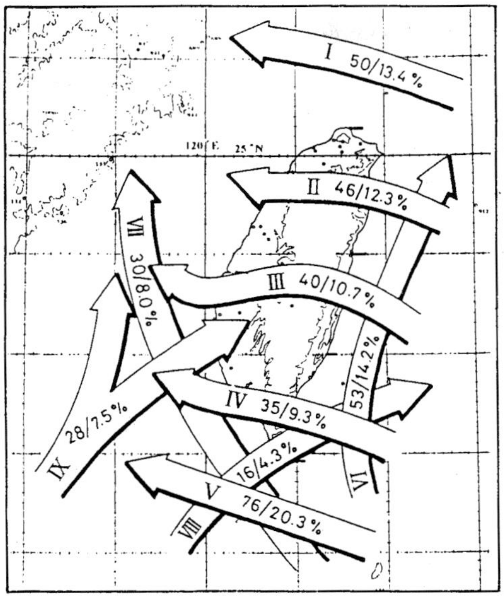

Fig. 1. The occurring number and frequency of the 9 sub-groups of the invading typhoons (in terms of their tracking patterns) in 1897-1996.

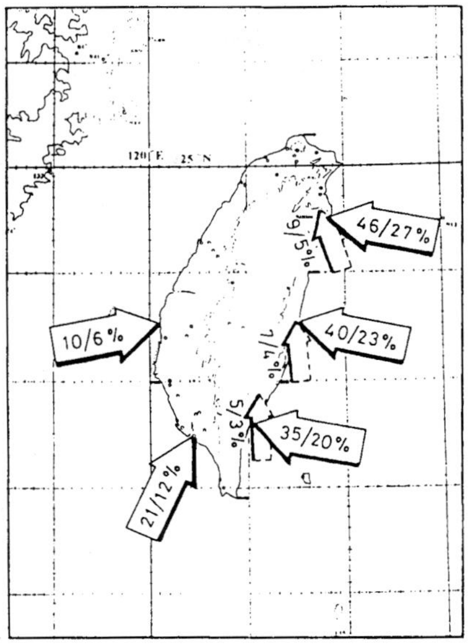

Fig. 2. The occurring number and frequency of typhoons mak-ing landfalls over different geographic locations in 1897-1996.

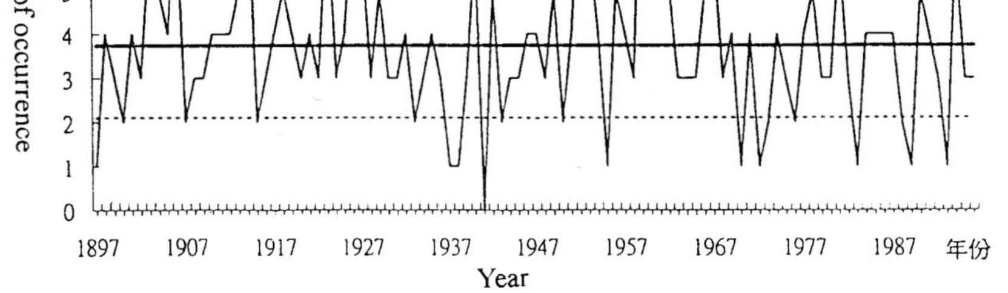

Fig. 3. Time series of the yearly total invading typhoons in 1897-1996. (The grand mean is indicated by the thick solid line, and the standard deviation the dashed lines.)

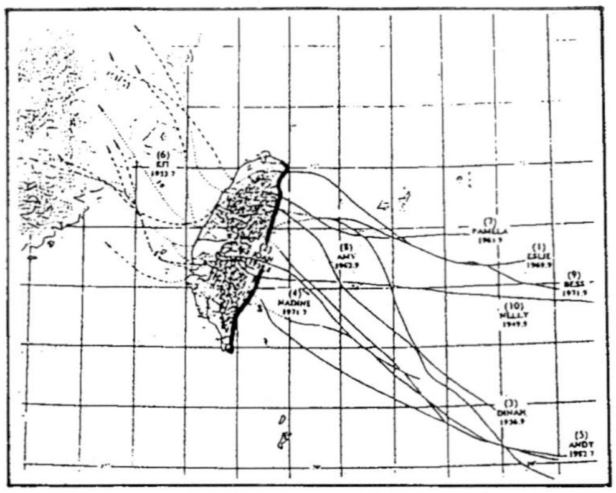

Fig. 4. The best tracks of the first 10 strongest invading typhoons (in terms of locally observed wind intensity) in 1949-1996.

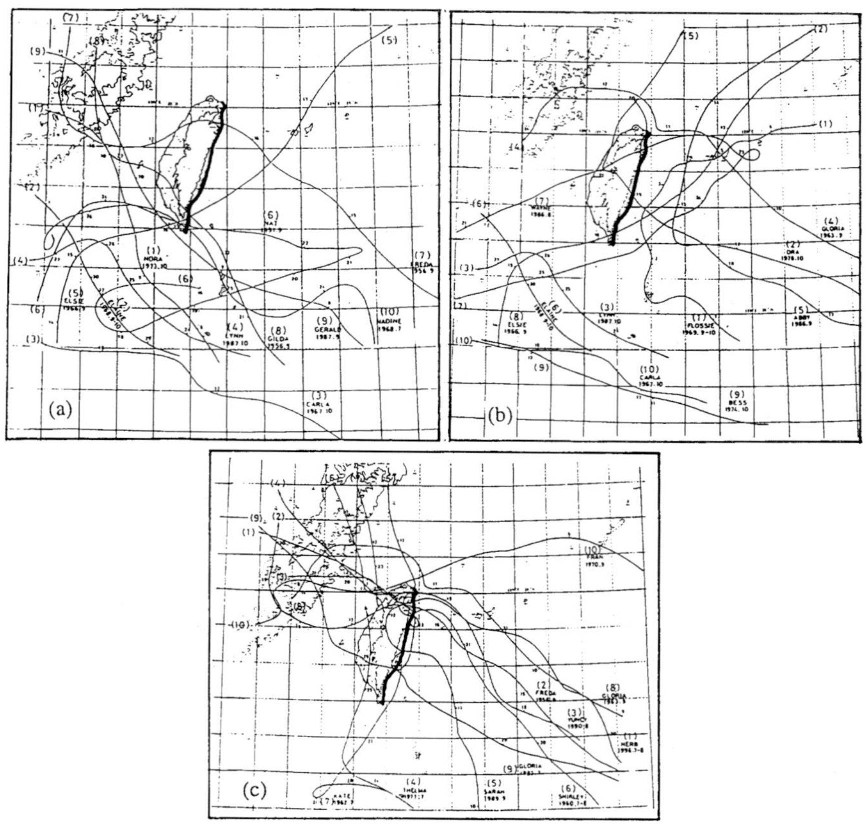

Fig. 5. The best tracks of the first 10 invading typhoons in 1949-1996 with most rainfalls over (a) the eastern, (b) the northern, and (c) the western Taiwan. The arabic number associated with each track represents the standing of that typhoon in terms of rainfall amount.

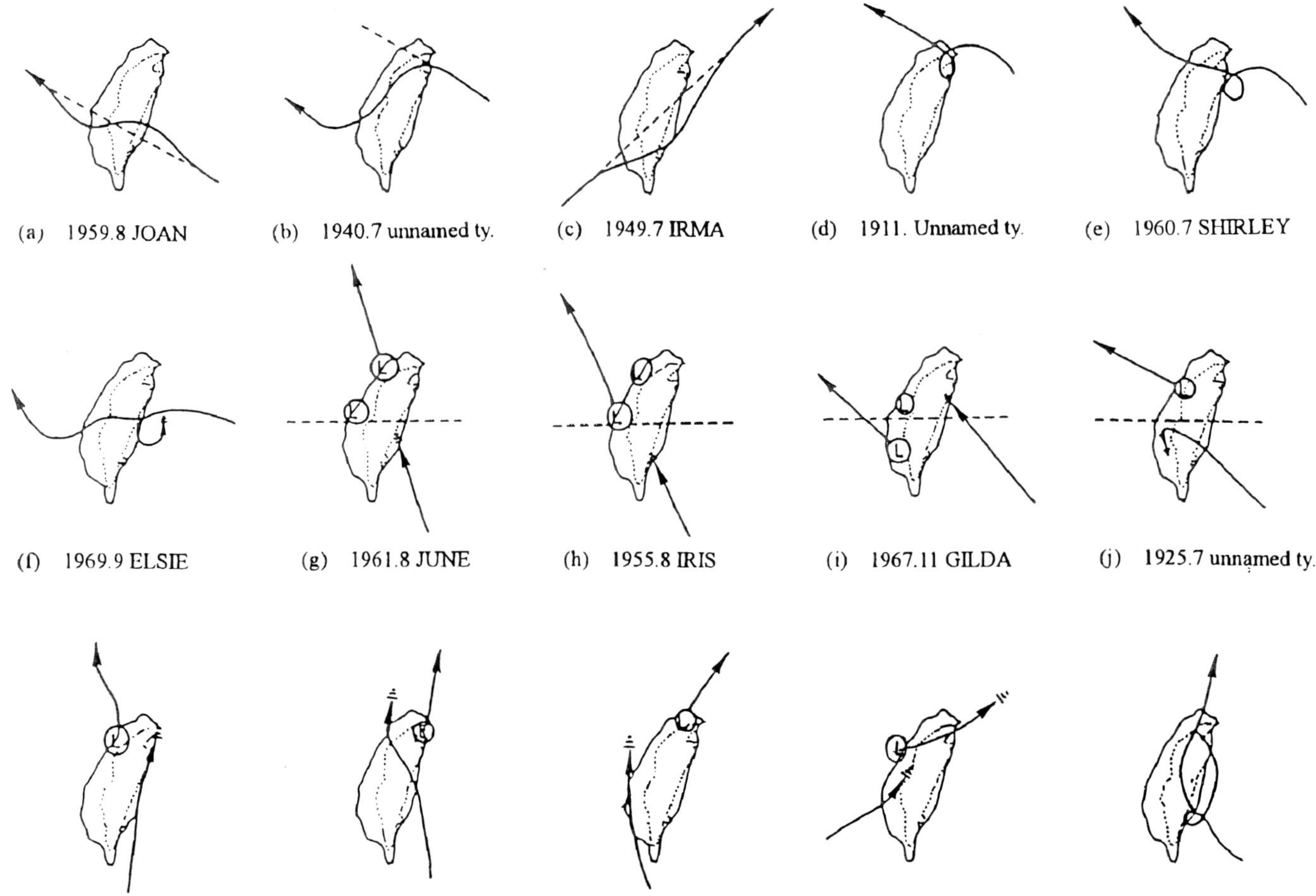

Fig. 6. The 15 conceptual models of track changes during the passage of typhoon centers over the Central Mountain Range in1897-1996.

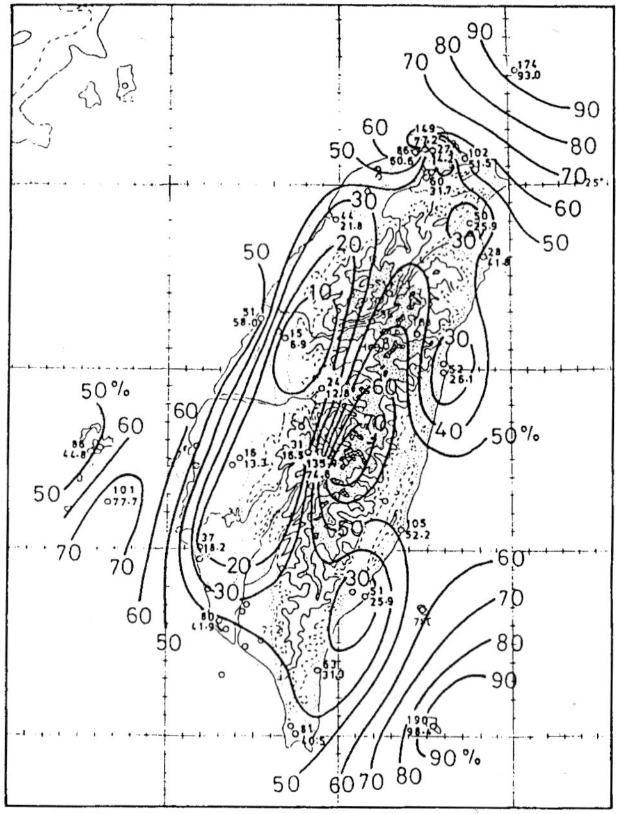

Fig. 7. The occurring frequency of the invading typhoons that reached the local office-closing intensity in 1949-1996.

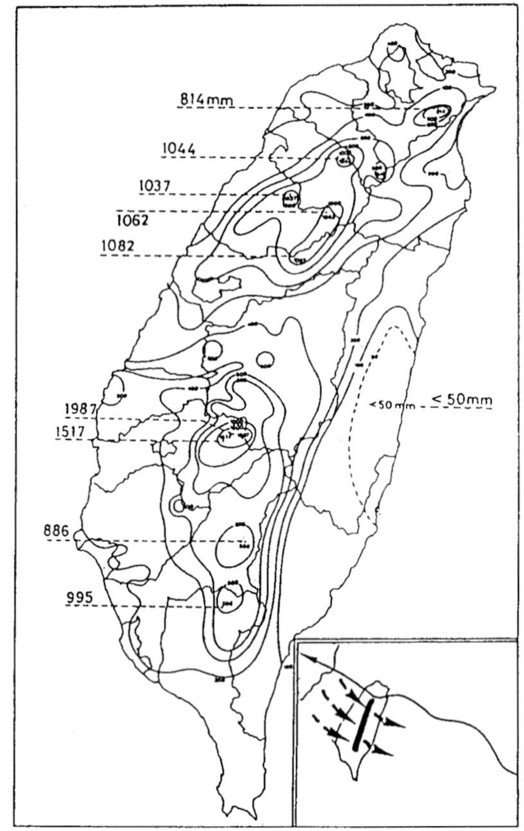

Fig. 8. The total rainfall (in mm) distribution associated with Typhoon Herb during its affecting period in July 31-August 1, 1996.

ON THE FORMATION OF THE HURRICANE EYE

Da-Lin Zhang
Department of Meteorology, University of Maryland, College Park, Maryland 20742
Yubao Liu
NCAR/Research Application Program, P.O. Box 3000, Boulder, CO 80307-3000
M. K. Yau
Department of Atmos. & Oceanic Sci., McGill University, Montreal, Quebec H3A 2K6

ABSTRACT

In this study, the vertical momentum budget is examined by inverting three-dimensional (3D) buoyancy and dynamic perturbation pressures using a high-resolution, explicit simulation of Hurricane Andrew (1992), in order to gain insight into the processes leading to the subsidence warming in the eye and the vertical lifting in the eye-wall.

It is found that a large portion of surface perturbation pressures is caused by the moist-adiabatic warming in the eye-wall and the subsidence warming in the eye. However, the associated buoyancy-induced pressure gradient force (PGF) is mostly offset by the buoyancy force, and their net effect is similar in magnitude but opposite in sign to the dynamically-induced PGF. Of importance is that the dynamically-induced PGF points downward in the eye to account for the maintenance of the general descent. But it points upward in the outer half portion of the eye-wall, particularly in the northern sector, to facilitate the lifting of high-θe air in the lower troposphere. Furthermore, this dynamic force is dominated by the radial shear of tangential winds. Based on this finding, a new theoretical explanation, different from previously reported, is advanced for the relationship among the subsidence warming in the eye, and the rotation and vertical shear in the eye-wall.

1. INTRODUCTION

Earlier theoretical studies suggested that the weak subsidence in the eye is caused by radially outward advection of the eye air into the eye-wall as a result of super-gradient flows (Malkus 1958; Kuo 1959). The existence of super-gradient flows, particularly near the radius of maximum winds (RMW), has been found by Gray and Shea (1973) from a composite analysis of many radial flight data. Smith (1980) demonstrated with an axi-symmetric vortex model that the subsidence warming in the eye is mechanically driven by the decreased vertical shear in tangential winds in the eye-wall as a consequence of thermal wind balance. This implicitly implies that the warm-cored eye is a result of the development of negative vertical shear in the eye-wall.

In contrast, Shapiro and Willoughby (1982) viewed the descent in the eye as compensational subsidence associated with vigorous convection in the eye-wall. The air mass in the eye-wall would detrain at the top, sink into the eye and flow back into the eye-wall at the bottom. The idea of subsidence caused by the eye-wall convection was questioned by Emanuel (1997). He used a balance model to demonstrate that the convectively induced subsidence cannot by itself raise the vertically averaged temperature in the eye to values greater than that inside the eye-wall. He then proposed that the amplification of the swirling flow is strongly frontogenetic, and it would lead to a thermal and momentum discontinuity at the inner edge of the eyewall. This discontinuity may collapse and cause the radial turbulent diffusion of momentum, thereby inducing mechanically a secondary circulation with the descending motion in the eye. Willoughby (1998) argued that such collapse may be prevented by mixing. Meanwhile, he modified the conceptual model of Shapiro and Willougby (1982), based on the analyses of soundings taken at the center of some hurricanes, and suggested that the descending air above an inversion (at an altitude of 2-3 km) in the eye has been there since the eye is first formed. As the storm deepens, the slow descent in the eye, occurring at a few cm s^{-1}, represents the response to the net mass loss caused by the moist downward cascade inside the eye and the convective updrafts in the eyewall above the inversion.

In this study, we attempt to understand the vertical motion in the hurricane eye and eyewall through vertical momentum budget. This is achieved by inverting three-dimensional (3D) perturbation pressures in the inner-core region of a hurricane using a 72-h, fully explicit, high-resolution ($\Delta x = 6$ km) simulation of Hurricane Andrew (1992) with MM5 (see Liu et al. 1997). The model reproduces well the track and intensity, the structures of the eye, eyewall, spiral rainbands, the radius of maximum wind (RMW), and other inner-core features as compared to available observations and the results of previous hurricane studies. The next section presents procedures for computing the vertical momentum budget in cylindrical coordinates. Section 3 describes the nature of dynamically driven vertical motion in the eye and eyewall. A summary and concluding remarks are given in the final section.

2. Analysis procedures

To compute the vertical momentum budget, a reference state pressure $\bar{p}$(z) is first defined as the horizontally averaged pressure over an area of 36 km x 36 km in the northeast of the storm. The mean virtual temperature $\hat{T}_v$(z) is then obtained from the hydrostatic equation using $\bar{p}$(z). Following Rotunno and Klemp (1982), the vertical momentum equation in cylindrical coordinates can be given by

$$\frac{dW}{dt} = (W_{CD} - \frac{1}{\rho}\frac{\partial P_d}{\partial z}) + (b - \frac{1}{\rho}\frac{\partial P_b}{\partial z}), \qquad (1)$$

where

$$b = g(\frac{\hat{\rho}}{\rho}\frac{T_v'}{T_v} - \frac{P'}{p}) - g(q_c + q_r + q_i + q_s + q_g), \qquad (2)$$

$$W_{CD} = 2\,\Omega\,u\cos\phi + D_w \qquad (3)$$

and P_d and P_b are the dynamic (P_d) and buoyancy (P_b) perturbation pressures, respectively. The variable P' (= $P_d + P_b$) is the total perturbation pressure from the basic state and almost equivalent to deviation pressure (see Zhang et al. 2000 for more details) All the other variables assume their typical meteorological meaning. Eq. (1) states that the vertical momentum acceleration is determined by the net dynamic force ($W_{ND} = W_{CD}$ + PGF_d) and the net buoyancy force (W_{NB} = b + PGF_b). To calculate PGF_d and PGF_b, we obtain P_d and P_b by inverting a 3D pressure equation. The derivation of the pressure equation and the inversion of P' are given in Appendix A.

The model-derived perturbation pressures (P'_m) exhibit a maximum surface pressure deficit of >80 hPa at the center of the storm and the magnitude decreases rapidly with height (see Zhang et al. 2000). Obviously, if the inverted perturbation pressure P' is accurate, it should be very close to the model-derived P'_m.

3. Dynamically-induced vertical motion

Figures 1a,b present cross sections of the azimuthally-averaged P_d and P_b, respectively, that are taken from 56-57 hour integration of the storm during its mature stage (see Liu et al. 1997). Although the error in the inverted P' is < 1% and amounts to only 0.1 to - 0.2 hPa in the eye (see Zhang et al. 2000), large errors can occur when computing the vertical acceleration because the magnitude of which is one to two orders smaller than the individual terms on the RHS of Eq. (1). Hence, it is not meaningful to discuss the residues between W_{ND} and W_{NB}.

It is evident from Fig. 1a that the buoyancy-induced P_b decreases rapidly with height, with a reversal of the sign of the horizontal gradients at 7 km where the peak warm core is located. However, the dynamically-induced P_d exhibits a "near-barotropic" structure in the vertical with only a slight tilt in the eye (cf. Figs. 1a,b). It is important to note that *a large portion of the buoyancy perturbation pressure is produced by the warming associated with moist-adiabatic ascent in the eyewall and the subsidence warming in the eye.* It accounts for more than 55 hPa of the pressure drop at the surface, or roughly 70% of the total deepening (Fig. 1a). By comparison, the dynamically-induced P_d is about – 22 to - 25 hPa at the eye's center. Such a dynamically-induced central

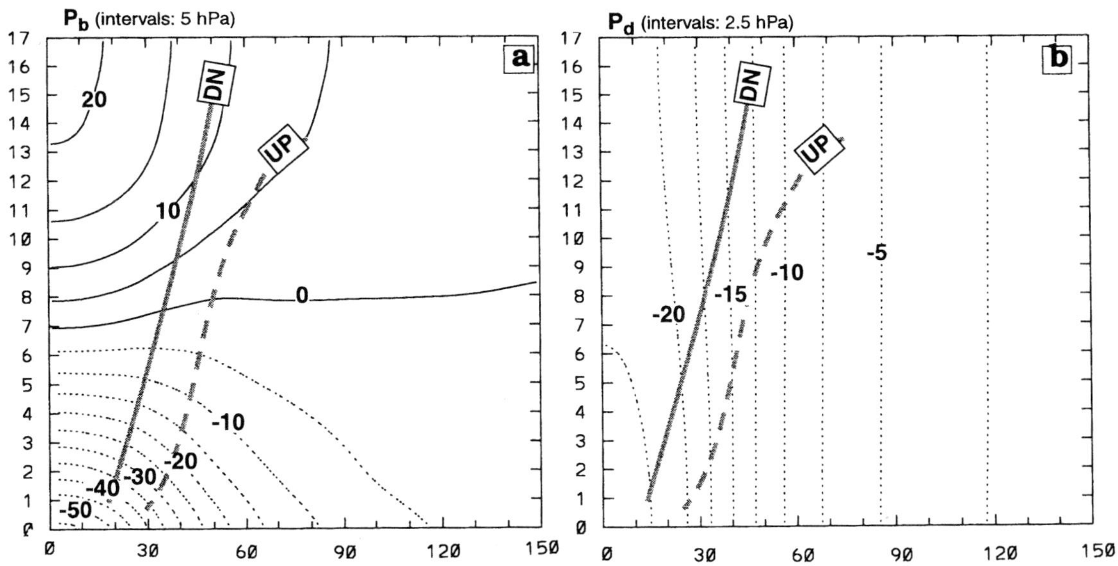

Figure 1. Radius-height cross section of the hourly and azimuthally-averaged (a) buoyancy-induced pressure perturbation (P_b) at intervals of 5 hPa; (b) the dynamically-induced pressure perturbation (P_d) at intervals of 2.5 hPa. They are obtained from 12 model outputs at 5-minute intervals during the 1-h budget period between 56-57 h integration ending at 2100 UTC 23 August 1992. Solid (dashed) lines are positive (negative) values.

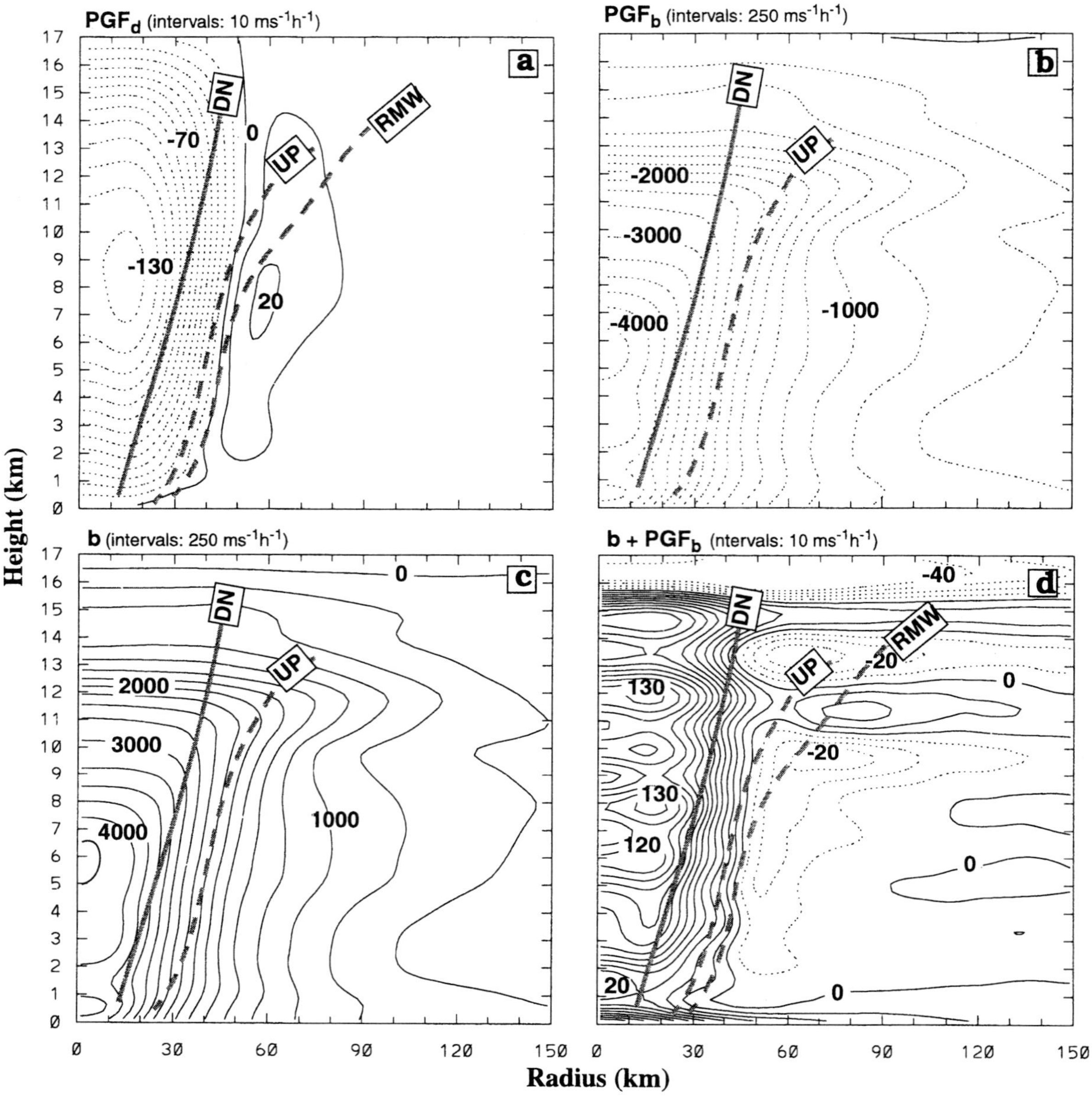

Figure 2. As in Fig. 1 but for (a) the dynamically-induced PGF_d; (b) the buoyancy-induced PGF_b; (c) the buoyancy force (i.e., b), and (d) the net buoyancy force (i.e., $W_{NB} = PGF_b + b$).

pressure drop finds analogy in a rotating cylindrical vessel filled with water. That is, the faster the rotation, the steeper is the parabolic free surface, and the lower the pressure at the bottom surface.

In spite of the near-barotropic structure, PGF_d points downward within the radius of the tilted updraft core but upward outside. Its value becomes negligible beyond R = 80 km (Fig. 2a). Since the azimuthally-averaged W_{CD} is small (see Zhang et al. 2000), the net dynamic force W_{ND} is nearly the same as PGF_d except in the boundary layer. Clearly, the downward PGF_d tends to suppress the upward convective current in the inner half portion of the eyewall and forces it to tilt outward. In the outer half of the eyewall, the upward PGF_d helps to lift low-level high-θe air to the condensation level to facilitate the development of convection. Note that both the upward and downward PGF_d are peaked in the midtroposphere at a radius of 12 km. The displacement of the peak downward PGF_d from the central axis appears to result from the azimuthally asymmetric distribution of tangential winds at the RMW (see Zhang et al. 2000).

In contrast, the large buoyancy-induced P_b results in extremely intense downward PGF_b, more than one order of magnitude larger than PGF_d, particularly in the eye (cf. Figs. 1a and 2b). The PGF_b is similar in structure and magnitude but opposite in sign to the buoyancy (cf. Figs. 2b,c). Thus, the net buoyancy force W_{NB} (Fig. 2d) is a small difference between two large terms, but it is responsible for various fluctuations in the vertical. On average, W_{NB} and the dynamically-induced PGF_d have similar magnitude but they oppose each other (cf. Figs. 2a,d). Since it is meaningless to examine the residues between W_{NB} and W_{ND}, we may state that *the dynamically induced PGF_d must overpower, but in close balance with the net buoyancy force W_{NB}, to maintain the general descent in a deep layer in the eye during the intensifying stage. The same should also be true for the lower-tropospheric ascent in the outer portion of the eyewall.* The inner half portion of the eyewall may be viewed as a transition zone where air parcels in both the updraft (UP) and the downdraft (DN) decelerate vertically. This result reveals that positive buoyancy may not be necessary for continued deepening of the hurricane as long as high-θe air is available in the vicinity of the eyewall.

To further understand the roles of the dynamic force W_{ND} in inducing vertical motion, we examine the dynamic sources in Eq. (A2). A scale analysis shows that PGF_d is dominated by the divergence of horizontal momentum advection, namely,

$$-\nabla\bullet(\rho\,\mathbf{V}\bullet\nabla\mathbf{V}) = -\rho\,[\nabla_h^2\frac{V_h^2}{2} - \nabla_h\bullet(V_h \times \zeta\,\mathbf{k})] + R \qquad (4)$$

and

$$R = -\rho\nabla\bullet(W\frac{\partial V}{\partial z} + \mathbf{V}_h\bullet\nabla W) - (\mathbf{V}\bullet\nabla\mathbf{V})\bullet\nabla\rho \qquad (5)$$

where the subscript 'h' denotes the horizontal component, and ζ is the vertical component of relative vorticity. In cylindrical coordinates, and after performing an azimuthal average, Eq. (4) can be written as

$$-\langle\nabla\bullet(\rho\,\mathbf{V}\bullet\nabla\mathbf{V})\rangle = -\frac{\rho}{r}\frac{\partial}{\partial r}[r\frac{\partial(V_h^2/2)}{\partial r}] + \frac{\rho}{r}\frac{\partial}{\partial r}[V_h\frac{\partial(rV_h)}{\partial r}] + R \qquad (6a)$$

$$\approx \frac{\rho}{r}\frac{\partial V^2}{\partial r} + R. \qquad (6b)$$

where the symbol < > denotes the azimuthal average operator and the contribution of the radial flows has been neglected in obtaining the final approximation. Eq. (6b) is essentially the cyclostrophic approximation, and it states that the azimuthally averaged dynamic source is dominated by the radial shear of V^2 weighted by the radial distance. Our result therefore differs from that of Rotunno and Klemp (1982) and Klemp and Rotunno (1983), where the important dynamic source in thunderstorm rotation and movement is the tilting of horizontal vorticity (associated with vertical wind shear) by non-uniform vertical motion. In our case, it is the low-level intense tangential winds at some distance from the circulation center, not their negative vertical shear, that is responsible for the general subsidence warming in the eye.

To substantiate the above analysis, Fig. 3 shows the azimuthally-averaged PGFs induced by the Laplacian of horizontal kinetic energy, and the radial gradient of "azimuthal vorticity advection" on the RHS of Eqs. (4) and (6a). One can see the opposing (or cancellation) effects between the first two components from their opposite signs. Obviously, the radial shear of the tangential winds dominates the dynamics of hurricanes (Fig. 3b). Since both V and ζ are intense at the RMW, the more compact the storm is, the stronger is the upward (downward) PGF_d at the outer portion of the eyewall (in the eye). *The net dynamic force is associated mostly with the radial gradient of V^2 weighted by radius*, similar to what is depicted in Fig. 2a. The residue source, dominated by the divergence of vertical advection of horizontal winds or the tilting of horizontal vorticity, indicates a deep layer of upward PGF_d within the RMW and it corresponds well to the strong vertical shear in the region (cf. Figs. 2b and 3a). This result is qualitatively in agreement with that of Rotunno and Klemp (1982) in their thunderstorm studies. However, the positive contribution of vertical shear is clearly too small compared to the large negative PGF_d generated by the radial shear within the RMW (not shown).

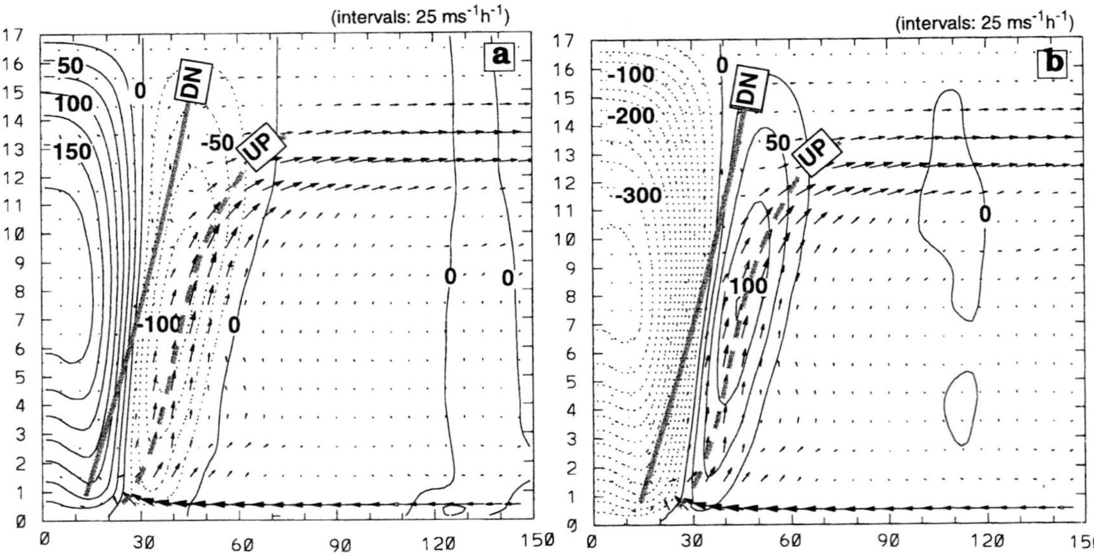

Figure 3. As in Fig. 1 but for the decomposed components of dynamic PGF_d associated with (a) $-\rho \nabla_h^2 V_h^2 / 2$; and (b) $\rho \nabla_h \bullet (V_h \times \zeta \mathbf{k})$ [see Eq. (4) for definitions].

4. Summary and conclusions

In this study, we examined the vertical momentum budget in the inner-core region using a high resolution ($\Delta x = 6$ km), explicit simulation of Hurricane Andrew (1992) to gain insight into the processes for subsidence warming in the eye and the vertical lifting at the outer edge of the eyewall. The results show that a large portion of the perturbation pressure is produced by the warming associated with moist-adiabatic ascent in the eyewall and subsidence warming in the eye. It is found that the net dynamic force points downward in the eye and accounts for the maintenance of the general descent in the eye. It, however, points upward in the outer portion of the eyewall, particularly in the northern sector, and facilitates the lifting of high-θe air from the lower troposphere into the storm.

Based on the above results, we propose a different theoretical explanation for the relationship among the subsidence warming in the eye, and the rotation and vertical shear in the eyewall. Specifically, as a tropical storm deepens, the lower-level tangential winds must intensify in accordance with gradient wind balance. The radial shear of the tangential winds in turn increases and induces a deep layer of downward PGF_d. This downward dynamic force could gradually overcome the positive buoyancy, thereby forcing the air to subside in the inner-core region. This forced descent results in warming and drying, and the formation of a hurricane eye. As a result of the middle- to upper-level warming, the tangential winds in the eyewall must decrease with height, as dictated by the thermal wind relation. Thus, *the negative vertical shear in the eyewall is caused by the forced subsidence warming, rather than the other way around as suggested by previous studies (e.g., Smith 1980).* Clearly, Smith's theory, could not explain how the initial subsidence is induced, because it requires a negative vertical shear in the eyewall to begin with. Our different explanation is based on the inverted PGF_d which shows that it is the low-level radial shear, rather than the vertical shear, of the tangential winds that drives the general descent in the eye. Of course, the dynamically-driven subsidence warming must occur at a rate to ensure near hydrostatic balance in a stratified fluid.

Acknowledgments

This work was supported by NSF grant ATM-9802391, NASA grant NAG-57842 and ONR grant N00014-96-1-0746. The computations were performed at the National Center for Atmospheric Research, which is sponsored by the National Science Foundation.

Appendix A: Inversion of three-dimensional perturbation pressures

The three-dimensional Euler momentum equation in Cartesian coordinates can be written as

$$\rho \frac{\partial V}{\partial t} = -\rho\, \mathbf{V} \bullet \nabla \mathbf{V} - \nabla P' + \rho\, b\, \mathbf{k} - 2\rho\, \Omega \times \mathbf{V} + \rho\, \mathbf{D_V} \qquad \text{(A1)}$$

where $\mathbf{V} = U\,\mathbf{i} + V\,\mathbf{j} + W\,\mathbf{k}$, ∇ is the three-dimensional gradient operator, P' is the perturbation pressure from the basic state $\bar{p}(z)$, $\mathbf{D_V}$ includes all the diffusive and boundary-layer effects, and b is the total buoyancy force defined in Eq. (2).

By applying $\nabla \bullet$ to Eq. (A1) yields

$$\nabla^2 P' = \frac{\partial \rho b}{\partial z} - \nabla \bullet (\rho\, \mathbf{V} \bullet \nabla \mathbf{V}) - 2\, \nabla \bullet (\rho\, \Omega \times \mathbf{V}) + \nabla \bullet (\rho\, \mathbf{D_V}), \qquad \text{(A2)}$$

where the first term on the RHS is referred to as the buoyancy (F_b) source and the remaining three terms the dynamic (F_d) sources. In deriving Eq. (A2), we have neglected the local tendency of $\nabla \bullet \mathbf{V}$ because its magnitude is very small relative to other terms during the mature stage. Assuming that all the RHS terms are known, and given proper boundary conditions, P' could be inverted from the 3D Poisson equation (A2).

The inversion of Eq. (A2) is carried out using the routine MUDPACK 4.0 on NCAR's CRAY Y-MP (see Adams 1993 for more details). To validate the accuracy of MUDPACK, we performed a preliminary inversion by specifying all the source terms on the RHS using a horizontal grid size of 6 km and a vertical grid size of 120 m. For this test case, we use a Dirichlet type boundary condition by specifying the model-calculated perturbation pressure at the top/bottom and lateral boundaries. It was found that when the P' obtained from the inversion is azimuthally averaged, it deviates from p_2' (Fig. 1a) by no more than 0.1 hPa (not shown). The maximum difference occurs at an altitude of 7.5 km in the eye where the warm core reaches its maximum value.

As mentioned in Introduction, we are mainly interested in the relative importance of the buoyancy- and dynamically-induced PGF in the development of vertical motion in hurricanes. However, we have little information to specify the individual *buoyancy and dynamic perturbation pressures* at the boundaries. As a first attempt, we follow Rotunno and Klemp (1982) by specifying the individual *perturbation pressure gradients* normal to the boundaries (the so-called Neumann-type boundary conditions). Specifically, we applied Eq. (A1) by ignoring the local momentum tendency and eliminating those terms which do not contribute to the pressure gradient normal to the boundaries. However, the results were not satisfactory as the azimuthally averaged inverted P' has a maximum error of about 0.8 hPa in the vicinity of V_{max} (not shown) when compared to p_2'.

Further experimentation was made and it was found that the following procedure and boundary conditions resulted in a more accurate solution. First, we extend the inversion domain from 360 km x 360 km to 480 km x 480 km and set the buoyancy source to zero in the extended region. Second, the boundary conditions for inverting the buoyancy-induced perturbation pressure P_b are specified as

$P_b|_X = 0; \; P_b|_Y = 0$ for lateral boundaries at X and Y, and
$\partial P_b/\partial z = \rho b$ for top and bottom boundaries. (A3)

By invoking (A3), the inverted P_b can be determined uniquely. Similarly, the dynamic sources are set to zero in the extended region when inverting the dynamically-induced P_d. The boundary conditions are

$P_d|_X = 0; \; P_d|_Y = 0$ for lateral boundaries at X and Y, and
$\partial P_d/\partial z = \rho (U\, \partial W/\partial x + V\, \partial W/\partial y + W\, \partial W/\partial z) + 2 \rho \Omega u \cos \phi + \rho D_w$
for top and bottom boundaries. (A4)

Zhang et al. (2000) showed that the difference between the inverted P' and the model-derived P'_m are very small, ranging from 0.1 to -0.2 hPa. Thus, we may state that our methodology for obtaining the PGFs are sufficiently accurate to allow an analysis of the relative importance of the dynamic and buoyancy forces in the development of secondary circulations in the inner-core region of the hurricane.

References

Adams, J., 1993: MUDPACK-2: Multigrid software for approximating elliptic partial differential equations on uniform grids with any resolution. *Appl. Math. & Comput.*, **53**, 235-249.

Emanuel, K.A.,1997: Some aspects of hurricane inner-core dynamics and Energetics. *J. Atmos. Sci.*, **54**, 1014-1026

Gray, W.M., and D.J. Shea, 1973: The hurricanes inner core region: Thermal stability and dynamic characteristics. *J. Atmos. Sci.,* **30**, 1565-1576.

Klemp, J.B. and R. Rotunno, 1983: A study of the tornadic region within a supercell thunderstorm. *J. Atmos. Sci.*, **40**, 359-377.

Kuo, H.-L., 1959: Dynamics of convective vortices and eye formation. *The Atmospheric and Sea in Motion*. B. Bolin, Ed., Rockefeller Institute Press, 413-424.

Liu, Y., D.-L., Zhang and M.K. Yau, 1997: A Multiscale Numerical Study of Hurricane Andrew (1992), Part I: An explicit simulation, *Mon. Wea. Rev.*, **125**, 3073-3093.

Malkus, J. S., 1958: On the structure and maintenance of the mature hurricane eye. *J. Meteor.*, **15**, 337-349.

Rotunno, R. and J.B. Klemp, 1982: The influence of the shear-induced pressure gradient on thunderstorm motion. *Mon. Wea. Rev.*, **110**, 136-151.

Shapiro, L.J. and H.E. Willoughby, 1982: The response of balanced hurricanes to local sources of heat and momentum. *J. Atmos. Sci.,* **39**,378-394.

Smith, R.K., 1980: Tropical cyclone eye dynamics. *J. Atmos. Sci.,* **37**,1227-1232.

Willoughby, H.E., 1998: Tropical cyclone eye thermodynamics. *Mon. Wea. Rev.* , **126**, 3053-3067.

Zhang, D.-L., Y. Liu and M.K. Yau, 2000: A Multiscale Numerical Study of Hurricane Andrew (1992), Part III: Dynamically-induced vertical motion. *Mon. Wea. Rev.*, in press.

SYSTEMATIC APPROACH FOR TROPICAL CYCLONE TRACK FORECASTING WITH DYNAMICAL MODEL SELECTIVE CONSENSUS

Russell L. Elsberry and Lester E. Carr, III
Department of Meteorology
Naval Postgraduate School
Monterey, California 93943-5114

1. Introduction

The long-range objective of the Systematic and Integrated Approach to Tropical (TC) Cyclone Track Forecasting (hereafter the Systematic Approach) technique (Carr and Elsberry 1999) is significant improvements in official TC track forecasts. Desired quantitative improvements include: lower average forecast track errors (FTE), official FTE's that are consistently better than the FTEs of the objective track forecast guidance available to the forecaster, and a reduction in the number of track forecasts that have very large FTEs (commonly referred to as "busts"). Unlike track forecasts provided by objective guidance (i.e., models), each official track forecast is accompanied by a subjective conceptual picture or "meteorological scenario" in the mind of the forecaster. The Systematic Approach is designed to help the TC forecaster develop a meteorological basis for the official track forecast that reflects dynamically-sound, state-of-the-science understanding of tropical cyclone motion and track prediction. A key component of the Systematic Approach is a forecast model traits knowledge base (hereafter Model Traits) that documents how models have performed in certain recurring meteorological scenairos.

To visualize how the forecast model traits knowledge base of the Systematic Approach is intended to assist the forecaster, consider the idealized distributions of model 72-h track forecasts in Fig. 1. Distributions of model forecasts can be viewed by the forecaster as constituting either: (i) a single cluster with a single associated meteorological scenario that explains all of the model tracks (Fig. 1a); or (ii) two or more clusters, with each cluster corresponding to a particular meteorological scenario that explains the model tracks in that cluster (Fig. 1b). It is emphasized that: (i) although very distinct clusters are shown in Fig. 1a and b for purpose of illustration, in reality the clusters can be quite "fuzzy," which may cause the forecaster to not assign a two-cluster classification to the distribution of model forecast tracks; and (ii) although clustering by forecast direction is illustrated in Fig. 1b, the clusters may be distinguished by any combination of speed and direction, and a cluster may contain only one forecast track.

In the case of a single cluster, an official track forecast that closely conforms to the consensus of all the model tracks (Fig. 1a; track C) will usually be the best option for the forecaster, and is the most

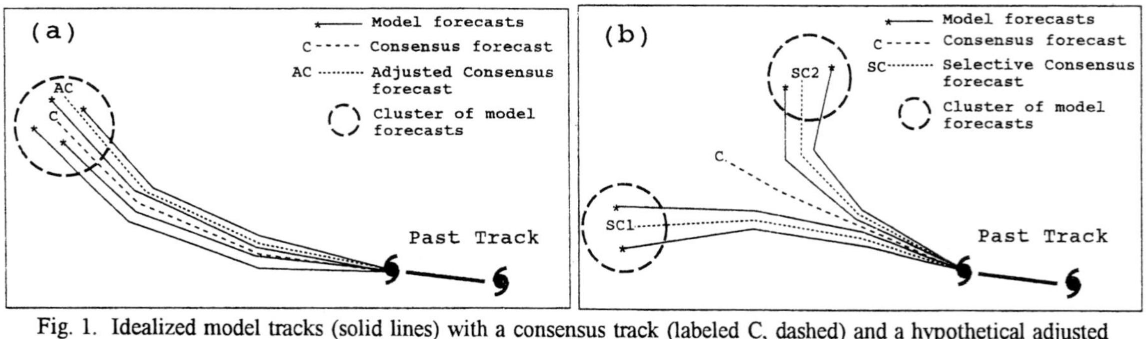

Fig. 1. Idealized model tracks (solid lines) with a consensus track (labeled C, dashed) and a hypothetical adjusted consensus (AC, dotted). In panel (a), the model tracks essentially are consistent and fall into a single cluster (heavy dashed circle). In panel (b), the tracks fall into two clusters in which the selective consensus is the average of the tracks in that cluster.

likely to result in the lowest errors on average (Goerss, personal communication, 1998). In this situation, application of the Model Traits knowledge base to discern subtle differences among the individual model forecast fields is not likely to consistently result in an improvement to a simple consensus forecast. In a minority of cases, model track forecasts in relatively good agreement may all have a qualitatively similar bias. In these situations, an adjusted consensus (AC) track forecast may make a modest improvement on the simple consensus forecast. Regardless of whether a simple consensus or adjusted consensus results from a single model track forecast cluster, the Systematic Approach process will still be qualitatively useful for formulating a consistent dynamically-informed explanation (conceptual picture) that accounts for the predicted TC motion.

In the case of two track clusters, a simple consensus average track (Fig. 1b; track C) down the middle may not be consistent with any of the meteorological scenarios represented by the individual clusters, and thus may not be meteorologically justifiable. One cluster usually represents the actual track better than the other cluster, and it is unlikely that both clusters will equally depart from reality, particularly if the separation distance between the cluster centroids is large. As a result, it is more likely that a *selective consensus* (Fig. 1b; tracks SC1 or SC2) of the members of one of the clusters will have lower forecast errors than the simple consensus of all the available track forecasts. It is especially in these cases that the Systematic Approach concept is expected to help the forecaster decide which selective consensus is the better choice on which to base the official track forecast. So long as the same meteorological conditions continue to apply, the official forecasts should follow this scenario so that shift-to-shift consistency in the official track forecasts is achieved. Only when it is clear that the actual meteorological condition differs from this scenario should the official track forecast be changed to the other track cluster.

The purpose of this research is to develop a Model Traits knowledge base that describes how objective TC track forecast guidance available to the forecasters at the Joint Typhoon Warning Center (JTWC) performs in the western North Pacific. In particular, this project focuses on the analysis of cases when highly erroneous forecasts have been made by the Navy Operational Global Atmospheric Prediction System (NOGAPS) and the U.S. Navy version of the Geophysical Fluid Dynamics Laboratory model (GFDN). Such documentation is a necessary step to successfully recognize highly erroneous numerical model track forecasts on a real-time basis, and either discard or greatly modify such objective forecasts when formulating the official track forecast. *That is, a knowledge of when specific model track forecasts are likely to be highly erroneous will permit the forecaster to formulate an official track forecast that reflects an informed adjusted or selective consensus of the remaining numerical model tracks (as in Fig. 1b), as opposed to using a simple consensus of all available track guidance.*

2. Summary of results

The results of the evaluation process for the large (> 300 n mi at 72 h) NOGAPS and GFDN errors are summarized in Table 1. Although each of these track error sources have been described in detail in Carr and Elsberry (1999), only an overview on the relative frequency from various error sources is presented here.

Erroneous binary cyclone interactions (Table 1; rows 1- 4) were by far the most frequent cause of poor NOGAPS and GFDN 72-h forecasts, and accounted for 42 (39%) and 36 (36%) of the large FTE cases, respectively. The three types of cyclone interactions listed are analogous to the three types of TC-TC interactions described in Carr et al. (1997), except that in the numerical model predictions the second cyclone was not always a TC. Excessive Direct Cyclone Interaction (E-DCI) was remarkably prevalent, and degraded more forecasts than any other single phenomenon by roughly a 2-to-1 margin for NOGAPS and 3-2 margin for GFDN. Those poor forecasts occurred during 18 separate periods involving 14 TCs in the western North Pacific in 1997. By contrast, evaluation of the TC tracks and NOGAPS analyses indicated that track-altering DCI actually occurred only twice in 1997; i.e., real DCI is a rare event in relation to these model-predicted events.

Table 1. Frequencies of the causes of large NOGAPS and GFDN 72-h track errors during 1997. The first (second) number indicates the times the phenomenon occurred excessively (insufficiently) in the model.

CAUSES OF NOGAPS OR GFDN 72-H FORECAST TRACK ERRORS GREATER THAN 300 N MI		NUMBER OF FORECASTS	
Phenomenon Name	Acronym	NOGAPS	GFDN
Direct Cyclone Interaction	DCI	39-0	31-0
Semi-direct Cyclone Interaction	SCI		
SCI on Western TC	SCIW	3-0	1-0
SCI on Eastern TC	SCIE	0-0	0-0
Indirect Cyclone Interaction	ICI		
ICI on Eastern TC	ICIE	0-0	3-0
ICI on Western TC	ICIW	0-0	0-1
Ridge Modification by TC	RMT	12-0	2-1
Reverse Trough Formation	RTF	10-0	2-0
Response to Vertical wind Shear	RVS	10-0	1-3
Baroclinic Cyclone Interaction	BCI	8-12	11-2
Midlatitude System Evolutions	MSE		
Midlatitude CycloGenesis	MCG	0-0	19-0
Midlatitude CycloLysis	MCL	0-0	0-0
Midlatitude AnticycloGenesis	MAG	3-0	0-0
Midlatitude AnticycloLysis	MAL	0-4	0-2
Tropical Cyclone Initial Size	TCS	5-0	9-0
Not discernable or explainable		2	2
Fields not available		0	9
All causes		108	99

Ridge Modification by a TC (RMT) and Reverse Trough Formation (RTF) are a manifestation of the beta effect and involve the development of an anticyclone eastward and equatorward of the TC (RMT) or TCs (RTF) that facilitates poleward motion (Carr and Elsberry 1997). Erroneous prediction of these two phenomena together degraded 22 (20%) NOGAPS track forecasts, but only 5 (5%) GFDN track forecasts.

Response to Vertical wind Shear (RVS) and Baroclinic Cyclone Interaction (BCI) represent the competing negative and positive dimensions of TC interaction with the midlatitudes. The RVS process tends to decrease the intensity of the TC, which usually results in a decrease in translation speed as the TC responds to environmental steering over a shallower layer. The BCI process tends to deepen, or at least maintain the strength, of the circulation via transition into an extratropical cyclone, which affects the motion of the TC. Erroneous prediction of these two phenomena together degraded 30 (28%) NOGAPS and 17 (17%) GFDN track forecasts.

The errors associated with the influences of Midlatitude Cyclogenesis (MCG), Midlatitude Anticyclogenesis (MAG), and Midlatitude Anticyclolysis (MAL), along with Midlatitude Cyclolysis (MCL) (which was not associated with any poor forecasts in 1997), relate the structure and movement of midlatitude features (i.e., troughs, ridges, cyclones, and anticyclones) to changes in environmental steering of the TC. That is, the model forecast has clearly misrepresented midlatitude features in a manner that is consistent with the character of the TC track error. Together these error mechanisms accounted for the degradation of 21 (21%) GFDN forecasts, but only 7 (6%) NOGAPS track forecasts.

Incorrect Tropical Cyclone initial Size (TCS) was responsible for the degradation of 5 (5%) NOGAPS and 9 (9%) GFDN track forecasts. All of these poor track forecasts were attributed to an overly large initial size of the TC in the models, which was possibly due to an overly large 35-kt radius in the JTWC warnings during a period when the TC was translating rapidly westward (up to 17 kt).

Only 2 (2%) NOGAPS and 2 (2%) GFDN forecasts with large 72-h errors were considered to have no discernable physical explanation. It is noteworthy that the largest of these four forecast errors was only 433 n mi. In other words, a discernable physical explanation could be provided for virtually all of the forecasts with 300 n mi FTEs. Since being able to readily discern a plausible reason for an erroneous track forecast is a necessary (but not sufficient) condition for anticipating such errors in an operational setting, this to considered to be a very encouraging result.

The error mechanisms in the first nine rows of Table 1 are related to TC-Environment Transformations in the Meteorological knowledge base, and in each case the TC circulation interacts significantly with the surrounding environment. By contrast, the error mechanisms in the next four rows involve large-scale midlatitude processes to which the TC is a comparatively passive respondent. Thus, a very important finding is that the vast majority (87%) of highly erroneous NOGAPS track forecasts and a significant majority (59%) of highly erroneous GFDN track forecasts may be attributed primarily to misrepresentations of phenomena that depend sensitively on the fidelity with which the structure of the TC is represented in the model. Furthermore, in the first nine rows of Table 1.3, it was usually excessive interaction of the TC with the environment that was the cause for the poor forecast. It is also significant that E-MCG was responsible for 19 (19%) of highly erroneous GFDN track forecasts, but no highly erroneous NOGAPS track forecasts. This difference suggests that perhaps some tuning of the GFDN model is necessary to function more effectively when using NOGAPS initial and boundary conditions.

These findings have important ramifications for operational TC forecasting and numerical TC forecast model development. For the forecaster, it means that particular attention should be paid to evaluating the numerical TC model forecasts for indications of excessive interaction with the environment. For the numerical modeler, it means that if continued improvements are to be achieved in the numerical prediction of western North Pacific TC tracks, then significant effort must be directed toward achieving improved (i.e., usually less vigorous) model representations of TC structure and interaction with the environment (particularly with other cyclones).

3. A preliminary test of large-error determination

Late in the 1999 typhoon season, a test of the ability to detect the likely presence of large track errors by five dynamical models was conducted at the Naval Postgraduate School (NPS). Although not done in real-time, the information used by the team of researchers (Mark Boothe, Pat Harr, and Grahame Reader assisted the authors) was almost the same as would be available to the Joint Typhoon Warning Center (JTWC) forecaster. The exception is that the NPS team did not have as good access to the variety of satellite imagery available at JTWC.

The primary objective of the test was to determine whether the NPS team could apply the large-error conceptual models of Carr and Elsberry (1999) to isolate likely erroneous tracks, and thus calculate a *selective consensus* (SCON) of the remaining models that would be more accurate than the *non-selective consensus* (NCON) forecast. Only in those cases in which one or more dynamical model tracks was considered to be in error would the SCON track differ from the NCON track. Thus, the two measures of accuracy are the improvement of the NCON track errors relative to a CLImatology and PERsistence (CLIPER) forecast, and whether the SCON forecasts had more skill relative to CLIPER when the SCON forecast was different from the NCON forecast. Although the consensus tracks were also calculated at 24 h and 48 h, the primary focus is the performance of the dynamical models at 72 h. It is noted that not all five dynamical models were available for each forecast -- the NCON track is the average of whatever dynamical model tracks that were available.

The steps in the construction of the NCON track are illustrated in Fig. 2, which is for a two-cluster scenario similar to Fig. 1. For this illustration, the three global models from Japan (JGSM), United States Navy (NOGAPS), and United Kingdom Meteorological Office (UKMO) that originated 12 h prior to the forecast time are predicting a left turn relative to a persistence-of-past-motion track. By contrast, the regional typhoon models from Japan (JTYM) and the U. S. Navy (GFDN) initiated 6 h prior to the forecast time are predicting a right turn. The first step is to identify the 12-h (6-h) positions in the global (regional) model predictions that verify at the forecast time. The second step is to translate the 12-h through 84-h (6-h through 78-h) global (regional) model tracks to originate from the forecast time position (see Fig. 2). When the global (regional) model tracks are only available to 72 h, the 84-h or 78-h positions are extrapolated. The final step to calculate a NCON track is to average the 24-, 48-, and 72-h positions from the available models. If only one global and one regional model track were available as in Fig. 2, the NCON positions would be half way between these two positions.

Such a large discrepancy between the predicted tracks from the three global models and the two regional models occurred for the 0000 UTC 14 November 1999 forecast of TC Gloria (Fig. 3). Given these five tracks, the NCON track is for a northwestward track for 24 h, and then to remain near-stationary for 48 h to 72 h. While such an intermediate track is possible, it is more likely that either the global model scenario or the regional model scenario will occur. The objective for the NPS team was to evaluate the dynamical model fields available to isolate the likely erroneous track scenario.

Applying the conceptual models in Carr and Elsberry (1999), a likely source of error for the cyclonically turning JGSM and NOGAPS tracks is the Excessive-Direct Cyclone Interaction (E-DCI) mechanism, which was the most common NOGAPS error during the 1997 season (Table 1). In the E-DCI conceptual model (Fig. 4), the second cyclone is usually a non-developing disturbance that is located to the west of the TC. Carr and Elsberry found the E-DCI mechanism is more likely to affect a weaker (< 50 kt) TC such as TC Gloria. One clear indication is a cyclonic rotation of the axis between the TC and the second cyclone with increasing forecast interval. In some cases, the two circulations merge to form a larger circulation than at the initial time. Both the onset and the cessation of the E-DCI may be rapid.

The application of the E-DCI conceptual model for the NOGAPS track in Fig. 3 is shown in the 700-mb streamlines after 36 h (Fig. 5a). Notice the large elliptical circulation in which Gloria is embedded with a marked trough to the southwest over the Philippines. As shown by the marked left turn of the NOGAPS track in Fig. 3 and typhoon symbols in Fig. 5a, Gloria is predicted to rotate cyclonically relative to the southwestern trough. Just as the predicted NOGAPS fields as in Fig. 5a agree with the E-DCI conceptual model in Fig. 4, the JGSM fields also agree with the E-DCI model after 24 h (not shown), which is consistent with the abrupt cyclonic turn of the JGSM track after 24 h in Fig. 3. Although these predicted fields confirm a consistency with the corresponding track prediction, it is possible that this is a real DCI case (although this was rare during 1997 -- see above) since all three global models suggest such a scenario.

The corresponding 700-mb GFDN prediction (Fig. 5b) at the same time as the NOGAPS prediction in Fig. 5a has a quite different scenario. Whereas a similar large elliptical cyclonic circulation is evident with a closed circulation over the Philippines, the GFDN model representation of Gloria is much smaller. Gloria is clearly predicted to be embedded in a north-northeasterly steering flow with an isotach maximum to the east, which is consistent with the GFDN track at 36 h in Fig. 3. It could been inferred that the turn of the other regional model (JTYM) to the north and east after 36 h may have resulted from a similar circulation scenario of a small TC that does not interact with the cyclonic circulation over the Philippines. If Gloria had interacted with the circulation to the southwest, the GFDN (and JTYM) model error would be a case of insufficient DCI; no I-DCI cases occurred for the GFDN model during 1997 (see Table 1).

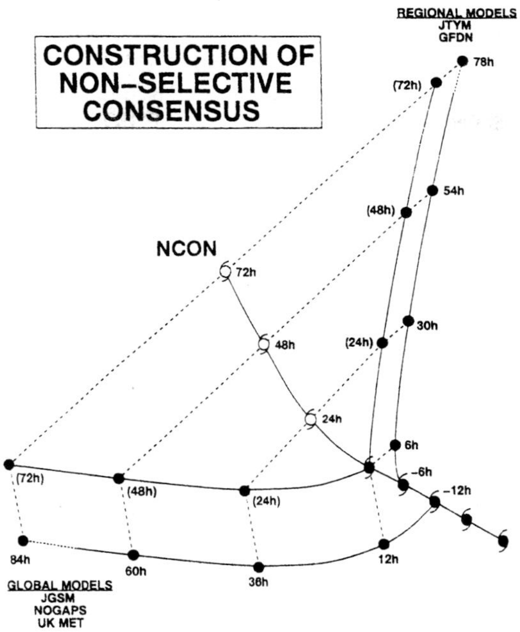

Fig. 2 Schematic of the construction of the non-selective (NCON) consensus of global model (left branch) and regional model (right branch) tracks originated 12 h (6 h) prior to forecast time that are matched with the forecast positions that verify at the forecast time, and then the 12-h through 84-h (6-h through 78-h) tracks of the global (regional) models are translated to begin at the forecast time. The NCON positions at 24, 48, and 72 h are the averages of the global and regional model predictions.

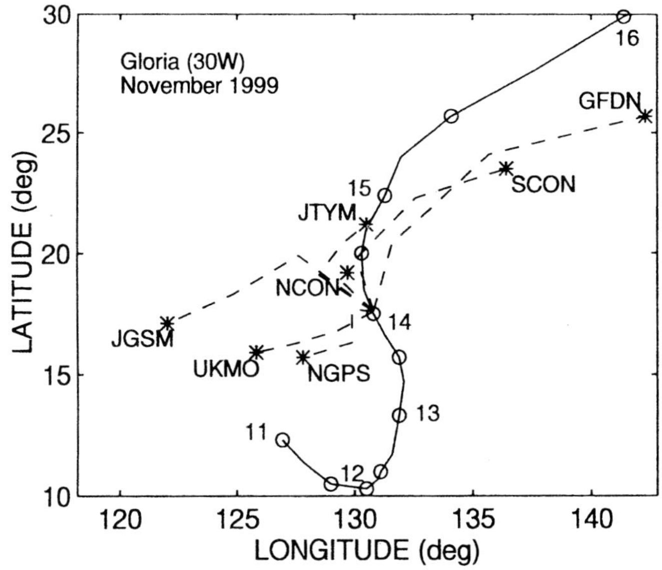

Fig. 3 Warning positions each 12 h and the five dynamical model (see insert) tracks with symbols at 24, 48, and 72 h for TC Gloria initiated at 0000 UTC 14 November 1999, the corresponding non-selective consensus (NCON) track, and a selective consensus (SCON) track in which the three global model (JGSM, UKMO, and NOGAPS) tracks are rejected and the other two regional model (JTYM and GFDN) tracks are retained. The 72-h verifying position at 0000 UTC 17 November is not shown as the storm had become extratropical at that time.

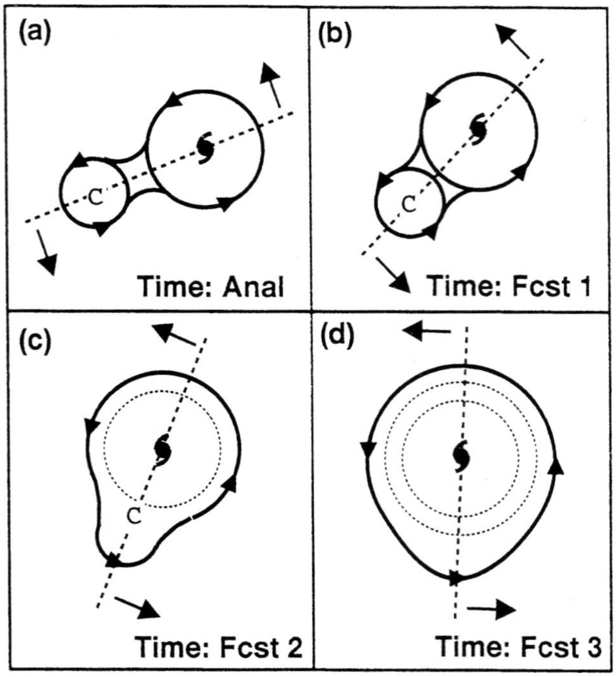

Fig. 4. Conceptual model by Carr and Elsberry (1999) of Excessive Direct Cyclone Interaction of a TC (symbol) and a second cyclone (C) at analysis and three forecast times. Note that the line connecting the TC and cyclone rotates cyclonically in time (arrows), and the size of the TC circulation increases during the merger with the cyclone (forecast times 2 and 3).

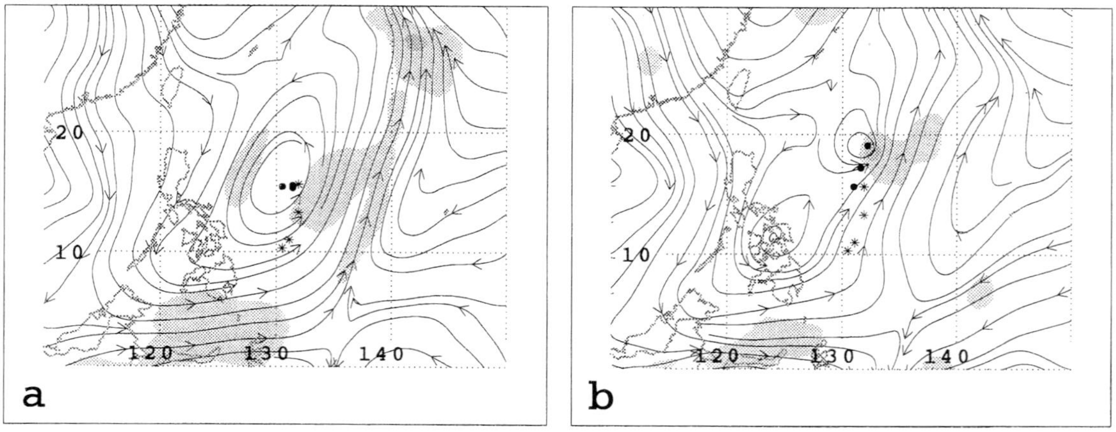

Fig. 5. Predicted 700-mb streamlines (solid) and isotachs (shaded, beginning at 20 kt and at 20-kt intervals) by the (a) NOGAPS model at 36 h and (b) GFDN model at 30 h after 0000 UTC 14 November 1999. Present and past TC Gloria positions (asterisks) and forecast positions (dots) are shown at 12-h intervals.

With track-circulation prediction consistency established for both the global (NOGAPS and JGSM) and the regional (GFDN) models, how may the forecaster decide which track prediction is likely to be erroneous? Examination of the satellite imagery (not shown) indicated that the predicted size and horizontal extent of Gloria in the NOGAPS field (Fig. 5a), and the strength of cyclonic circulation over the Philippines, were not likely to verify. Given the weak intensity of Gloria, it was more likely that the GFDN prediction (Fig. 5b) had a better representation. Because of the frequent occurrence of E-DCI in the NOGAPS model, and the absence of any I-DCI cases for the GFDN model during 1997 (Table 1), the likely erroneous track prediction was by the NOGAPS model. After also rejecting the other two global models for the same reason, the SCON track in Fig. 3 is the average of the two remaining regional model tracks. As shown by the subsequent track, this decision was basically correct, although Gloria actually accelerated rapidly to the northeast so that the SCON track has a slow bias. Notice that the SCON track was considerably more accurate than the nearly stationary NCON track.

A summary of the SCON versus NCON test is given in Fig. 6 as a skill comparison relative to CLIPER. Notice that the number of 72-h forecasts that could be verified is only about half of the 24-h forecasts, which is an indication that many of the TCs (numbers 19 through 30 in the JTWC listing) were weak and did not have long tracks. It is emphasized that all forecasts for which the TC at the initial and verifying times had an intensity equal to or greater than 25 kt are included, whereas other forecast centers and researchers use a 35-kt criterion. Although the dynamical model and CLIPER track forecast errors are typically larger for the weaker TCs, the 305 n mi 72-h error for CLIPER does not indicate that this sample of forecasts was unusually difficult.

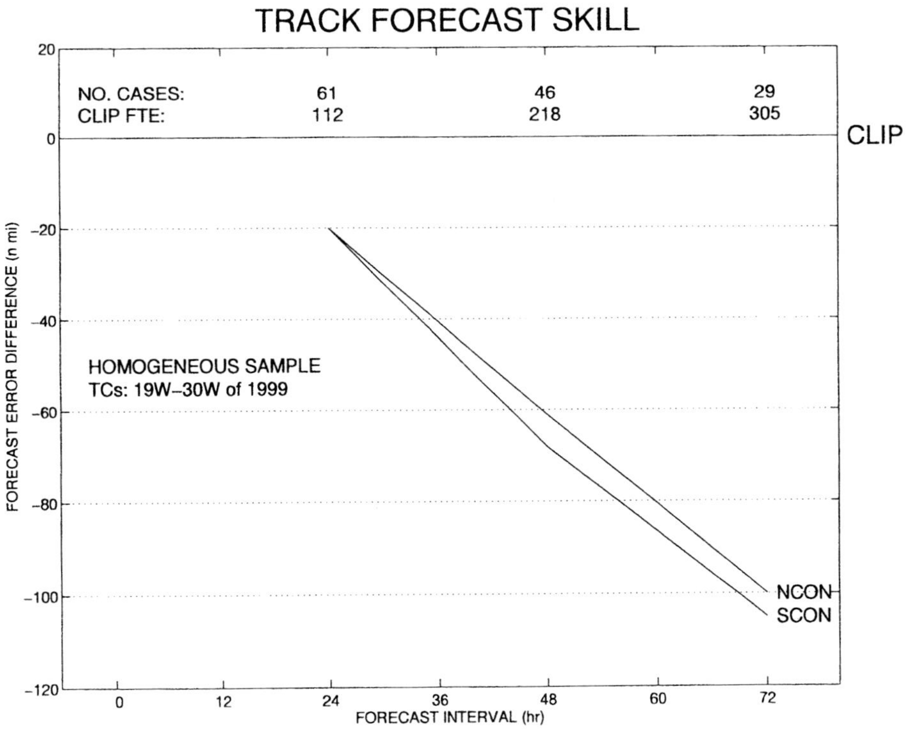

Fig. 6. Track forecast skill relative to CLImatology and PERsistence (CLIP) forecasts (error magnitudes along top) as a function of forecast interval for the non-selective consensus (NCON) and selective consensus (SCON) forecasts for TCs 19W-30W during August-November 1999 with both initial and verifying intensities equal or greater than 25 kt. The numbers of forecasts for this homogeneous sample is indicated along the top.

Notice that both the NCON and SCON track errors are substantially smaller than CLIPER at 48 h (statistically significant at 95% confidence) and at 72 h. The 105 n mi improvement relative to CLIPER represents an impressive 34% gain. Although the improvement of 5 n mi at 72 h for SCON relative to NCON may not seem that good, in 9 of the 29 cases the SCON was equal to the NCON value because no clearly erroneous dynamical model tracks would be isolated. If only the remaining 20 cases in which SCON departed from NCON are considered, the improvement of SCON over NCON at 72 h is 7 n mi. One insight gained during the test was that our rejection of models tracks in the small consensus spread cases (less than 250 n mi at 72 h) failed in 9 out of 10 cases to make an improvement over the NCON forecast. Had this consensus spread condition been applied during the test, only in 14 of the cases would the SCON track have been different from the NCON track. However, the improvement of the SCON over the NCON for these 14 cases would have been 23 n mi at 72 h! This improvement is quantitative evidence that the NPS team could apply the large-error conceptual models of Carr and Elsberry (1999) to define a SCON that has smaller errors, i.e., the forecaster can add value relative to a non-selective consensus (NCON). Of course, this potential improvement needs to be demonstrated by operational forecasters!

Acknowledgments. This research has been sponsored from the beginning by the Office of Naval Research Marine Meteorology Program, and more recently by the Space Warfare Systems Command. The manuscript was expertly prepared by Mrs. Penny Jones.

REFERENCES

Carr, L. E., III, and R. L. Elsberry, 1997: Models of tropical cyclone wind distribution and beta-effect propagation for application to tropical cyclone track forecasting. *Mon. Wea. Rev.*, **125**, 3190-3209.

Carr, L. E., III, and R. L. Elsberry, 1999: Systematic and integrated approach to tropical cyclone track forecasting. Part III: Traits knowledge base for JTWC track forecast models in the western North Pacific. Tech. Rep. NPS-MR-99-002, Naval Postgraduate School, Monterey, CA 93943-5114, 227 pp.

Carr, L. E., III, M. A. Boothe, and R. L. Elsberry, 1997: Observational evidence for alternate modes of track-altering binary tropical cyclone scenarios. *Mon. Wea. Rev.*, **125**, 2094-2111.

PERFORMANCE OF NOGAPS ON THE PREDICTIONS OF TROPICAL CYCLONES USING DIFFERENT CONVECTIVE PARAMETERIZATION SCHEMES

Melinda S. Peng and Timothy F. Hogan
Naval Research Laboratory, Monterey, CA

ABSTRACT

The performance of U.S. Naval Operational Global Atmosphere Prediction System (NOGAPS) on the predictions of tropical cyclones in August of 1998 using two different convective parameterization schemes, the Arakawa-Schubert scheme and the Emanuel scheme, is evaluated. The Emanuel scheme provided superior track forecasts than the Arakawa-Schubert scheme did and the structures and precipitation were also better simulated by the Emanuel scheme.

1. INTRODUCTION

As computer resource increases along with the advancement of modern technology, the resolutions in global forecast systems have been increasing steadily to levels that could only be reached by regional models in the past. The global forecast models can now provide reliable forecasts of synoptic and sometimes mesoscale phenomena in the five to seven day range. One method to evaluate the prediction skill of global forecast systems is the tracking of tropical cyclones in the model in which synthetic data are included to represent the cyclone. NOGAPS has demonstrated its capability of predicting tropical cyclone motions and is providing objective guidance for the Navy's tropical cyclone prediction mission (Fiorino et al. 1993, Goerss and Jeffries 1994).

The main source of energy for tropical cyclones comes form underlying warm oceans and is released through latent heat of condensation. Therefore, one of the most important sub-grid physical parameterizations for the simulation of tropical cyclone is the convective parameterization. Early convective parameterization schemes designed for use in numerical models include schemes such as moist convective adjustment by Manabe et al. (1965) and determination of final cloud state by large-scale moisture convergence by Kuo (1965, 1974). The Arakawa and Schubert scheme (Arakawa and Schubert 1974) is unique in its determination of the spectrum of the cloud base mass flux by demanding that convection dissipates instability at the same rate as which it is generated by large-scale forcing. The main deficiency of the scheme is the lack of downdrafts in its original design. Betts (1986) introduced an alternative relaxed adjustment scheme with the consideration of convective instability. A more recently developed scheme that became very popular is the Kain-Fritsch scheme (Fritsch and Chappell 1980, Kain and Fritsch 1990) in which the whole cloud is treated as an entraining plume as the other schemes and convective instability is eliminated by removing convective available potential energy (CAPE). However, aircraft observations reveal a large degree of inhomogeneity in individual convective clouds characterized by sub-cloud scale drafts. Emanuel (1991), based on these observations, designed a convective scheme that includes the effect of the updrafts and downdrafts in sub-cloud scale on the order of 100 m. The scheme takes the approach of buoyancy sorting in determining the subsequent ascent or decent of individual air parcel to reach a level where the liquid water potential temperature equals to that of its environment. The cloud mass flux at each level can thus be contributed from all levels above or below.

The relaxed Arakawa Schubert scheme (Moorthi and Suarez 1992, denoted as RAS) has been the one chosen for the operational NOGAPS with a crude consideration of saturation downdraft. Recently, the Emanuel scheme has been implemented in NOGAPS as an alternative scheme for convective parameterization. During the testing period of August 1998, the synoptic forecasts with Emanuel scheme showed an improved forecast skill compared to forecasts using the RAS scheme, indicated by the statistics both regionally and globally. For example, the 500 mb height anomaly correlation for the northern hemispheric at 120-h forecast is increased from 0.75 using the RAS scheme to 0.77 using Emanuel Scheme. Furthermore, the average forecast distance errors of the six tropical cyclones occurring during that period was also reduced by using the Emanuel scheme. In this study we analyze and compare the forecasts of the six storms by NOGAPS using these two different convective parameterization schemes.

2. The model

The NOGAPS is a global forecast system with a spectral representation in the horizontal plane and a finite-difference approximation in the terrain-following σ vertical coordinate. The spectral representation in a global model has the advantage of avoiding singularity at the poles and having continuity of the differential variables. The operational NOGAPS has a resolution of T159 with 24 vertical levels. In our study, T79 and 24 levels are used due to computer resource limitations. A semi-implicit scheme is used for the time integration. The model's sub-grid physical parameterizations include: surface fluxes parameterization by Louis (1979) and Louis et al. (1982) with vertical diffusive processes approximated by the K-theory, radiation by Harshvardhan et al. (1987), shallow convection by Tiedtke (1984), and deep convection either by RAS or by Emanuel (1991). Other physical processes include large-scale precipitation and gravity wave drag. A detailed listing of these references can be found in the technical report of NOGAPS (Hogan et al. 1991).

3. Predictions of Tropical Cyclones

During the month of August in 1998, there were six tropical cyclones that had reached cyclone strength, they include Bonnie and Danielle in the Atlantic; Howard, Estelle, and Georgette in the eastern Pacific; and Rex in the Western Pacific. The characteristic of the best tracks for these storms is distinguished by their geographic locations. Namely, the two storms in the Atlantic made recurvature, the three storms in the eastern Pacific moved west by northwest. The only typhoon in the western Pacific during that period, Rex, had an unusual north-south oscillation in its generally eastward movement.

All forecast integrations are carried out to 120 h. The average 120 h forecast distance error from the control run using RAS scheme is 416 NM for 35 cases verse 354 NM from the 27 cases using the Emanuel scheme. The greater number of cases in the RAS scheme group compared to those of the Emanuel scheme is due to the fact that the former scheme sustained a stronger cyclone intensity than the latter. In a homogeneous sample that included all the cases of Bonnie, Howard, Estelle, and Georgette, the forecast distance errors are depicted in Fig. 1, showing a consistently superior forecasts using the Emanuel scheme.

For each forecast of the storm, we examine the synoptic-scale flows within an area covering 100 degrees in longitude and 50 degrees in latitude, centered roughly near the storm track in the five-day forecast period. For the two recurving hurricanes in the Atlantic, the intensity of the mid-latitude trough over the U. S. Continent is over-predicted by the RAS scheme and is better predicted by the Emanuel scheme. This leads to a better prediction of the tracks for Hurricane Bonnie and Danielle. For the three more westward moving hurricanes in the eastern Pacific, there appears to be no significant synoptic system that affects the movement of the storms. The more realistic structure of the storm simulated by the Emanuel scheme, as will be described later, may contribute to the better track prediction. The most challenging task is for Typhoon Rex in the western Pacific. Even though the Emanuel scheme produces smaller forecast distance error, the irregular motion of Rex is not captured by either scheme. The mechanism associated with Rex's oscillatory motion remains undetermined.

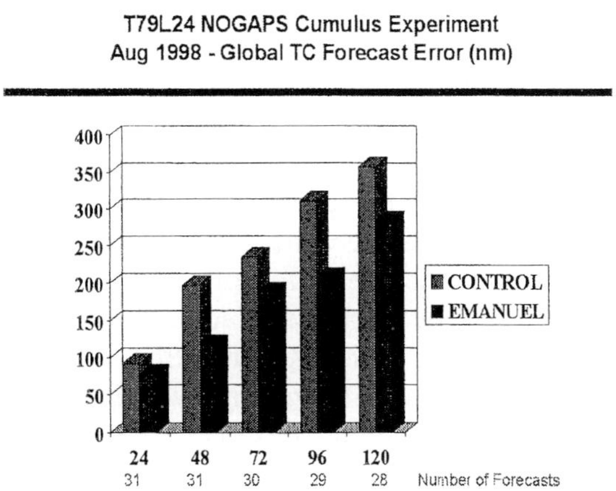

Figure 1. Mean tropical cyclone track error for the tropical storms Bonnie, Howard, Estelle, and Georgette vs. forecast hour for the month of August 1998 using T79 NOGAPS with the RAS scheme (control) and the Emanuel scheme.

In global average, the Emanuel scheme consistently produces larger amounts of precipitation than the RAS scheme does. But the RAS scheme generates more intense precipitation in smaller area over ocean as well as over heated land. The horizontal scales of the 24-h accumulated precipitation patterns from forecasts using the Emanuel scheme are smaller in general than those using the RAS scheme. The threat score computed against the NCEP rain analysis indicates that the Emanuel scheme is superior to the RAS scheme for all scales over the tropics and over the continental United Stated, the score is higher for light rain events but lower in the heavy rain events (Fig. 2). The localized high-intensity rainfalls in the RAS scheme was eliminated by the Emanuel scheme and replaced by more scattered precipitation. The quasi-stationary rain belt in the monsoon trough during the forecast period, extending from the Pacific Ocean to the southern part of North America, is also better predicted by the Emanuel scheme.

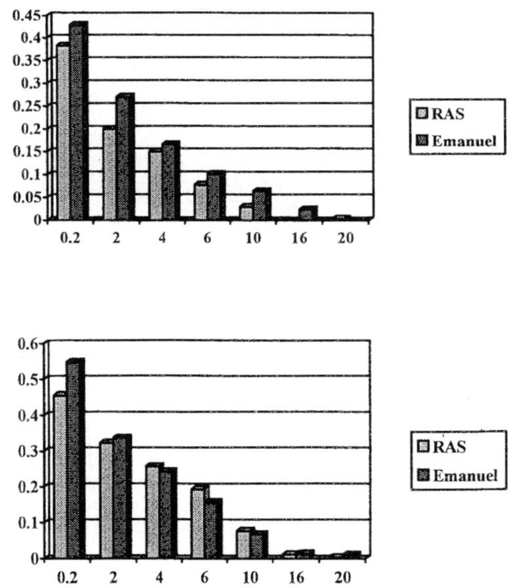

Figure 2. Threat score for the first week of August 1998: a) over tropics; b) over continental United States.

In most cases, the rainfall pattern generated by RAS scheme surrounding a cyclone is concentric with its maximum near the storm center, while the rainfall amounts using Emanuel scheme show more of an asymmetric annulus with a minimum at the center and several maxims surrounding it. We use Hurricane Estelle in the eastern Pacific to demonstrate the characteristic differences of the prediction by these two schemes. The 12h-accumulated rainfall amounts at 48 h forecast for Estelle on July 31 are depicted in Fig. 3.

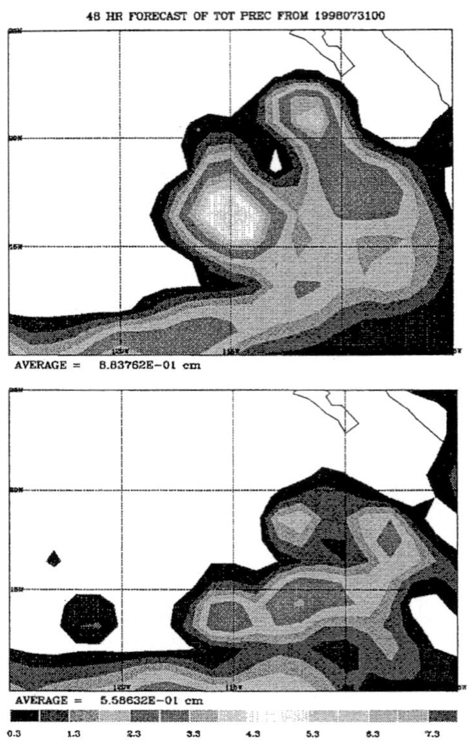

Figure 3. 12h accumulated rainfall amount at 48h for Hurricane Estelle on 31 July 1998: a) RAS scheme, and b) Emanuel scheme.

There is no indication of any eye-wall structure in the simulations using the RAS scheme, while the simulations using the Emanuel scheme certainly conform better with the conventional understanding of the rain band structure of tropical cyclones where the eye is dominated by downward motion with little rain (Anthes 1982). The intense precipitation predicted for tropical cyclones by the RAS scheme likely accounts for some spurious tropical genesis in NOGAPS. The overall intensities of the storms are weaker in the experiments using the Emanuel scheme compared to those using the RAS scheme

In the simulation of Hurricanes Estelle by RAS scheme, the vertical circulation in the downward branch of the vertical circulation surrounding the storm breaks into two parts in early stage. Later on, the upper part of the downward motion propagates over the storm center and caps the storm with a large downward motion (Fig.4). The large descending motion above the storm center causes distortion of the horizontal circulation at upper levels so that centers, determined either by vorticity maximum or pressure minimum, at lower levels are offset horizontally from those at higher levels (Fig. 5). On the other hand, the predicted cyclones using the Emanuel scheme have a well-organized vertical circulation extending to the top of the atmosphere (Fig.6) and the structure of the storm remains barotropic (Fig. 7).

Some experiments are carried out in an attempt to improve the RAS scheme in simulating the structure of tropical cyclones, especially the vertical circulation. The first attempt is to include a downdraft in the current version of the RAS scheme. This, however, does not alleviate the problem. Other experiments include adjusting the vertical weighting of the work function, modifying the reference profile of the work function. None of these led to any significant improvement of the tropical cyclone simulations. This study raises the hypothesis that the superiority of the vertical circulation structure in simulated tropical cyclones using the Emanuel scheme may be due to its more accurate treatment of the sub-cloud drafts and the microphysics processes.

These encouraging results prompted further test of the Emanuel scheme using

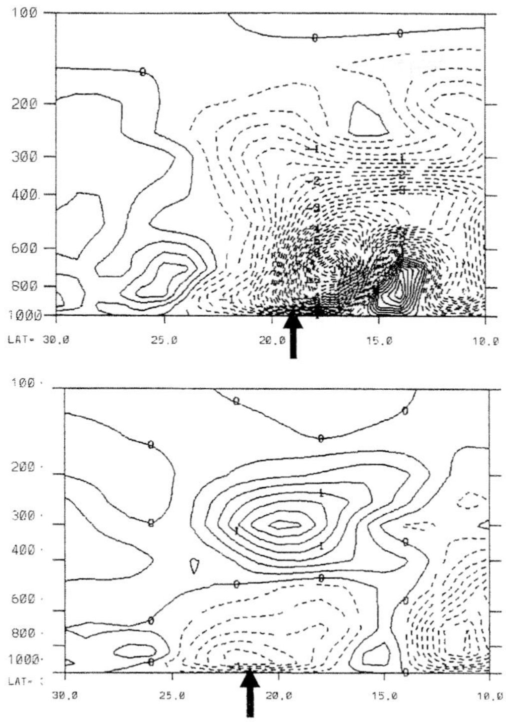

Figure 4. Vertical motion across the center of the storm for Estelle using the RAS scheme at a) 48h and b) 06 h. Dashed lines indicate upward motion and solid lines indicate downward motion. The center of the storm is marked by an arrow.

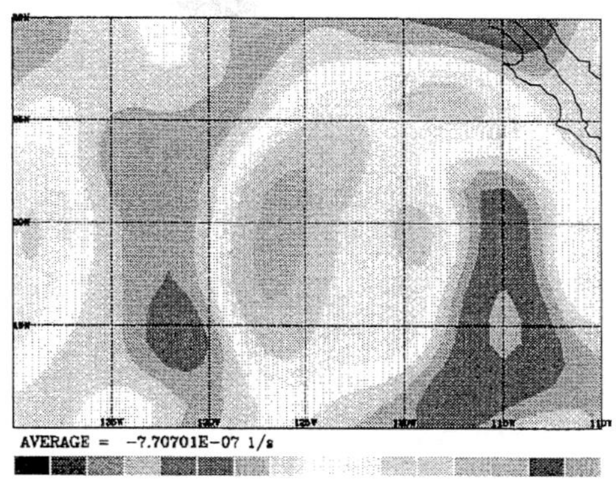

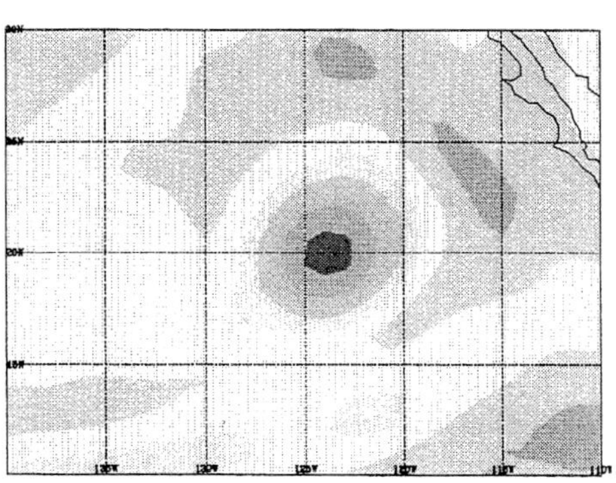

Figure 5. Relative vorticity for Estelle at 96h simulated using the RAS scheme: a) on the 1000 mb surface, b) on the 500 mb surface.

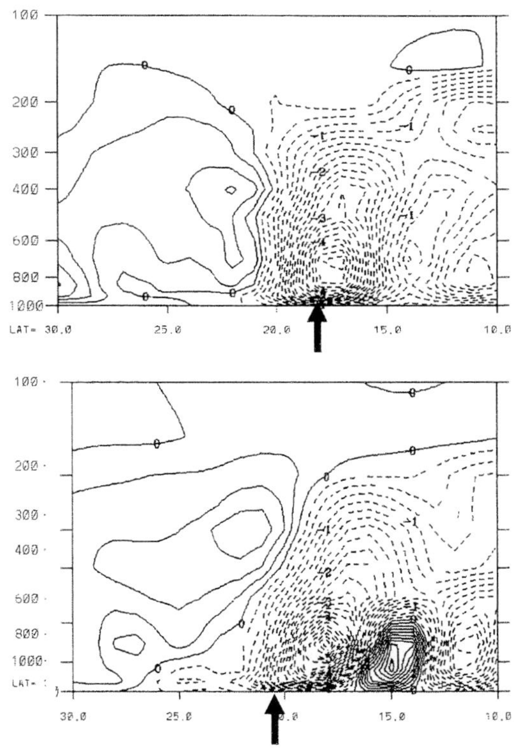

Figure 6. Same as Fig. 4 except using the Emanuel scheme.

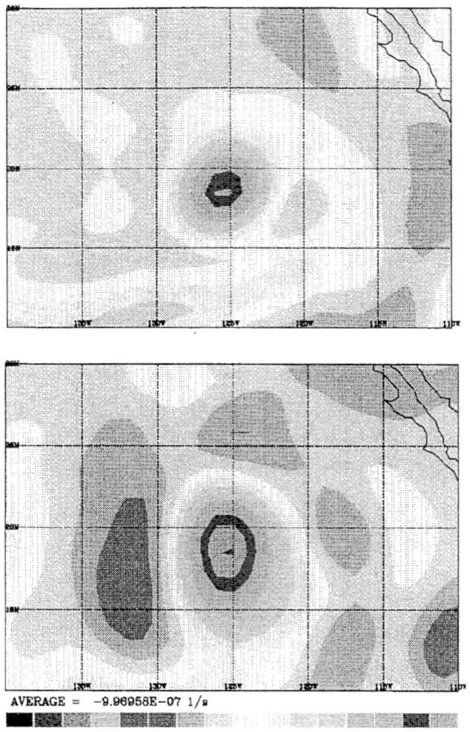

Figure 7. Same as Fig. 5 except using the Emanuel scheme.

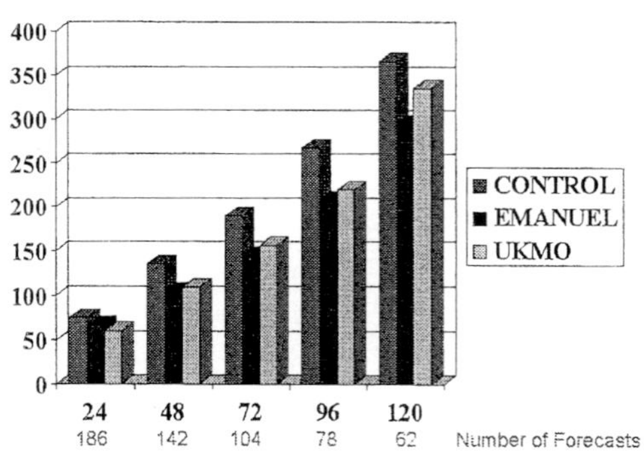

Figure8. Mean tropical cyclone track error vs. forecast hour for the months of August-September using the T159 NOGAPS with the RAS scheme, the Emanuel scheme, and data provided by UKMO.

higher resolution T159, which is the same as in the operational NOGAPS. The track distance errors for the summer of 1999 are shown in Fig.8. For all the forecast time, the Emanuel beats the RAS scheme.

4. Summary

This study analyzes and compares the tropical cyclone simulations of NOGAPS using the Arakawa-Schubert scheme and the Emanuel scheme for convective parameterization over the period August 1998. The Emanuel scheme is found to perform better than the RAS scheme in the track forecast for all six tropical cyclones that occurred within that period. The average 120h forecast distance error improved from 416 NM to 354 NM. Detailed analysis indicates that unrealistic vertical structures of tropical cyclones exist in simulations using Arakawa-Schubert scheme that do not occur in those using the Emanuel scheme. Even with its modest resolution of T79, the precipitation pattern produced by Emanuel scheme suggests that an eye structure may be generated. The Emanuel scheme also produces more scattered small-scale precipitation over a large area outside of tropical storm region. Instead, the Arakawa-Schubert scheme generates very intense rainfall near the storm center for both tropical cyclones as well as convective systems over land.

Examination of the vertical circulation associated with simulated tropical cyclones indicates that the Arakawa-Schubert scheme produces de-coupled vertical circulation on the downward branch surrounding the storm. The upper part of the downward motion moves toward the center of the storm and caps the storm with strong downward motion. This downward motion distorts the horizontal circulation in the upper part of storm and the center is de-coupled from lower levels. These unrealistic characteristics associated with tropical cyclones simulated by Arakawa-Schubert scheme suggests that there is an inherited problem in the Arakawa-Schubert scheme that is not easily resolved. The superior performance of Emanuel scheme leads to the adoption of it in NOGAPS in the daily operational forecasts in April 2000.

REFERENCES:

Anthes, R. A., 1982: Tropical cyclones. Their evolution, structure and effects. American Meteorological Society. 203pp.

Arakawa, A., and W. H. Schubert, 1974: Interaction of a cumulus cloud ensemble with the large-scale environment, Part 1. *J. Atmos. Sci.*, **34**, 674-701.

Betts, A. K., 1986: A new convective adjustment scheme. Part I: Observational and theoretical basis. *Quart. J. Roy. Meteor. Sco.*, **112**, 677-691.

Emanuel, K. A., 1991: A scheme for representing cumulus convection in large-scale models. *J. Atmos. Sci.*, **48**, 2313-2335.

Fiorino, M., J. S. Goerss, J. J. Jensen, and E. J. Harrison, 1993: An evaluation for the real-time tropcial cyclone forecast skill of the Navy Operational Global Atmospheric Prediction System in the western North Pacific., *Wea. Forecasting*, **8**, 3-24.

Fritsch, M., and C. F. Chappell, 1980: Numerical prediction of convectively driven mesoscale pressure system. Part I: Convective parameterization. *J. Atmos. Sci.*, **37**, 1722-1733.

Goerss, J. S., and R. A. Jeffries, 1994: Assimilation of synthetic tropical cyclone observations into the Navy Operational Global Atmospheric Prediction System., *Wea. Forecasting*, **9**, 557-576.

Hogan, T. F., T. E. Rosmond, and R. Gelaro, 1991: The NOGAPS forecast model: A technical description. NOARL Report 13. Naval Research Laboratory (formerly NOARL), Monterey, CA.

Kain, J. S., and M. Fritsch, 1990: A one-dimensional entraining/detraining plume model and its application in convective parameterization. *J. Atmos. Sci.*, **47**,2784-2802.

Kuo, J.-L., 1965: On formation and intensification of tropical cyclones through latent heat release by cumulus convection. *J. Atmos. Sci.*, **22**, 40-63.

-----------, 1974: Further studies of the parameterization of influence of cumulus convection on large-scale flow. *J. Atmos. Sci.*, **31,** 1232-1240.

Louis, J. F., 1979: A parametric model of vertical eddy fluxes in the atmosphere. *Boundary Layer Meteor.*, **17**, 187-202.

Manabe, S., J. Smagorinsky and R. F. Strickler, 1965: Simulated climatology of a general circulation model with a hydrologic cycle. *Mon. Wea. Rev.*, **93**, 769-798.

Moorthi, S., and M. Suarez, 1992: Relaxed ArakawRASchubert: A parameterization of moist convection for general circulation models. *Mon. Wea. Rev.*, **120**, 978-1002.

AN EMPIRICAL ORTHOGONAL FUNCTION ANALYSIS OF THE CHARACTERISTICS OF TYPHOON RAINFALL IN THE VICINITY OF TAIWAN

Tien-Chiang Yeh
Central Weather Bureau, 64 Kung Yuan Road, Taipei
Email: yeh@cwb.gov.tw

ABSTRACT

The typhoon rainfall distributions over Taiwan area are investigated through empirical orthogonal function analysis. The data set includes the hourly rainfall observed at 23 stations when a typhoon was inside the region between 18°N and 28°N, 116°E and 126°E from 1970 to 1996. The rainfall pattern of the leading EOF modes shows that the Taiwan topography is a dominant factor to the distribution of typhoon rainfalls. The rainfalls are enhanced on the up-slope side of the mountain range and the rainfalls are suppressed on the down-slope side of the mountain. By investigating the relationship between the amplitude of each EOF mode with the typhoon intensity, Julian day, typhoon moving speed and direction, some favourable conditions for enhancing or suppressing the rainfalls with the selected mode are found.

1. INTRODUCTION

The torrential rainfalls associated with typhoons often caused severe damages over Taiwan area. For an example, the daily rainfalls more than 1,000 mm and two-day accumulated rainfalls of nearly 2,000 mm were observed at Alishan station during the invasion of typhoon Herb in 1996. Although the Central Weather Bureau (CWB) had issued skilful track forecasts of typhoon Herb, however the damages were still not preventable. Fifty-one people were killed and the losses were more than NT$35 billions.

Wu and Chi (1973) have shown the typhoon rainfalls over Taiwan area are correlated with the location of the typhoon centres. Climatology averaged rainfall (based on the spatial distribution of typhoon centres) was then developed and applied (Wu and Shieh 1990) by the CWB as one of the operational objective typhoon rainfall forecast aids in CWB. On the other hand, Chang et al. (1993) displayed the typhoon hourly rainfall distributions over Taiwan area as a function of typhoon-centre location. Several modes of rainfall patterns were shown associating with the dominated empirical orthogonal function (EOF) pressure modes. This work extends their analysis to study the typhoon rainfalls over Taiwan area based on EOF modes.

2. DATA

A 20 year data set, in 3-hour intervals, from 1971 to 1990 was used by Chang et al. (1993) to show the typhoon surface structure over Taiwan area. To extent the data set, the hourly data and the JTWC's best tracks from 1970 to 1996 were prepared. The cases were selected when the typhoon centre were inside the region between 18°N and 28°N, 116°E and 126°E. During this period of the selected 27 years, 241 typhoons were found within the selected region. The database includes a total of 13,036 cases of hourly rainfalls at 23 surface stations over Taiwan area. Since three stations, of station numbers 46690, 46691 and 46693, are located very close to each other, instead of including all the three observations in the EOF analysis, only the averaged values were used. Therefore, the total number of the EOF modes is 21.

The mean rainfalls are in a range from 0.4 to 1.8 mm/h. Smaller values are over western Taiwan. Larger values are over eastern and northern Taiwan. The maximum value is over the central mountain area. The mean rainfalls are relatively small compare to what we expecting of the rainfalls during the invasion of typhoons. The smaller values are due to that of many cases included are the typhoon centres still located few degrees away from Taiwan. For those cases, no or very small rainfall was observed.

3. HOURLY RAINFALL EOF MODES

The percentage of the rainfall variances represented by the first eight hourly rainfalls EOF modes are 23.6%, 14.7%, 9.1%, 6.4%, 5.5%, 5.0%, 4.2% and 4.2%, respectively. Those values are smaller than that of the variances represented by the leading EOF modes in the surface pressure analysis (Chang et al. 1993). Typhoon rainfall distributions are more complicated than that of the surface pressure distributions.

The pattern of the hourly rainfall EOF mode 1 (Fig. 1a) is very similar to that of been shown in Fig. 12 of Chang et al. It has a uniform sign over the entire domain. The maximum value of about 6 mm/h locates over the central Taiwan. Another local maximum value is over the northern Taiwan. Chang et al. (1993) shown the amplitudes of the leading surface pressure modes depend mainly on the 'spatial distribution' according to the typhoon centre locations. Similar spatial distribution of the mean amplitudes of mode 1 (Fig. 2a) shows a maximum centre over northeastern Taiwan. When typhoon centres move to this region, the mode will have the largest amplitude. The maximum rainfall rate, of about 15 mm/h, contributed from this mode is located over the western side of the Central Mountain Range (CMR). Here large rainfall is expected as the typhoon cyclonic flow is encountering with the steep topography.

The EOF mode 2 (Fig. 1b) has a pattern with out-of-phase rainfall distributions between northern and southern Taiwan. The maximum of 3.4 mm/h is over southwestern Taiwan and the minimum value of –2.0 mm/h is over the central mountain area. The spatial distribution of the amplitudes associated with mode 2 (Fig. 2b) shows a positive/negative centre over southern/northern tip of Taiwan. When a typhoon moves to the southern Taiwan, the maximum/minimum centre of the EOF mode 2 will be in the up-/down-stream side of the typhoon cyclonic flow over mountain. The topography is again a dominant factor to the rainfall distributions of EOF mode 2. When a typhoon moves to the northern Taiwan, the rainfall from both of mode 1 and mode 2 will have a same sign over central and northwestern Taiwan. The combination of these two modes will reach to a value of about 17 mm/h rainfall over the central mountain area. On the other hand, the rainfall contributes from mode 1 and mode 2 will have opposite signs for area over southern and eastern Taiwan. The rainfalls become very small over the downwind side of CMR.

The pattern of EOF mode 3 (Fig. 1c) has an out-of-phase rainfall distribution between northern and southern Taiwan. The amplitude associated with the mode has a positive/negative centre over east/south of Taiwan. Right over the Taiwan, where the amplitudes associated with mode 1 and mode 2 are large, the amplitudes associated with mode 3 are relatively small. The spatial distribution of the amplitude associated with mode 3 (Fig. 2c) shows that when typhoon centres move to the northeast of Taiwan, a maximum of about 3 mm/h rainfall is associated with this mode. Meanwhile, over the central mountain area, the mode will have a negative contribution. The rainfall will be reduced by a value of about 2 mm/h. For the negative phase of mode 3, the rainfall over the central mountain area will be enhanced from this mode by a value of about 1.5 mm/h when typhoon centres move to the northern part of the Taiwan Strait. The strong southwesterly flow associated with the storm when the typhoon centres move to the northeast of Taiwan can explain the enhancement of rainfall upstream of the central mountain. On the other hand, suppressions of rainfall are at the downstream side of the mountain. The maximum value of the drying effect is about 2 mm/h over Taipei area. The topography is again a dominant factor to the rainfall distributions of EOF mode 3.

Both rainfall patterns of mode 4 and mode 5 have two node-lines over Taiwan to separate the rainfall distributions into three parts. For mode 4, the sign over central Taiwan is out-out-phase with the signs over the northern and the southern Taiwan. This mode has an effect of enhancing the rainfall near the typhoon centre. For mode 5, the structures of the rainfall distributions are even more complicated. In general, the signs over the Taiwan are separated into two pairs. A pair, of positive sign, is over the southeastern and the northwestern Taiwan and the other pair, of negative sign, is over the southwestern and the northeastern Taiwan. The amplitude of this mode has a maximum when the typhoon centres move to the central Taiwan. This mode will increase/decrease the rainfall in northeast/southeast and in southwest/northwest of the centre by a value of about 2.4/1.5 mm/h. Here the cyclonic flow associating with the typhoon is in up-/down-slope of the mountain. For the negative phase, the typhoon centres spread around Taiwan with relatively small amplitudes. Therefore, it is likely be a result of computational.

4. 12-HOUR ACCUMULATED RAINFALL EOF MODES

Each of the hourly rainfall EOF leading modes is physically meaningful. Those EOF patterns show the feature of redistributing rainfalls by the effect of topography. However, the accumulated percentage of variances represented by those leading modes is still small. Pronounced short period variations of rainfalls may cause such a result. To eliminate the short period variations, accumulated rainfalls are analyzed. The pattern of each leading accumulated rainfall EOF mode and its associated spatial distribution of amplitudes are very similar to those from hourly rainfalls EOF modes. Some differences are that the accumulated percentage of the variances represented by each of the leading modes is increased with longer period of rainfall accumulation. For example, the accumulated percentage of variances represented by the leading 5 modes increase from 59% (hourly rainfalls) to 68%, 74%, and 80% for 3, 6, and 12 hours accumulated rainfalls, respectively. For 12 hours accumulated rainfall, the percentage of variances represented by the leading three modes are 40%, 20%, and 13%. About 3 quarters of 12-hour accumulated rainfalls can be explained by these three modes.

For typhoon rainfall forecast, the climatology average according to the spatial distribution of typhoon centres is developed (Wu and Shieh 1990) for operational use at CWB. Chen et al. (1992) included more predictors, such as typhoon intensity, typhoon moving speed and Julian day, to develop a typhoon rainfall forecast model based on regression formulas. However, no significant improvements were reported, which may indicate that the typhoon rainfalls over Taiwan area do not depend on the typhoon intensity, the typhoon moving speed and the Julian day in a simple way. To evaluate whether each mode is related to those 'predictors' in a systematic way, Figs. 3, 4 and 5 show the scatter distributions of the 'amplitudes deviation' and the 'predictors' for modes 1, 2 and 3 respectively. The 'amplitude deviation' is defined as the difference between the observed amplitude and the mean that is averaged according to the spatial distribution of the typhoon centres. The figures show the amplitude deviations are also not dependent on the 'predictors' in a simple way. Nevertheless, some guidance can be found. For example, below-normal rainfalls from mode 1 are usually found for cases with weaker intensity (<60 knots central winds), in earlier (before June) or later season (after September), faster moving (>7.5 m/s), and the moving direction between 0° (eastward) to about 30° (northeastward) and 210° (southwestward) to 360° (eastward).

For the negative phase of mode 2, when typhoon centres are located at north of about 23.5°N, the mode is suppressed for cases with weaker intensity (<50 knots), in earlier season (before middle of June) or later season (after September), with smaller (<2.5 m/s) or larger (>7.5 m/s) translation speed, and the moving direction are between 30° to 60° and between about 200° to 300°. For the positive phase of mode 2, when typhoon centres are located at south of about 23.5°N, the rainfall of the mode is suppressed for cases with weaker intensity (<30 knots), very small moving speed (<1 m/s) or very large moving speed (>9 m/s), and the moving direction between about 220° to 360°.

For mode 3, the systematic relations between the rainfall of the mode and the 'predictors' become less obvious. For the negative phase of mode 3, when the typhoon centres are over west of about 122°E, the rainfall of the mode is suppressed for cases in earlier (before July) season or later season (after September). For the positive phase of mode 3, when the typhoon centres are over east of Taiwan, the rainfall of the mode is suppressed for cases with very strong intensity, in earlier season (before middle of June), very fast moving (>10 m/s), and the moving directions are between 320° to 70° when the typhoons are moving away from Taiwan.

5. SUMMARY

Typhoon rainfall forecast is a crucial issue to the weather service in Taiwan. More studies are required to improve the forecasts. Based on the empirical orthogonal function (EOF) analyses of the typhoon rainfall distributions over Taiwan, we found the topography is a major factor to redistribute typhoon rainfalls. For hourly rainfalls, the total percentage of variances represented by the leading five EOF modes is about 60%. Meanwhile, the percentage increases to 81% for the 12 hours accumulated rainfall modes. Among the modes, the leading three modes can explain 73% of the rainfall variances.

The EOF mode 1 represents the increase of rainfalls over the whole island when typhoon moving into Taiwan area. The EOF mode 2 reflects the topography enhancement of rainfalls due to the up-slope flow impending the north-south oriented Central Mountain Range. The EOF mode 3 reflects the heavy rainfalls over northern/southern Taiwan when the typhoons move to northeast/northwest of Taiwan, where the induced strong northwest/southwest flow encounters the mountain.

The amplitude of each leading modes is mainly dependent on the spatial distribution of typhoon centre location. The dependence of the amplitude deviation, from the spatial mean, to the typhoon intensity, the Julian day, the typhoon moving speed and direction has been investigated. The results show the relations are not linked in a simple way. However, some conditions favourable for enhancing or suppressing the rainfall of selected modes are found.

ACKNOWLEDGMENTS. This study is supported by the Central Weather Bureau and the National Science Council under Grant NSC87-2621-P052-012.

REFERENCES

Chang, C.-P., T.-C. Yeh, and J. M. Chen, 1993: Effects of terrain on the surface structure of typhoons over Taiwan. *Mon. Wea. Rev.*, **121**, 734-752.

Chen, G. T., C.-S. Li, S.-T. Wang, S.-S. Chi, B. C. Chou, 1992: *The development of the statistics forecast model for typhoon rainfall prediction.* Chapter 2, Studies On The Improvement Of The Rainfall Forecast: Typhoon Rainfall Forecast, Research Report of Year 1992, Central Weather Bureau, 133pp.

Wu T.-Y., and K.-H. Chi, 1973: A study on the prediction of rainfall over Taiwan during typhoon invasion by analogue techniques (II). *Meteorological Bulletin*, **19-3**, 10-19.

Wu T.-Y., and S.-L. Shieh, 1990: *On the applications of the scientific results in meteorological research to typhoon forecast operations (III).* Technical Report 79-37, National Science Council, 137pp.

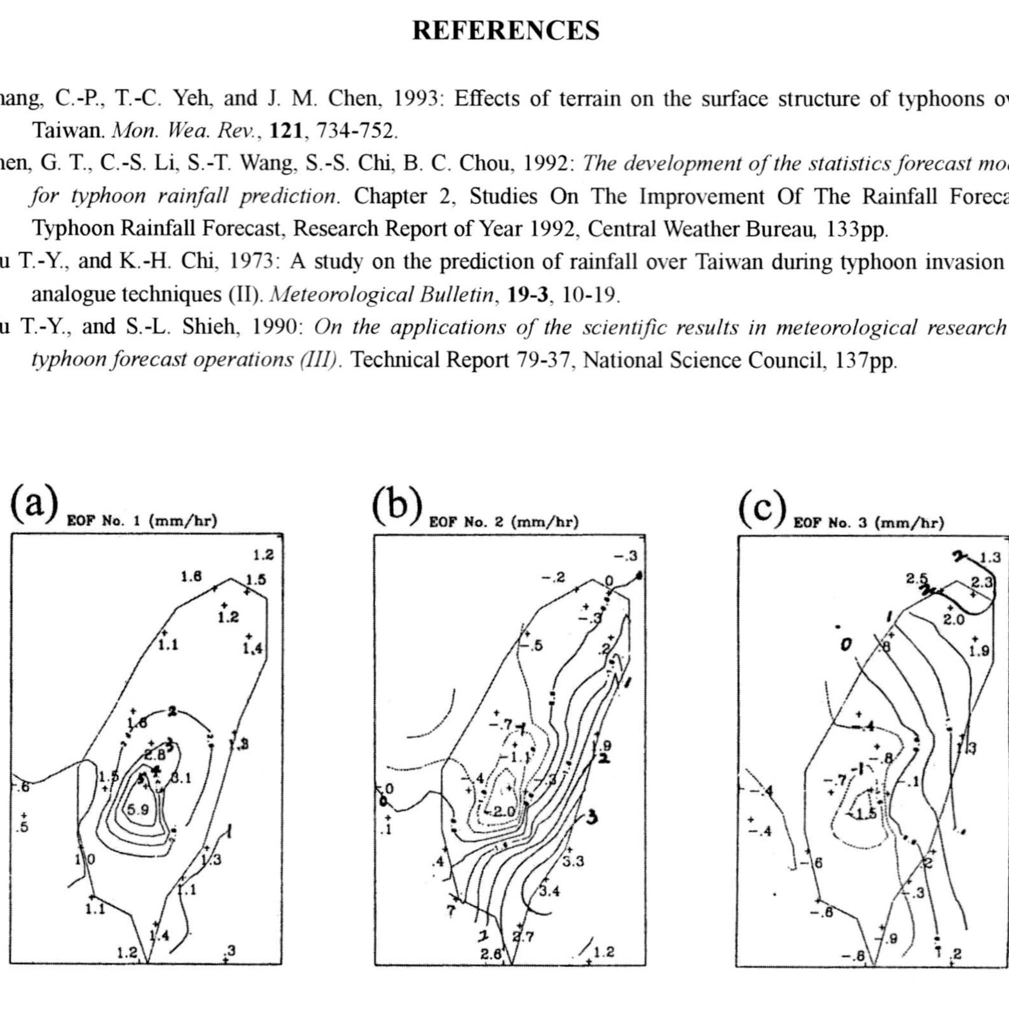

Fig. 1. The rainfall pattern of EOF (a)mode 1, (b)mode 2, and (c)mode 3. Contour intervals are (a) 1 mm/h, (b) and (c) 0.5 mm/h.

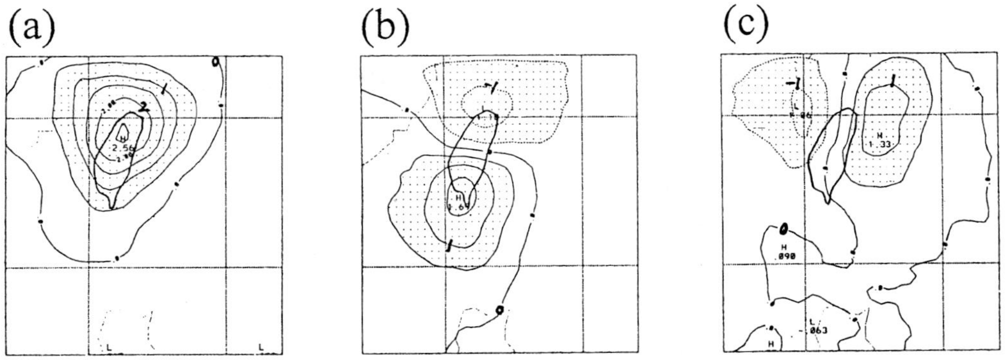

Fig. 2. The spatial distribution of the amplitude for EOF (a)mode 1, (b) mode 2, and (c) mode 3. Contour intervals are 0.5.

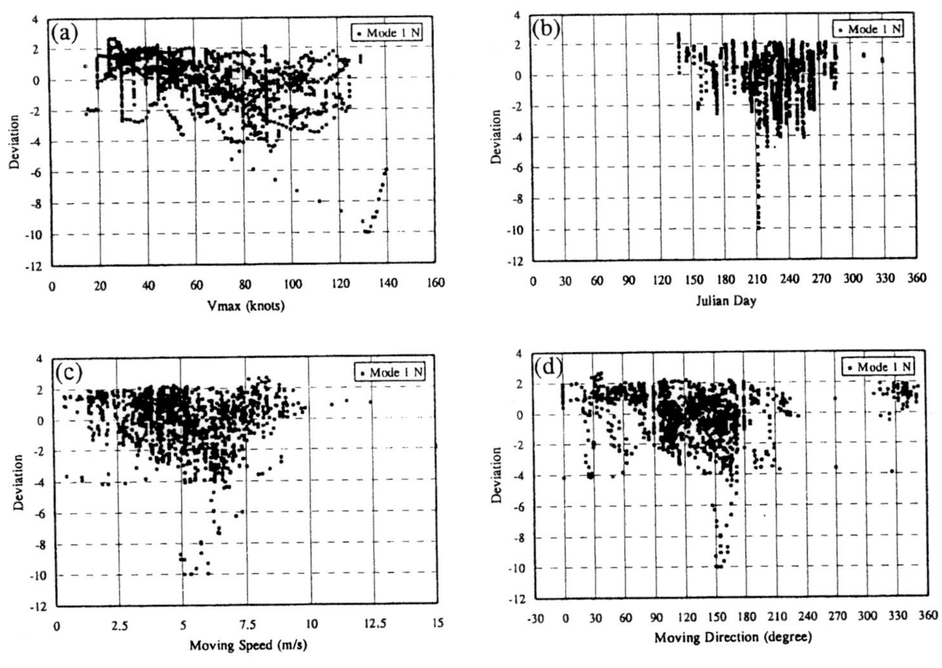

Fig. 3. The amplitude deviation of EOF mode 1 as function of (a) typhoon intensity measured in the maximum wind speed (knots), (b) Julian day, (c)typhoon moving speed, and (d)typhoon moving direction.

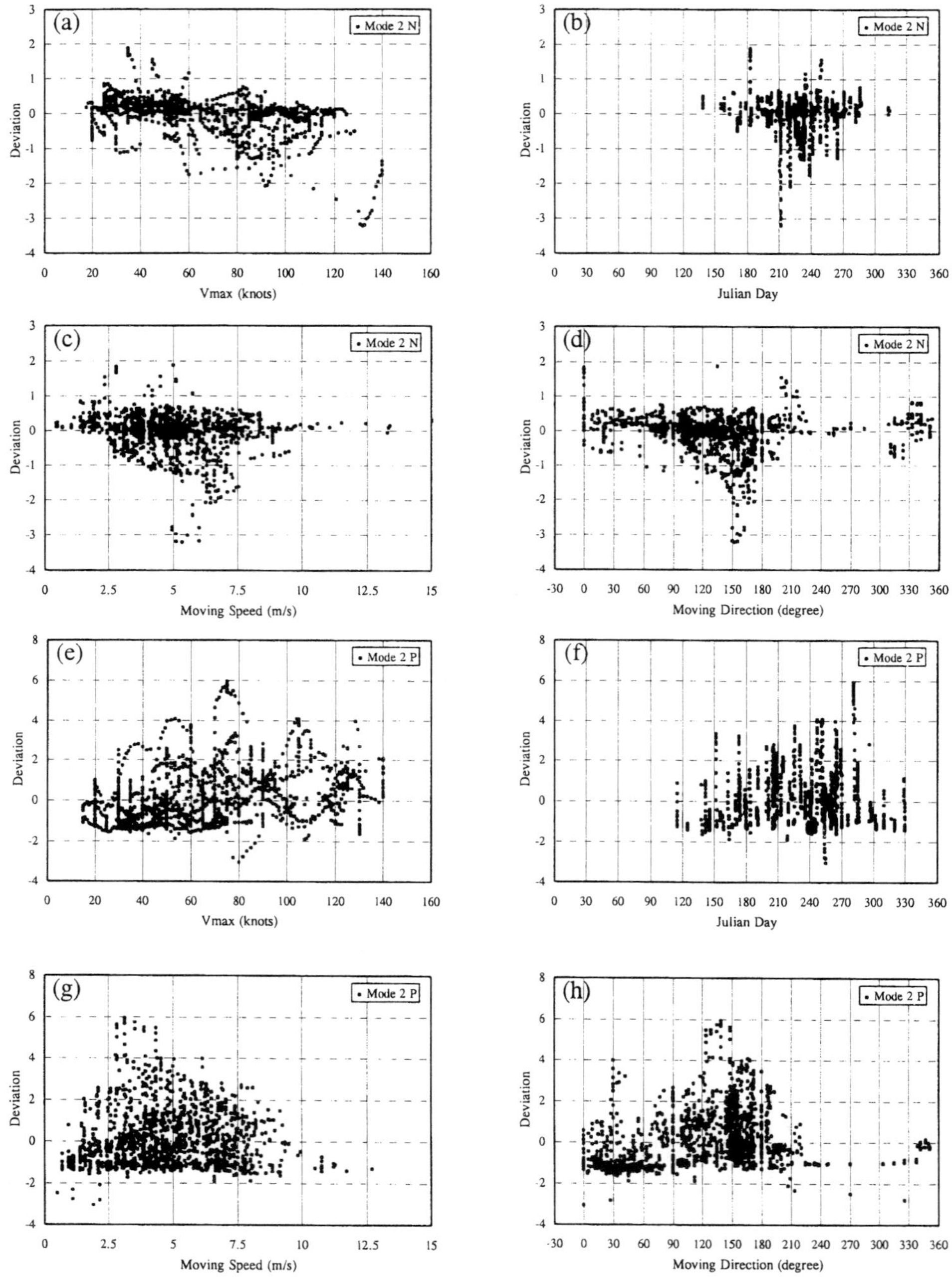

Fig. 4. Same as Fig. 3, except for the amplitude deviation of EOF mode 2 in negative phase. (e) to (f) in positive phase.

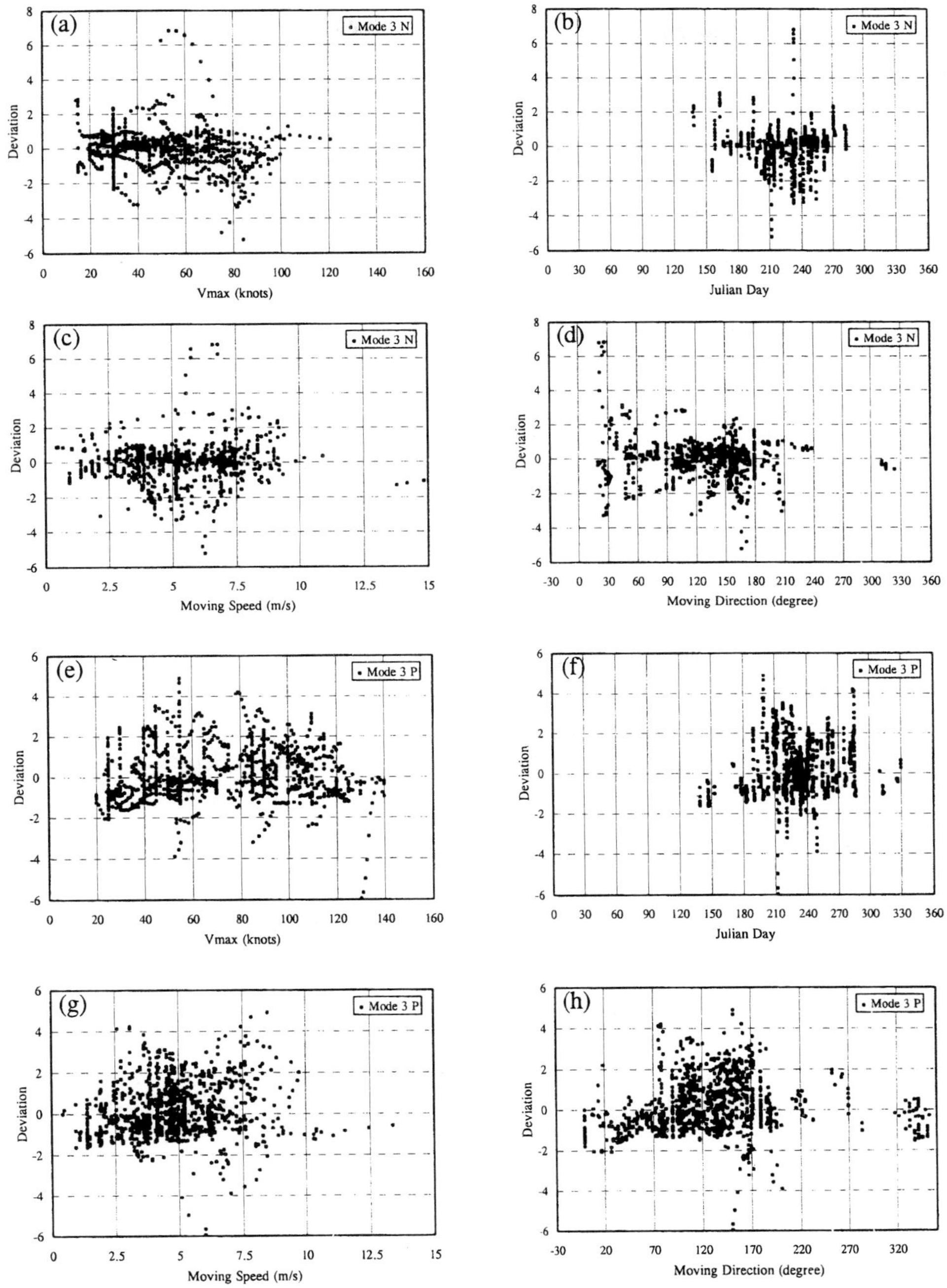

Fig. 5. Same as Fig. 4, except for the amplitude deviation of EOF mode 3.

A NUMERICAL SIMULATION OF THE RAINFALL ASSOCIATED WITH TYPHOON HERB (1996) USING THE PSU/NCAR MM5

Chun-Chieh Wu, Tzuu-Hsiung Yen
Department of Atmospheric Sciences, National Taiwan University, Taipei
and
Ying-Hwa Kuo and Wei-Wang
National Center for Atmospheric Research, Boulder, CO

ABSTRACT

High-resolution numerical simulations and detailed analyses are carried out to examine the ability of the model in simulating the track, intensity change, and the detailed mesoscale wind and precipitation distributions associated with Typhoon Herb (1996). The 2.2 km MM5 successfully simulates the mesoscale rainfall distribution associated with Herb, although the predicted maximum 24-h rainfall of 1290 mm was about 75% of the observed amount. It is shown that the ability of the model in successfully simulating the observed rainfall hinges on the model's horizontal grid resolution in describing the Taiwan terrain. The presence of the Central Mountain Range plays a key role in substantially increasing the total rainfall amounts associated with Typhoon Herb by lifting the moisture-laden air over the upslope regions of the mountains. Overall, Herb is a case in point to indicate the intimate relation between the rainfall distribution and the Taiwan topography.

1. INTRODUCTION

Typhoon Herb (1996) made landfall at I-Lan in the northeast Taiwan at 1444 UTC July 31, 1996, and became one of the most damaging tropical cyclones in the recent history of Taiwan (Wu and Kuo 1999). Besides the intense gusty wind associated with Herb, the most interesting aspect of Typhoon Herb is the torrential rainfall it produced over the Central Mountain Range (CMR). As Typhoon Herb traversed northern Taiwan, the airflow over the main body of Taiwan changed from northwesterly to westerly, and then to southwesterly. The moisture-laden typhoon circulation was forced to move over the CMR, and the lifting associated with the upslope motion produced extremely heavy precipitation, which resulted in numerous devastating landslides and debris flows in central Taiwan and caused heavy loss of both life and property. Fig. 1 shows the rainfall accumulated during the passage of Herb from 0000 UTC July 31 to 0000 UTC, August 1, and the corresponding Taiwan terrain. Two major precipitation centres are identified: one at Mountain Nio-Dray in northern Taiwan, with a total precipitation of 858 mm, and the other at Mount A-Li in central Taiwan, with a record-breaking total accumulation of 1736 mm over the one-day period. In sharp contrast, downslope flow persisted over eastern Taiwan, and little precipitation was recorded. The downslope flow also produced a topographically induced meso-low over eastern Taiwan near Hualien.

In terms of track prediction, Herb is a relatively easy case for most dynamic and statistical models (Wu and Kuo, 1999). For a storm that produced such a severe damage to Taiwan, the interest goes beyond an accurate track forecast. For evacuation, flood mitigation, and decisions related to all public activities, accurate forecasts of mesoscale wind and precipitation distribution are needed. The challenging question is: can a numerical prediction model provide a good rainfall and wind field forecast for a landfalling typhoon with such intensity? Wang

(1980; 1989) showed that the mesoscale precipitation and wind field distributions during typhoons affecting Taiwan are strongly modulated by the topography of Taiwan. Wu and Kuo (1999) indicated that a high-resolution numerical model is needed to resolve the detailed topography of the CMR and to evaluate how it affects the track, intensity, and mesoscale wind and precipitation distribution as Herb approached and crossed northern Taiwan.

Previous work (Liu et al., 1997) have demonstrated the capability of the PSU-NCAR non-hydrostatic mesoscale model (MM5) in providing a multiscale numerical study of Hurricane Andrew. Following Wang and Kuo (1997), in this work, we experimented with a high-resolution (2.2 km) MM5 on the prediction of Typhoon Herb. Our primary objective is to investigate the ability of the model in simulating the track, intensity, as well as the detailed precipitation distribution associated with Typhoon Herb, and to assess the impact of the model physics and the model's horizontal resolution and detailed topography on rainfall simulation. The model description and experiment design are presented in section 2. Section 3 provides the results of the model simulations. The summary is outlined in section 4.

2. MODEL DESCRIPTION AND EXPERIMENT DESIGN

This study makes use of the Penn State--NCAR mesoscale model version 5, MM5 (Dudhia 1993; Grell et al. 1995). The MM5 model is a non-hydrostatic, primitive-equation model in sigma coordinates. A suite of model physical parameterization schemes is available for subgrid-scale convection, grid-resolvable scale microphysical parameterization, the planetary boundary layer forcing, and radiation. The particular version of MM5 used here in the control experiment includes the following physics options: (1) the Betts and Miller (1986) cumulus parameterization, (2) the Blackdar PBL scheme, (3) the atmospheric radiation scheme of Dudhia (1989), and (4) the simple ice physics scheme of Dudhia (1989). The initial condition was based on the ECMWF global analysis. We first perform a 60-km/20-km MM5 simulation starting at 1200 UTC 29 July with a bogus Rankine vortex according to the characteristics of the storm estimated by the best-track analysis. As the typhoon vortex is well developed in 12 h (0000 UTC 30 July), the fully developed typhoon vortex is then lifted from the model simulation to merge with the environmental flow filed (using the filtering method from Kurihara et al., 1995) derived from EC/TOGA analysis at 0000 30 July. The spun-up typhoon vortex is placed at the observed storm location according to the best track analysis. The forward model simulation then commences at 0000 UTC 30 July. The model is integrated for 48 hours, ending at 0000 UTC 1 August.

We have conducted a series of experiments with different horizontal resolutions. All these experiments used the same PBL, radiation, and cumulus parameterization schemes. Experiment E60 is a single mesh experiment with a horizontal resolution of 60 km. E20 is the control experiment with the nested 60/20 km mesh. E6.7 is a single-mesh simulation with a horizontal resolution of 6.7 km, and is run in one-way mode with initial and boundary conditions provided from the output of E20. E2.2 is the doubly-nested experiment of E6.7, with the inner grid having a horizontal resolution of 2.2 km (the configuration of the model nests is shown in Fig. 2). The experiment, E6.7NT, is similar to E6.7 except that the topography of Taiwan is removed. The purpose of this experiment is to show the impact of the CMR on the rainfall prediction and wind field distribution. Other simulations which test the sensitivity of the model results to different initial conditions, cumulus parameterization schemes and microphysics are also performed, and will be discussed in section 4.

3. RESULTS

(a) Track and intensity

Fig. 3 shows the comparison of forecast track with the best track analyses from the Joint Typhoon Warning Center (JTWC). The position error as compared to the JTWC analysis is 80 and 160 km for the 24 and 48h forecasts, respectively. Overall, the control experiment shows a fairly accurate track simulation for Gladys, although the simulated storm moved slightly too fast, and thus made landfall at 32h and left Taiwan at 38h, as compared with observed time of 38 and 43h, respectively. In the experiment with no Taiwan topography (E6.7NT), the track was very similar to the E6.7, indicating that the track of Herb was not much affected by the presence of the CMR. This result is likely resulted from the nature that Herb was approaching the lower and the northern end of the CMR.

As to the intensity simulation, the comparison of the central mean sea level pressure from the model output and the JTWC (Fig. 4) indicates that the intensity agreed well with the observation during most of the integration period. Although the model simulation slightly underestimates Herb's intensity (by about 5 hPa) throughout the forecast period, the model can capture the evolution of Herb's intensity, especially the steadiness and weakening before and during landfall. In the experiment E6.7NT, the model was able to maintain the same intensity for longer time period until it made landfall near southern China at 40h.

(b) Rainfall

Fig. 5 shows the daily (24-48h) rainfall prediction and the corresponding model terrain (from 0000 UTC 31 July to 0000 UTC 1 August) from four experiments with resolutions of 60 km (E60), 20 km (E20), 6.7 km (E6.7) and 2.2 km (E2.2), respectively. In general, all the four experiments produce similar 48h tracks (figures not shown). Comparison of Figs. 5a and b with Fig. 1a clearly indicates that both E60 and E20 produce heavy rainfall over the western side of the model terrain, but underestimate the rainfall amount and cannot simulate the major rainfall pattern as observed. Compared to the actual Taiwan terrain (Fig. 1b), the model terrains (Figs. 5a and 5b) appear to be too coarse to produce the detailed observed rainfall.

On the other hand, with a better horizontal resolution, the daily rainfall in E6.7 (E2.2) captures the two rainfall maxima, one over northern Taiwan with a peak amount of 816 (1002) mm, and the other over Mountain A-Li with 772 (1290) mm. The model also predicts a band of precipitation over the southern CMR with an amount exceeding 649 (1046) mm, which is in agreement with the observation. Meanwhile, the model is very successful in capturing the rain shadow over eastern Taiwan, which is on the Lee side of the mountains.

A careful comparison of the model terrain and daily rainfall in all four experiments shows that the rainfall takes place mainly along the western slopes of the CMR. On the other hand, results from Experiment E6.7NT (figures not shown) indicate broader and weaker rainfall distribution over Taiwan. For experiments with Taiwan terrain, the maximum rainfall tends to occur immediately to the west of the mountain peaks. The importance of upslope flow in producing the extreme heavy rainfall over the CMR is evident. Comparison of the rainfall amounts and distributions from the above experiments clearly indicates that the ability of the model in successfully simulating the observed rainfall degrades with the reduction of horizontal grid resolution.

One might ask whether the improvement in rainfall simulation in E6.7 against E20 is due to the improvement in resolving the storm structure or is due to the improvement in describing the model terrain. To attest this, we conduct another experiment, which is identical to E6.7, but with model terrain interpolated from E20. Even though the model resolution is 6.7 km, yet with a coarser 20-km model terrain, the rainfall distribution (figures not shown) becomes similar to E20 with only about 10% increase in the rainfall amount. This result strongly suggests that the better representation of the model terrain is a key for producing more realistic rainfall pattern associated with Herb.

Sensitivity experiments with various cumulus parameterization schemes (e.g., Kain and Fritsch, 1993, and Grell 1993) are also performed. Results (figures not shown) indicate that similar track and precipitation are produced, and that 90% of the rainfall is resolved on the grid-scale due to the strong topographic forcing, and therefore the rainfall is less sensitive to the choice of cumulus parameterization. Meanwhile, instead of using the simple-ice microphysics, we also perform a 6.7-km simulation using the Goddard scheme with graupels (Tao and Simpson, 1993). The results (figures not shown) indicate a dramatic increase in rainfall amount near Mountain A-Li from 772 mm in E6.7 (Fig. 5c) to 1294 mm. This result points out that for such extreme rainfall event with strong topography-induced upward motion, the consideration of the graupel process is effective in producing hydrometers with larger fall-speed, and increases the precipitation efficiency. Detailed analyses still needs to be performed to evaluate the impact of different microphysics schemes on the rainfall simulation associated with typhoons near the CMR.

4. SUMMARY

Typhoon Herb made landfall at northeastern Taiwan in July 31 1996, and is the most damaging tropical cyclones in the recent history of Taiwan. In this study we have performed a series of simulation experiments on Typhoon Herb using the MM5 model with different horizontal resolutions of 60, 20, 6.7 and 2.2 km. Our primary interest is to examine the ability of the model in predicting the detailed precipitation distribution associated with Typhoon Herb, and to assess the impact of horizontal resolution and detailed topography on rainfall prediction. The results show that a high-resolution mesoscale model, such as MM5, can be a very useful tool in forecasting the detailed mesoscale precipitation and wind distribution associated with typhoons near Taiwan. It is also shown that the ability of the model to successfully simulate the observed rainfall degrades with the reduction of horizontal grid resolution. Not surprisingly, the presence of CMR plays a key role in substantially increasing the total rainfall amounts associated with Typhoon Herb by lifting the moisture-laden air over the upslope of the mountains.

Nevertheless, questions remain unanswered regarding the uncertainty and limitation of the model to the quality of the initial data, the model initialization, the options of the model physics. As indicated in Wang (1980; 1989) and Wu and Kuo (1999), although the rainfall distribution for typhoons affecting Taiwan is often phased-locked with the Taiwan terrain, giving the uncertainty in simulating the landfall location and typhoon intensity, the mesoscale precipitation distribution associated with a typhoon making landfall in Taiwan remains a big challenge for numerical models. Therefore, additional research is required to access the potential of such high resolution mesoscale or typhoon models in order to improve the detailed wind and rainfall forecasts for typhoons near Taiwan.

It is hoped that improved observations from the satellites (e.g. Merrill and Velden, 1996), Doppler radars (e.g. Marks et al., 1996; 1998), reconnaissance (e.g. the impact of Omega Dropwindsonde, Franklin et al., 1996), unmanned aircraft (e.g. Holland et al., 1992; Langford and Emanuel, 1993), as well as specific field experiments and the GPS/MET system (Ware et al., 1996; Kou et al. 2000), can all be incorporated into high-resolution numerical models through advanced data assimilation techniques. By combining the observational analyses and numerical model experiments, we hope to advance our knowledge of the dynamics of the mountain-typhoon interaction, and to use such knowledge to aid typhoon forecasting near Taiwan. It is also hoped that more basic and applied typhoon research can be carried out to improve our understanding of typhoon dynamics, as well as to improve typhoon forecasting in the Taiwan area.

REFERENCES

Betts, A. K., and M. J. Miller, 1986: A new convective adjustment scheme. Part II: Single column tests using GATE wave, BOMEX, ATEX and Arctic air-mass data sets. *Quart. J. Roy. Metero. Soc.*, **112**, 693-709.

Dudhia, J., 1989: Numerical study of convection observed during the Winter Monsoon Experiment using a mesoscale two-dimensional model. *J. Atmos. Sci.*, **46**, 3077-3107.

Dudhia, J., 1993: A non-hydrostatic version of the Penn State--NCAR mesoscale model: Validation tests and simulation of an Atlantic cyclone and the cold front. *Mon. Wea. Rev.*, 121, 1493-1513.

Franklin, J. L., S. E. Feuer, J. Kaplan, and S. D. Aberson, 1996: Tropical cyclone motion and surrounding flow relations: Searching for beta gyres in Omega Dropwindsonde Datasets. *Mon. Wea. Rev.*, **124**, 64-84.

Grell, G. A., 1993: Prognostic evaluation of assumptions used by cumulus parameterizations. *Mon. Wea. Rev.*, **121**, 764-787.

Grell, G. A., J. Dudhia, and D. R. Stauffer, 1994: *A description of the fifth generation Penn State/NCAR mesoscale model (MM5)*. NCAR Tech. Note, NCAR/TN-398+STR, 138 pp.

Holland, G. J., T. McGeer, and H. Youngren, 1992: Autonomous aerosondes for economical atmospheric sounding anywhere on the globe. *Bull. Amer. Meteor. Soc.*, **73**, 1987-1998.

Kain, J. S., and J. M. Fritsch, 1993: Convective parameterization for mesoscale models: The Kain-Fritsch scheme. The representation of Cumulus Convection in Numerical Models, *Meteor. Monger.*, No. **46**., Amer. Meteor. Soc., 165-177.

Kuo, Y.-H., and W. Wang, 1997: Rainfall prediction of Typhoon Herb with a mesoscale model. *Preprints, Workshop on Typhoon Research in the Taiwan Area*. Boulder, CO, National Science Council. May 17-18, 1997. 35-45.

Kuo, Y.-H., S. Sokolovskiy, R. A. Anthes, and F. Vandenberghe, 2000: Assimilation of GPS radio occultation for numerical weather prediction. *Terr. Atmos. and Oceanic Sci.* (in press).

Kurihara, Y., M. A. Bender, R. E. Tuleya, and R. J. Ross, 1995: Improvements in the GFDL hurricane prediction system. *Mon. Wea. Rev.*, **123**, 2791-2801.

Kurihara, Y., M. A. Bender, R. E. Tuleya, and R. J. Ross, 1998: The GFDL hurricane prediction system and its performance in the 1995 hurricane season. *Mon. Wea. Rev.* (in press).

Langford, J. S., and K. A. Emanuel, 1993: An unmanned aircraft for dropwindsonde development and hurricane reconnaissance. *Bull. Amer. Meteor. Soc.*, **74**, 367-375.

Liu, Y., D.-L. Zhang, and M. K. Yau, 1997: A multiscale numerical study of Hurricane Andraw (1992). Part I:

Explicit simulation and verification. *Bull. Amer. Meteor. Soc.*, **125**, 3073-3093.

Marks, F. D., R. A. Houze, and J. F. Gamache, 1992: Daul-aircraft investigation of the inner core of Hurricane Norbert. Part I: Kinematic structure. *J. Atmos. Sci.*, **49**, 919-942.

Marks, F. D., et al., 1998: Landfalling tropical cyclones: Forecast problems and associated research opportunities. *Bull. Amer. Meteor. Soc.,* **79**, 305-323.

Merrill, R. T., and C. S. Velden, 1996 : A three-dimensional analysis of outflow layer of Super-typhoon Flo (1990). *Mon. Wea. Rev.*, **124**, 47-63.

Shieh, S.-L, S.-T. Wang, M.-D. Cheng, and T.-C. Yeh, 1997: *User's guide (2) for typhoon forecasting in the Taiwan area.* Res. Rep. CWB85-1M-01, 381 pp. (in Chinese.)

Tao, W.-K., and J. Simpson, 1993: The Goddard cumulus ensemble model. Part I: Model description. *Terr. Atmos. and Oceanic Sci.*, **4**, 35-72.

Wang, S.-T., 1980: *Prediction of the behavior and intensity of typhoons in Taiwan and its vicinity.* Res. Report 018, National Science Council, NSC-67M-0202-01 (01), 100 pp.

Wang, S.-T., 1989: *Track, intensity, structure, wind and precipitation characteristics of typhoons affecting Taiwan.* Disaster Mitigation Res. Report 80-73, National Science Council, NSC 80-04140-P052-02B, 285 pp.

Ware, R., et al., 1996: GPS sounding of the atmosphere from low Earth orbit: Preliminary results. *Bull. Amer. Meteor. Soc.*, **77**, 19-40.

Wu, C.-C., and Y. H. Kuo, 1999: Typhoons Affecting Taiwan – Current Understanding and Future Challenges. *Bull. Amer. Meteor. Soc.* **80**, 67-80.

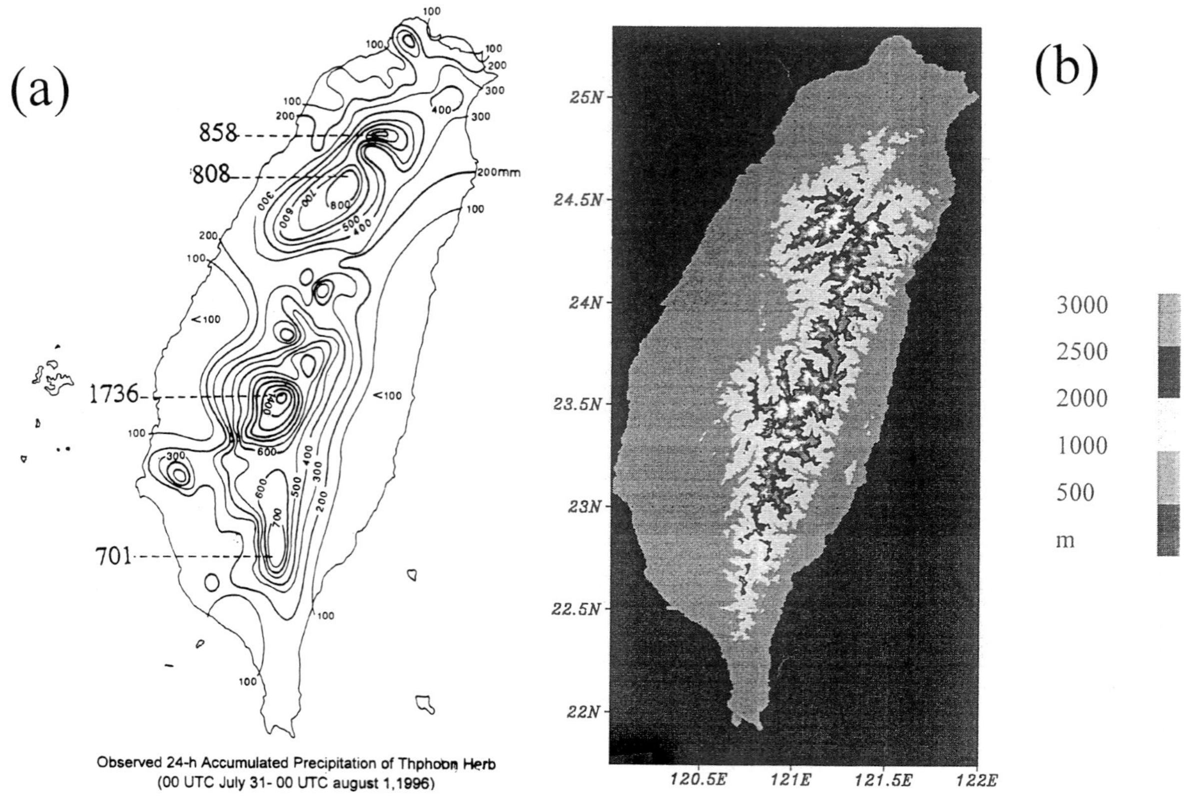

Fig. 1 (a) 24h (0000 UTC, July 31 to 0000 UTC August 1, 1996) accumulated rainfall (contour interval of 100 mm) during passage of Typhoon Herb. (b) Taiwan terrain shown in shading for every 500 m.

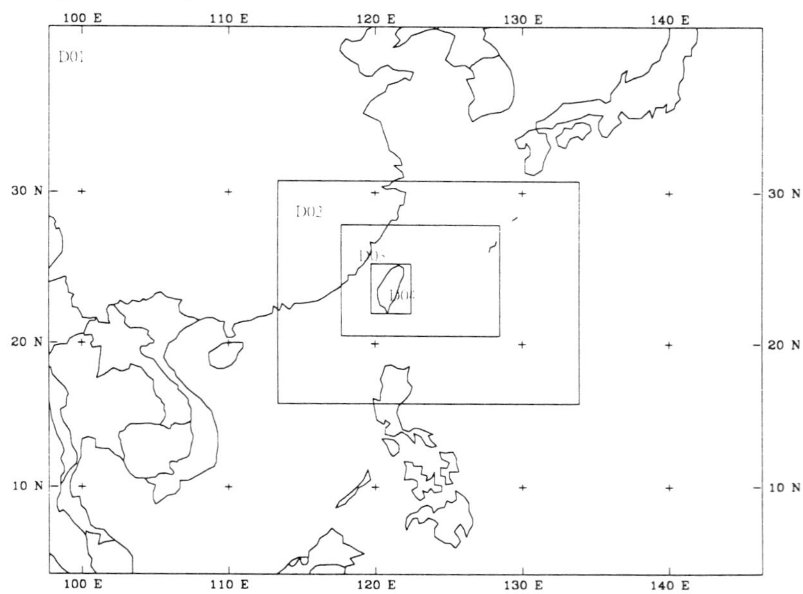

Fig. 2 Configuration of the model nests.

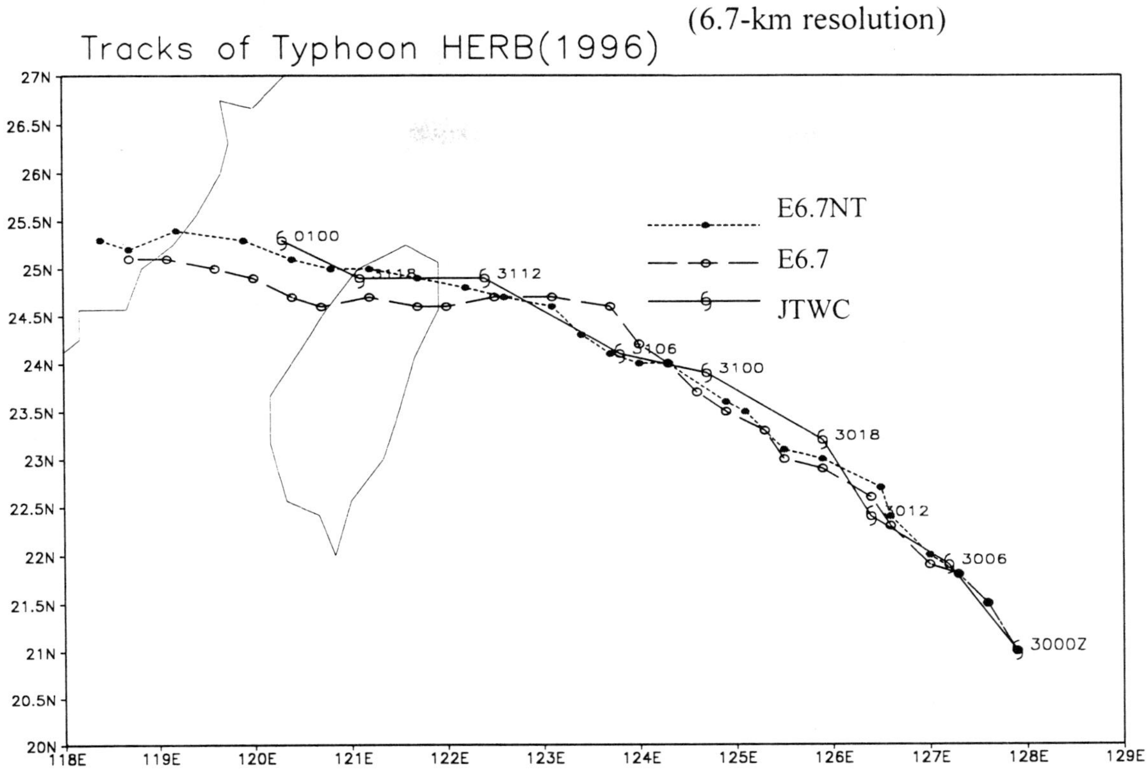

Fig. 3 Tracks of Typhoon Herb (1996) from best tracks of JTWC and from the model experiments. The storm symbol indicates the best track of Herb for every 6 hours. The solid (open) circle indicates the model track for the experiment E6.7 (E6.7NT) for every 2 hours from 0000 UTC, July 30 to 0000 UTC August 1, 1996.

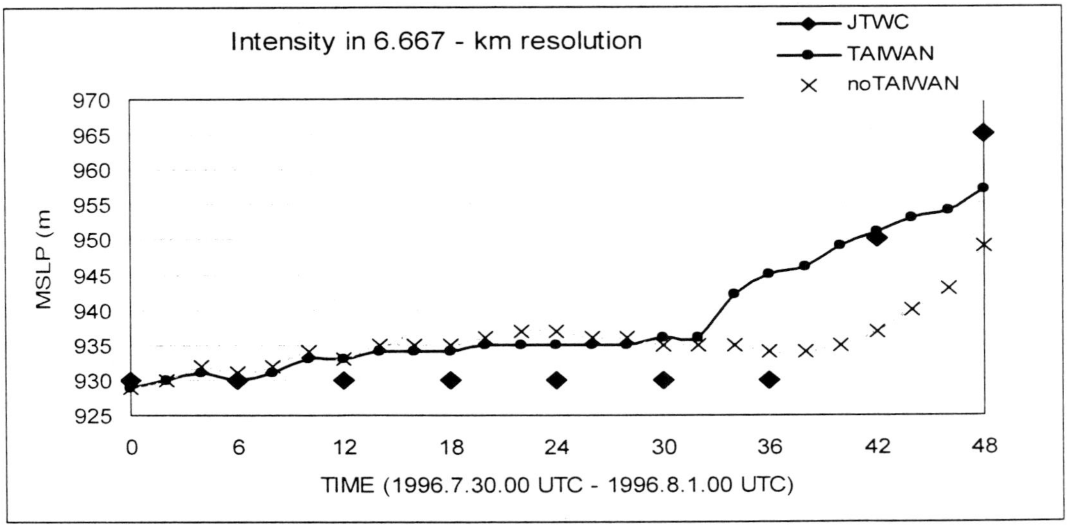

Fig. 4 Time evolution of the central minimum mean sea level pressure of Herb from JTWC. E6.7, and E6.7NT from 0000 UTC, July 30 to 0000 UTC August 1, 1996.

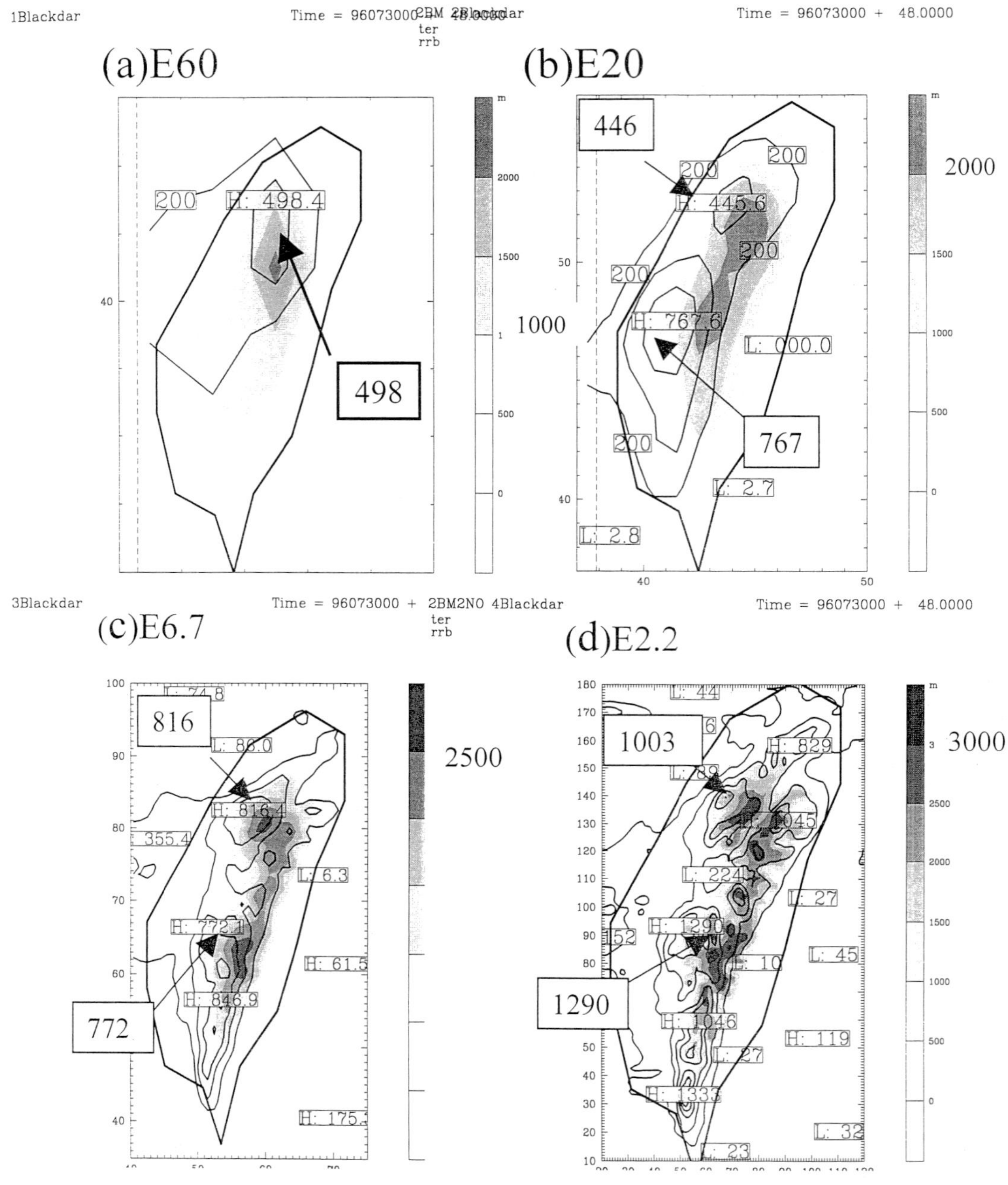

Fig. 5 24-h rainfall (contour interval of 200 mm) ending at 0000 UTC August 1, 1996 from MM5 simulations with highest horizontal resolutions of (a) 20 km; (b) 20 km; (c) 6.7 km; and (d) 2.2 km. Corresponding model terrains (m) are shown in shading for every 500 m

A PRELIMINARY STUDY OF ENVIRONMENTAL MOMENTUM INFLUENCES ON TROPICAL CYCLONE FORMATION

Cheng-Shang Lee and Chii-Horng Lee

Department of Atmospheric Sciences, National Taiwan University, Taipei

ABSTRACT

During the formation of typhoon Mike (1990), the trade wind to the north of the system was very strong over a very large domain. However, there were evident cross-equatorial flows associated with the formation of typhoon Betty (1987). The intensities of both systems increased greatly after they reached tropical storm intensity. To help understand the influences of the environmental momentum field on the formation of typhoon, we used the PSU/NCAR MM5 to simulate the formation process of Betty and Mike. The ECMWF grid point data were used as the model initial field. Results showed that the strong trade wind associated with Mike's formation and the cross-equatorial flows associated with Betty's formation were simulated reasonably well. The intensities of model typhoons were also close to the observations for Betty. However, the model vortex developed faster than the observation for Mike.

Both the tangential momentum and the vorticity budgets were analyzed using model-simulated data. Results showed that the low-level eddy horizontal angular momentum transport caused by the strong trade winds played a significant role to the spin-up of the model vortex, thus was favorable for the formation of Mike. The horizontal vorticity advection caused by the cross-equatorial flow associated with Betty's formation did not contribute significantly to the vorticity generation of model vortex. However, the low-level convergence enhanced by the cross-equatorial flow might have caused the vorticity generation near system centre and increased heating efficiency. Such effect might have played an important role on the organization of convection and helped the formation of tropical cyclone.

Keywords: typhoon, tropical cyclone formation, numerical simulation

1. INTRODUCTION

The physical processes that lead to tropical cyclone formation are still not fully understood. Observations showed that among numerous tropical cloud clusters, only very few developed to tropical cyclone stage with well-defined structures (Gray, 1981). However, about half of tropical cyclones developed to typhoon intensity with maximum sustained surface wind of 64 knots. In addition, some tropical storms (maximum intensity between 34 and 64 knots) never reached typhoon intensity, while some cloud clusters developed very fast and reached typhoon intensity in a short time period. These observation evidences reveal that the physical processes determining tropical cyclone formation are likely quite different from those affecting the intensification of tropical cyclone. Observations also showed that there were often significant large-scale wind field characteristics when a tropical disturbance developed to a tropical storm. These wind field characteristics included the cross-equatorial flow (Love, 1985), the strong trade wind (Lee, 1986) and the upper level tropospheric trough.

Due to the limitation of the observation, it is very difficult to study the dynamics of each tropical cyclone formation process using conventional data or grid point data. Therefore, this study attempts to simulate those tropical cyclone formation cases, with evident large-scale wind field characteristics, using NCAR/PSU MM5 (The Fifth-Generation NCAR/PSU Mesoscale Model). The model-simulated results will be diagnosed to study the large-scale momentum forcing on tropical cyclone formation.

In section 2, the model design, the initial conditions and the case selection procedures are discussed. The cases simulated in this study are described briefly in section 4 together with the model simulation results. Section 5 discusses results of the vorticity and tangential momentum budget analysis. Finally, the summary and conclusions are presented in section 6.

2. MODEL DESIGN AND INITIAL CONDITIONS

This study uses a two-way interactive nested two-layer model with 60 and 20-km grid spacing as well as 120 and 40 seconds time-steps. The model domains are 7260km x 7980km (121 x 133 grid points) and 3620km x 3620km (181 x 181 grid points). In the vertical, there are 23 σ levels. The radiative boundary condition and the high resolution Blackadar parameterization scheme are used for the upper and lower boundaries, respectively. For the lateral boundary, the nudging and the time-dependent boundary are used for the coarse and fine grids, respectively. The Betts-Miller cumulus parameterization (Betts and Miller, 1986) with modified parameters is used in our simulation. The ECMWF (European Centre for Medium-range Weather Forecasts) basic 2.5 $^\circ$x 2.5 $^\circ$global analysis is used as initial condition and time-dependent boundary. For the sea surface temperature, the monthly mean sea surface temperature analyzed by NCEP is used.

The focus of this study is on the environmental momentum influence on the formation of tropical cyclone, specifically the effects of the strong trade wind and the cross-equatorial surge. Previous analysis indicated that these systems often experienced rapid intensification after becoming tropical cyclones. Therefore, we used the JTWC (Joint Typhoon Warning Center, Guam) best track data to select those tropical cyclones in the Western North Pacific whose intensity increased significantly after formation. It has to be noted that tropical cyclone formation is a continuous process as pointed out by Ooyama (1982). There is no genesis point. However, for the convenience of discussion, we used the term formation when the JTWC post-analysis (or the best track) could determine the minimum central pressure and the maximum surface wind speed for the disturbance. The system should have been undergoing the formation process and have a well-organized tropical cyclone structure at this stage.

The ECMWF global analysis was used to examine the 850 hPa wind field associated with these systems to identify if there was large-scale environmental momentum influence during formation. Although observations are very sparse over the western North Pacific, the wind field feature we are interested in is over a very large domain, thus should be resolved in ECMWF global analysis. Next, we used MM5 to simulate those cases with significant large-scale environmental momentum influence during formation stage. Up to now, we have successfully simulated the formation processes of typhoons Betty (1987) and Mike (1990). It is not surprising that only few cases can be simulated reasonably well because a minor error in the initial field can lead to large errors after time-integration. The ECMWF global analysis often can not resolve well the circulation pattern and structures associated with a tropical disturbance due to lack of the observation there.

3. CASE DESCRIPTION AND SIMULATION RESULTS

a. Super typhoon Betty (1987)

According to the JTWC annual report, Betty was first detected as a tropical disturbance embedded in the monsoon trough on August 7, 1987. The system cloudiness developed rapidly early on the 8th and JTWC issued the Tropical Cyclone Formation Alert (TCFA) at 0300UTC. The system moved northwestward slowly and reached tropical storm intensity on 9 August. Betty experienced a rapid intensification in the next 60 hours and reached super typhoon intensity (minimum central pressure of 891 hpa) on the 11th. Figure 1 shows the maximum surface wind speed and the minimum central pressure up to 1200UTC, August 9, according to the JTWC estimates (or the best track). Betty was also characterized by the large size of the area of intense convection and the associated strong low-level southwest monsoon inflow.

It is obvious that pre-Betty disturbance started to develop some time around the 7th and was undergoing the formation process when JTWC issued the TCFA at 0300 UTC, August 8. Therefore, the numerical simulation starts at 1200 UTC, August 6, or 39 hours prior to the TCFA and runs for 72 hours to 1200 UTC, August 9. The sea-surface minimum central pressure and the maximum sustained surface wind speed of the model vortex are shown in Fig. 1.

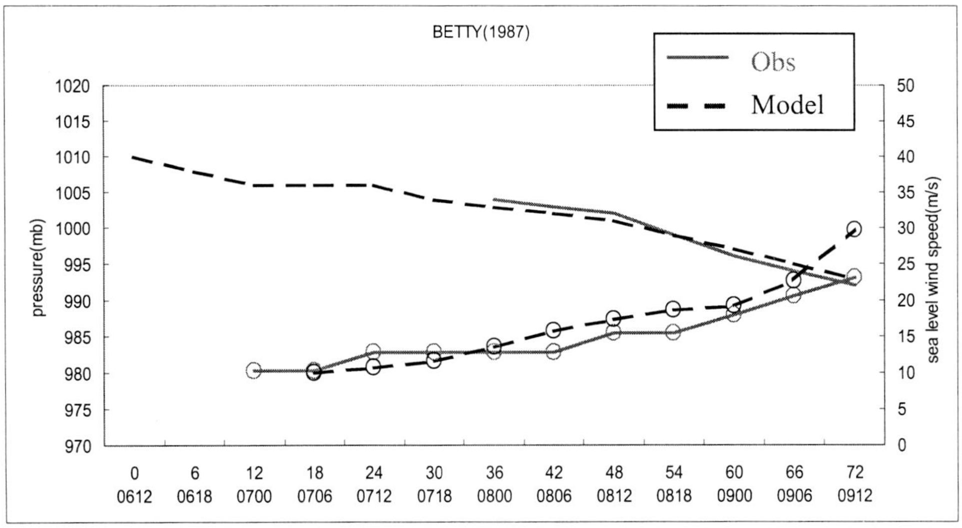

Fig. 1. The observed and the simulated minimum central pressure and maximum surface wind speed of typhoon Betty (1987).

Results show that a well-defined tropical cyclone with minimum central pressure of 1003 hpa is observed after 36 hours of integration (or at 0000 UTC, August 8). This minimum central pressure is slightly lower than the observed 1004 hpa. However, the maximum surface wind speed (13 m/s) is identical to that observed. The model vortex develops to tropical storm intensity at 1200 UTC, August 8 (or after 48 hours of integration), with the maximum surface wind speed of 17 m/s, and continues to intensify afterward. In summary, the intensity of model vortex is about the same as that observed during the first 60 hours of integration and is stronger than that observed after 66 hours of integration.

The cross-equatorial flow associated with Betty's formation is also simulated well as shown in Fig. 2, the 850 hpa wind field at integration times of 00, 12, 24 and 36 hours. Two cross-equatorial streams are located to the southwest and southeast of vortex centre at 12 hours. The strong wind regions (wind speed ≧ 10 m/s) have crossed the equator at this time. The strong wind region of the west-bond cross-equatorial flow reaches the southwest vicinity of vortex centre at 18 hours and merges with the vortex circulation at 24 hours. The east-bond cross-equatorial flow also reaches the southeast-eastern vicinity (but at larger radii) of vortex circulation at 24 hours. The structure of the model vortex becomes better organized during the next 12 hours of integration. In short, the cross-equatorial flow might have played an important role on the formation process of Betty and such progressing sequence in the flow pattern is well simulated.

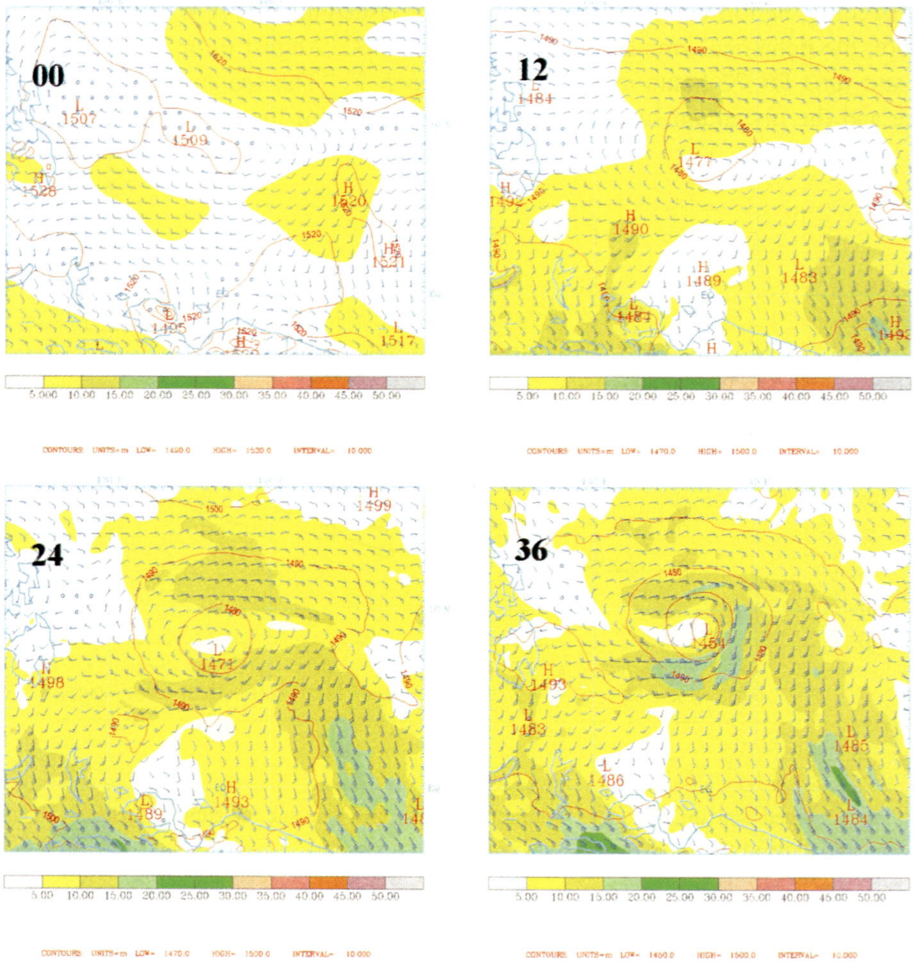

Fig. 2. The 850 hpa wind field and geopotential contours at integration times of 00, 12, 24 and 36 hours for typhoon Betty. The gray area shows wind speed ≧ 5 m/s (see the gray-scale bar).

b. Super typhoon Mike (1990)

The pre-Mike disturbance was first mentioned on Significant Tropical Weather Advisory as an area of persistent convection on 0600 UTC, November 6, 1990, according to the JTWC tropical cyclone report. JTWC then issued the Tropical Cyclone Formation Alert at 1530 UTC, based on rapidly improving outflow and curvature and an increase in central convection. The system was upgraded to tropical storm at 0000 UTC, November 8. The satellite images show that the convection associated with pre-Mike disturbance was not very active especially when compared to that of Betty.

It is obvious that the developing speed of Mike is faster than that of Betty. It also has to be noted that although the pre-Mike disturbance was first mention at 060600 UTC, the JTWC best track estimated intensity starts from 060000 UTC. Our simulation starts at 0000 UTC, November 5, or 39.5 hours before TCFA, and runs for 72 hours to 0000 UTC, August 8. Fig. 3 shows the observed (or JTWC estimated) and simulated minimum central pressure and maximum surface wind speed. Results show that the minimum central pressure of the model vortex is lower than the observation at the same time period. Apparently, the model vortex develops faster than the observation and reaches tropical storm intensity at 061800 UTC (maximum wind speed of 17 m/s). However, the developing trend of the model vortex is similar to the observation. It has to be noted that the observed system intensity actually was estimated based on satellite images.

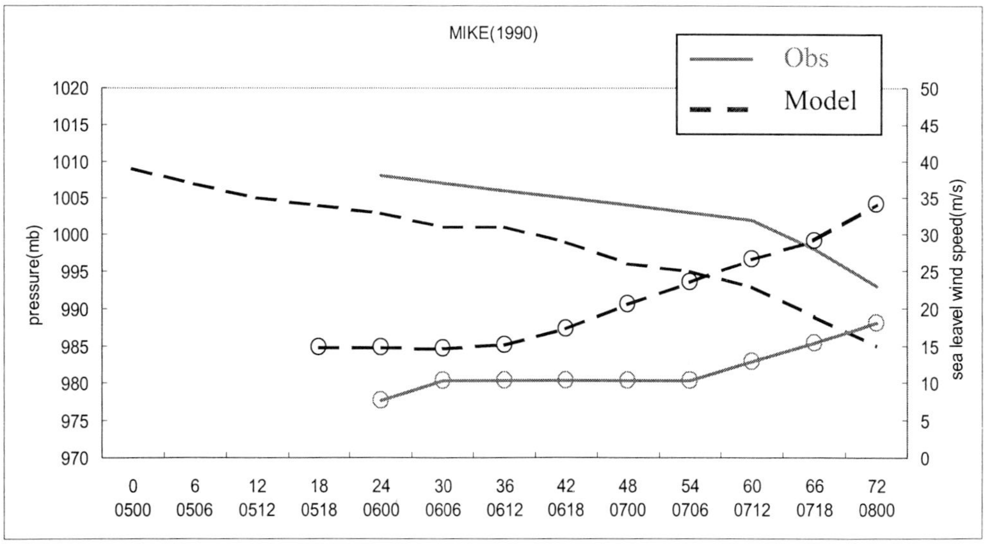

Fig. 3. The observed and simulated minimum central pressure and maximum surface wind speed of typhoon Mike (1987).

The initial 850 hPa wind field (Fig. 4) shows a broad region of strong easterly trade wind to the north of the region where pre-Mike disturbance will develop. At 12 hours, a long and narrow strong wind region (with wind speed ≧ 15 m/s) extends from the east-northeastern outer region toward the northwestern inner region of the vortex. The maximum wind speed of this strong wind region exceeds 20 m/s at 18 hours. The vortex structure becomes better organized in the next 12 hours. Apparently, the strong trade winds

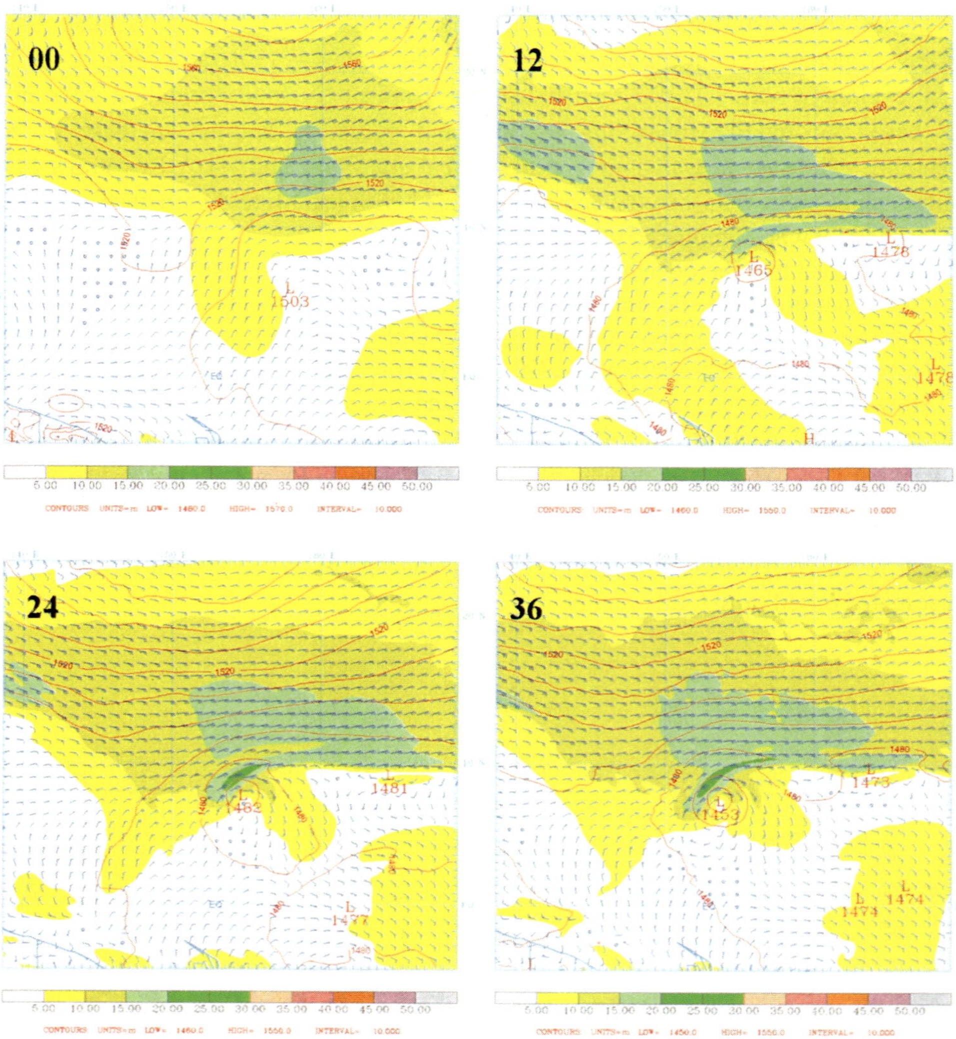

Fig. 4. The 850 hpa wind field and geopotential contours at integration times of 00, 12, 18 and 24 hours for typhoon Mike. The gray area shows wind speed ≧ 5 m/s (see the gray-scale bar).

to the north of the disturbance have played an important role in the formation process of Mike.

c. Comparisons

Our analysis has revealed some important features associated with the formation of Betty and Mike. Our simulation has also reproduced well the changes in the environmental wind field and the evolution of the system (although the model vortex develops at a speed faster than observation for Mike case). In summary, there are cross-equatorial flow and strong trade winds associated with Betty and Mike's formation, respectively. The convection is very active for pre-Betty's disturbance but not so active for pre-Mike's disturbance. The moving speed is much faster for Mike than that for Betty. In addition, the pre-Mike disturbance develops faster than the pre-Betty disturbance does.

Further analysis of the axisymmetric tangential wind reveals that the low-level cyclonic circulation of Mike extends beyond 14 ° radius and also increases significantly over a very large domain (radius ≧ 14 ° latitude) during the first 12 hours of integration. The cyclonic spin-up is confined to within 10 ° radius afterward. For Betty, the cyclonic spin-up occurs only within 8 ° radius. At radius greater than 8-9 °, the wind speed even decreases slightly during the first 24-hour integration leading to anticyclonic circulation beyond 10-11 ° radius. Such results seem to indicate that the strong trade winds can cause an effective vortex spin-up over a larger domain while the direct spin-up caused by the cross-equatorial flow is not so significant. Results also show that for Betty, the 950 hpa vorticity and convergence to the south-southwest of the system centre increases significantly from 12 hours to 24 hours, or after the west-bond cross-equatorial flow reaches the southwest vicinity of the vortex centre. The 3-hourly accumulated rainfall also increases greatly from 12 hours to 24 hours at this region.

4. BUDGETS ANALYSIS

Lee (1986) used the rawinsonde composite data to study the tangential momentum budget during tropical cyclone formation process. Results showed that the low-level eddy angular momentum transport was very important for vortex spin-up before formation. Similar analysis is performed in this study. Results show that the mean angular momentum transport is large below 900 hpa with maximum value located at about 200-km radius at 12 hours of integration for both cases. As the time integrated, the radius of maximum flux decreases and the magnitude increase. However, the increase of mean angular momentum transport is more significant for Mike than that for Betty, after 18 hours of integration.

Results also show that Mike has significant eddy angular momentum transport below 850 hpa at 12 hours (Fig. 5). The radius of maximum eddy angular momentum transport also decreases with time. On the other hand, the low-level eddy angular momentum transport is negative for Betty. These results reveal that the strong trade wind can provide large amount of eddy angular momentum for pre-Mike tropical disturbance to spin-up while the direct effect of the cross-equatorial flow on the spin-up of pre-Betty disturbance is negative (since f of southerly flow is smaller).

To further examine the processes affecting vortex spin-up, the low-level vorticity budget is analyzed. Results show that both the divergent effect and the horizontal advection are important to vortex vorticity generation at 950 hpa for Mike. However, the horizontal advection term dominates at 850 hpa with the maximum vorticity tendency located to the west-northwest of system centre as shown in Fig. 6 for 24 hours of integration. Such horizontal advection effect would have caused a fast west-northwestward movement as was observed for Mike. It has to be noted that this strong horizontal advection effect is associated with the strong trade winds as Fig. 6b and 6c show. For Betty, the contribution of the horizontal advection on the vorticity generation during formation appears to be negligible. The divergent term plays the major role in the increase of the low-level vorticity. However, the divergent effect is confined to below 850 hpa and is mostly balanced by the vertical transport, resulting in a relatively smaller tendency term.

In summary, the physical processes that lead to vorticity generation during the tropical cyclone formation stage is different for Betty and for Mike. Fig. 7 shows the idealized depiction concerning the vorticity generation of these two systems. During the early formation stage of Mike, the strong trade winds to the north-northeast of the system centre (region A) moves cyclonically (form the northwestern side) toward system centre and brings in strong cyclonic shear vorticity to spin-up vortex (region B). At levels near surface (about 950 hpa), the divergent terms also has significant positive effect (but the vorticity generation by the flow convergence tends to be balanced by the vertical advection). However, when the strong trade wind flow reaches the inner region, the horizontal advection effect appears to dominate the vorticity generation. The cyclonic shear vorticity associated with the trade winds to the northeast of system centre is also enhanced when the southerly merges with trade winds at region C.

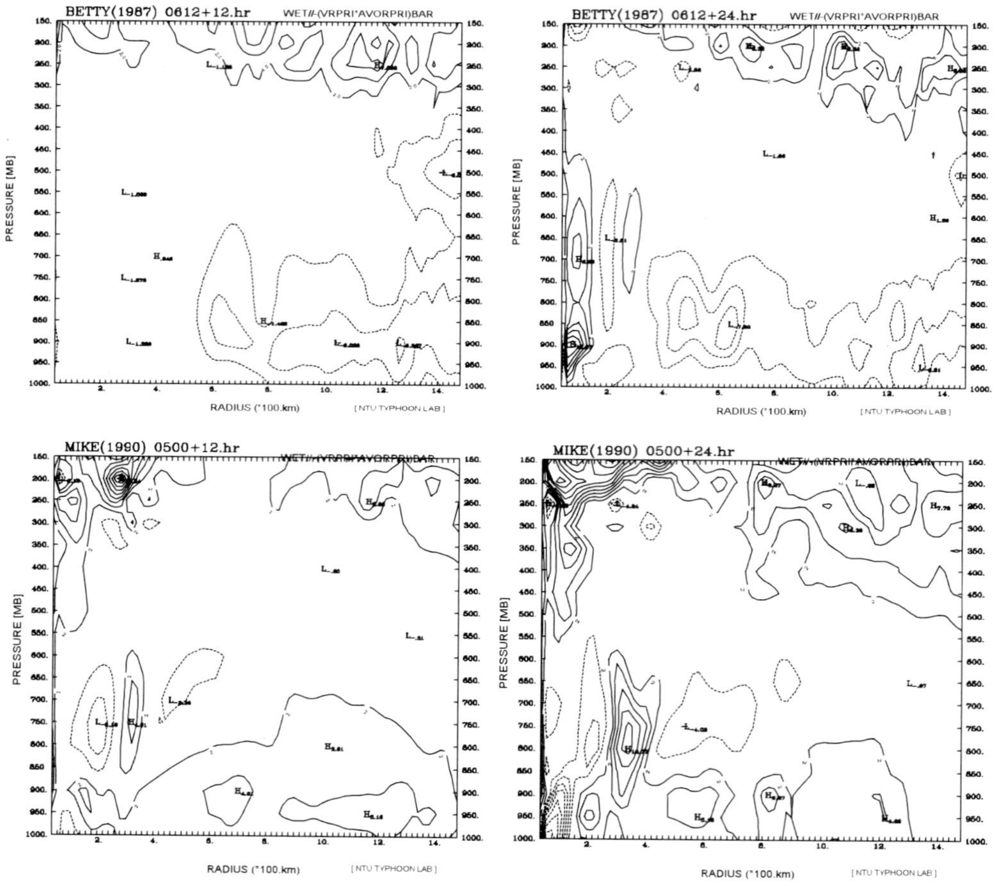

Fig. 5. The eddy angular momentum transport at 12 (left) and 24 (right) hours of integration for Betty (top) and Mike (bottom).

For Betty, the horizontal vorticity advection associated with the cross-equatorial flow does not seem to contribute directly to the vortex spin-up. The advection term only contribute to the vorticity generation at several smaller regions as shown in Fig 7. However, the cross-equatorial flow might have enhanced the convergence at the west and southwest vicinities of system centre (as shown by the confluent flow in Fig. 7) and increased local vorticity. The increase of local vorticity at where convection is enhanced might lead to an increase in local heating efficiency and help organize convection that was favorable for the formation of Betty.

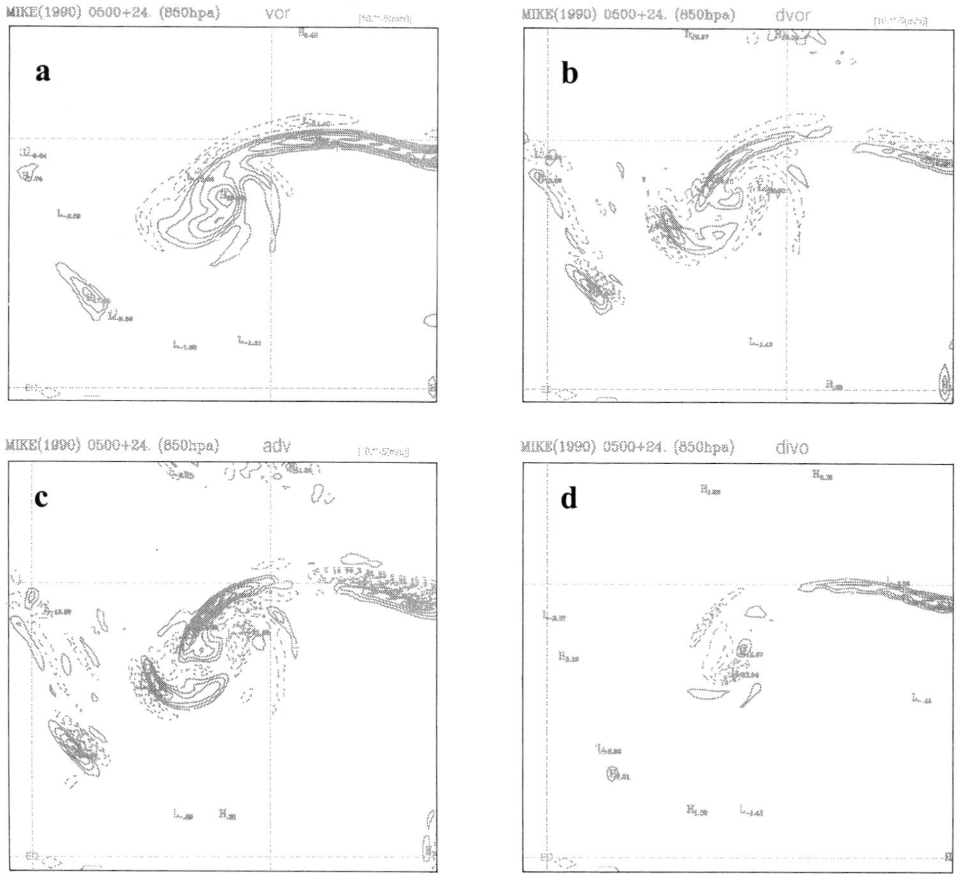

Fig. 6. The 850-hpa-vorticity (a) and the vorticity budget terms (b, c and d) for Mike at 24 hour of integration. Diagram b is the vorticity tendency, c, the advection term and d, the divergent term. The negative contours are dashed.

5. SUMMARY AND CONCLUSIONS

This study aimed to help our understanding in the environmental momentum influence on the formation of tropical cyclone. The environmental flow features associated with the formation of typhoons Betty (1987) and Mike (1990) were analyzed and the MM5 was used to simulate the formation processes. Results showed that MM5 could simulate reasonably well the tropical cyclone formation process and the changes in the environmental circulation patterns, namely the strong trade wind surge and the cross-equatorial flow. The analysis of model results and observation had revealed some important features associated with the formation of Betty and Mike as summarized in Table 1.

Our analysis has pointed out that the strong trade wind is the key effect to the vorticity

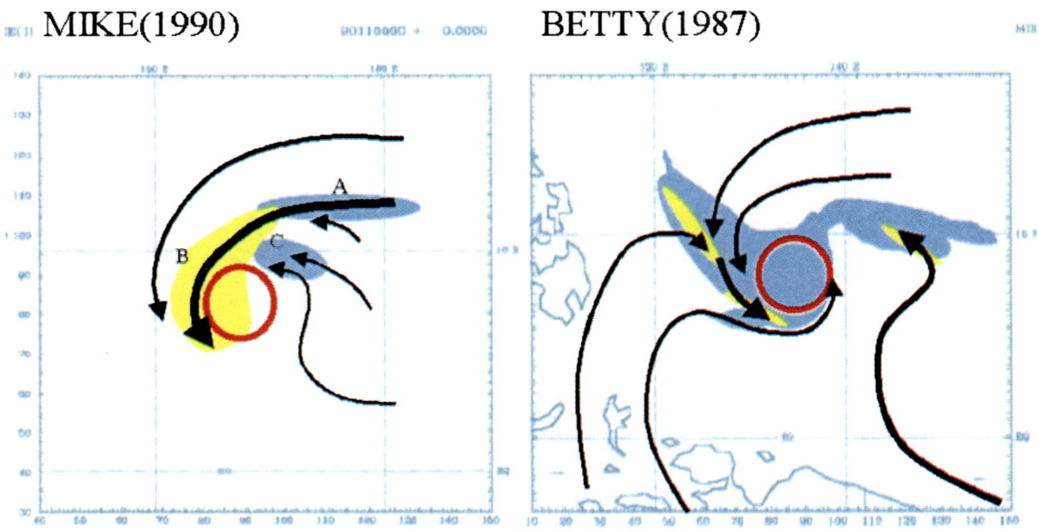

Fig. 7. Idealized depiction showing the environmental flow influences on the vorticity generation of pre-Mike (left) and pre-Betty (right) disturbances. The lighter (darker) shading areas show where the divergent term (advection term) is important to vorticity generation.

generation that leads to the formation of Mike. The strong cyclonic shear vorticity associated with strong trade winds, also enhanced by the merging of the southerlies, is advected towards the west-northwest inner region of disturbance. This horizontal advection effect leads to a significant increase of cyclonic vorticity to the west-northwest of the original centre. Such vorticity tendency would make the system move fast westnorthwestwards and develop.

However, the role that the cross-equatorial flow plays on the formation of Betty is not as obvious. It is speculated that the cross-equatorial flow can enhance the local vorticity and convection at the west and southwest vicinities of the system centre. Such effect might lead to the increase of local heating efficiency and help organize cyclone scale convection. Nevertheless, more studies are still needed to address fully such type of tropical cyclone formation process.

Table 1. A summary of some important features associated with the formation of typhoons Betty (1987) and Mike (1990).

	Betty (1987)	**Mike (1990)**
Large scale flow pattern	Cross-equatorial flow	Strong trade wind
Convection before formation	Very active	Less active
Motion	Very slowly	Fast west-northwestward
Low-level azimuthal average tangential wind	Increases within 8 ° radius, decrease slightly outside 8-9 ° radius	Increases over 14 ° radius
Low-level angular momentum transport	Negative	Positive, maximum area moves inward
Vorticity generation at 850 hpa	~	Advection term dominates
Vorticity generation at 950 hpa	Divergent term dominates	Both divergent term and advection term are important

Acknowledgement. The authors would like to thank Miss Ching-yi Tsai for her help in manuscript preparation. This research is supported by the National Science Council under grant no. NSC 89-2111-M-002-004-AP1 and NSC 88-2111-M-002-006-AP1.

REFERENCES

Betts, A. K., and M. J. Miller, 1986: A new convective adjustment scheme. Part II: Single column tests using GATE wave, BOMEX, ATEX, and Arctic air-mass data sets. *Quart. J. Roy. Meteor. Soc.*, **112**, 693-709.

Gray, W. M., 1981: *Recent advances in tropical cyclone research from rawinsonde composite analysis.* Paper prepared for the WMO Committee of Atmospheric Science. Geneva, Switzerland, 407pp.

Lee, C.-S., 1986: *An observational study of tropical cloud cluster evolution and cyclogenesis in the Western North Pacific.* Dept. of Atmos. Sci. paper No.403, Colorado State Univ., Fort Collins, Co. 250pp.

Love, G., 1985: Gross equatorial interactions during tropical cyclone genesis. *Mon. Wea. Rev.*, **113**,1499-1509.

Ooyama, K., 1982: Conceptual evolution of the theory and modeling of the tropical cyclone. *J. Meteor. Soc. Japan*, **60**, 369-379.

~ Mesoscale Meteorology ~

CONVECTION AND LOCAL CIRCULATIONS UNDER THE INFLUENCE OF MEIYU FRONT OVER NORTHERN TAIWAN

George Tai-Jen Chen
Department of Atmospheric Sciences, National Taiwan University

ABSTRACT

The Meiyu fronts affecting northern Taiwan during the period of May-June 1991-1996 were categorized into the moving, the stationary, and the retreating types. Radar echoes observed at the CKS International Airport and surface observations from the Central Weather Bureau for the same period were analyzed to reveal the relationship between daytime land convection and local circulations under the influence of different types of fronts over northern Taiwan.

Results show that the sea breeze and localized vortices prevailed over northern Taiwan. A convergent line was located over the northwestern slope of the Snow Mountain Range with localized cyclonic vortices along the line. Convective activities were modulated strongly by the sea breeze and the convergent line for different types of fronts. More widespread convection was triggered on the day when front entered northern Taiwan for the stationary type as compared to the moving and the retreating types. Convection was enhanced on the day when front entered northern Taiwan and suppressed on the following day for the moving type and the reverse was true for the stationary type. On the other hand, it was enhanced and organized consecutively from the day when front entered northern Taiwan to two days after for the retreating type.

1. INTRODUCTION

There are about 4-5 frontal systems affecting Taiwan during the Meiyu season each year (Chen 1988; Chen and Chi 1978; 1980). The Meiyu front is usually accompanied by a nearly continuous cloud band which produces the stratiform and /or convective precipitation. The secondary circulation across the synoptic-scale Meiyu front was found to be thermally direct with warm air rising to the south and cold air sinking to the north of front (Chen and Chang 1980). Using composite technique, Chen and Chi (1978) was able to deduce two secondary circulation cells across the Meiyu front in northern Taiwan with one cell across the front and the other one to the south of the front. Convective activities were enhanced over the area of ascending part of the cell and were suppressed over the area of descending part. Besides the frontal forcing, local circulations played important roles on modulating the diurnal variation of convective rainfall and heavy rain event in the Meiyu season (Chi and Chen 1989b; Chen and Yang 1988; Johnson and Bresch 1991; Chen 1992). The land-sea breezes and island circulations usually become primary forcings for convection under weak synoptic-scale control (Chen et al., 1989; Chi and Chen 1989a; Chen and Hui 1990; Lin and Sheng 1990; Chen 1992). Apparently, the convective rainfall during Meiyu season over northern Taiwan was strongly modulated by the Meiyu front and the local circulations. The purpose of this paper is to study the spatial and temporal variations of convective activities over northern Taiwan as revealed by radar reflectivity under the influence of different types of Meiyu fronts and local circulations.

2. DATA AND ANALYSIS

The surface Meiyu fronts affecting northern Taiwan during the period of May-June 1991-1996 were categorized into three types based on their movements. Day zero (T0) is the day when the front moved southward towards northern Taiwan within a distance of 200 km from the CKS International Airport Doppler radar site of the Civil Aeronautics Administration (CAA) at 0000 UTC. T-1 is the day before and T1 is the day after T0. When two fronts consecutively affected northern Taiwan, the day between T0s of the two fronts was considered to belong to the second front. The moving type (M type) is defined as the front moved continuously toward south and the stationary type (S type) as the front remained within a distance of 200 km from CKS Airport for at least two days. While the retreating type (R type) is defined as the front retreated northward away from the CKS Airport with a distance greater than 200 km after T0. There were a total of 14, 8 and 10 cases for the M, S and R types, respectively. Table 1 shows dates of T0 for different types of fronts. While fronts were evenly distributed in May and June for the moving and stationary types, there were only two out of ten fronts occurred in May for the retreating type.

Radar echoes observed at the CKS International Airport in 1100-1800 LST for these frontal cases were used to reveal the characteristics of daytime convection over land. In order to study the role of local circulations in the formation and development of land convection, the convection embedded in the Meiyu front cloud band was excluded in this study. Spatial distribution of the frequency of convection initiation and convection occurrence was composited for different types of fronts. In order to compare the behaviour of convective activities over the plain, the mountain slope and the high mountain areas, cross-sections of different altitudes to the northwest and north of the Snow Mountain Range (SMR) were chosen (Fig. 1, Table 2). Snow Mountain Range is oriented in a northeast-southwest direction over northern Taiwan, and the high mountain range oriented in a north-south direction to the south is the Central Mountain Range (CMR). Temporal variations of the convection frequency at various locations with different altitudes along each cross-section were analyzed during the period of frontal influence. Surface hourly observations of the Central Weather Bureau (CWB) during the period of 1100-1800 LST were used to composite wind field during the influence of different types of fronts.

3. SPATIAL DISTRIBUTION OF CONVECTION

Frequency distribution of the convection initiation revealed by the first radar echo of ≧ 15 dBZ at day T0 for different types of fronts is presented in Fig. 2. In general, convection was first initiated mainly over the mountain slope and the high mountain areas of the SMR for all types of fronts, although the detail distribution was quite different. More widespread convection was triggered for the stationary type front as compared to the moving type and retreating type fronts. For the moving type front, the first convection tended to occur over an elongated area over the northwestern slope and the mountain areas of the SMR. For the stationary type front, it appeared to be initiated over the northwestern mountain slope and the high mountain areas of the SMR and the CMR with a higher frequency than that for the moving type. For the retreating type front, it tended to occur mainly along the SMR.

Fig. 3 illustrates the frequency distribution of convection revealed by radar echo of ≧

30 dBZ for the moving type front from one day before (T-1) to two days after (T2) the day zero. At day T-1, convection mainly occurred over the SMR. It tended to become more active and more organized over the northwestern slope of the SMR at day T0. Convection was suppressed at day T1 and was mainly confined over the northwestern slope of the SMR. This is consistent with the less unstable air mass to the rear of Meiyu front over northern Taiwan at day T1. Convection became more active over the northwestern slope and the high mountain areas of the SMR at day T2, suggesting that the static instability already recovered to the level equivalent to that before the frontal passage.

The frequency distribution of convection for the stationary type front is presented in Fig. 4. Overall, convection was much more active over northern Taiwan for the stationary type than for the moving type front during the period of analysis. At day T-1, convection prevailed over most of the northern Taiwan area with an active region to the northwest of Taiwan. At day T0, convection seems to be suppressed somewhat. At day T1, convection was enhanced with maximum activity over the SMR and its northeastern slope. It became less active at day T2 with an elongated area of maximum along the northwestern slope of the SMR.

Fig. 5 illustrates the frequency distribution of convection for the retreating type front. In general, convection was less active than the stationary type and more active than the moving type. At day T-1, convection was active and organized along the northwestern slope of the SMR. At day T0, convection was more active only over the northeastern part of the SMR. At day T1, convection became more active with maximum activity over the northeastern part and southwestern part of the SMR. At day T2, convection was enhanced and an area of maximum activity became organized and elongated along the northwestern slope of the SMR.

From this analysis, it becomes clear that the daytime convection over northern Taiwan was modulated by different types of fronts. Convection was enhanced and organized in an elongated area along the northwestern slope of the SMR at day T0 by the moving type front and at day T2 by the retreating type front. It was enhanced and was widespread over and to the northeast of SMR at day T1 by the stationary type front.

4. TEMPORAL VARIATIONS OF CONVECTION

Temporal variations of convection frequency revealed by radar echo of $\geqq$30 dBZ along different cross-sections during the frontal influence from day T-3 to day T5 for the moving type front are presented in Fig. 6. To the northwest of the SMR, the Chupei-Neiwan-Shiehpaishan cross-section shows no systematic tendency of convection enhancement during the frontal passage. The Chupei-Kuangshi-Takuangshan cross-section seems to indicate convection enhancement over the mountain slope (Kuangshi) during the frontal passage and over the mountain area (Takuangshan) before the front entered northern Taiwan. To the north of the SMR, two cross-sections show convection enhancement mainly over the mountain slope (Shientien) during the frontal passage. It is clear that convection was enhanced over the mountain slope during the frontal passage and over the mountain area before the front entered northern Taiwan for the moving type front.

Fig. 7 illustrates the frequency distribution of convection from day T-3 to day T5 for the stationary type front. To the northwest of the SMR, two cross-sections show a tendency of convection enhancement over the plain area (Chupei) and the mountain areas (Shiehpaishan, Takuangshan) at day T1 and T2 and over the mountain slope (Neiwan, Kuangshi) from day T-1 to day T2. To the north of the SMR, two cross-sections indicate a tendency of convection enhancement at day T1 over the Taipei Basin, the mountain slope and the mountain areas. It is clear that convection enhancement was much stronger at different altitudes to the north than to the northwest of SMR after the front entered northern Taiwan for the stationary type.

For the retreating type front, frequency distribution of convection from day T-3 to day T5 is presented in Fig. 8. To the northwest of the SMR, two cross-sections seem to indicate that convection was somewhat enhanced over the mountain slope (Kuangshi) at day T1 and over the mountain area (Takuangshan) at day T2. To the north of the SMR, two cross-sections show that convection was enhanced over the Taipei Basin (CWB, Sungshan) from day T0 to day T2 and over the mountain slope (Shientien) at T2. Convection was enhanced from day T-1 to day T0 over northern tip of the SMR (Pinglin) and at day T2 over mountain area of Takuangshan. It is clear that the convection was enhanced by the retreating type front over the mountain slope at day T1 to the northwest and at day T2 to the north of the SMR. Convection was more enhanced over the Taipei Basin to the north as compared to the northwest of the SMR from day T0 to day T2. Over the mountain areas, it was enhanced either at day T1 or day T2.

5. LOCAL CIRCULATIONS

Streamline analyses from day T-1 to day T2 for the moving type front are presented in Fig. 9. In general, local circulations such as sea breeze, upslope wind, localized vortices, and convergent lines existed during the frontal passage. At day T-1, onshore flow prevailed along coastal area indicating that sea breeze circulation developed well during the daytime period. There were three cyclonic vortices formed along the convergent line over the northwestern slope of the SMR. At day T0, sea breeze again prevailed with an increase of onshore flow along the northern coastal area. This was the time that convection tended to be enhanced and organized along the convergent line over the northwestern slope of the SMR. At day T1, northerly flows dominated over the coastal area to the north and northwest of Taiwan. This together with the decrease in temperature and the increase in dew point depression suggested that front had moved to the south of northern Taiwan. Although the convergent line still existed along the northwestern slope of the SMR, convection was suppressed over the same area under the influence of less unstable air mass. At day T2, north-northeasterly flows intensified over most of the coastal area such that the orographic lifting was enhanced on the windward side of the SMR, and convergent line was intensified along the northwestern slope of the SMR. This was the time that convection was enhanced over the same area. From this streamline analysis, it seems clear that convection tended to develop and organize along a convergent line on the northwestern slope of the SMR. The area of maximum frequency of convection seems to correlate well to the localized cyclonic vortices that formed along the convergent line.

Fig. 10 illustrates the streamline distribution from day T-1 to day T2 for the stationary type front. The local circulations appear to be quite similar to that of the moving type except that sea breeze circulations prevailed throughout the entire period of analyses. A convergent line together with cyclonic vortices again existed on the northwestern slope of the SMR at day T-1 and T0. The winds increased at day T1 over the area to the immediate northwest of the convergent line such that the convergent line and cyclonic vortices were intensified. This was the time that convection was enhanced over the same area. At day T2, cyclonic vortices weakened along the convergent line and convection was also suppressed over the same area as compared to that at day T1.

Streamline analyses for the retreating type front from day T-1 to day T2 are presented in Fig. 11. At day T-1, sea breeze prevailed all over the coastal area. The convergent line and cyclonic vortices again existed on the northwestern slope of the SMR. At day T0, winds weakened over the area immediate to the northwest of the southwestern part of convergent line such that convergent line and cyclonic vortex also weakened. This was the time that convection was suppressed over the same area. At day T1, sea breeze increased such that convergent line was also intensified and convection was enhanced over the same area. At day T2, sea breeze continued to increase and again convergent line was intensified. Thus, convection was also enhanced and organized along the intensified convergent line on the northwestern slope of the SMR.

6. CONCLUDING REMARKS

The Meiyu fronts affecting northern Taiwan during the period of May-June 1991-1996 were categorized into three types based on their movements. Radar echoes observed at the CKS International Airport in 1100-1800 LST were used to study the characteristics of daytime land convection under the influence of different types of fronts. Surface observations during the same period were used to reveal the relationship between daytime convection and local circulations. The results can be summarized as follows.

1) More widespread convection was triggered at the day when front entered northern Taiwan for the stationary type front as compared to the moving type and retreating type fronts.
2) For moving type front, convection tended to be more active and more organized at day T0 and suppressed at day T1 over the northwestern slope of the Snow Mountain Range. It was enhanced over the mountain slope during the frontal passage and over the mountain area before front entered northern Taiwan.
3) For stationary type front, convection was suppressed at day T0 and was enhanced at day T1 over the Snow Mountain Range and its northern slope. Convection enhancement was much stronger at different altitudes to the north than to the northwest of the Snow Mountain Range after the front entered northern Taiwan.
4) For the retreating type front, convection was enhanced continuously from day T0 to day T2 over the northwestern slope of the Snow Mountain Range. During this period, convection was much more enhanced over the Taipei Basin as compared to the area northwest of the Snow Mountain Range.
5) A convergent line formed over the northwestern slope of the Snow Mountain Range with localized cyclonic vortices along the convergent line for all types of fronts.
6) The intensification of sea breeze circulation appears to be related to the intensification of convergent line and cyclonic vortices which in turn tended to be related to the convection enhancement over the same area.

ACKNOWLEDGMENTS. We wish to thank Mr. H. C. Chou, Mr. J. S. Yang, and Ms. T. C. Chang for preparing the manuscript. This work was supported by the National Science Council Grant NSC89-2111-M002-015.

REFERENCES

Chen, G. T. J., 1988: On the synoptic-climatological characteristics of the east Asian Mei-Yu front. *Atmos. Sci.*, **16,** 435-446 (in Chinese with English abstract).

Chen, G. T. J., 1992: Mesoscale features observed in the Taiwan Mei-Yu season, *J. Meteor. Soc. Japan,* **70,** 497-516.

Chen, G. T. J., and C. P. Chang, 1980: The structure and vorticity budget of an early summer monsoon trough (Mei-Yu) over southeastern China and Japan, *Mon. Wea. Rev.*, **108,** 942-953.

Chen, G. T. J., and S. S. Chi, 1978: On the mesoscale structure of Mei-Yu front in Taiwan. *Atmos. Sci.*, **5**, 1, 35-47 (in Chinese with English abstract).

Chen, G. T. J., and S. S. Chi, 1980: On the frequency and speed of Mei-Yu front over southern China and the adjacent areas. *Papers. Meteor. Res.*, **3, 1 & 2**, 31-42.

Chen, G. T. J., and J. S. Yang, 1988: On the spatial and temporal patterns of heavy rainfalls in Taiwan Mei-Yu season. *Atmos. Sci.,* **16**, 151-162 (in Chinese with English abstract).

Chen, Y. L., and N. B. F. Hui, 1990: Analysis of a shallow front during Taiwan Area Mesoscale Experiment. *Mon. Wea. Rev.*, **118,** 2649-2667.

Chen, Y. L., Y. X. Zhang, and N. B. F. Hui, 1989: Analysis of a surface front during the early summer rainy season in Taiwan. *Mon. Wea. Rev.*, **117**, 909-931.

Chi, S. S., and G. T. J. Chen, 1989a: A study of the MCS case of TAMEX IOP-13. *Proc. Workshop on TAMWX Preliminary Scientific Results, Taipei, 22-30 June*, 14-21. [Available from the Dept. of Atmospheric Science, National Taiwan Univ., Taipei].

Chi, S. S., and G. T. J. Chen, 1989b: Case study of the MCSs and rainfall during TAMEX/Phase I. *Atmos. Sci.*, **17,** 59-75 (in Chinese with English abstract).

Johnson, R. H., and J. F. Bresch, 1991: Diagnosed characteristics of precipitation systems over Taiwan during the May-June 1987 TAMEX. *Mon. Wea. Rev.*, ***119***, 2540-2557.

Lin, P. L., and Y. V. Sheng, 1990: A preliminary study on the characteristics of sea and land breeze in Taiwan area during TAMEX period. *Proc. Conference on Weather Analysis and Forecasting, Central Weather Bureau, Taipei,* 133-144. (in Chinese with English abstract) [Available from Central Weather Bureau, 64 Kung Yuan Rd., Taipei].

Table 1. Dates of T0 for different types of fronts affecting northern Taiwan during May-June 1991-1996.

Types	Dates
Moving	14 May 1991, 27 May 1991, 4 June 1991, 8 May 1992, 11 June 1992, 16 June 1992, 3 May 1993, 2 June 1993, 17 May 1994, 26 May 1994, 18 June 1994, 3 May 1995 8 June 1995, 28 June 1995
Stationary	6 June 1991, 20 June 1991, 25 June 1991, 15 May 1992, 22 May 1993, 22 May 1994, 15 June 1994, 27 May 1996
Retreating	18 June 1991, 5 June 1992, 15 May 1993, 11 June 1993, 23 June 1993, 21 June 1994, 27 May 1995, 8 June 1995, 5 June 1996, 22 June 1996

Table 2. Locations selected for the plain, the slope and the mountain areas along each cross-section to the northwest and north of the Snow Mountain Range.

Location / Cross-section	Plain	Slope	Mountain
Northwest A	Chupei	Neiwan	Shiehpaishan
Northwest B	Chupei	Kuangshi	Takuangshan
North A	CWB	Shientien	Pinglin
North B	Sungshan	Shientien	Takuangshan

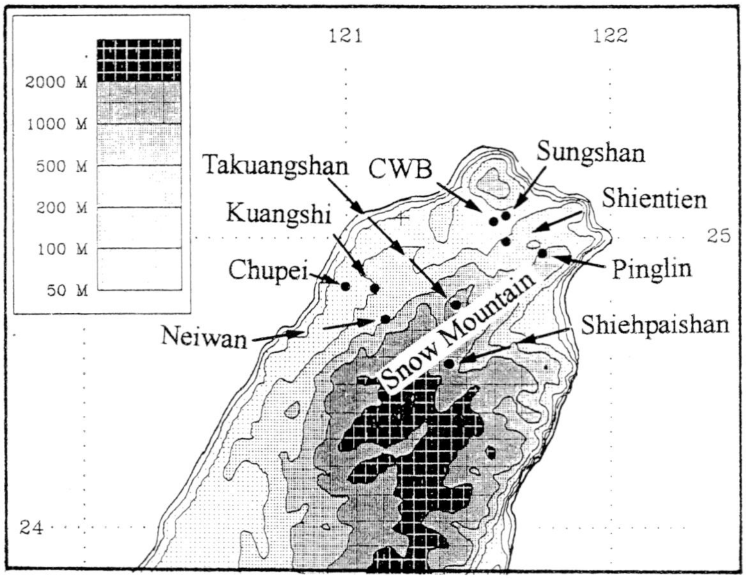

Fig.1 Smoothed terrain heights (m) over a 5 km by 5 km grid square using 1 km grid data. Locations of interest are indicated. + sign indicates the CKS Doppler radar site.

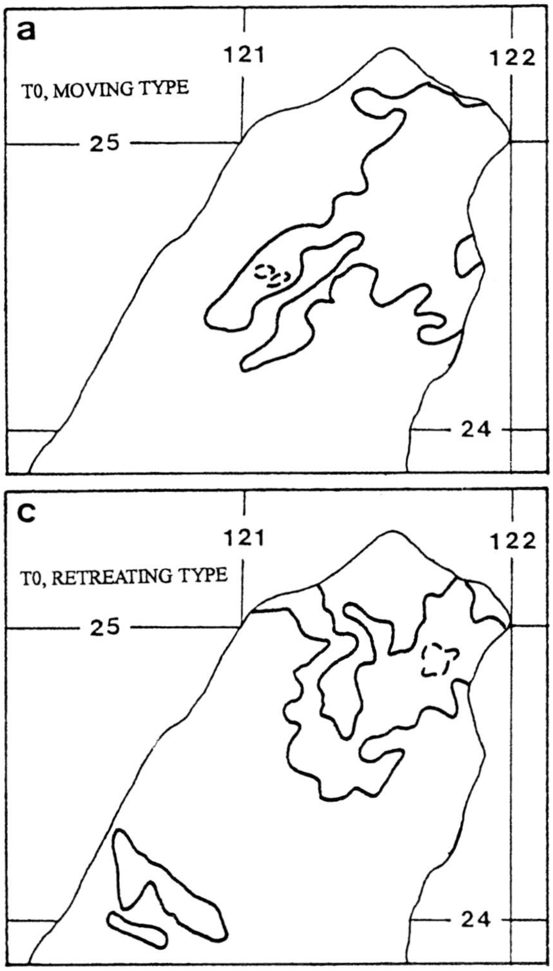

Fig.2 Frequency of convection initiation revealed by the first radar echo of ≧ 15 dBZ at day 0 for fronts of (a) moving type, (b) stationary type, and (c) retreating type. Isolines indicate frequency of 1~2 (solid) and 3~5 (dashed).

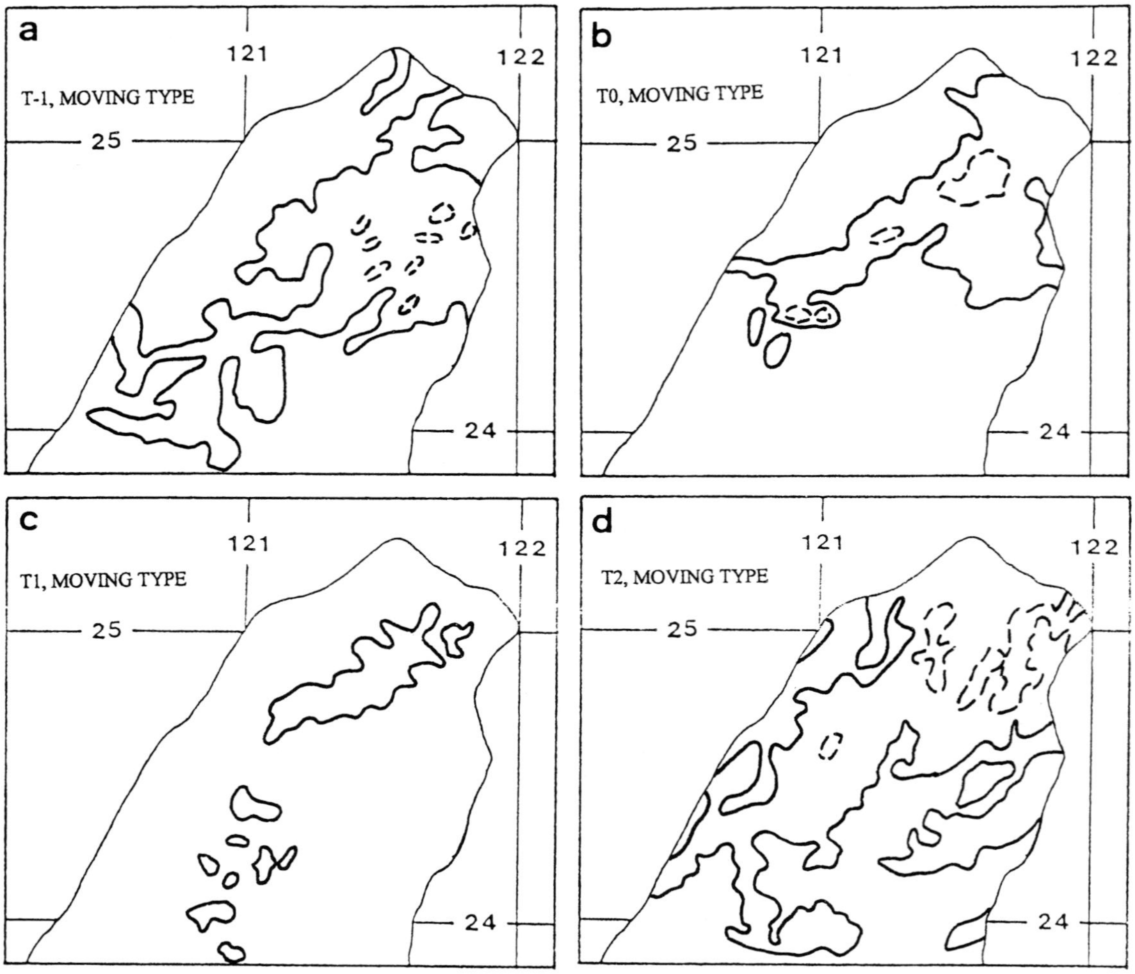

Fig.3 Frequency of convection revealed by radar echo of ≧ 30dBz for moving type front on day (a) T-1, (b) T0, (c) T1, and (d) T2. Isolines indicate frequency of 1~3 (solid), 4~7 (dashed).

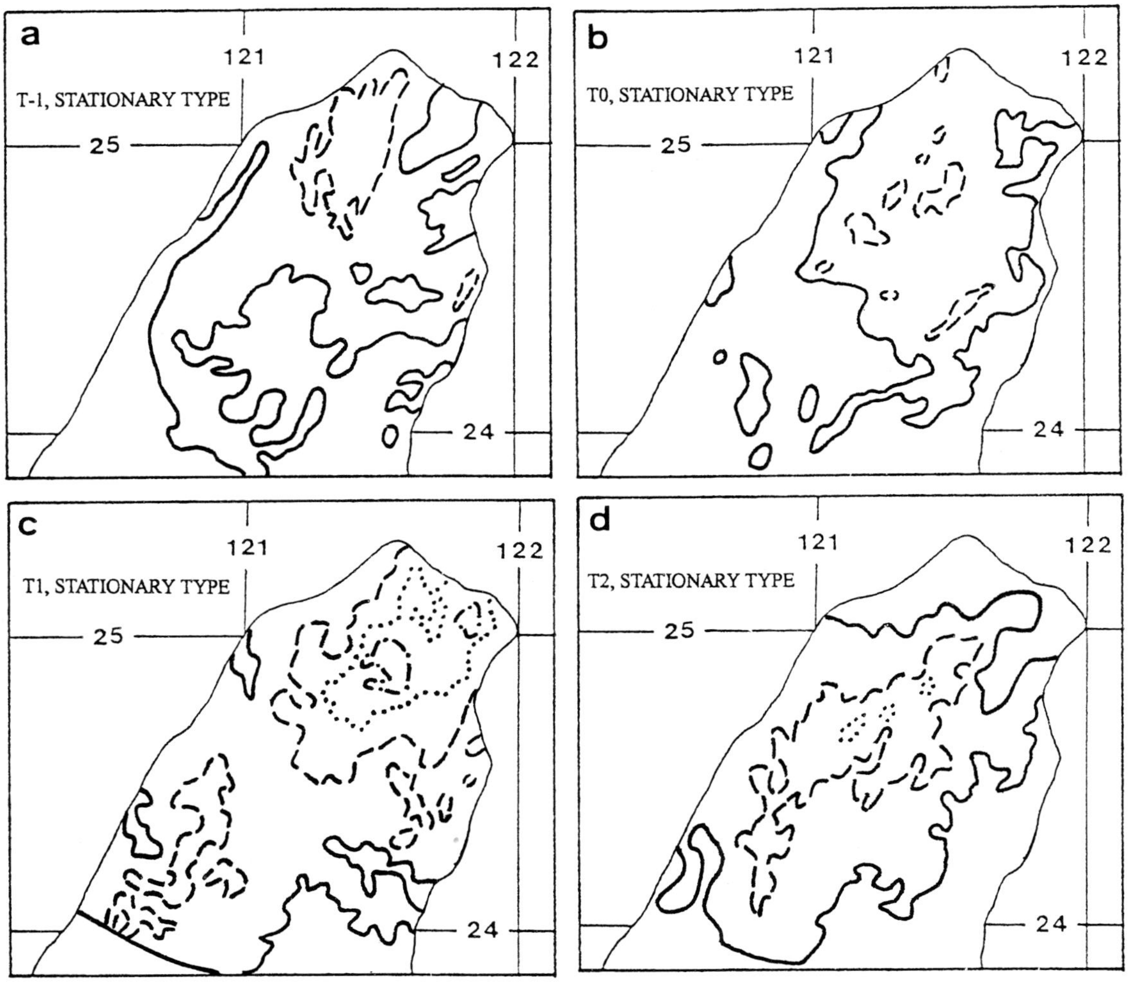

Fig.4 Same as in Fig.3, except for stationary type front. Isolines indicate frequency of 1~3 (solid), 4~7 (dashed), 8~11(dotted) ,and ≧12(dash-dotted).

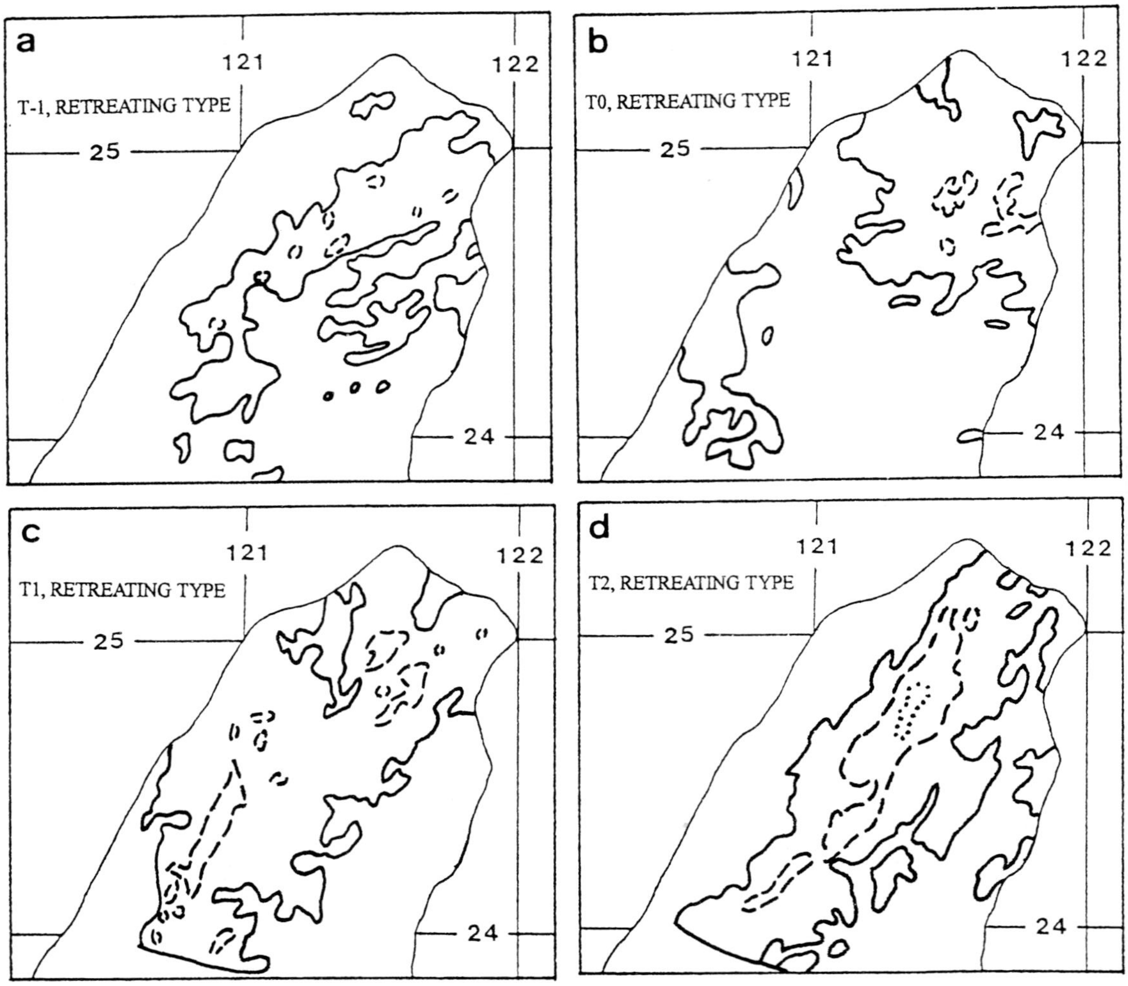

Fig.5 Same as in Fig.3, except for retreating type front. Isolines indicate frequency of 1~3 (solid), 4~7(dashed), and 8~11(dotted).

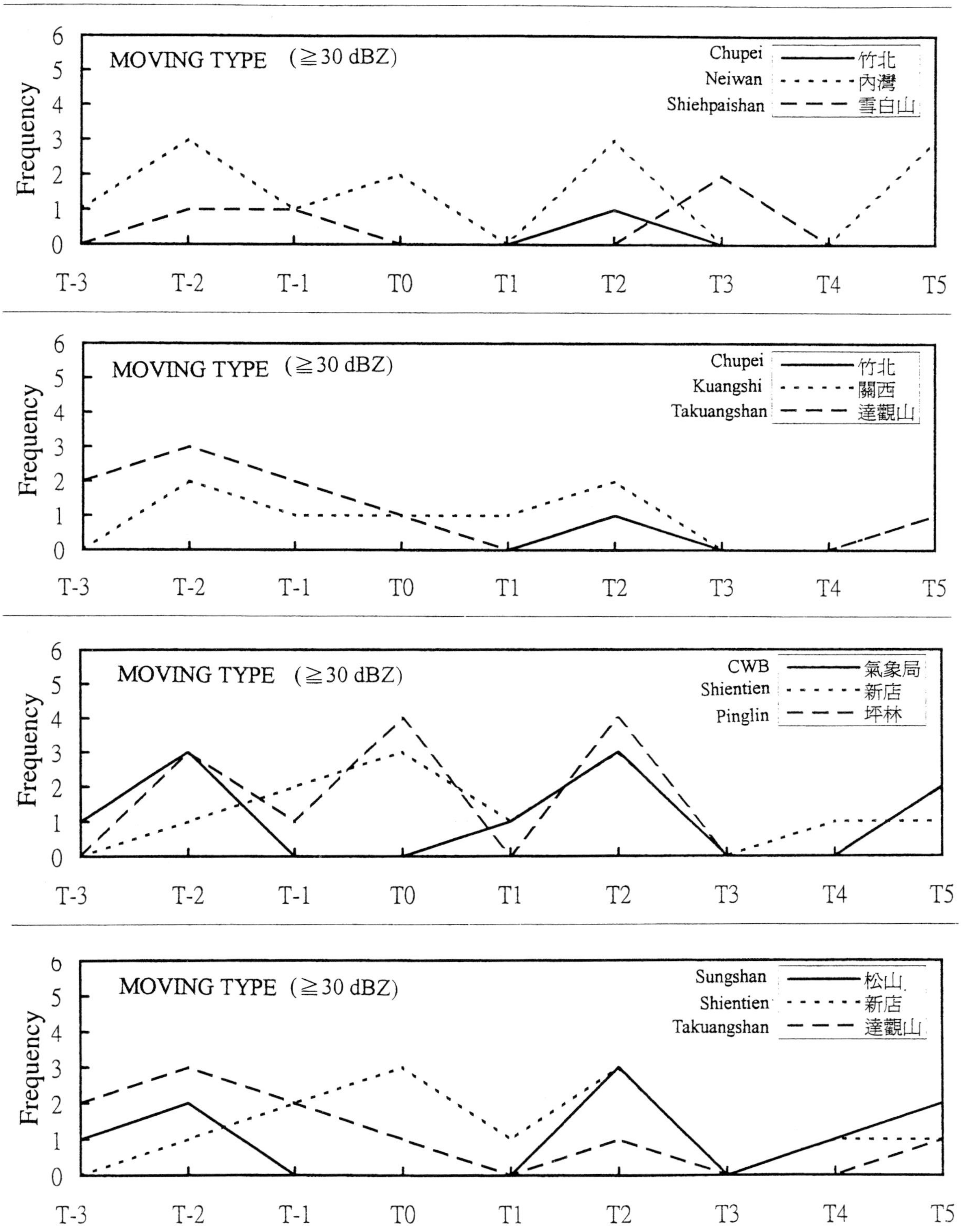

Fig.6 Temporal variations of convection frequency revealed by radar echo of ≧30 dBZ for moving type front at different locations from day T-3 to day T5.

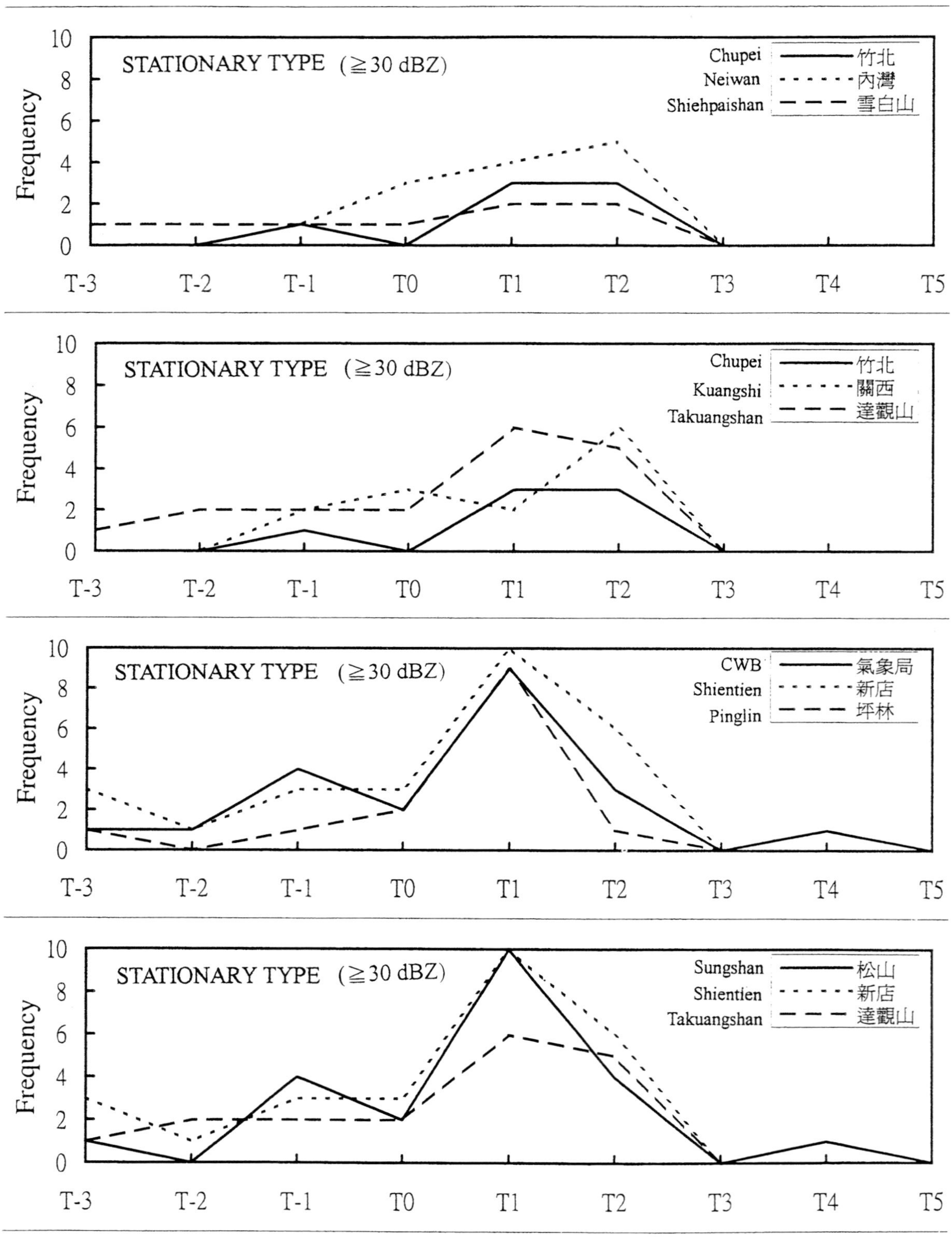

Fig.7 Same as in Fig.6, except for stationary type front.

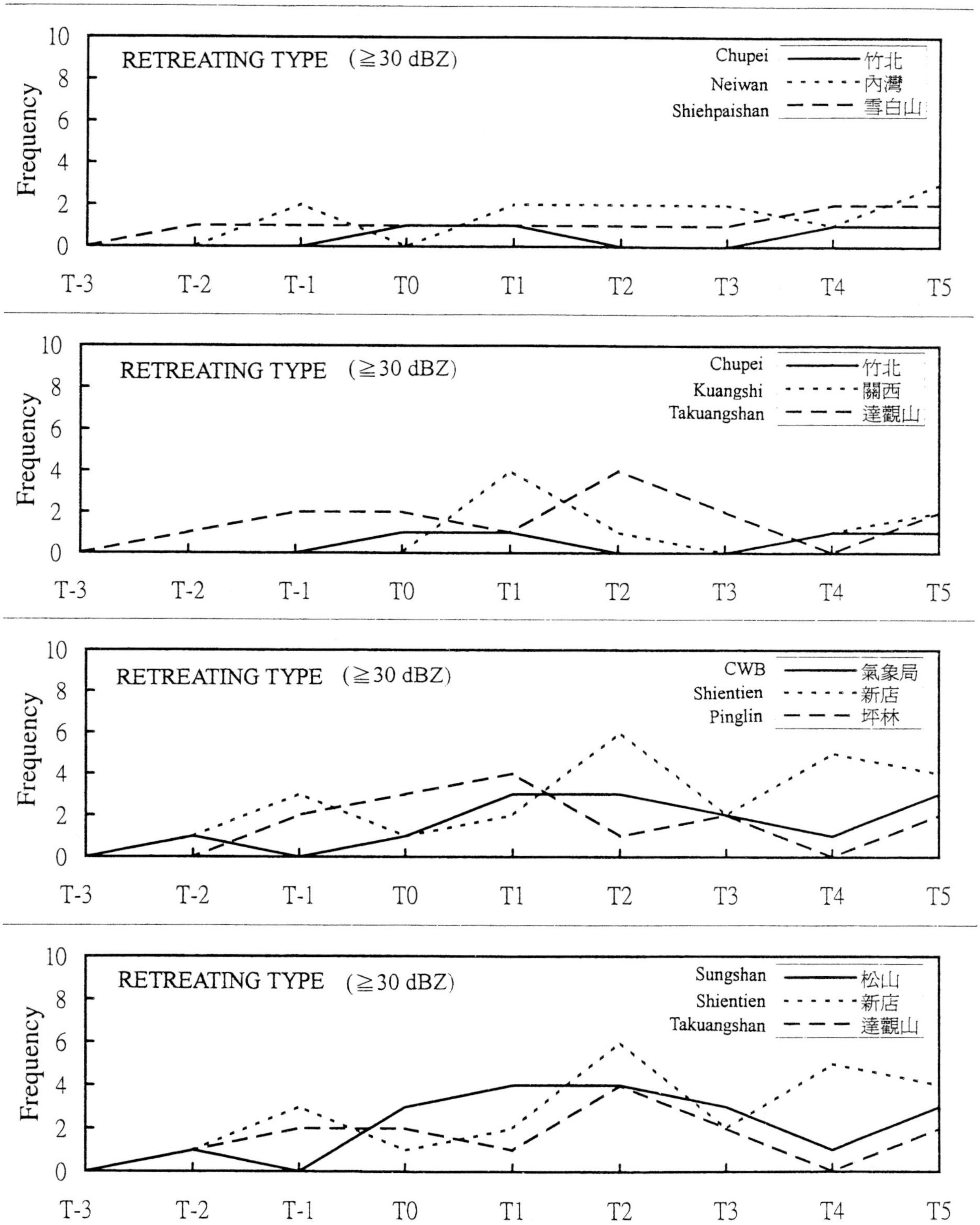

Fig.8 Same as in Fig.7, except for retreating type front.

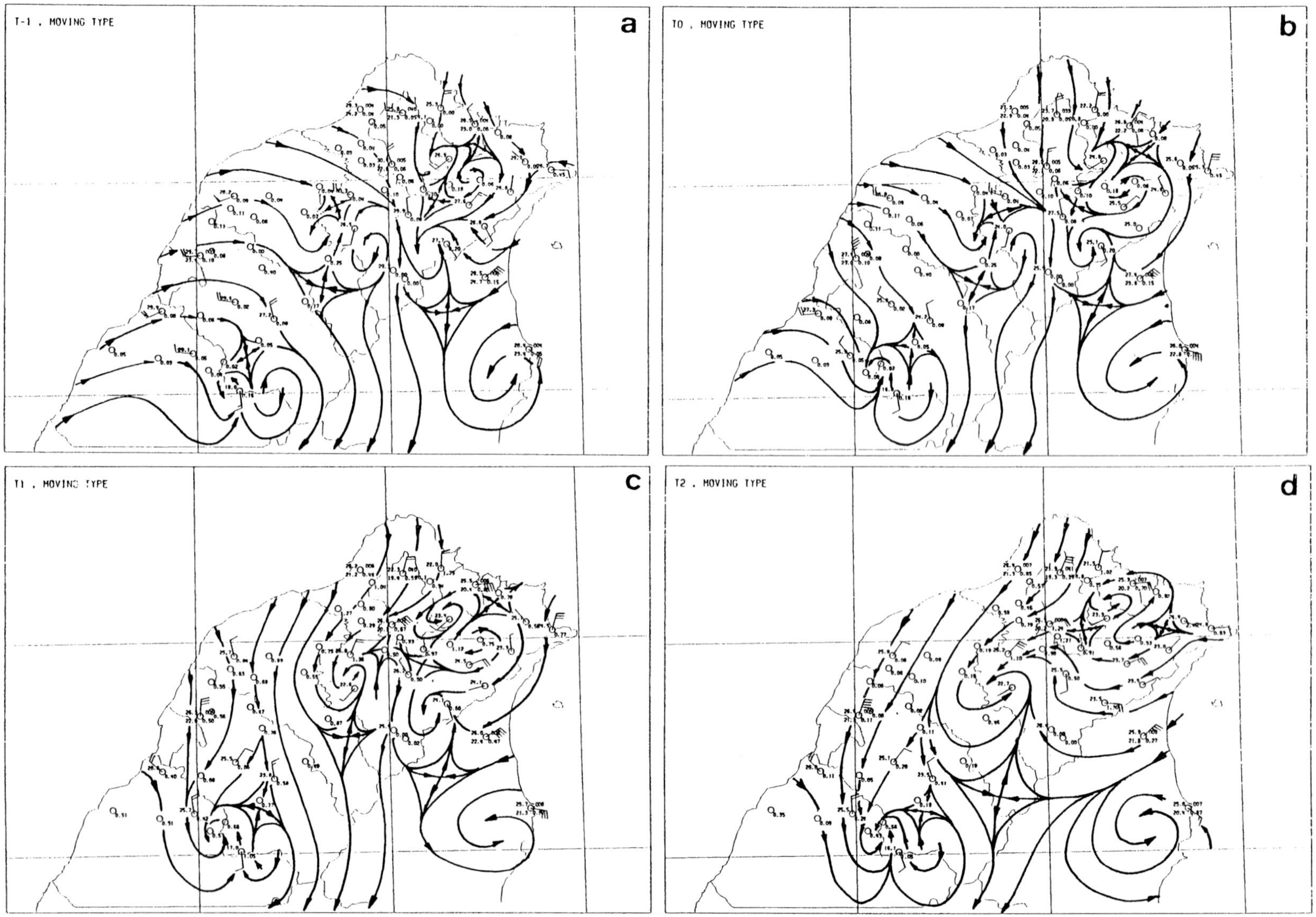

Fig.9 Surface streamlines from day T-1 to day T2 for moving type front. Station plots are conventional, except wind speed. Triangle flag, full barb, and half barb represent 5, 1, and 0.5 ms^{-1} , respectively.

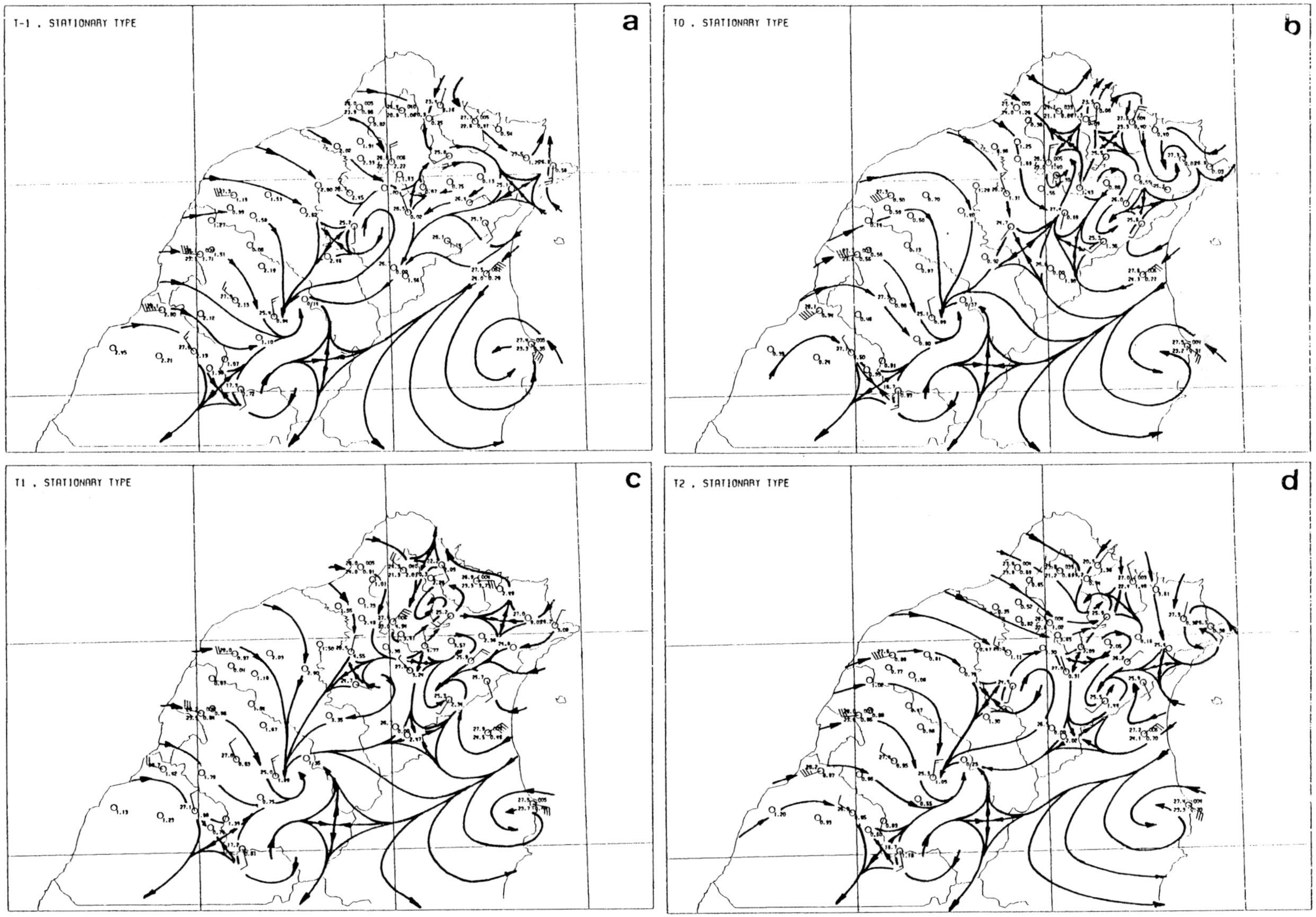

Fig.10 Same as in Fig.9, except for stationary type front.

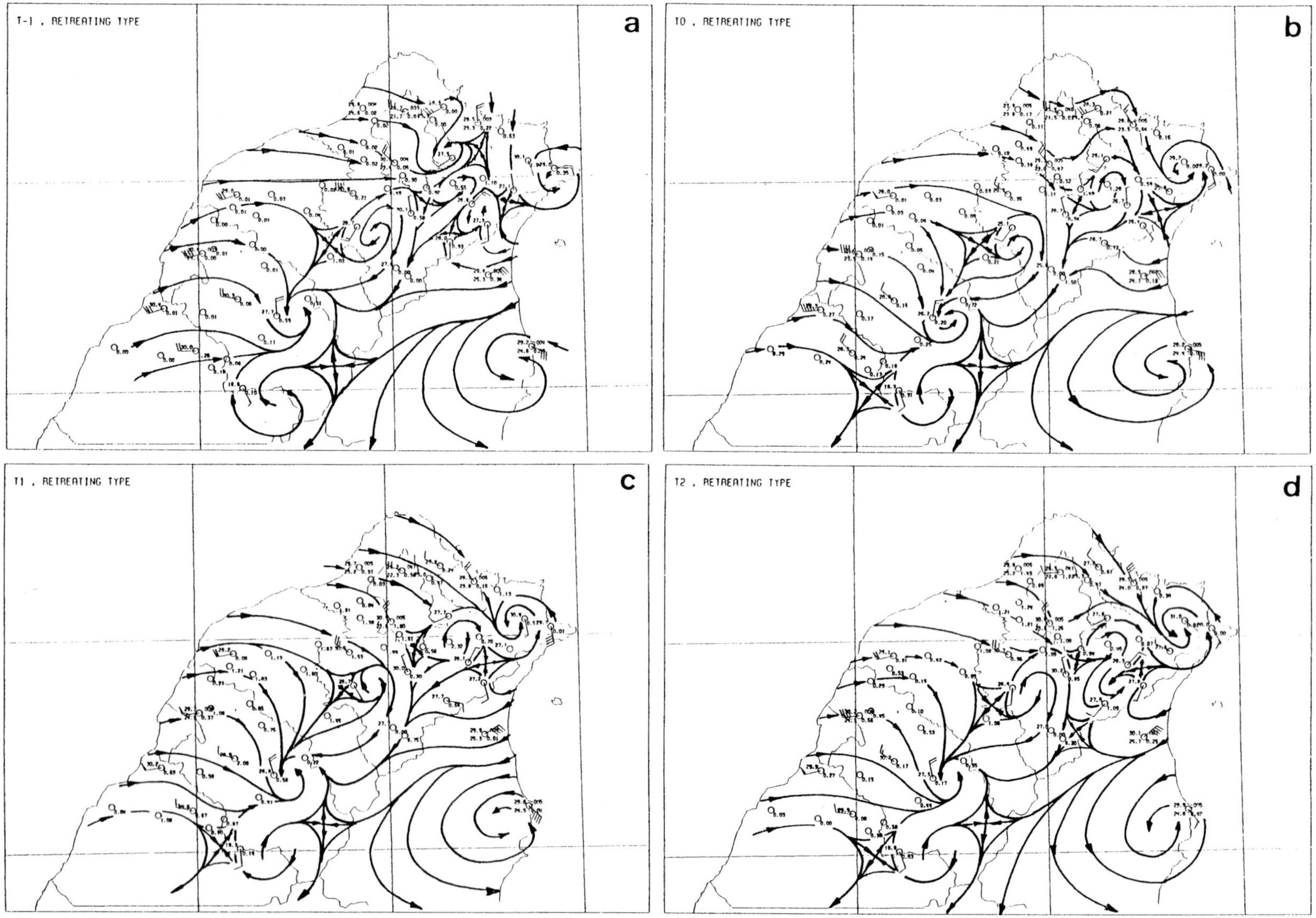

Fig. 11 Same as in Fig. 10, except for retreating type front.

NUMERICAL SIMULATION OF THE MESOSCALE CONVECTIVE SYSTEM OBSERVED OVER TAIWAN STRAIT ON 7-8 JUNE 1998

Zhang Qinghong*, Kai-Hon Lau, Chen Jay-chung and Chen Shoujun
Center for Coastal and Atmospheric Research, The Hong Kong University of Science and Technology
Laboratory for Severe Storm Research, Department of Geophysics, Peking University

ABSTRACT

A mesoscale convective system (MCS) was generated over the South China Sea at around 115° E, 21° N on 0000 UTC 7 June 1998. Riding along the Meiyu front, the system moved through the Taiwan Strait and brought over 300 mm of rain to parts of Taiwan in the 24-hour period after hitting the southern part of the island. Torrential rain resulted in severe flooding. We found that the Penn State-NCAR Mesoscale Model Version 5 (MM5) did quite well in simulating the evolution of the MCS, including the shape of clouds and the amount of precipitation over Taiwan island.

In the mature phase of the MCS, a mesoscale low, a mesoscale low-level jet (mLLJ) and a mesoscale upper-level jet (mULJ) could be clearly identified in the numerical simulations. At the 850 hPa level, the mLLJ was located to the southwest of the MCS, which was also associated with a mesoscale low generated by convection. In the 300 hPa level, the mULJ was found due east of the MCS (and a upper-level mesoscale high). Budget studies were used to investigate the physical processes associated with the maintenance of the mLLJ and mULJ. Horizontal advection of momentum acted to accelerate the jet at the exit of the mLLJ but decelerate the jet at the entrance of the mLLJ where the pressure gradient force acted to compensate this loss and maintain the strength of the jet in its mature phases. For maintenance of the mULJ, vertical advection of momentum and the pressure gradient force were both found to be important.

1. INTRODUCTION

The Meiyu front is one of the most significant circulation systems in the East Asia summer monsoon region. It is characterized by a rather weak temperature gradient but a strong moisture contrast. The Meiyu front first appears in southern China in May and migrates to central China and Japan from mid-June to mid-July each year. Although the Meiyu front can be several thousand kilometers long, the cross-front dimension is within the mesoscale range (Chen 1992). Many observational studies reveal features of mesoscale convective systems (MCSs) (some of them are long-lived) developing and propagating along the Meiyu frontal zone (Ninomiya *et al.* 1988a, b). Intense and concentrated rainfall events in the Meiyu season are often associated with these organized MCSs.

One of the very important features accompanying the Meiyu front is the synoptic low level jet (LLJ) to the south of the shear line in the 850-700 hPa levels. It has been suggested that the LLJ transports warm, moist air northward in the lower levels, creates a convectively unstable layer, and induces a second circulation with ascent at the northern tip of the LLJ. Although many studies have concentrated on the dynamic processes of the acceleration and maintenance of LLJ and its relationship with Meiyu frontogenesis (Chou 1986, Matsumoto and Ninomiya 1971, Nagata and Ogura 1991, Chen *et al.* 1994, Kato 1998), results are still

---

* Present affiliation: The National State Key Laboratory for Severe Storm Research, Peking University, Beijing 100871, Email address: zqh@pku.edu.cn

inconclusive owing to the lack of mesoscale observations.

Chen *et al.* 1998 recently observed mesoscale disturbances in the wind fields, in the synoptic LLJ (referred to as mLLJ) and also in the 200-300 hPa level jet (referred to as mULJ), in the simulation of the rainstorm case along the Meiyu front of 12-13 June 1991.

This study simulates another mesoscale convective system (MCS) generated over the South China Sea near 115° E, 21° N at 0000 UTC on 7 June 1998. Riding along the Meiyu front, the system moved across the Taiwan Strait and hit the southern part of the Taiwan island at about 1200 UTC that day. During the next 24 hours, the MCS brought more than 300 mm of rain to parts of Taiwan. Massive flooding followed, causing severe and widespread damage to thousands of homes, roads, bridges, cultivated lands and crops.

The objectives of the study are: (1) to test the ability of the Pennsylvania State University-NCAR Mesoscale Model Version 5 (MM5); (2) to study the mesoscale features of the MCS (such as mLLJ, mULJ, and meso-low) during its mature stage; (3) to investigate the dynamic mechanism of the maintenance of mLLJ and mULJ through momentum budget studies. Section 2 gives an overview of the model results. Section 3 describes the detailed mesoscale features of the MCS. The momentum budget analysis is presented in Section 4. A summary including a conceptual model of the MCS along the Meiyu front is given in Sections 5.

2. OVERVIEW OF THE MODEL RESULTS

With only two sounding observations at the initial time (0000 UTC 7 June) over the South China Sea, MM5 (three nested domains with the resolutions of 54, 18 and 6 km) simulated the evolution of this MCS quite well. Over the ocean, the shape and position of simulated radar reflectivity at 500 hPa (t = 13, 15 and 17 h; Fig.1a, c and e respectively) compared well with the observed satellite images (Fig. 1b, d and f respectively). The 24-hr rainfall amount simulated in the model (Fig. 2a) also matched with the observation (Fig. 2b), although the center of the rain area was located about 30 km more to the north. In addition, there was also a delay of about 3-6 hours in the evolution of the MCS.

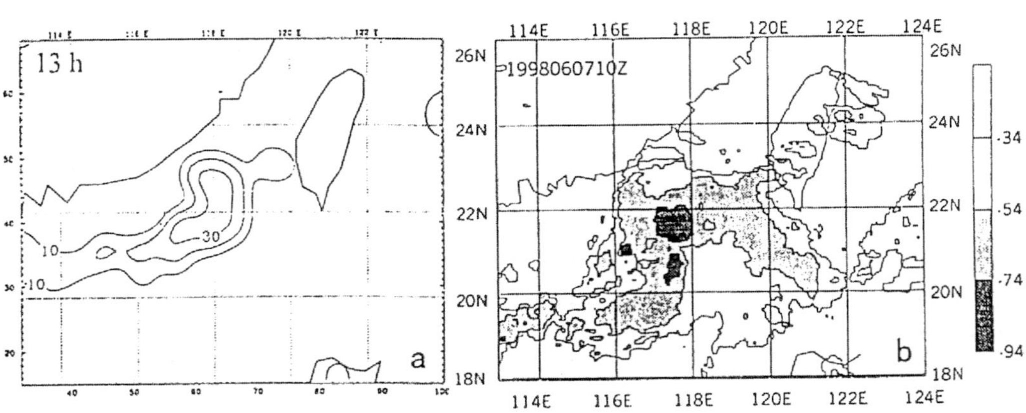

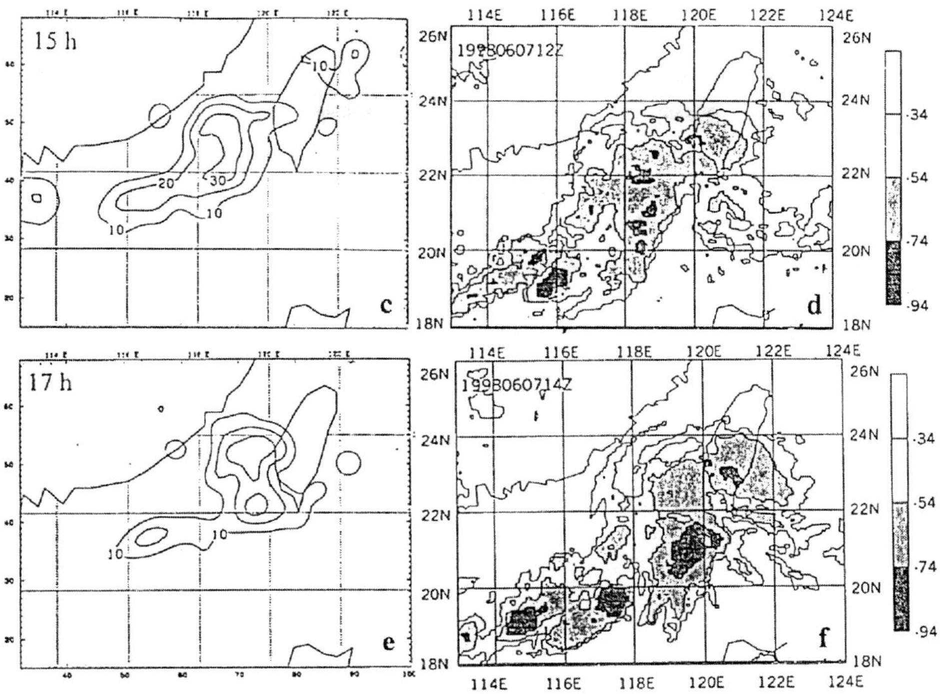

Fig. 1. 500-hPa simulated radar reflectivity (every 10 dBz) in Domain 2 at: (a) t = 13h, (c) t = 15 h, and (e) t = 17 h. Also shown alongside are Tbb from satellite images observed at: (b) 1000 UTC, (d) 1200 UTC, and (f) 1400 UTC (first level of shading is for -54 to -74° C, second is for less than -74° C).

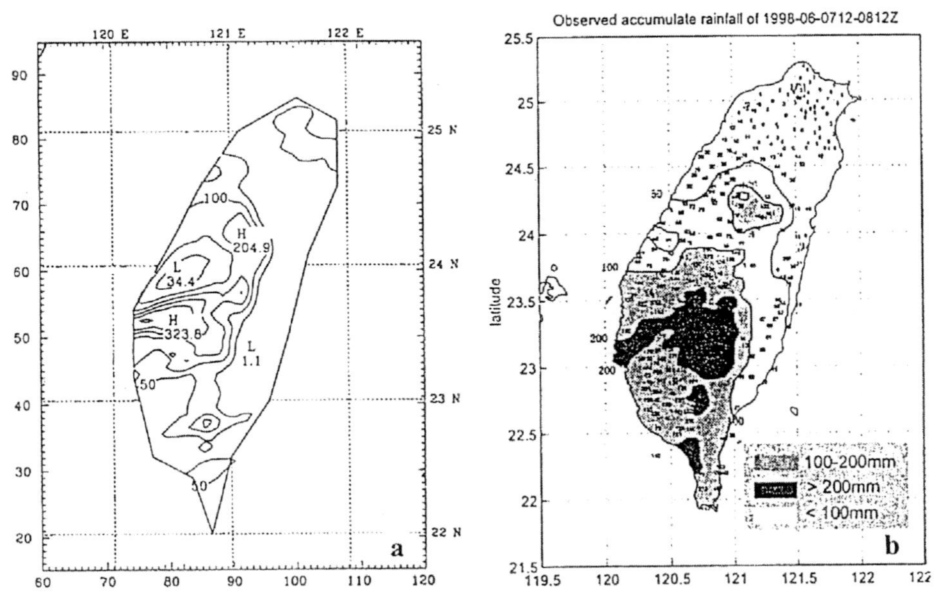

Fig. 2. Accumulated 24-hour rainfall from 1200 UTC 7 June to 1200 UTC 8 June over Taiwan: (a) simulated (every 50 mm); (b) observed (first level shading is for 100-200 mm, second is for greater than 200 mm).

3. STRUCTURE OF MATURE MCS

In the mature phase of the MCS, a mesoscale low, mLLJ and mULJ were clearly depicted in the numerical simulations. At the 850-hPa level, the mLLJ was located to the southwest of the MCS (Fig. 3a). A mesoscale low generated by convection was also found. Strong convergence happened at ahead of the mLLJ (Fig. 3b). In the 300-hPa level, the mULJ was found due east of the MCS (Fig. 3c), coupled with mesoscale high at the upper levels caused by convective heating. Strong divergence occurred to the west of mULJ (Fig. 3d).

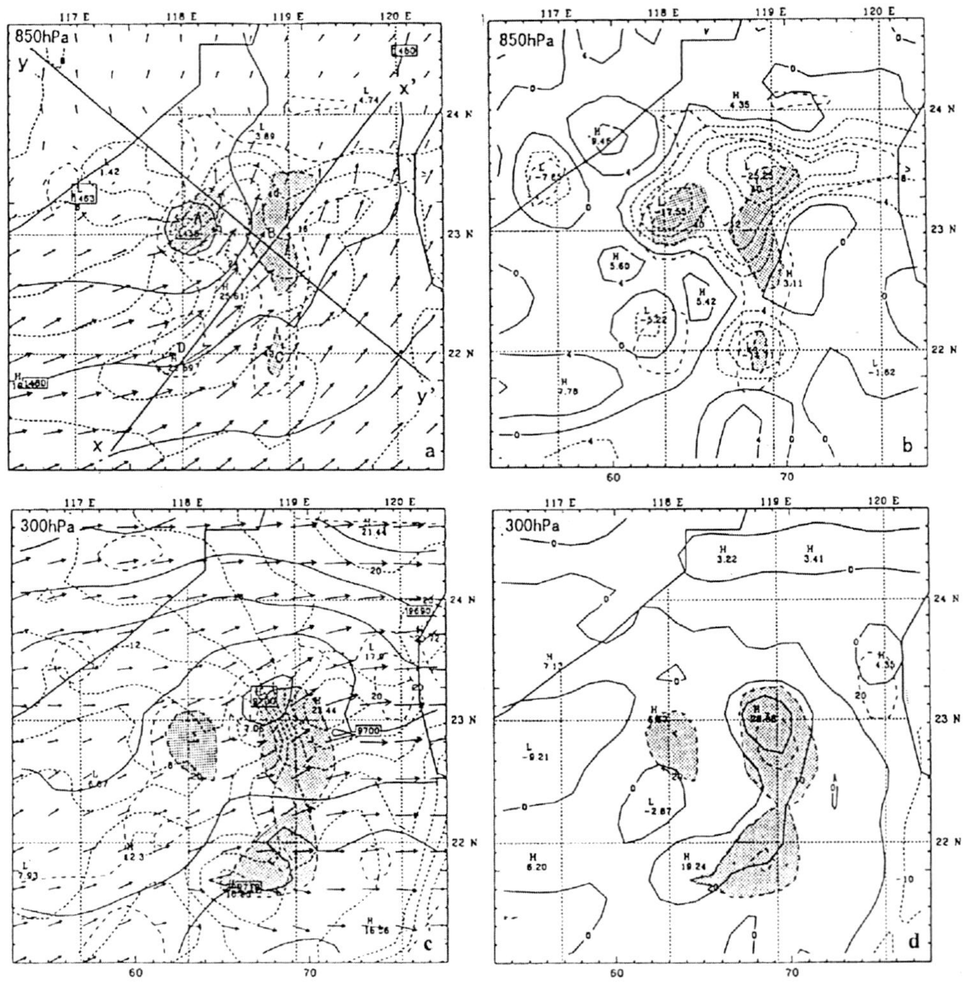

Fig. 3. (a) Simulated geopotential height (every 10 gpm, solid lines), wind vector, horizontal wind speed (every 4 ms^{-1}, short dashed lines) and radar reflectivity (every 10 dBz, long dashed lines, shading area is for greater than 40 dBz) at 850-hPa level at t = 15 h; (b) horizontal divergence (every 4.0×10^{-5} s^{-1}) and radar reflectivity (every 10 dBz, long dashed lines, shading area is for greater than 40 dBz) at 850-hPa level; (c) same as (a) but for 300 hPa; (d) same as (b) but for 300 hPa.

The mLLJ and mULJ could also be seen on the cross section XX' through the centre of maximum winds at 850 hPa (Fig. 4a). The MCS associated with strong upward motion is located between mLLJ and mULJ. Interestingly, mLLJ was associated with a relatively dry core in the mature stage, different from the moist synoptic low–level jet (LLJ) identified in earlier studies. Similar to these studies, moisture convergence was found to be important for the development of this MCS (Fig. 4b). However, we found that moisture convergence in this case was generated mainly through strong horizontal convergence in the wind field associated with mLLJ and not through down-gradient moisture transport across the humidity gradient.

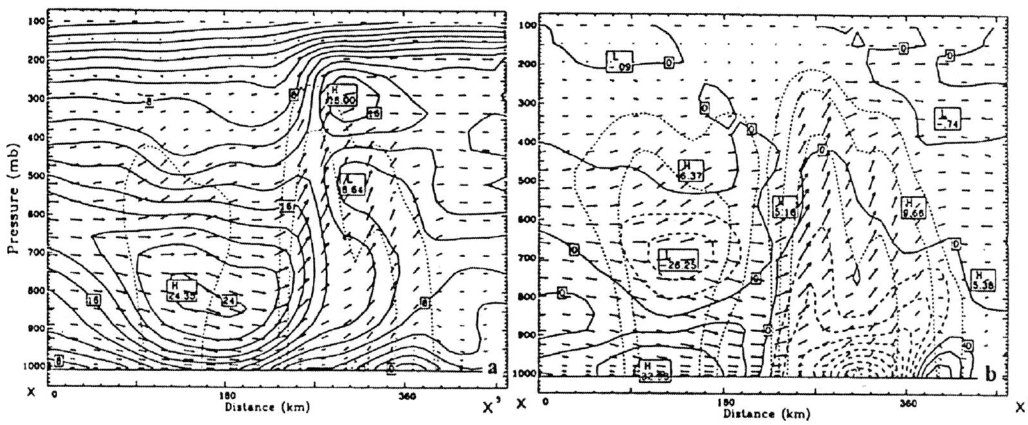

Fig. 4. (a) Simulated radar reflectivity (every 10 dBz, dashed line), wind vector and horizontal wind speed (every 2 ms^{-1}, solid line) along XX' shown in Fig. 3a at t = 15 h; (b) same as (a) but for divergence of water vapor flux (every 10^{-5} $gkg^{-1}m^{-1}s$).

4. MOMENTUM BUDGET RESULTS

Budget studies were used to investigate the physical processes associated with the maintenance of mLLJ and mULJ. To simplify the computation, we define the x- coordinate to be along the southwest-northeast orientation, with positive direction (referred to as S) being from southwest to northeast in the direction of mLLJ and mULJ.

We calculated eight momentum budget terms along S in MM5 and found only four of them to be of importance, namely pressure gradient force (SPGR), advection of momentum flux divergence (SADV), vertical advection of momentum flux divergence (SVAD), and Coriolis term (SCOR). Figs. 5 and 6 plot the above terms (except SCOR) and momentum tendency (STOT) at 850 hPa and 300 hPa respectively. All the terms appeared to exhibit mesoscale characteristics and, to a greater or lesser extent, contributed to the momentum budgets at the entrance and exit of mLLJ and mULJ. Average momentum budgets at the entrance and exit of mLLJ and mULJ are tabulated in Table 1.

Ahead of the mLLJ, horizontal advection of momentum acted to accelerate the jet. At the rear, momentum advection acted to decelerate the jet, but the pressure gradient force acted to compensate this loss and maintain its strength (during the developing and mature phases).

For maintenance of the mULJ, vertical advection of momentum and pressure gradient force were both found to be important.

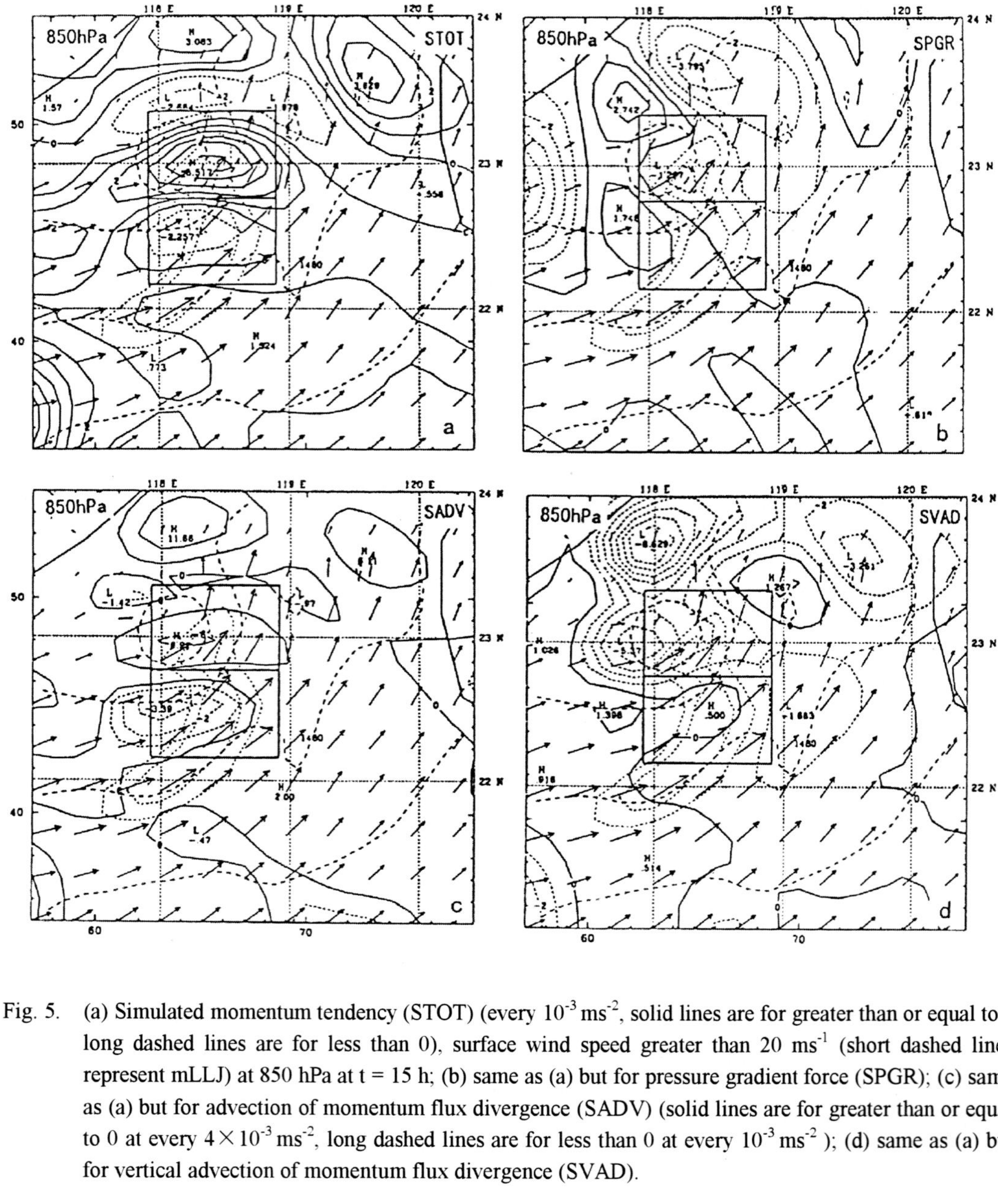

Fig. 5. (a) Simulated momentum tendency (STOT) (every 10^{-3} ms^{-2}, solid lines are for greater than or equal to 0, long dashed lines are for less than 0), surface wind speed greater than 20 ms^{-1} (short dashed lines represent mLLJ) at 850 hPa at t = 15 h; (b) same as (a) but for pressure gradient force (SPGR); (c) same as (a) but for advection of momentum flux divergence (SADV) (solid lines are for greater than or equal to 0 at every 4×10^{-3} ms^{-2}, long dashed lines are for less than 0 at every 10^{-3} ms^{-2}); (d) same as (a) but for vertical advection of momentum flux divergence (SVAD).

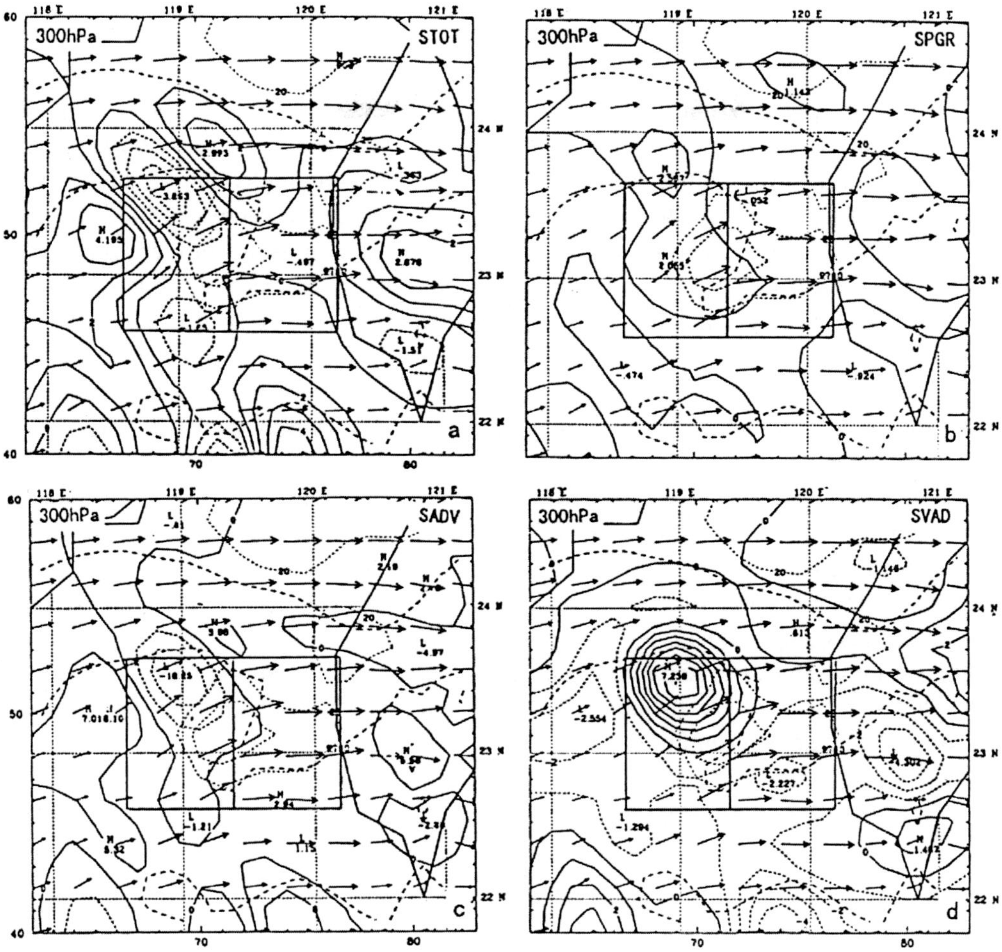

Fig. 6. Same as Fig. 5, but for 300 hPa.

Table 1. Momentum budget area averages for mLLJ (850 hPa) and mULJ (300 hPa).

	850-hPa mLLJ average			300-hPa mULJ average		
	Exit	Entrance	Total	Exit	Entrance	Total
STOT(10^{-3}m s^{-2})	2.30	-0.79	0.76	-0.02	-0.31	-0.17
SPGR(10^{-3}m s^{-2})	-0.48	0.23	-0.13	0.43	1.35	0.89
SADV(10^{-3}m s^{-2})	4.41	-0.93	1.74	0.97	-2.46	-0.75
SVAD(10^{-3}m s^{-2})	-1.88	-0.41	-1.15	-0.74	1.63	0.45
SCOR(10^{-3}m s^{-2})	0.44	0.29	0.37	-0.66	-0.96	-0.81

5. SUMMARY

A conceptual model for the self-development mechanisms of the MCS, mLLJ and mULJ under typical synoptic environment along the Meiyu front in southern China is proposed in Fig. 7. Convective heating associated with moisture convergence in the synoptic-scale LLJ induces mesoscale lows in the lower levels and mesoscale highs in the higher levels. Subsequently, the pressure gradient force associated with the meso-lows in the lower-levels acts to generate the mLLJ. The mLLJ then tends to increase moisture flux convergence and induces upward motion in the cloud cluster. In addition, the convection-induced meso-high in the upper level generates a mULJ downstream of the MCS. Strong divergence at the rear of the mULJ helps to enhance the upward motion within the MCS, which in turn helps to channel horizontal momentum from the mLLJ to the mULJ. Our results also suggest that evaporative cooling, which tends to inhibit vertical motion within the MCS, may act as a damping factor in the aforementioned mechanisms. Sensitivity experiments with and without latent heating show that latent heating is crucial for the development of the rainstorm. In an experiment without latent heating (FAKE DRY), mLLJ, mULJ and the meso-low were all missing. In contrast, in an experiment where evaporative cooling was excluded in cloud and rainwater, the simulated mLLJ, mULJ and the meso-low were enhanced and more rainfall was predicted (as compared with the control experiment).

This is only a semi-idealized hypothesis as a result of our modelling study. It would be of interest to check whether the aforementioned mechanisms can be identified through a denser observation network in future studies.

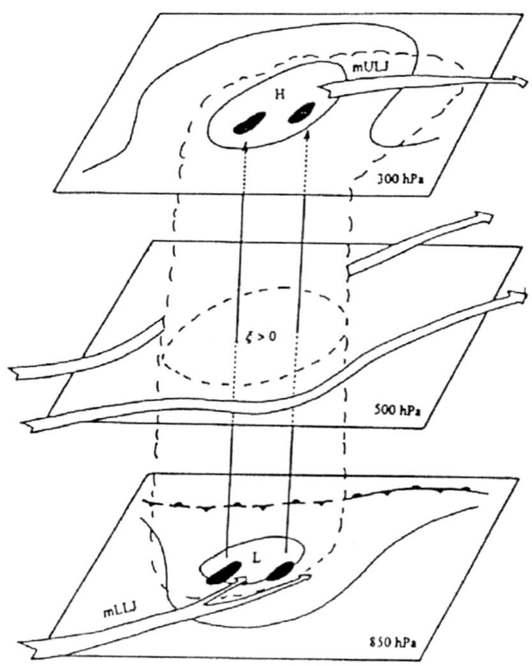

Fig. 7. Conceptual model of MCS along the Meiyu front in southern China. Dark shading areas are strong convergent areas in MCS where heavy rain occurs. White vectors are horizontal mesoscale streamlines at 850 hPa and 300 hPa, and synoptic streamlines at 500 hPa. Dark vectors indicate upward motion. The quasi-stationary front represents the Meiyu front.

REFERENCES

Chen, G.T.J., 1992: Mesoscale features observed in the Taiwan Mei-Yu Season. *J. Meteor. Soc. Japan,* **70**, 497-515.

Chen, S.-J., Y.-W. Kuo, W. Wang, Z.-Y. Tao, and B. Cui, 1998: A modeling case study of heavy rainstorms along the Mei-Yu Front. *Mon. Wea. Rev.*, **126**, 2330-2351.

Chen, Y.-L., X.-A. Chen, and Y.-X. Gang, 1994: A diagnostic study of the low-level jet during TAMEX IOP5. *Mon. Wea. Rev.*, **122**, 2257-2284.

Chou, L. C., 1986: *A numerical simulation of the Mei-Yu front and the associated LLJ.* Ph.D. Thesis, Naval Postgraduate School, 157pp.

Kato, T., 1998: Numerical simulation of the band-shaped torrential rain observed over southern Kyushu, Japan on 1 August 1993. *J. Meteor. Soc. Japan*, **76**, 97-126.

Matsumoto, S., and K. Ninomiya, 1971: On the mesoscale and medium-scale structure of a cold front and the relevant vertical circulation. *J. Meteor. Soc. Japan*, **49**, Special issue, 648-662.

Nagata, M., and Y. Ogura, 1991: A modeling case study of interaction between heavy precipitation and a LLJ over Japan in the Baiu Season. *Mon. Wea. Rev.*, **119**, 1309-1336.

Ninomiya K., T. Akiyama, and M. Ikawa, 1988a: Evolution and fine structure of a long-lived meso-α-scale convective system in a Baiu front zone. Part I: Evolution and Meso-β-scale Characteristics. *J. Meteor. Soc. Japan*, **66**, 351-371.

Ninomiya, K., T. Akiyama, and M. Ikawa, 1988b: Evolution and fine structure of a long-lived meso-α-scale convective system in a Baiu front zone. Part II: Meso-γ-scale characteristics of precipitation. *J. Meteor. Soc. Japan*, **66**, 331-350.

DOPPLER RADAR OBSERVATIONS OF A QUASI-STATIONARY MESOSCALE CONVECTIVE SYSTEM AND THE ASSOCIATED VORTEX

Ben Jong-Dao Jou
Department of Atmospheric Sciences
National Taiwan University, Taipei

ABSTRACT

Doppler radar observations of a mesoscale vortex associated with a quasi-stationary mesoscale convective system in the Taiwan Mei-yu season is presented. By using GBVTD technique, the mean tangential winds are retrieved. The results showed the vortex had a dipole structure in perturbed Doppler radial winds embedded within the prevailing southwesterly winds. Heavy rainfall occurred at downstream of the enhanced southwesterly flows induced by the mesoscale vortex.

1. INTRODUCTION

Observational studies show that a variety of mesoscale convectively generated vortices (MCVs) with horizontal scales ranging from tens to hundreds of kilometers and time scales ranging from hours to days can be generated within mesoscale convective systems (MCSs). These MCVs play a variety roles, from locally enhancing the strength of rear-inflow jets, which may contribute to the production of damaging surface winds, to initiating the development of long-lived, balanced mid-tropospheric circulations, which can help trigger new convective episodes on subsequent days (Smull and Houze 1985; Stirling and Wakimoto 1989; Brandes 1990; Johnson and Bartels 1992; Jorgensen and Smull 1993; Bartels et al. 1997). In most above-mentioned studies, observations indicated that mesoscale vortices are frequently found in the low to mid-troposphere of MCS stratiform precipitation regions. Stretching of planetary vorticity in response to latent heat release in the stratiform region was thought to be important mechanism for the formation and maintenance of the meso-vortex (Skamarock et al. 1994). On the other hand, Yu et al. (1999) showed that, with an exceptionally moist atmosphere and moderate ambient vertical shear through a deep layer with much weaker shear and winds aloft, a meso-vortex was formed within the convective precipitation region of a developing MCS. By using airborne Doppler radar data, they showed that vertical stretching in the convective region was a primary mechanism contributing to low-level development of the vortex and then vertical transport of low-level convectively generated vorticity acted to strengthen and extend the more widespread cyclonic vorticity at upper levels. In addition, the cyclonic vorticity originated along the mesoscale shear/convergence zone resulted from interaction of the large-scale southwesterly inflow with the relatively cool orographically-trapped northeasterly low-level flow aided the MCV's spin up at lower levels.

An early study on the meso-low in the Taiwan Mei-Yu season by Chen (1978) showed that a positive correlation existing between rainfall amount and meso-low frequency to the west of the Central Mountain Range (CMR). Chen and Chi (1980) further suggested that meso-low over the northwestern Taiwan probably served as a mechanism for producing heavy rainfall through enhancing southwesterly flows. However, the questions such as detailed structure and dynamic processes relating to the formation of the meso-low as well as the specific dynamic link between the convection and the meso-low remain unanswered. A recent study conducted by Chen and Wang (1992) investigated the composite mesoscale structure of the meso-low at different life stages. The results showed that the deep convection and thus the heavy rainfall to the south of the low centre were closely associated with the formation and intensification of the meso-low through the enhanced southwesterly flows and the boundary layer convergence. It was suggested the behavior of convection is heavily modulated by the surface boundary layer convergence.

The objective of this paper is to present observations of a mesoscale vortex embedded within a quasi-stationary MCS as captured by the C-band Doppler radar at CKS International airport. In May and June 1997, an observational and forecasting experiment was conducted to study the heavy rainfall events over the Taiwan area during the Mei-yu season. Intensive observations of meteorological data of 7 heavy rainfall cases were collected in the experiment. In late afternoon of 2 June, a heavy rainfall event occurred in Hsin-Chu area. Mesoscale surface pressure analysis showed that a meso-low was found over the region. A cyclonic circulation was also identified through surface wind analysis. From radar echo, it was shown that the convection was first found over the coast of central Taiwan and propagated northward along the sloping area west of CMR. The convection slowed down and became quasi-stationary while approached Hsin-Chu county. From 2 to 5 p.m., more than 120 mm rain was collected by an automatic rain gauge in the area.

2. DATA

During the period of heavy rainfall, intensive observations were conducted by a C-band Doppler radar located about 40 km north of MCS. Precipitation reflectivity and Doppler radial winds were obtained by the radar in a horizontal resolution of 1km X 1^0 and with 13 elevation angles in each volume scan and in time interval of 15 minutes. In this study, the Doppler radial winds were analyzed to reveal the mesoscale flow structure associated with the meso-vortex by using GBVTD (ground-based velocity track display) technique developed by Lee et al. (1999).

3. RESULTS AND DISCUSSION

A sequence of radar reflectivity maps from 11:47-15:47 LST June 2 1997 taken by CKS Doppler radar is given in Fig. 1. The closed contours indicate the isolines of radar reflectivity factor of 40 dBZ at elevation angle 2.4^o. Each closed contour is 15 minutes apart. It is shown that the convective system formed over the sloping area of central Taiwan in late morning and then gradually moved northward along the flat plateau at a height 200-500 m west of CMR. The convective system slowed down and became quasi-stationary near Hsin-Chu county in early afternoon. The convective system developed and became an organized mesoscale convective system (MCS) in this region. Fig. 2 showed the detailed terrain feature in the Hsin-Chu area. The numbers indicated in the figure are accumulated hourly rainfall amount (2:01-3:00 p.m.) for each automatic rainfall stations. Hourly rainfall of 50 mm was measured at the station Da-Tei-Kun (DTK) and this station accumulated more than 100 mm rain between 2-5 p.m. It is worth to mention that the MCS slowed down when it approached the concave terrain feature in the county of Hsin-Chu. It seems that the MCS was trapped and became stationary because of the concave shape of the complex terrain. This feature is especially clear when mapping the centres of the associated vortex in sequence of time.

Sounding of Pan-Chiao on 00 Z June 2 1997 is given in Fig. 3. The well-mixed layer and high moist content below 950 hPa is clearly shown. Pronounced inversion layer at 800 hPa prohibited large area convective activity development. However, the huge amount of CAPE (~2800 m^2s^2) indicated great potential for convection development in this area. Vertical profiles of horizontal wind derived from CKS Doppler radar using VAD technique is given in Fig. 4. The 15 minutes interval of vertical profiles of horizontal wind showed there is a moderate vertical wind shear in the lower atmosphere. A pronounced wind shift zone existed below 1 km height with southwesterly prevailed above and northeasterly down below, indicated the approach of surface frontal system. This feature is further confirmed by the wind measurements at the surface stations on the northwest coast of Taiwan. Surface streamline analysis (not shown) indicated the existence of mesoscale cyclonic circulation over the Hsin-Chu and Miao-Li area. Surface pressure analysis also showed the existence of a meso-low accompanying with the mesoscale cyclonic circulation as revealed in the local wind analysis. This result is similar to the results shown in Chen and Wang (1992) analyses. In their study, Chen and Wang (1992) showed the major precipitation and cloudiness distributions located in the area where enhanced southwesterly flow prevailed. As will be shown later, similar organization was also observed in this study however in a much finer scale.

Radar reflectivity and Doppler radial wind at height 2 km on 2 p.m. is given in Fig. 5. The Doppler winds in this plot are the perturbation winds. The mean southwesterly flow component determined from the VAD analysis has been deduced from the total Doppler radial winds in order to show the disturbed wind pattern. In the figure, the dipole structure of the perturbed wind is vividly shown. This disturbance embedded within the prevailing southwesterly flows. This signature was not showed in the original Doppler wind pattern. The "+" sign at the top of the figure indicates the location of CKS radar. The perturbed winds showed a maximum value of 6 m/s both on the approaching and the receding components. This dipole pattern of Doppler wind indicates a signature of cyclonic vortex motion in this region. The dipole pattern of the perturbed radial winds was not only observed at height 2 km but up to 5 km indicating the meso-vortex is not only a boundary layer phenomenon.

The size of the vortex (~20 km in diameter) is much smaller compared to that of midlevel mesoscale vortex in the trailing-stratiform rain region of an asymmetric squall line (Houze 1993). The major reflectivity field was found to locate over the down stream of the southwesterly flow. In the region of perturbed northeasterly flow, the reflectivity field was a minimum. This organization indicates the importance of the intensifying warm and moist southwesterly flow on the development and maintenance of this mesoscale convective system.

By using pattern recognition method, Lee et al. (1999) demonstrated that GBVTD technique successfully reconstructs the horizontal tropical cyclone primary circulation from single Doppler radar radial winds. Detailed mathematical derivation and interpretation of the method can be found in Lee et al. (1999) and will not be repeated here. In this paper, we applied the GBVTD technique to analyze the mean and the disturbed winds associated with the vortex. In Fig. 6, a sequence of mean tangential winds from 1302-1432 LST at height 2 km is given. The evolution of the mean tangential winds at 2km height showed the mean tangential wind increased its intensity from 2 m/s to 4 m/s in the early development stage and the radius of the maximum wind shrink from 6km to 3 km at the same time. The mean tangential wind intensified to 6 m/s and the meso-vortex reached its mature stage. During this period, no apparent change of the radius of the maximum wind (RMW) was observed. The vortex dissipated rapidly after its mature phase. In Fig. 7, evolution of the mean tangential winds at various height and radius were displayed in time sequence starting from 1302 to 1432 LST June 2, 1997. It is interested to note that orientation of the axis of the maximum mean tangential wind radius has been gone through a dramatic change from a straight-up-to-outward pattern (small RMW at the bottom and larger RMW at the top) to a downward-outward (reversed) pattern. In the intensifying stage, the RMW at upper portion of the vortex (at height 4 km, for example) shrunk from10 km down to 3 km. The mean tangential wind was strengthened rapidly from 3 m/s to 6 m/s at height 2 km and from near zero to 4 m/s at 5 km height. These features can be interpreted with the conservation law of angular momentum. In the decaying stage, the strong mean tangential wind at upper level the vortex dissipated quickly and the vortex collapsed afterwards.

4. CONCLUDING REMARKS

It is interested to note the centre of the vortex at different heights tilted toward southeast with altitude as indicated in Fig. 2. The tilt is consistent with the direction of shear between 700 and 500 hPa. It is also interested to note that the maximum rainfall measured by the surface rain gauge station was at DTK and was located at downstream of the enhanced southwesterly flows induced by the mesoscale vortex. It is known the precipitation echoes of radar reflectivity were changing rapidly due to complex precipitation mechanisms involved. It is very difficult to trace back the history of each individual echo and make accurate precipitation forecast. However, the radial wind pattern with dipole signature has a much longer time memory. This dipole signature can be served as a pre-cursor of the heavy rain as indicated in this case study. More studies are needed to clarify the relationship between the mesoscale vortex and the MCS in which the vortex is embedded within.

ACKNOWLEDGEMENTS This study is partially supported by National Science Council, under Grant NSC89-2111-M-002-019-AP6. Figure drafting of Ms. Rita Chou is highly appreciated.

REFERENCES

Bartels, D. L., J. M. Brown, and E. I. Tollerud, 1997: Structure of a midtropospheric vortex induced by a mesoscale convective system. *Mon. Wea. Rev.*, **125**, 193-211.

Brandes, E. A., 1990: Evolution and structure of the 6-7 May 1985 mesoscale convective system and associated vortex. *Mon. Wea. Rev.*, **118**, 109-127.

Chen, G. T.-J., 1978: On the mesoscale systems for the Mei-yu regime in Taiwan. *Proc. Conf. Severe Weather in Taiwan Area,* NSC and Academia Sinica, 150-157.

Chen, G. T.-J., and S.-S. Chi, 1980: On the mesoscale rainfall and mesolow in the Mei-yu season in Taiwan. *Atmos. Sci.*, **7**, 39-48.

Chen, G. T.-J., and C.-C. Wang, 1992: The composite structure of mesolows accompanying heavy rainfall in the Taiwan Mei-yu season. *TAO*, **3**, 533-556.

Jorgensen, D. P., and B. F. Smull, 1993: Mesovortex circulation seen by airborne Doppler radar within a bow-echo mesoscale convective system. *Bull. Amer. Meteor. Soc.*, **74**, 2146-2157.

Johnson, R. H., and D. L. Bartels, 1992: Circulations associated with a mature-to-decaying midlatitude mesoscale convective system. Part II: Upper-level features. *Mon. Wea. Rev.*, **120**, 1301-1320.

Jou, B. J.-D., and C.-K. Yu, 1992: Airborne measurement of an oceanic mesoscale convective system during TAMEX. Proc. *Second Int. Conf. On East Asia and W. Pacific Meteorology and Climate, Hong Kong*, Meteorological Society of Hong Kong, 453-462.

Lee, W.-C., B. J.-D. Jou, P.-L. Chang, and S.-M. Deng, 1999: Tropical cyclone kinematic structure retrieved from single Doppler radar observations. Part I: Interpretation of Doppler velocity patterns and the GBVTD technique. *Mon. Wea. Rev.*, **127**, 2419-2439.

Skamarock, W. C., M. L. Weisman, and J. B. Kelmp, 1994: Three-dimensional evolution of simulated long-lived squall lines. *J. Atmos. Sci.*, **51**, 2563-2584.

Smull, B. F., and R. A. Houze, Jr., 1985: A midlatitude squall line with a trailing region of stratiform rain: Radar and satellite observations. *Mon. Wea. Rev.*, **113**, 117-133.

Stirling, J., and R. M. Wakimoto, 1989: Mesoscale vortices in the stratiform region of a decaying midlatitude squall line. *Mon. Wea. Rev.*, **117**, 452-458.

Yu, C.-K., B. J.-D. Jou, and B.F. Smull, 1999: Formative stage of a long-lived mesoscale vortex observed by airborne Doppler radar. *Mon. Wea. Rev.*, **127**, 838-857.

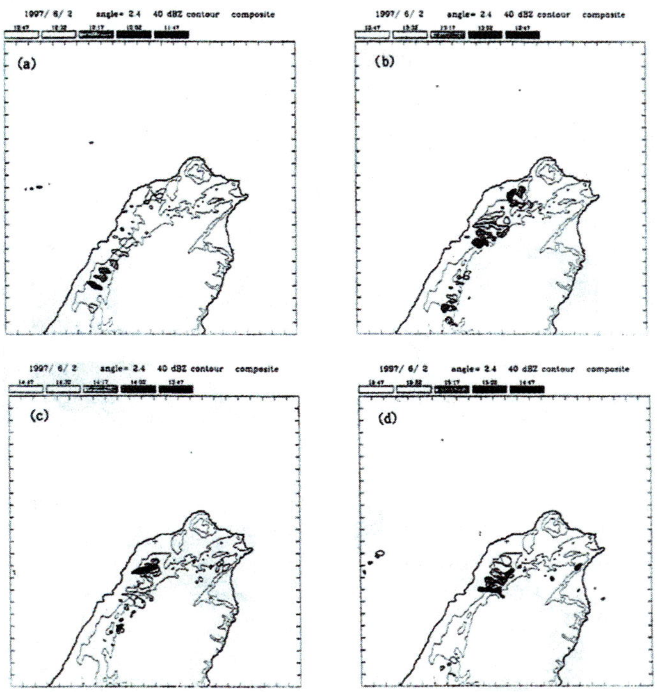

Fig.1. Radar reflectivity summary maps of (a) 11:47-12:47, (b) 12:47-13:47, (c) 13:47-14:47, and (d) 14:47-15:47 LST on June 2, 1997 observed by CKS Doppler radar. The elevation angle is 2.4^0. The closed contours indicate the isolines of reflectivity intensity of 40 dBZ.

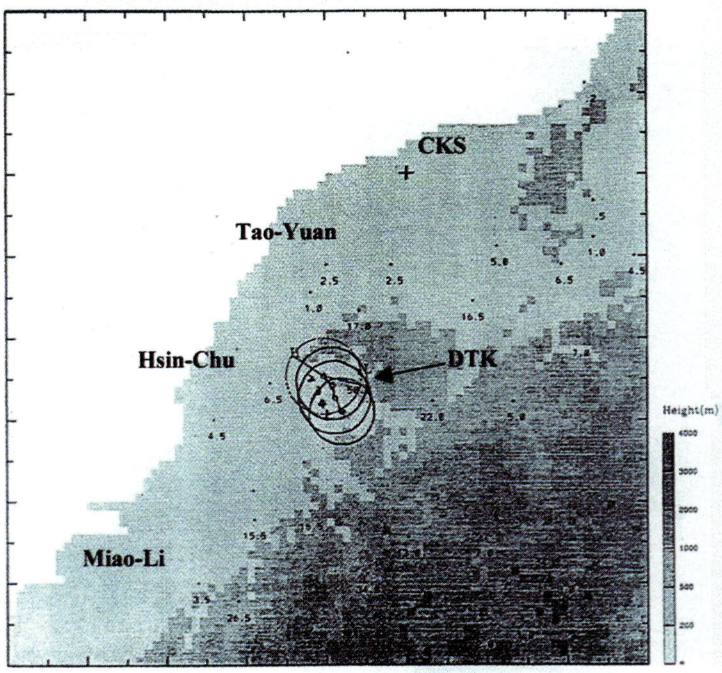

Fig.2. Detail terrain feature around Hsin-Chu area is given. The numbers in the figure showed the hourly rainfall of 1500LST for the stations in the area. Rainfall station Da-Tei-Kun is marked with DTK. The small circles indicated the radius of maximum wind and the center of the mesoscale vortex at different altitudes (from 2-5 km).

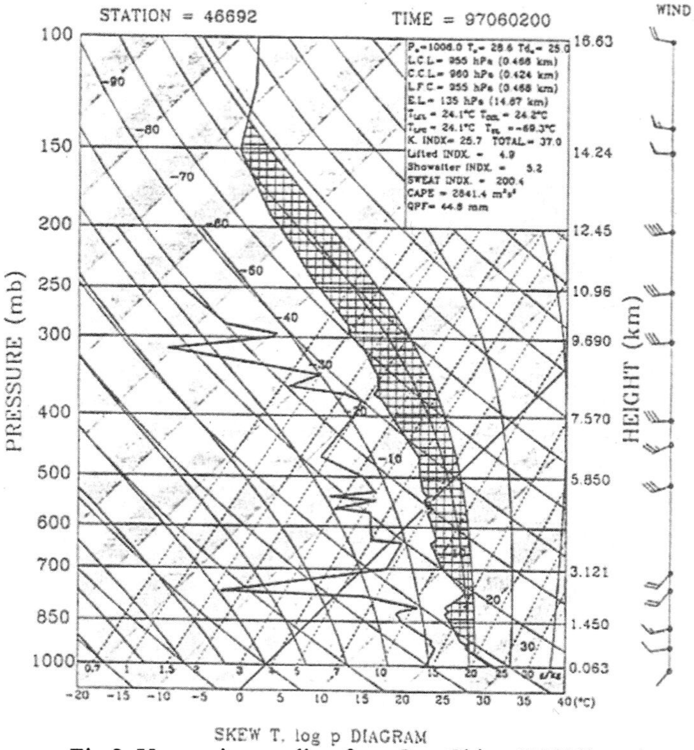

Fig.3. Upper air sounding from Pan-Chiao (46692) station on 00Z June 2 1997.

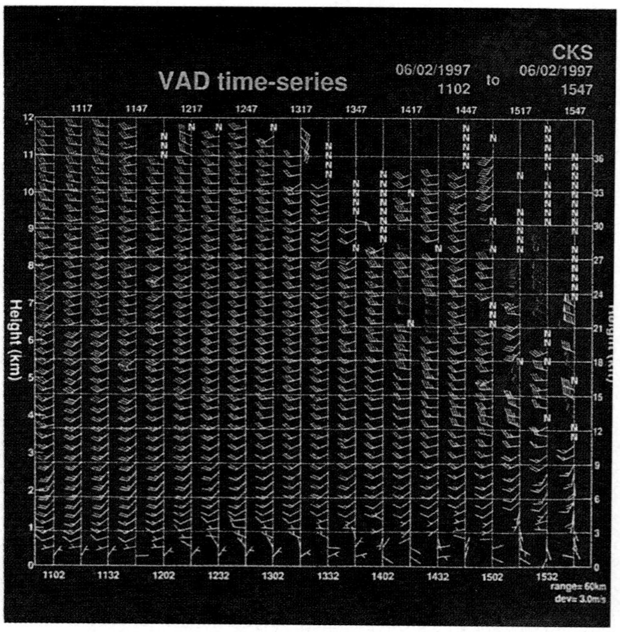

Fig.4. Time series of the vertical profiles of horizontal winds with 15 minutes interval derived from CKS Doppler radar using VAD technique.

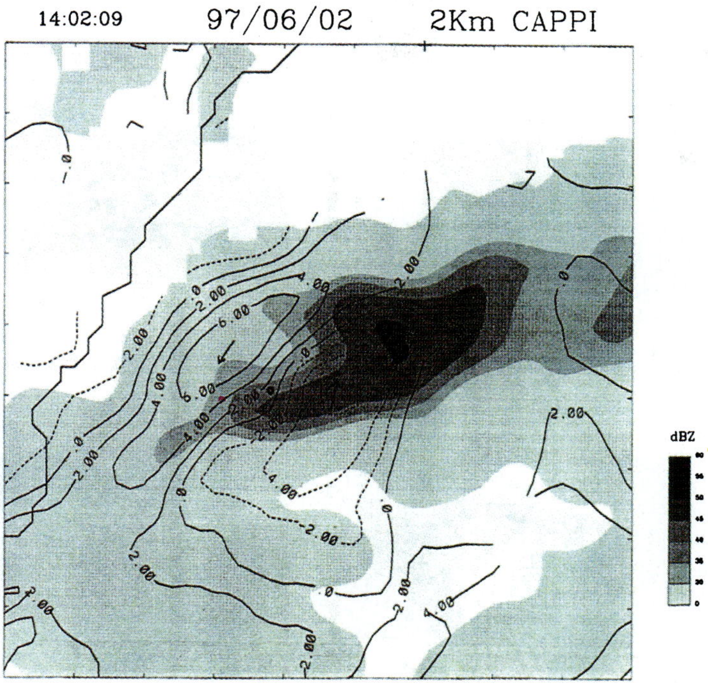

Fig.5. Radar reflectivity and perturbed radial winds at 2 km height on 1402LST June 2 1997. Each tick mark is 10 km in length.

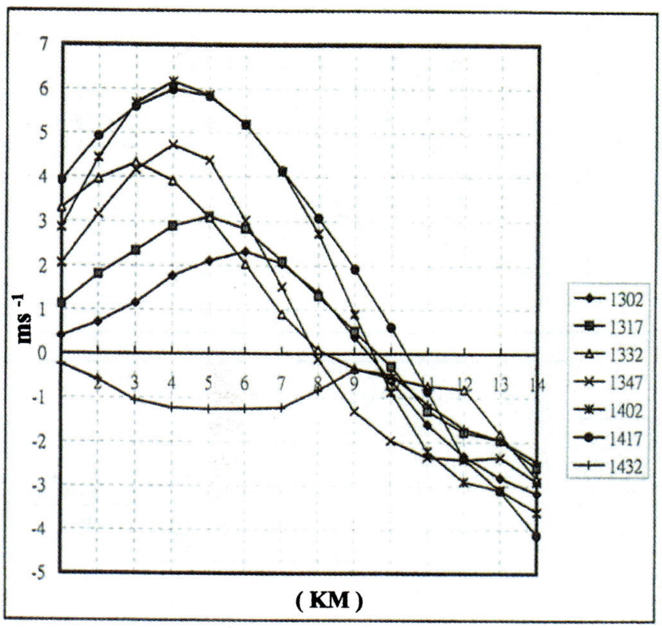

Fig.6. Time sequence of the mean tangential winds at 2 km height derived from GBVTD technique.

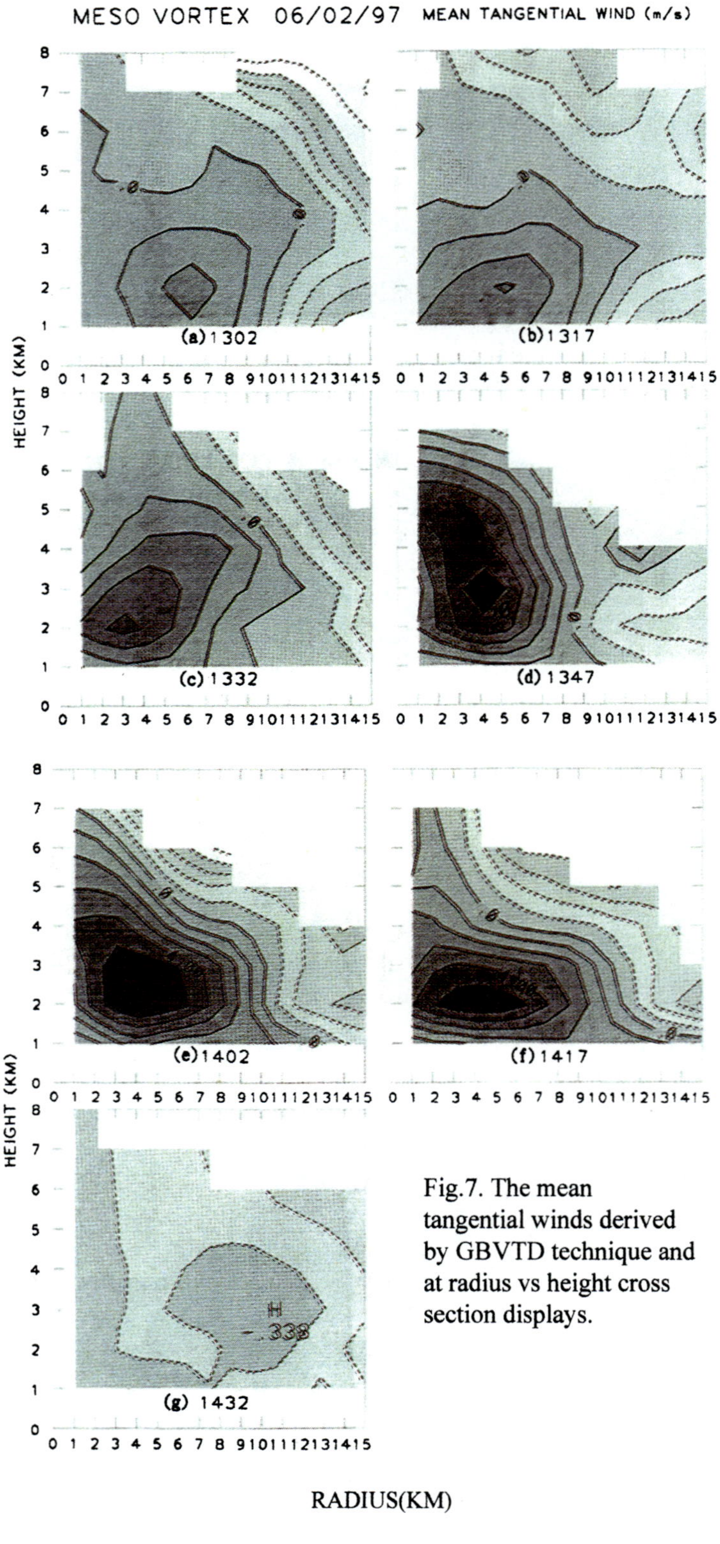

Fig.7. The mean tangential winds derived by GBVTD technique and at radius vs height cross section displays.

MESOSCALE PHENOMENA OBSERVED BY WIND PROFILER DURING HUAMEX IOP608

Li Chenguang*, Li Yaping, Tao Zuyu
Laboratory for Severe Storm Research, Dept. of Geophysics, Peking University, Beijing 100871

ABSTRACT

Based on the wind observations of the wind profiler located in Hong Kong, the mesoscale phenomena in the lower troposphere for the severe rainstorm case of 7-9 June 1998 were analyzed. The wind profiler revealed an easterly jet stream in the boundary layer bringing a cold air mass from the East China Sea prior to the occurrence of the rainstorm. The development of the South China Sea monsoon was also shown in details by the wind profile. The mesoscale phenomena of the jet stream were found in the low levels and the boundary layer during the period of torrential rain.

From 18 UTC 8 June to 12 UTC 9 June, torrential rain of 403 mm fell over Hong Kong. Based on the wind observations from two wind profilers located at Sha Lo Wan and Sham Shui Po of Hong Kong, the evolution and mesoscale structure of the wind field was analyzed. The data had a time resolution of 10 minutes and a vertical spatial resolution 60 m or 200 m. The time cross section of the wind profiles revealed an easterly jet stream in the boundary layer bringing a cold air mass from the East China Sea prior to the occurrence of the torrential rain. The development of the South China Sea monsoon was also shown by the wind profiles. Mesoscale phenomena of the jet stream were found in the low levels and the boundary layer during the period of torrential rain.

1. SYNOPTIC STUDY

Tables 1 and 2 summarize the surface weather phenomena and the 6-hourly precipitation in Hong Kong. Before 12 UTC 7 June, rain was rather light. The surface easterlies and the inversion (Fig. 1) indicated that the light rain was caused by the low clouds above the cold air mass (Tao et al. 1999). After 12 UTC, light showers occurred and a number of weak convective cells were observed around the Pearl River delta in the infrared satellite images. Severe thunderstorms occurred at 22 UTC 8 June, producing torrential rain in excess of 100 mm for each of the three successive 6-hourly periods that followed. The satellite images indicated that the Pearl River delta was covered by strong mesoscale convective systems at the time (Fig. 2).

2. WIND PROFILE ANALYSIS

In the following sections, winds for this heavy rain case measured by the wind profiler located at Sham Shui Po, Hong Kong, were analyzed to reveal the mesoscale phenomena in the wind field.

The wind profiler had two modes. The high mode sampled the atmosphere every 10 minutes at 30 fixed altitudes, beginning at around 200 m above ground level (AGL) and extended at a regular interval of 202 m to a maximum height of 6240 m AGL.

* Current Tel: 010-62751132

The time-height cross section (Fig. 3), with time increasing from right to left, provided a detailed view of the wind profile from the wind profiler measurements (horizontal coordinate is time at 10-minute interval and vertical coordinate is height). In this figure, the blanks indicate that the values at those data points were invalid, the winds flagged with '_' indicate that the values were suspect, while the rest are valid. It showed that the information from the profiler measurements was rather inconsistent with no complete coverage over time and height. A large number of winds were invalid and data marked with '_' differed remarkably from the neighboring measurements in both time and height.

Statistics on the effective rates of the profiler measurements were computed over height and time (Fig. 4 and Fig. 5 respectively). It clearly showed that the effective rates (the number of valid data/the totals) of the wind profiler measurements was about 60% on average, but it was much higher (approaching 90%) at the lower levels. When compared against the 6-hourly precipitation trends (Table 2), it was not obvious that the weather environment exerted any influence on the capability of the wind profiler during the period of heavy rain.

In Figs. 6, 7 and 8, winds displayed at each height were one-hour average. For example, 10-min measurements acquired between 00 and 01 UTC were averaged and displayed with a time stamp of 0100 UTC (Weber et al. 1983). Winds marked with '_' had been eliminated and only valid data remained. In these figures, the hourly averaged winds showed better consistency with more coverage over time and height. They presented a clearer picture of the mesoscale structures in the wind field before, during, and after the severe rainstorm event. From the foregoing statistics (Weber et al. 1983), it was found that the lower-level wind field could be much better represented by the hourly averaged wind profiler measurements.

Based on the observations from the Hong Kong profiler, time-height sections of the wind profile were drawn (Figs. 6, 7 and 8; horizontal coordinate is time increasing from right to left at hourly interval). It could be seen in Fig. 6 that there was an easterly jet stream with a maximum wind of 14 m/s below 1.5 km from 01 UTC to 06 UTC on 7 June, indicating the presence of a cold air mass in the boundary layer flow from the East China Sea (Tao et al. 1999). After 07 UTC, the intensity and thickness of the easterly layer reduced. During this period, only light rain occurred over the Pearl River delta. Above the easterly layer, the southwest monsoon developed after 12 UTC and a jet streak first appeared at 3000 m AGL. After 18 UTC, the southwest monsoon controlled the mid-lower troposphere above 1500m AGL over the Pearl River delta as light showers developed.

On 8 June (Fig. 7), the easterly cold air mass in the boundary layer disappeared and four intensity cores of the monsoon were found in typical meso-timescale (3 - 8 hours). During the period of 18 UTC to 24 UTC in particular, the appearance of an intensity core with maximum winds of 20 m/s of in the monsoon coincided with a high rainfall amount of 160 mm at the surface.

From the wind cross-section in Fig. 8 for 9 June, the southwesterly jet stream extended to the boundary layer between the levels of 200 m and 1000 m. During the torrential rain period of 9 June, the wind profile revealed two intensity cores in the boundary layer with maximum winds of 16-18 m/s.

3. CONCLUSIONS

The wind profiler measurements with high resolution over time and height could provide detailed vertical structure and time evolution of winds associated with the mesoscale systems. But owing to data discontinuity and the incomplete coverage of the 10-min wind measurements, the data had to be treated with hourly-averaged quality control before they could reveal the mesoscale features on the time-height cross sections clearly.

From the wind profiler measurements, we could clearly see the mesoscale structures in the southwest monsoon and the boundary layer jet stream, features that could not be identified from conventional soundings. The development of the southwest monsoon, as well as the relationship between the severe rainstorms and the strong mesoscale activity in the southwest monsoon, could also be observed.

Note: There are two wind profilers (Sham Shui Po and Sha Lo Wan) in Hong Kong. Having analyzed the data from the two profilers, we found that the results of the two data sets were very similar (Fig. 9), except that the wind directions at Sha Lo Wan did not change as much as those of Sham Shui Po. Because of a distance of over 30 km between the two wind profilers, data obtained from one profiler could not be used to improve upon those of the other in describing the mesoscale phenomena observed in the southwest monsoon jet, even though both might be able to depict separately the time-height structure of the wind field rather well. As such, only results from one data set, i.e. those from the profiler at Sham Shui Po, were presented to describe the wind structures during the period of heavy rain.

ACKNOWLEDGEMENT. This project is supported by Pandeng-95-zhuan-03 and Chinese Natural Science Foundation under Grant 49735180.

REFERENCES

Tao Zuyu, Wang Hongqing, Zheng Yongguang, and Wang Likun, 1999: Mesoscale convective system during the IOP609, Preprint of the workshop on disaster weather on both sides of Taiwan Strait, Taibei.

Weber, B. L., D. B. Wuertz, D. C. Welsh, and R. McPeek, 1993: Quality controls for profiler measurements of winds and RASS temperatures. J. Atmos. Oceanic Tech., **10**, 452-464.

Table 1. Hourly weather phenomena of Hong Kong during 7-10 June 1998.

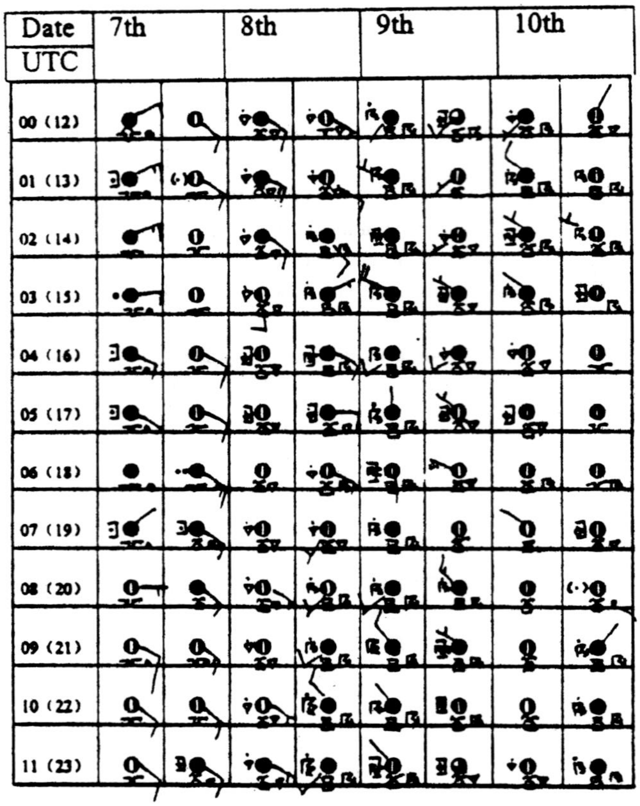

Table 2. 6-hourly precipitation of Hong Kong.

Date	7th				8th				9th				10th			
R	T	T	3	3	5	6	24	160	123	120	5	14	11	T	5	23

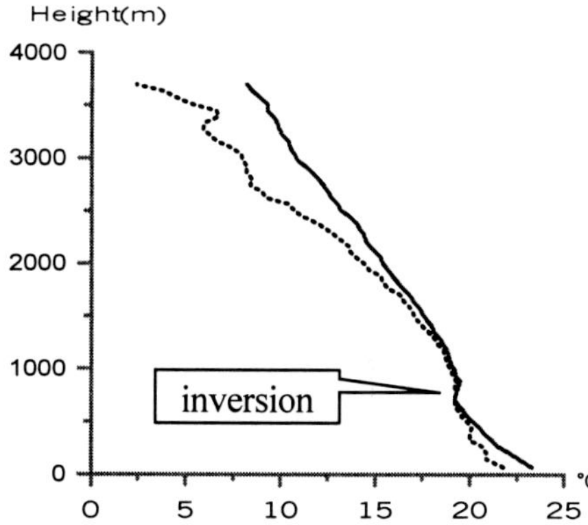

Fig. 1. Hong Kong sounding at 06 UTC 7 June (solid: temperature ; dashed: dew point).

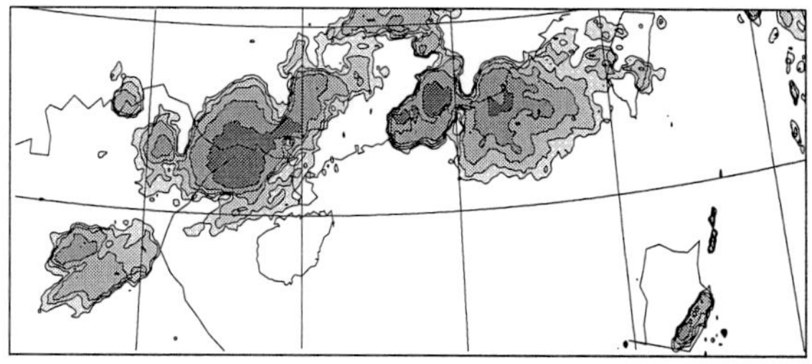

Fig. 2. TBB of infrared satellite images (isoline from –32 ℃, interval 10℃) on 8 June.

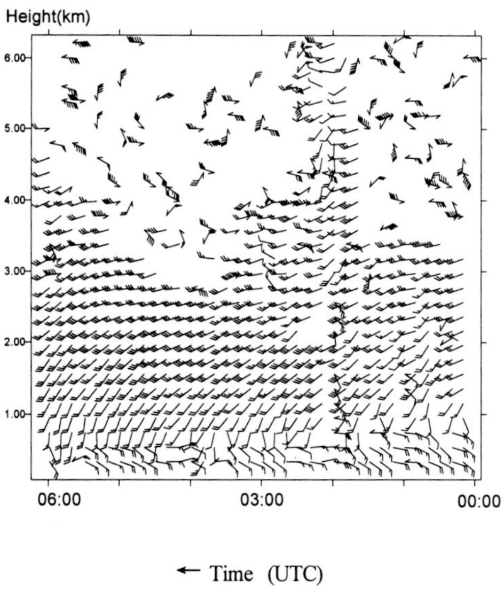

Fig. 3. 10-min wind from profiler measurements without quality control (horizontal coordinate stands for time).

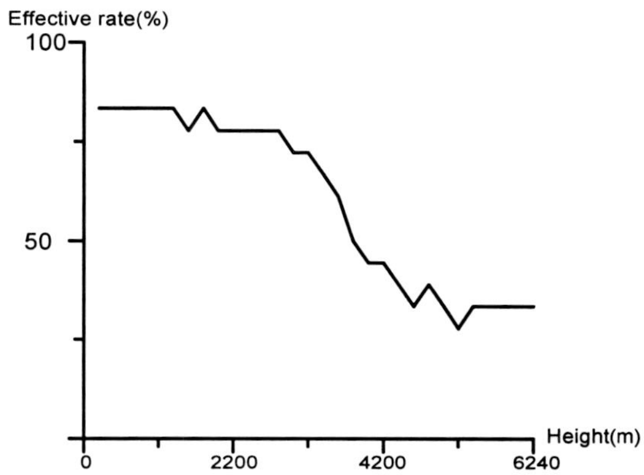

Fig. 4. Effective rates of the profiler measurements at different heights.

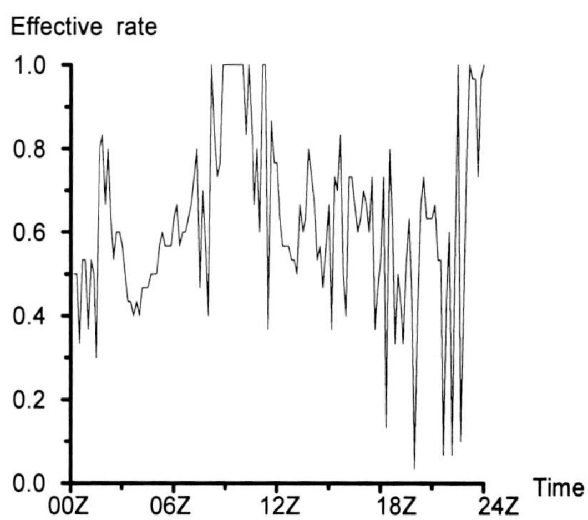

Fig. 5. Effective rates of the profiler measurements at different times.

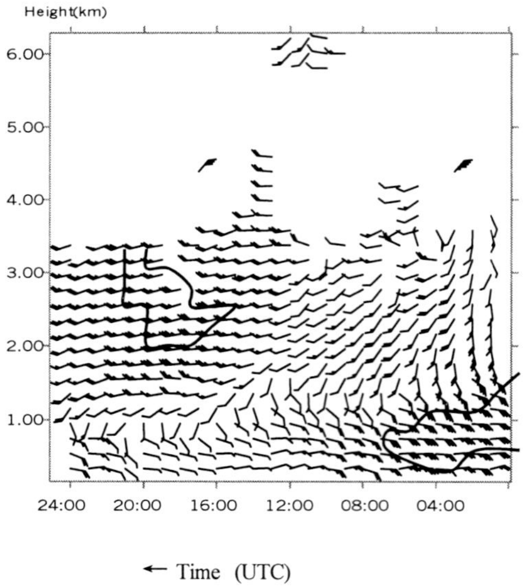

Fig. 6. Time cross section of wind profile at Sham Shui Po, Hong Kong on 7 June (wind speeds > 12m/s were encircled by heavy lines).

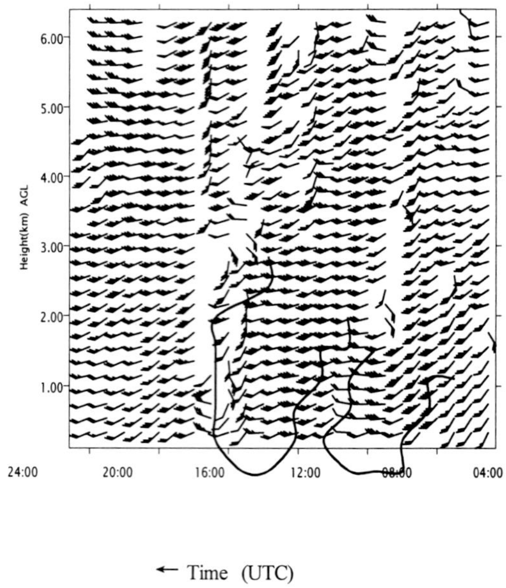

Fig. 8. Same as Fig. 6, but for 9 June.

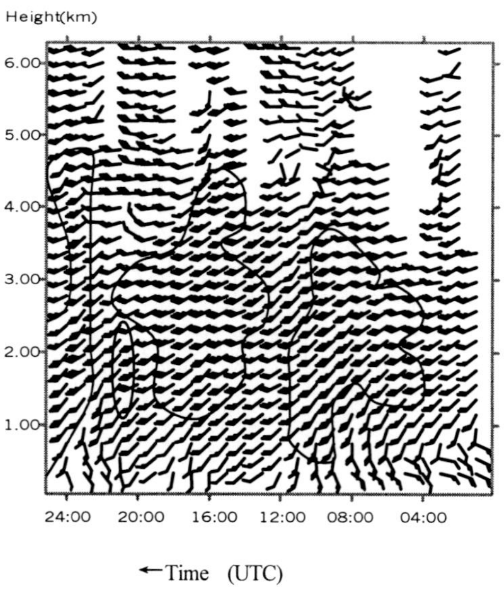

Fig. 7. Same as Fig. 6, but for 8 June.

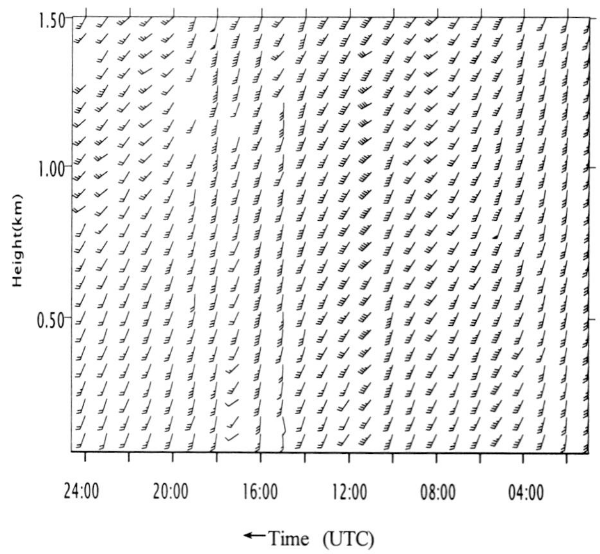

Fig. 9. Same as Fig. 8, but for Sha Lo Wan.

THE APPLICATION OF GPS DATA TO SYNOPTIC ANALYSIS*

Yang Hongmei   Xu Baoxiang   Zhou Xiuji

Chinese Academy of Meteorological Sciences, Beijing 100081

ABSTRACT

Atmospheric water vapor data were obtained by two Global Positioning System (GPS) receivers at Shantou and Yangjiang during the Huanan (southern China) Torrential Rainfall Experiment from 5 May to 26 June 1998. The ground-based GPS technique was tested to assess its accuracy for meteorological applications. Integrated water vapor was estimated through zenith wet delay and the results were compared with routine radiosonde data (at mandatory and significant levels). The trends shown by the two methods were generally consistent. It was demonstrated that GPS sounding signals were sensitive to convective systems and could be used to measure atmospheric water vapor. As such, it would be feasible to make use of GPS data for monitoring thunderstorm development associated with mesoscale convective systems.

Key words: GPS sounding technique, integrated water vapor, torrential rainfall forecast.

1. INTRODUCTION

Water vapor is one of the primary conditions for precipitation processes. How to evaluate changes of water content within an air column objectively and quantitatively remains an unsolved problem that has been attracting the attention of meteorologists. In recent years, satellite sounding has shown great progress. Measurements of atmospheric water vapor with GPS have been carried out (Bevis *et al.* 1992; Rocken *et al.* 1995; Rocken *et al.* 1993). Chinese meteorologists (Mao 1993; Peng *et al.* 1997; Li *et al.* 1998) have also done a lot of research in this respect.

As GPS signal is sensitive to the index of atmospheric refraction, which is a function of pressure, temperature, and moisture, GPS can be used directly for sensing of the atmosphere. Based on this principle, atmospheric water vapor was measured with two GPS dual-frequency receivers at Shantou and Yangjiang during the Huanan Torrential Rainfall Experiment. The wet delay can be determined directly from GPS observations. Precipitable water (PW) can be derived from the wet delay multiplied by 0.156 (empirical constant). PW is the depth of water that would result if all the atmospheric water vapor within a vertical column of air were to be condensed to liquid. In this study, GPS data at Haikou and Tianjin were also used.

2. PRINCIPLE FOR GROUND-BASED GPS REMOTE SENSING OF ATMOSPHERIC WATER VAPOR

Because the GPS signals are sensitive to the amount of moisture in the atmosphere, the time delay in GPS signals reaching ground-based receivers due to the presence of water vapor

---

* supported by Special Project of National "Climb Plan " (85-Special-03) and National Key Basic Research and Development Project (G1998040909).

is nearly proportional to the quantity of water vapor integrated along the signal path. As discussed in Rocken *et al.* (1995), it only requires a small amount of atmospheric water vapor to significantly affect GPS signal propagation velocities. Thus, GPS is especially suitable for remote sensing of atmospheric water vapor.

To estimate the zenith wet delay (ZWD) and precipitable water (PW), knowledge of the station position and the satellite orbits is required. In general, the integrated water vapor can be expressed in terms of PW, the height of an equivalent column of liquid water, i.e.

$$PW = \int_0^{\infty} \gamma \frac{\rho_d}{\rho_w} dz \quad , \qquad (1)$$

where γ is the mixing ratio, ρ_d is the density of dry air, and ρ_w is the density of liquid water. It can then be written that

$$PW = \Pi \cdot ZWD \quad , \qquad (2)$$

where ZWD is the zenith wet delay in units of mm, and

$$\Pi = 10^6[\rho_w R_v (k_3/T_m + k_2')]^{-1} \quad , \qquad (3)$$

Here Π is a dimensionless constant and is a function of the empirical constants related to the refractivity of moist air (k_3 and k_2'), the gas constant for water vapor (R_v), and the mean temperature of the atmosphere (T_m), which is defined as

$$T_m = \frac{\int (e/T) dz}{\int (e/T^2) dz} \quad , \qquad (4)$$

where e is the water vapor pressure and T is the absolute atmospheric temperature.

Equation (3) can be re-written as

$$\Pi = k_2' + \frac{k_3}{T_m} \quad , \qquad (5)$$

where $k_2' = k_2 - k_1 \frac{R_d}{R_v}$, k_1 and k_2 are empirical constants related to the refractivity of moist air, and R_d and R_v are the gas constants of dry air and water vapor respectively.

In this study, Π is calculated by using two sets of retrieval coefficients (K_1, K_2 and K_3) and an empirical constant as follows:

a. $K_1 = 77.6$, $\quad K_2 = 70.4$, $\quad K_3 = 3.739 \times 10^5$

b. $K_1 = 77.6$, $\quad K_2 = 64.79$, $\quad K_3 = 3.776 \times 10^5$

c. Empirical constant = 0.156 (6)

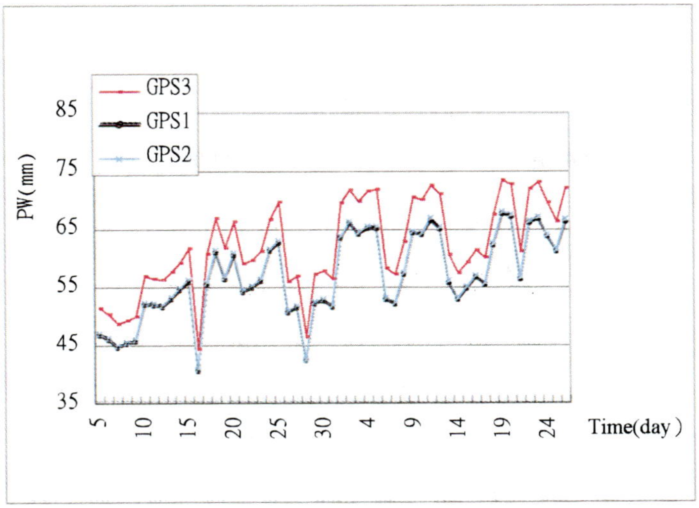

Fig. 1. Shantou PW curves of three sets of retrieval coefficients from GPS data.

3. ANALYSIS OF GPS DATA

Daily PW results (at 0000UTC) at Shantou computed from the three sets of empirical coefficients, GPS1, GPS2 and GPS3 are shown in Fig. 1. GPS1 and GPS2 curves are approximately the same with differences confined between 0.3 and 0.4 mm. GPS3 curve plots the results calculated from the empirical constant of 0.156. The values thus obtained are found to be higher than the others.

To check for consistency, the following analyses are made:

(a) Contrast of total water content in different areas within the same season
Climatologically from late spring to early summer, the monsoon rain belt is located in southern China where the water content of air column is relatively high. Fig. 2 shows the curves of daily PW values (at 0000UTC and with empirical constant = 0.156) for Yangjiang, Shantou, Haikou and Tianjin during the experiment period. Apart from Tianjin, which is at a higher latitude in northeastern China under the control of the westerly circulation regime, all the other three stations are in southern China under the influence of the monsoon rain belt. Not surprisingly, PW at Tianjin is the lowest among the four, consistent with the characteristic north-dry and south-wet pattern.

(b) PW in relation to weather systems
Atmospheric water vapor is quite variable irrespective of how dense the spatial and temporal resolutions are. Fig. 3 shows the histogram of PW values at Shantou plotted every two hours on 14-15 May 1998. Thunderstorms were approaching Shantou when PW values increased rapidly from 58.9 mm to 74.7 mm between 16 and 18 UTC on the night of 14 May. PW values then decreased gradually as rain ceased at 06 UTC the next day.

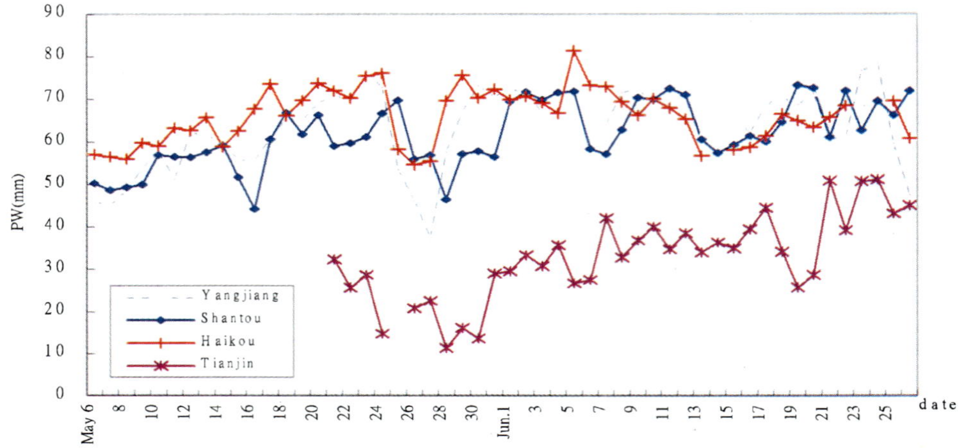

Fig. 2. PW curves of retrieved coefficient from GPS data of four stations.

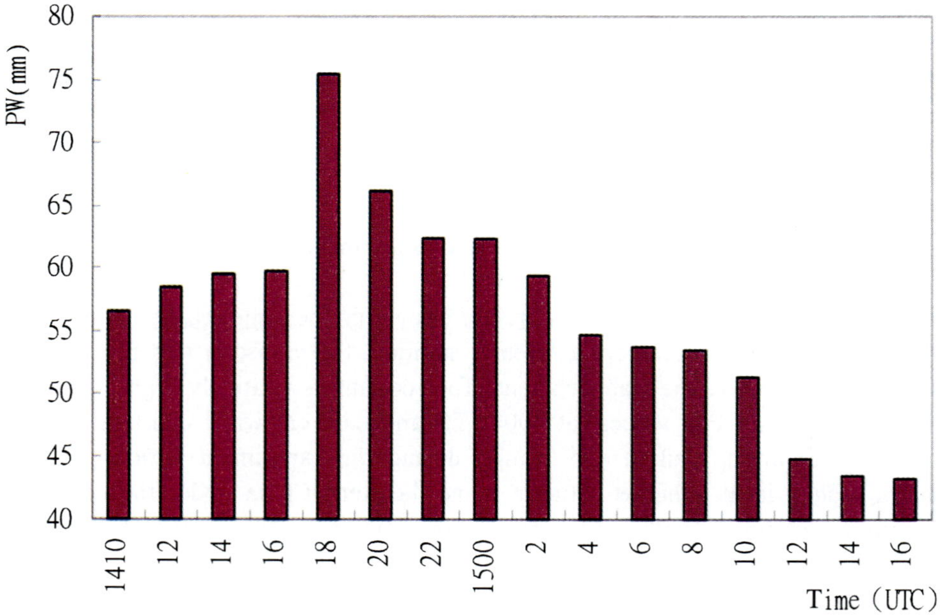

Fig. 3. Shantou PW histogram of GPS data May 1998.

(c) <u>GPS-derived PW compared to conventional radiosonde data</u>

According to Eq. (1) and the equation of static equilibrium, precipitable water is then

$$\mathrm{PW}=\frac{1}{g}\int_{p2}^{p1}\gamma\,dp \qquad , \qquad (7)$$

where γ is mixing ratio, and subscripts 1 and 2 denote the top and bottom of each pressure layer respectively.

Apart from the routine soundings four times a day, radiosonde ascents were made every three hours for enhanced observations in the events of severe weather during the experiment period. In this study, radiosonde-derived PW values at Yangjiang are computed four times a day from the mandatory and significant level data. Time series for radiosonde-derived and GPS-derived PW values over the experiment period are plotted in Fig. 4. It can be seen that the two curves are generally similar with nearly identical increasing and decreasing trends.

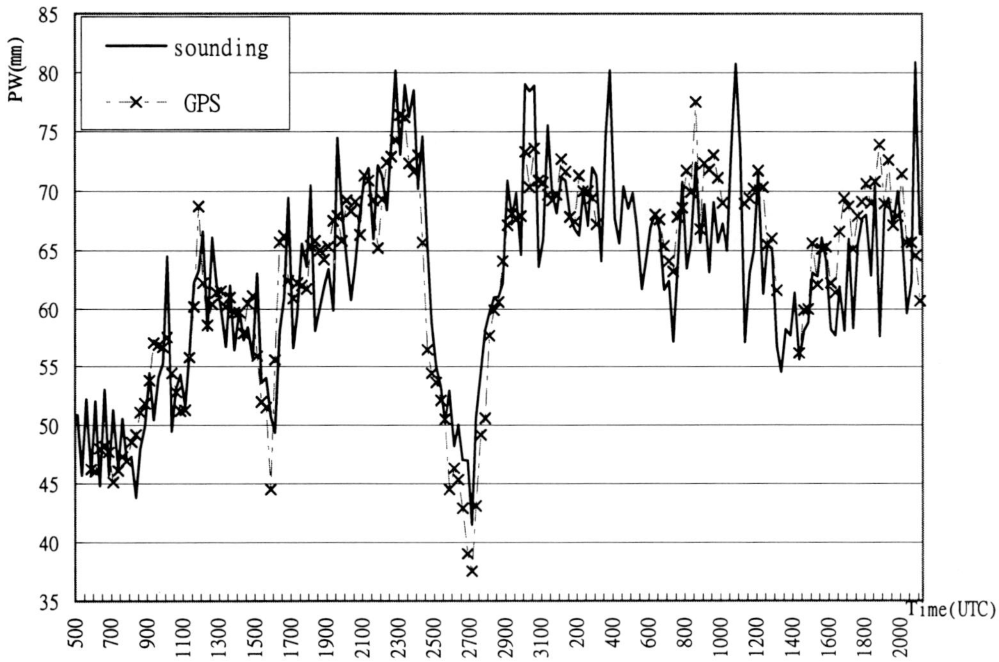

Fig. 4. Yangjang PW curves from GPS and radiosounding data during experiment.

The difference distribution between the two sets of PW values is tabulated in Table 1. In the majority of cases (i.e. 112 occurrences or 67.5%), the differences are less than 5 mm which is within the acceptable error limits. As such, GPS-derived values are shown to be in good agreement with those obtained from sounding data. It demonstrates that GPS-based techniques are capable of monitoring atmospheric water vapor distribution at increased temporal resolution.

Table 1. Distribution of PW differences between sounding and GPS data at Yangjiang.

Absolute value (mm)	>≥15	10 -≥15	5 -≥10	< 5
No. of Occurrences	2	2	50	112
Percentage (%)	1.2	1.2	30.1	67.5

4. CONCLUSIONS

In this paper, the results clearly demonstrate that GPS-derived data are reliable. Atmospheric water vapor data obtained by ground-based GPS receivers during the Huanan Torrential Rainfall Experiment are shown to be comparable, and in some respects superior, to those measured by radiosondes. Increased temporal resolution of the water vapor distribution from the GPS data would be most valuable in monitoring thunderstorm development.

GPS data have the potential to enhance the analysis and prediction of weather. To fully utilize such information, further investigation should be made.

REFERENCES

Bevis, M., S. Businger, T.A. Herring, C. Rocken, R. A. Anthes, and R.H. Ware, 1992: GPS meteorology: remote sensing of atmospheric water vapor using the global positioning system. *J. Geophys. Res.*, **97,** 1578–1580.

Li Chengcai *et al.*, 1998: Analysis for remote sensing of atmospheric precipitable water using ground-based GPS receiver. *Quarterly Journal of Applied Meteorology*, **9(4)**, 470-477 (in Chinese).

Mao Jietai, 1993: Meteorological application of GPS. *Qixiang Keji*, **4,** 45-49 (in Chinese).

Peng Hong *et al.*, 1997: The application of GPS in atmospheric sounding. *Collected papers of atmospheric remote sensing technology,* China Meteorological Press, 110-118 (in Chinese) .

Rocken, C. *et al.*, 1993: Sensing atmospheric water vapor with the Global Positioning System. *Geophysical Research Letters*, **20(23)**, 2631-2634.

Rocken, C. *et al.*, 1995: GPS/STORM - GPS sensing of atmospheric water vapor for meteorology. *J. Atmos. Ocean Technol.*, **12(3)**, 468-478.

TEST OF THE APPLICATION OF SATELLITE REMOTE SENSING LAND VEGETATION DATA IN MESOSCALE SIMULATION OF HEAVY RAINFALL OF SOUTHERN CHINA*

Wang Pengyun
Chinese Academy of Meteorological Sciences, Beijing
Xiao Qianguang
National Satellite Meteorological Center, Beijing

ABSTRACT

Land vegetation categories and physical parameters retrieved from NOAA satellite remote sensing data in the HUAMEX project during May and June 1998 was tested by using the PSU/NCAR mesoscale model MM5 to simulate an IOP case of heavy rainfall during 23-24 May 1998 and compared with that by using NCAR land vegetation data-set. Results show that with our land vegetation data, which represents in more detail the complex land surface characteristics of China and near area, the simulated 24h accumulated rainfall was increased about 12-14% for grid spacing 5-15km compared with that using NCAR data. The dynamic structures on the low levels are also affected. For larger grid spacing, 45km, the effects are not obvious.

Since the retrieve of land vegetation, especially for the associated physical parameters, is hardly accurate, it needs more studies to enhance the accuracy of the retrieved data and test in various area and synoptic situations.

Key words: land vegetation, mesoscale simulation.

1. INTRODUCTION

Mesoscale weather processes could be induced by either the instability within larger scale system or the local topography forcing or thermodynamic action of surface, or the interaction of these two scales, i.e. induced by the former then enhanced further by the later. The interactions among various scales are more often and make more serious damages. Therefore both initial data and the terrain and land vegetation data and analyses are important to the simulation and forecast for the mesoscale processes.

The analysis of initial field includes first-guess from global model and objective analysis of observational data. The emphases are currently concentrated on 3-D or 4-D assimilation by using radar and satellite remote sensing data in order to involving more mesoscale information in initial field. There have been more studies on the effects of topography to the local weather processes. Currently most mesoscale models have considered the effects of terrain by using the real terrain data with high resolution (e.g. 5' of latitude/longitude). Under some special PBL processes even need consider the effects of slope and direction of topography to the thermodynamic processes. The characteristics of surface affect both the dynamic processes (e.g. effects of roughness length on momentum diffusion), and thermodynamic processes

* Supported by 95 Special Topic of Key National Project (95-special-03), National Natural Science Foundation of China (49735180) and Research on the Formation Mechanism and Prediction Theory of Severe Synoptic Disasters in China (G1998040907).

(e.g. effects of albedo, emissivity, thermal inertial on thermal diffusion). Moreover, the wet of surface (moisture availability) also directly affects the water cycle in atmosphere.

The land vegetation categories and physical parameters retrieved from NOAA satellite remote sensing data during the HUAMEX project in May and June 1998 make it possible to test their effects on mesoscale simulation of heavy rainfall. In this paper we will present the main results of the test.

2. LAND VEGETATION CATEGORIES AND PHYSICAL PARAMETERS RETRIEVED FROM SATELLITE REMOTE SENSING

Land vegetation categories and physical parameters retrieving from satellite data has been studied abroad since 1980th (e.g. Tucker et al., 1985; Townshed et al. 1987; Norwine and Greegor, 1993; Derrien, 1993). In China, this work was started in 1990. Shen et al. (1995) classified the land vegetation types on China continent into 18 categories by using NDVI satellite data in 1990.

The global land vegetation is divided into 13 types in the PSU/NCAR mesoscal model MM5 with resolution of 10' latitude/ longitude grids. Since the data are considered better to fit for North American and global distribution of vegetation, it seems too rough for other local area. Also, their physical parameters associated with each type of land vegetation are classified for only summer and winter season (Grell et al. 1994).

However, the vegetation is quite diverse in the much broader continental area of our country. In order to improve the effects of land characteristics in mesoscale model, we reclassified land vegetation categories into new 13 types according to the distribution of vegetation in our country and near area, and retrieved the data in the area from 74-135.44°E of longitude and 12-56.88°N of latitude, with resolution of 0.8°, by using NOAA satellite NDVI data in 1997-1998. We also estimated the physical parameters for each type of land vegetation using satellite data in May and June 1998 respectively (table 1).

3. NONHYDROSTATIC MESOSCALE MODEL AND EXPERIMENT CASE

PSU/NCAR nonhydrostatic mesoscale model MM5(V2) was used to test the effects of land vegetation data. The physical process parameterization schemes we used include nonhydrostatic version; Blackadar high-resolution PBL scheme; four types of explicit moisture schemes: warm rain; Reisner mixed-phase; Reisner graupel; and Goddard microphysics; for sub-grid cumulus parameterization schemes: Kain-Fritsch and Grell scheme (Grell et al, 1994).

The model was initiated using the grid data (grid spacing 1.125° of latitude/longitude) of NMC medium range spectral model T106 as a "first guess" field, supplemented by operational surface and rawinsonde data, and sea surface temperature and land use data. The time-dependent inflow-outflow conditions were used for lateral boundary, and the data at boundary were linearly interpolated at 12h intervals in time by using T106 analyses and observational data.

Table 1. Description of NSMC 13-category land vegetation categories and physical parameters for May and June 1998.

Landuse integer identification	Land vegetation description	Albedo (%)		Moisture avail. (%)		Emissivity (%)		Roughness length (cm)		Thermal inertial (calcm^{-2}k^{-1}s$^{-1/2}$)	
		May	June	May	June	May	June	May	June	May	June
1	Gebi, desert	25	25	3	2	84	86	9	10	0.02	0.02
2	Agriculture	18	16	65	70	91	92	14	15	0.03	0.04
3	North grass land	17	15	70	65	92	91	12	11	0.03	0.03
4	Tundra	58	60	95	95	95	95	5	5	0.05	0.05
5	Subtropical shrub	15	15	70	75	94	94	40	40	0.03	0.03
6	Tropical forest	12	13	85	80	95	95	50	50	0.05	0.05
7	Water	8	8	100	100	98	98	0.0001	0.0001	0.06	0.06
8	Dry grass land	19	18	46	55	91	92	11	12	0.03	0.03
9	Deciduous forest	16	17	70	75	90	95	60	65	0.04	00.4
10	Coniferous forest	13	12	50	60	95	95	50	50	0.04	0.05
11	High and cold wilderness	27	26	15	18	90	90	15	15	0.03	0.03
12	Sandy waste	22	22	6	7	88	88	12	12	0.03	0.03
13	Mixed deciduous/ coniferous forest	13	14	65	70	94	95	55	60	0.03	0.03

Three nested domains are set (Fig. 1): the out-large domain (A) contains 61x61 grids centred at 114°E and 25°N with grid spacing of 45km; the middle domain (B) contains 61x61 grids centred at 113.1 °E and 24.2 °N with grid spacing of 15km; the inner-small one (C) contains 73x73 grids centred at 111.9 °E and 22.5 °N with grid spacing of 5km. The model atmosphere is divided into 25 layers from surface to 10hPa.

The terrain is represented by actual height on a 10'x10' latitude/longitude grid, and analyzed on grids by using 16 points and two dimension parabolic interpolations.

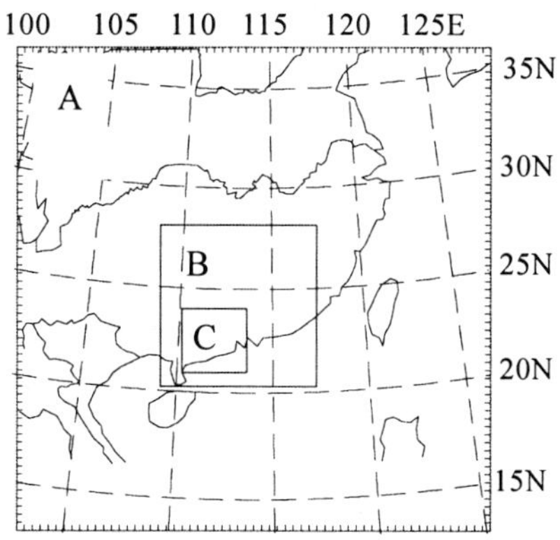

Fig. 1. Three domains in simulations.

The PSU/NCAR 13-categories land vegetation data and associated physical parameters (known as NCAR data hereafter) were used as comparison to our data set (knows as NSMC data hereafter).

The land vegetation distributions of NCAR data and NSMC data in domain B are shown in Fig. 2a and 2b, respectively. Comparing Fig. 2a and 2b we can see that the NSMC land vegetation distribution is in more detail than NCAR data. The land vegetation in Fig. 2a contains only three categories: coniferous forest, tropical or subtropical forest and small part of savannah; but in Fig. 2b there are agriculture, subtropical shrub, deciduous forest, coniferous forest and mixed deciduous/coniferous forest etc.

Clearly, these land vegetation represent better the distribution on the continent of our country and near area.

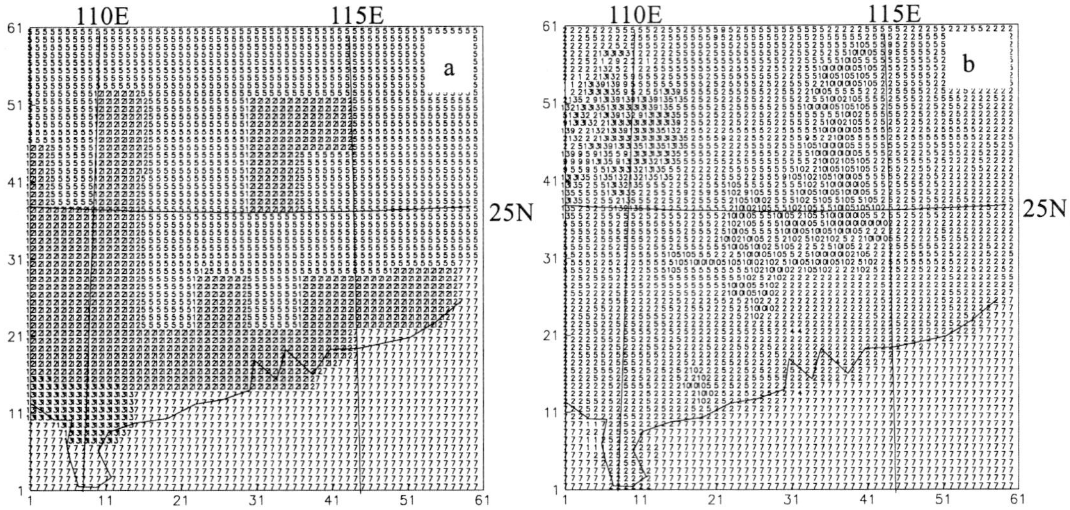

Fig. 2. Land vegetation distribution in domain B. (a) NCAR data; (b) NSMC data.

A case of heavy rainfall storm in South China from 23rd to 24th May 1998 was used as the simulation test. The heavy rainfall was produced associated with a southward moving cold air mass. Cold front entered into Guangdong province in 23rd. During southward movement of the cold front, strong convective cloud mass was formed near shore to the west of Zhujiang river. Satellite infrared cloud pictures showed the convective cloud was developing at 0000 UTC 23rd, stayed in the area to the west of Zhujiang river during 0600-1800 UTC 23rd, moved into sea after 0000UTC 24th. Associated with the southward moving cold front and convection, heavy rain fell at the mid-part of Guangdong province. In the Xijiang basin and Zhujiang delta, 14 observatories had more than 50mm of rainfall in 24h. In Sihui, 24h rainfall reached 233.7mm. All three domains covered the heavy rainfall area.

4. TEST RESULTS

This case was mainly characterized by the cold front passing southward and convective rainfall ahead of the front, we shall examine the effects of the land vegetation data on rainfall amount and frontogenesis function simulation.

(1) 24h rainfall simulation

MM5 model simulated 24h rainfall in three domains by using the scheme of Reisner graupel and Kain-Fritsch cumulus parameterization with NCAR land vegetation data as landuse input data shows (figures omitted) 135mm, 162mm, and 172mm of maximum rainfall for domain A, B and C, respectively. The fine mesh grids domain simulated not only better rainfall amount maximum than larger grids domain but also showed in more detail mesoscale structures. The location and scope of rainfall were simulated similar using NSMC data, but the maximum was 133mm, 183mm and 197mm for three domains, respectively, i.e., near the same as with NCAR data for domain A, but increased about 13% and 15% for domain B and C, respectively by using NSMC data. Fig. 3 shows the differences of rainfall in three domains by using NCAR and NSMC data. It clearly shows that the differences are small for domain A but obvious for domain B and C.

Table 2. Schemes of Moisture parameterization and land vegetation data used and resulted maximum rainfall (mm) for three domains.

Experiment index	Moisture parameterization scheme	Land-use data	Grid spacing		
			45km	15km	5km
EXP1A	WARM-KF	NCAR	130	159	171
EXP1B	WARM-KF	NSMC	131	170	183
EXP2A	MIX-KF	NCAR	125	142	147
EXP2B	MIX-KF	NSMC	124	170	184
EXP3A	RESG-KF	NCAR	135	162	172
EXP3B	RESG-KF	NSMC	133	183	197
EXP4A	GODG-KF	NCAR	167	220	239
EXP4B	GODG-KF	NSMC	169	256	281
EXP5A	MIX-GR	NCAR	129	156	162
EXP5B	MIX-GR	NSMC	127	161	173

Rainfall amount simulated by mesoscale model is somewhat related with the option of grid scale moisture schemes and sub-grid scale cumulus parameterization schemes, therefore we tested the case with five moisture parameterization schemes, and use NCAR data or NSMC data in each scheme respectively to compare their effects. The experiments are shown in Table 2.

In Table 2, WARM represents warm rain scheme for grid scale, MIX represents Reisner mixed phase scheme for grid scale, RESG represents Reisner Graupel scheme for grid scale, GODG represents Goddard Graupel scheme for grid scale, and KF represents Kain-Fritsch cumulus parameterization for sub-grid scale, GR represents Grell cumulus parameterization for sub-grid scale, respectively (Grell et al., 1994).

From Table 3 we can see that 24h rainfall maximum in all experiments for domain A with either NSMC or NCAR data was similar (difference less than 1%). However, rainfall maximum was increased obviously for domain B and C using NSMC data comparing with that by using NCAR data: 3-20% with average of 12% for domain B, 7-25% with average of 14% for domain C. Considering the fact that model simulated rainfall generally less than observations, for example 233mm observed at Sihui, but all simulated rainfall was less than it except by GODG-KF for domain B and C. Therefore, increasing rainfall by using NSMC land vegetation data may improve the rainfall simulation and forecast.

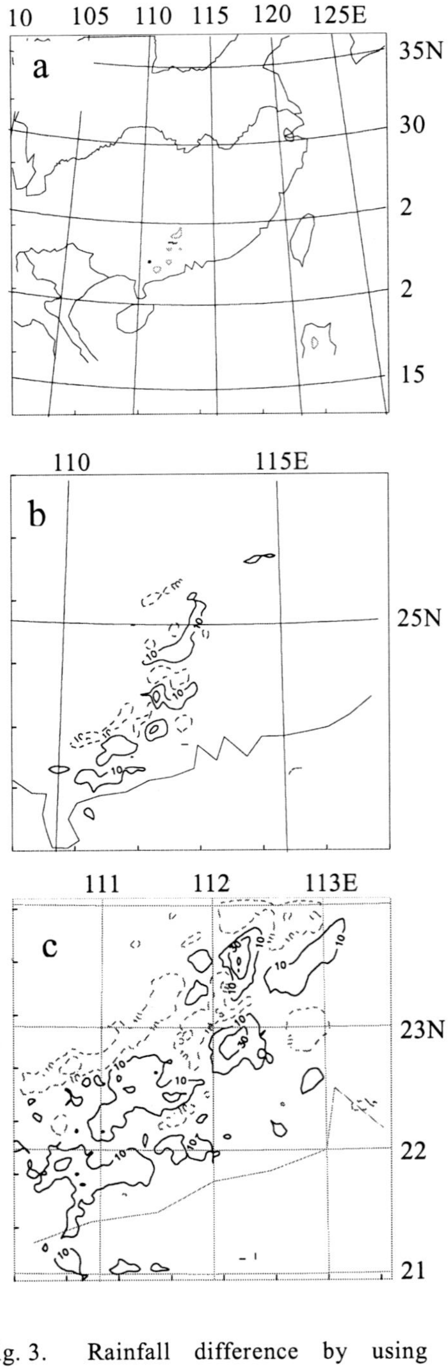

Fig. 3. Rainfall difference by using NCAR data and NSMC data. (a) domain A; (b) domain B; (c) domain C. Counter units: mm.

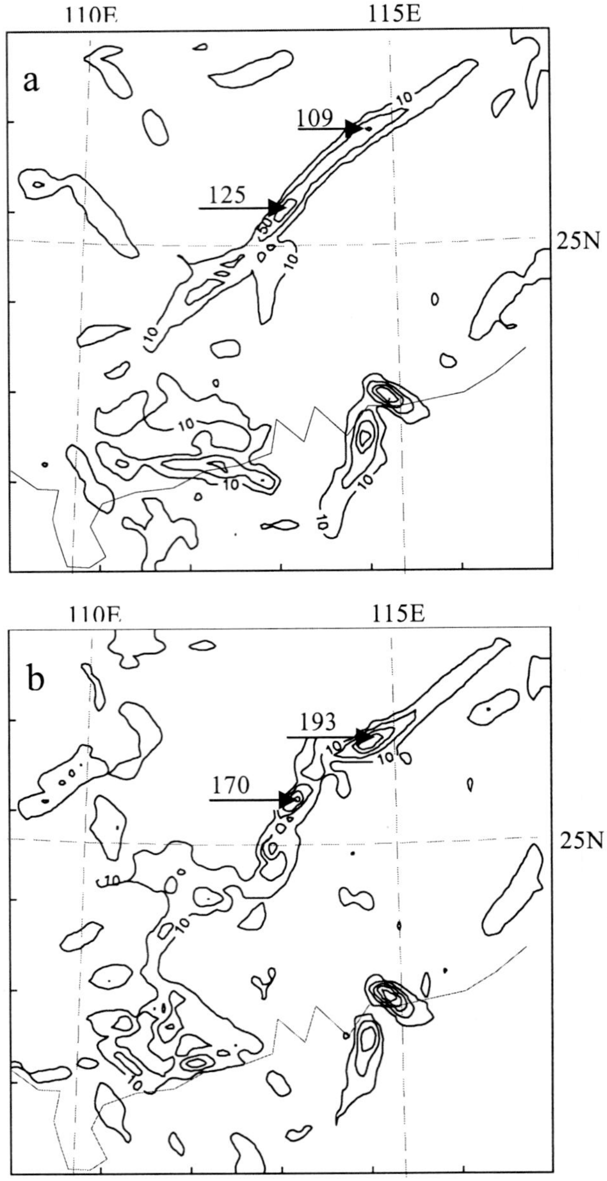

Fig. 4. Frontogenesis function at 950 hPa. (a) domain A; (b) domain B. Counter units: $KS^{-1}M^{-1}$

(2) Frontogenesis function near the cold front

Considering the deformation and convergence of horizontal wind, according to Ogura and Portis (1982), frontogenesis function F is expressed as:

$$F = \frac{d}{dt}\left|\nabla \theta\right| = \frac{\left|\nabla \theta\right|}{2}\left|def_r \cos 2\beta - \frac{\left|\nabla \theta\right|}{2} D\right|$$

where $$def_r = \left[\left(u_x - v_y\right)^2 + \left(v_x + u_y\right)^2\right]^{1/2}$$

β is the angular from the axis of dilatation to the potential temperature lines, D is horizontal divergence. The subscripts x and y denote derivatives with respect to x and y respectively.

The frontogenesis was clear near the cold front. The main part of cold front could be covered in the domain A and B. The domain C was too small to express the long and thin characteristics of the front. So we will compare the frontogenesis function in domain A and B by using NCAR and NSMC land vegetation data.

The frontogenesis function simulated by EXP3A and EXP3B at 1200UTC 23rd May (12h integrate) shows they are similar for domain A (figure omitted). For domain B, the frontogenisis at low levels using NSMC data was stronger than that by using NCAR data. Fig. 4 shows the frontogenisis function at 950hPa. There are some mesoscale larger values on the long and thin line of frontogenesis function along the cold front. The maximum of frontogenesis function by using NCAR data was 125×10^{-10} $KS^{-1}M^{-1}$ and 109×10^{-10} $KS^{-1}M^{-1}$. At the same site, the values were 179×10^{-10} $KS^{-1}M^{-1}$ and 193×10^{-10} $KS^{-1}M^{-1}$ by using NSMC data, increased 36% and 77%, respectively.

5. CONCLUSIONS

The land vegetation and associated physical parameters with 0.08° latitude/longitude resolution retrieved from satellite remote sensing data collected in the HUAMEX project represent better characteristics of land cover of China and East Asia than the NCAR 10' land vegetation data. Using NSMC and NCAR vegetation data in mesoscale model MM5 to simulate the 23rd - 24th may 1998 South China heavy rainfall event, results show that the difference are small for larger grid size (45km), but obvious for small grid size (15-5km): increased 12-14% by using NSMC data for 24h rainfall maximum; and also clear different for dynamic characteristics at low levels. Therefore, more fine and accurate land vegetation data is an important factor for improving the mesoscale simulation and forecast.

The difficulties are accurately determining the physical parameters for each type of land vegetation. Our experiments here show that only the land vegetation data and associated physical parameters can affect the quality of mesoscale simulations, but may not conclude that our retrieved data are exactly correct and accurate. Further study and test are needed for various synoptic situations at various areas.

REFERENCES

Grell, G.A., 1994: A description of the fifth-generation Penn State / NCAR mesoscale model (MM5). *NCAR/TN-398+STR NCAR technical note*, 138.

Ogura, Y., And D. Portis, 1982: Structure of cold front observed in SESAME-AVE III and its comparison with Hoskins-Bretherton frontogenesis model. *J. Atmos. Sci.*, **39**, 2773-2792.

Sheng, Y.-W., 1995: Vegetation classification of China using satellite data. *Chinese Science Bulletin*, **40**, 68-71.

PRELIMINARY STUDY ON THE FLOW STRUCTURE OF THE STRONG PRECIPITATION ON JUNE 9 IN CHANGLE AREA DURING HUAMEX IOP#6

Jiang Haiyan Ge Runsheng Zhu Xiaoyan
Chinese Academy of Meteorological Sciences, Beijing 100081

ABSTRACT

During the sixth intensive observation period (IOP#6) of Torrential Rainfall Experiment over the Both Sides of the Taiwan Strait and Adjacent Area (HUAMEX), localized intense precipitation occurred on June 9,1998 in Changle area in Fujian province. The primary objective of this paper is to analyze this case using single-Doppler weather radar data, in particular the wind field structures retrieved by VAD (Velocity Azimuth Display) and V-D (Vorticity-Divergence) methods. The results showed that the event was related to wind shear at 3-5 km height and the activity of a mesoscale cyclone wave as seen on the radar echo images.

1. INTRODUCTION

With ample supply of water vapour, torrential rain affects the coastal region of southern China every year. Studies on the precipitation processes in the pre-flood season within the region tend to focus on the synoptic background and numerical modeling analyses. Analyses of the mesoscale and microscale flow structures in and around the rainstorms are often handicapped by the lack of high-resolution observational data.

Seven heavy precipitation events were observed during the IOPs of HUAMEX in 1998. Doppler radar data obtained from the field experiment provided information on the evolution of echo intensity and wind field in the rain areas. From the methods and softwares developed by the Chinese Academy of Meteorological Sciences for retrieving Doppler winds, the evolution of average wind profiles in time was derived from the VAD method (Browning and Wexler 1968) while the two-dimensional and three-dimensional wind fields in rain clouds were acquired by using V-D method (Jiang and Ge 1997, Wang 1998).

An intense precipitation event brought 200 mm of rain to Changle and the adjacent areas on June 9,1998. The rain area was less than 100 km in diameter. Intense radar echoes at the low levels had a wave-like shape within a zonal band and flow convergence was suggested from the retrieved wind field. The localized heavy rain was mainly related to a wave perturbation along a stationary front and the activity of a mesoscale cyclone.

2. GENERAL PATTERNS OF PRECIPITATION

The range of rainfall in both Fujian and Guangdong from 08:00 H on June 8 to 08:00 H on June 11 (all times in Beijing Local Time) was large and the rainfall distribution was rather uneven. Rain in Fujian province was mainly in the northern and central parts. Changle, in central-eastern Fujian and where the Doppler radar station was located, had a rainfall total of 75 mm. But rainfall amounts exceeded 200 mm in the surrounding areas. Fig. 1 shows the distribution of 8-hour precipitation (15:00-22:00 H on June 9) of 9 stations near Changle. Rainfall at Minhou about 30 km northwest of Changle was 220 mm and at Fuqing about 20 km southwest of Changle was 260 mm, much higher than those in other areas during the 8-hour period. The sequence of events associated with this short-lived, localized yet intense rainstorm took place within the effective scanning range of the Doppler weather radar at Changle.

3. ANALYSIS OF SYNOPTIC BACKGROUND AND RADAR ECHO EVOLUTION

From the synoptic chart, the rain event on June 8-11 was set against a frontal background, enhanced by the combined effects of frontal trough, shear line and southwesterly jet flow. The frontal zone moved southwards to Guangdong and Fujian under general conditions of instability ahead of a 500-hPa westerly trough.

As seen from the continuous evolution of echoes on the radarscope, rain development near Changle was mainly associated with local convective systems. There were several cycles of storm genesis, intensification, dissipation and regeneration as the convective cells went through different phases of development within a mesoscale convergence zone from 08:00H on June 8 to 08:00H on June 11. The most intense episode occurred at 16:00H-21:00H on June 9 and lasted for about 5 hours. The depth of the convective system was about 7-8 km.

Radar images of Changle in Fig. 2 show the rain echoes in different phases of evolution. There were signs of a mesoscale cyclone wave along a zonal echo band. At 17:46 H, the east-west echo band was joined by a northeast-southwest oriented short echo band northwest of Changle radar station. The short echo band was less than 100 km in length with central intensity of about 45 dbz. At 18:45 H, the short echo band moved southwest towards Changle and developed a mesoscale cyclone wave structure. It intensified significantly and rain fell between 18:00 H and 19:00 H. At 19:24 H, the east-west echo band weakened while the short echo band continued to intensify, bringing heavy rain to Minhou and Minqing northwest of Changle and to Fuqing southwest of Changle. At 20:16 H, the length of the echo band was reduced to about 50 km, but intensity was maintained at 45 dBz. Precipitation data showed that rain was heaviest at 20:00 H and lasted till 22:00 H.

4. ANALYSIS OF VAD WIND PROFILE

The evolution of vertical wind profiles with time at Changle obtained using VAD method derived from the Doppler radar data from 10:00 H to 22:00 H on June 9 is shown in Fig. 3. Below 3-km level, winds were mostly westerly or southwesterly, favouring transport of moisture from the sea areas. Wind shears detected at 16:00 H and 19:00 H were apparently linked directly to the rain event. At 16:00 H, wind shear was located between 3 km and 5.5 km levels. At 19:00 H, wind shear was at the 2 – 3 km levels, and most prominent at an altitude of 3 km. These suggested that the rain processes were caused by mid-level systems.

5. ANALYSIS OF TWO-DIMENSIONAL WIND FIELD RETRIEVED BY V-D METHOD

The low-level two-dimensional horizontal wind fields near Changle along with echo intensity contours (Fig. 4a) were retrieved by V-D method from the Doppler radar data at 17:46 H, 18:45 H, 19:24 H and 20:35 H on June 9. It was found that changes in the wind fields were related to echo intensity in the developing phases of the convective cell. At 17:46 H, there were two air currents at the low levels: one was southwesterly and strong; the other was weak and located in the northeast-southwest short echo band. At that time, the echo band had just formed and its inner flow led to weak convergence with the background wind within a cyclonic system. At 18:45 H, the echo band and the westerly flow within it both intensified. Strong speed convergence suggested that new echoes would continue to develop ahead of the band. At 19:23 H, the band had an obvious mesoscale cyclone wave and became very severe. The inner flow showed cyclonic rotation and had westerly, northwesterly and southwesterly winds. Both speed and directional convergences were strong. By that time, rain had started. At 20:16 H, the echo band diminished in size but kept its intensity. The most intense echo coincided with strongest convergence and heaviest precipitation.

In addition, a mesoscale cyclone was clearly depicted in the PPI scan (elevation angle at 3.4°) at 20:35 H on June 9. It suggested that the mesoscale cyclone wave had evolved into an occluded cyclone by this time, bringing rainfall of about 200 mm to the Changle areas in the process. Fig. 4b shows the two-dimensional horizontal wind fields, with echo intensities in different colours, at 20:35 H on June 9.

6. ANALYSIS OF THREE-DIMENSIONAL WIND FIELD RETRIEVED BY V-D METHOD

To analyze the flow structure associated the rain event in more details, the three-dimensional wind fields at seven levels (1 km, 2 km, 3 km, 4 km, 5 km, 6 km, 7 km above ground) were retrieved by the V-D method from the volume scan (VOL) data at 20:16 H. As shown in Fig. 5, the arrows represent the horizontal components of winds and vertical velocities are shown in colours. Rising motion as a result of convergence occurred mainly at the contour surfaces of 3 km, 4 km and 5 km. The system was strongest at the 5-km surface.

The two small diagrams on the bottom-right corner of Fig. 5 show the two-dimensional vertical-section wind fields at the point of intersection of the two red lines. The y-axis represents height. The x-axis is along east-west direction in the upper diagram and along north-south direction in the lower diagram. Rising motion due to convergence between easterly and westerly flows was obvious in the upper diagram, especially at the 3-5 km surfaces. Rising motion above the 4-km surface in a northerly flow was depicted in the lower diagram.

The observed evidences suggested that convection affecting the Changle areas on June 9 was mainly related to mid-level systems at the 3-5 km levels, with strong precipitation linked to an occluded cyclone.

7. CONCLUSIONS

(1) Intensive observation during HUAMEX provided important data for studying the torrential rain in the pre-flood season over southern China in 1998. Rain events on June 9 were due to convection enhanced by the combined influences of frontal trough, shear line and southwesterly jet flow. The systems moved southwards to Guangdong and Fujian in

generally unstable conditions ahead of a 500-hPa westerly trough. Analyses of the wind fields for this case, which should be typical of the storm environment for the region in general, allowed us to better understand the mechanisms involved in the triggering of torrential rain in the pre-flood season of southern China.

(2) Torrential rain in the pre-flood season over southern China was related to wind shear at the 3-5 km levels associated with the activities of mesoscale cyclone waves, especially occluded cyclones.

(3) This paper successfully tested and applied some single-Doppler radar data retrieval techniques to obtain two-dimensional and three-dimensional wind fields from PPI and VOL data. With such helpful tools, structure and evolution of airflow in clouds could be analyzed.

ACKNOWLEDGMENT. We wish to express our gratitude to Senior Engineer Que Qiufang for helpful discussions on the manuscript.

REFERENCES

Browning, K. A., and R. Wexler, 1968: The determination of kinematic properties of a wind field using Doppler radar. J. Appl. Meteor., 7, 105-113.

Jiang Haiyan, and Ge Runsheng, 1997: A new retrieval technique for single–Doppler radar. Quarterly Journal of Applied Meteorology, **8(2)**, 219-223.

Wang Dongfeng, 1998: Studies on the three-dimensional wind field retrieval technique from single-Doppler radar. Master thesis, Chinese Academy of Meteorological Sciences.

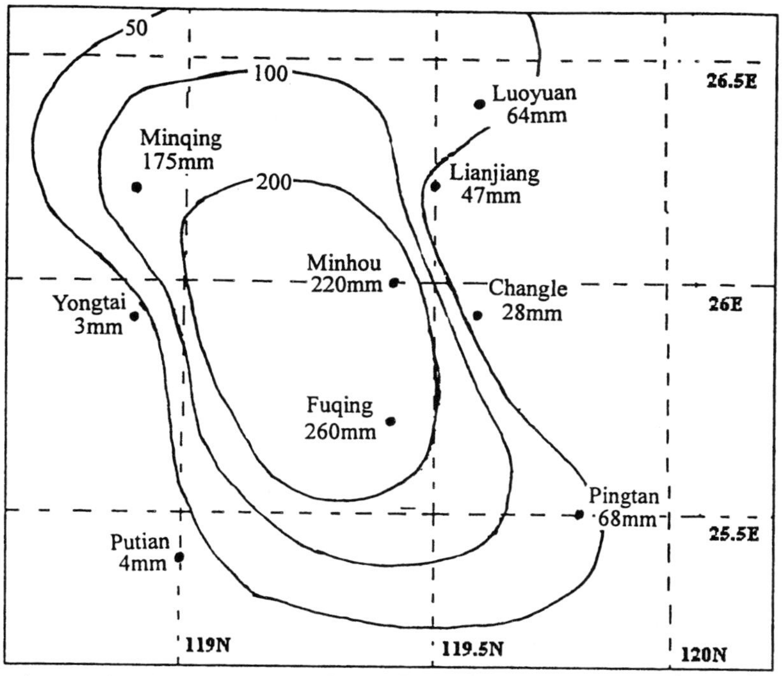

Fig.1 The distribution of precipitation over Changle and adjacent area at 15:00—22:00 on June 9, 1998

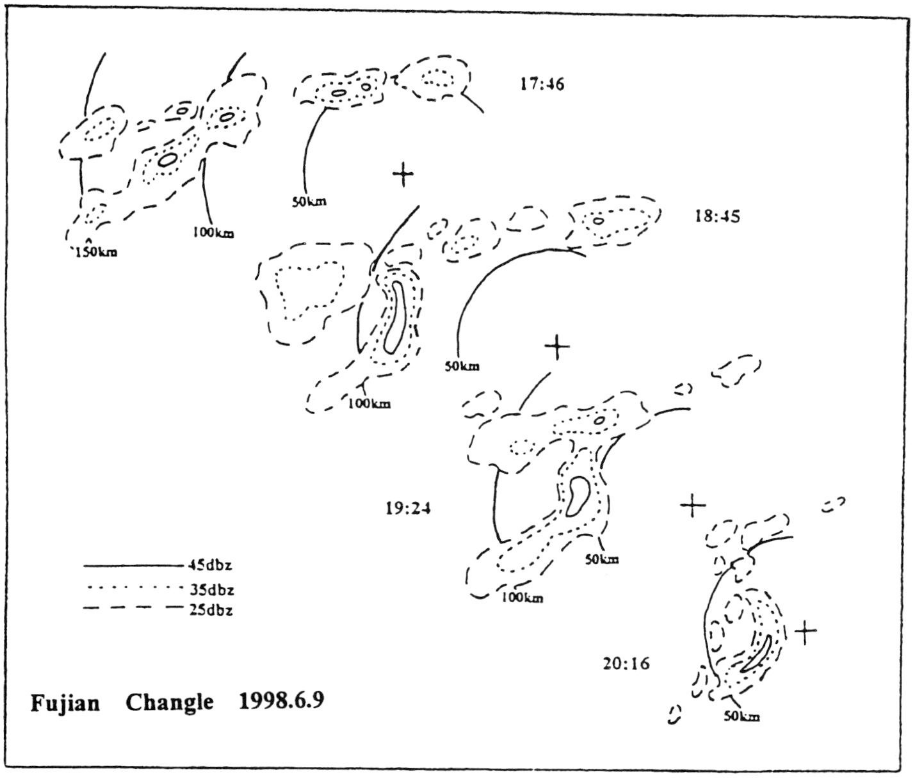

Fig.2 The evolution of radar echo over Changle area during IOP#6

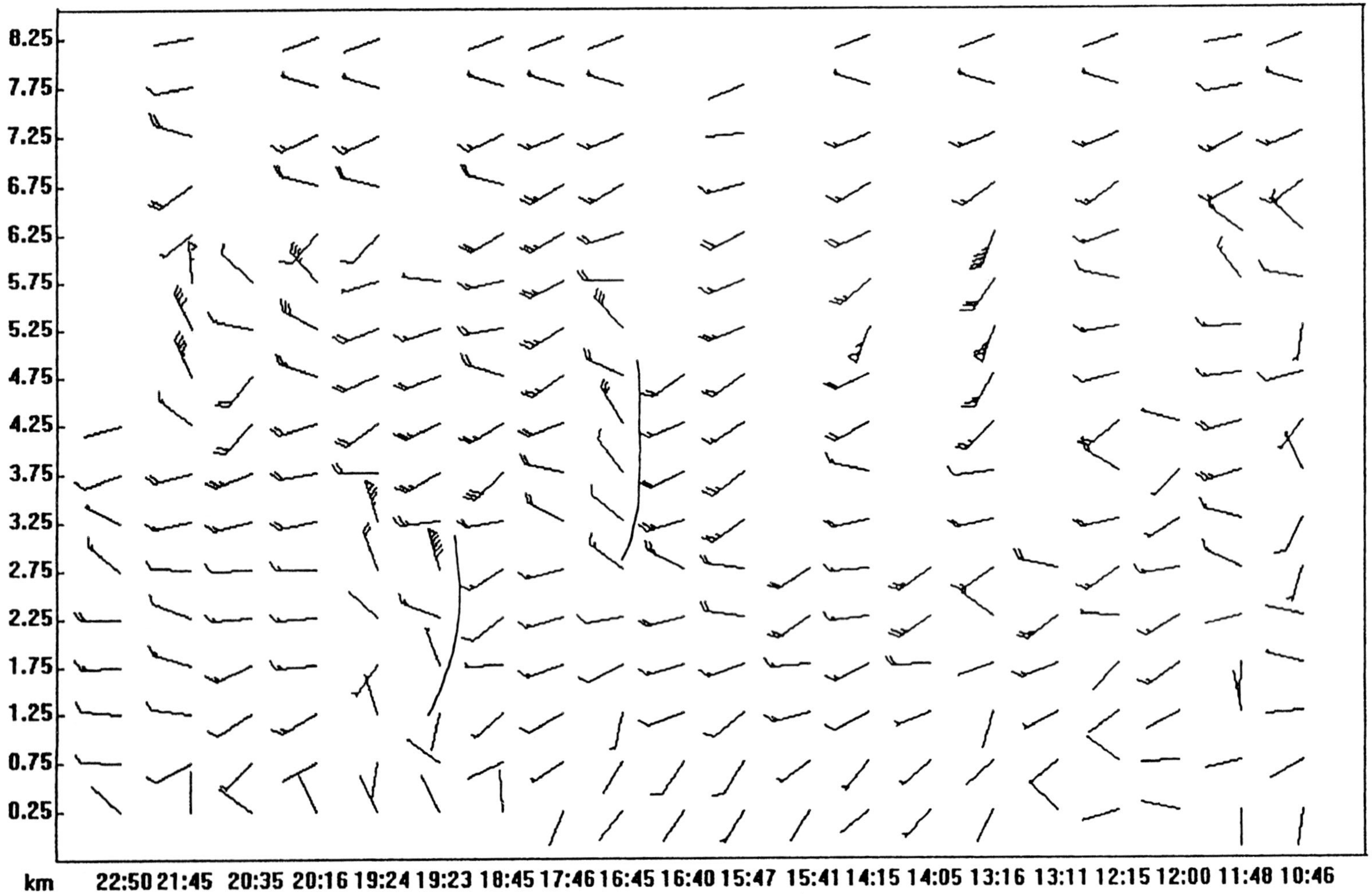

Fig.3 The evolution of VAD wind profiles at 10:00-22:00 on June 9, 1998

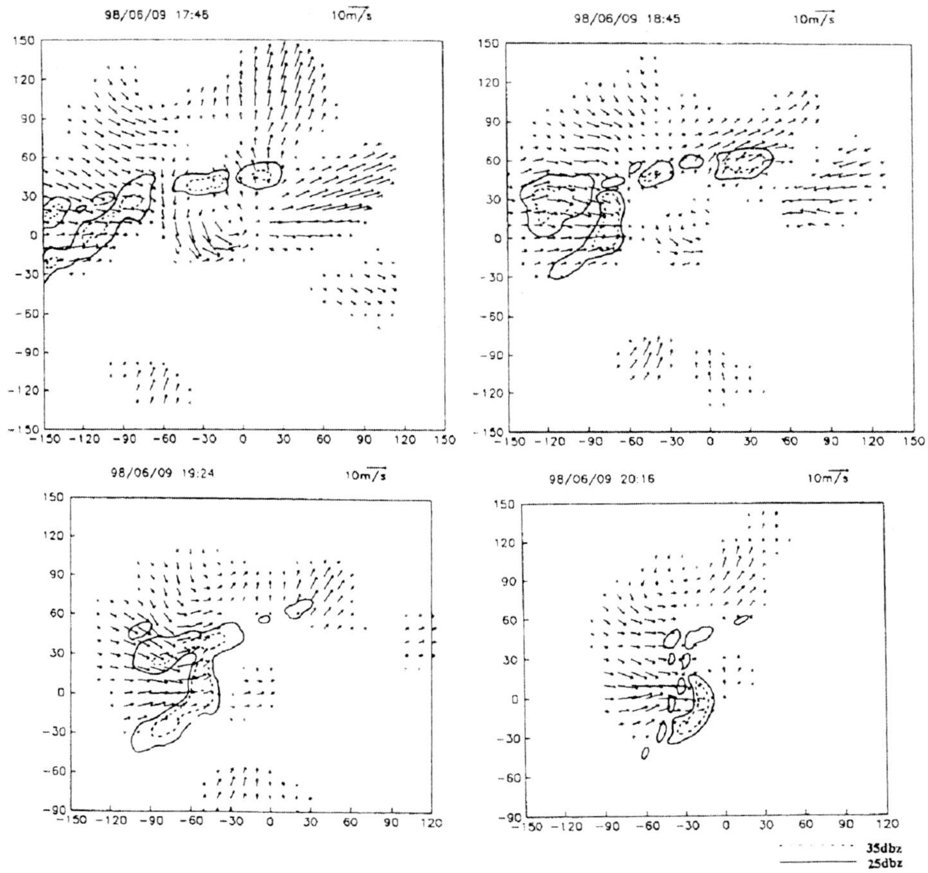

Fig.4a The evolution of low-level wind field from Changle Doppler radar at 17:00---20:00 on June 9, 1998

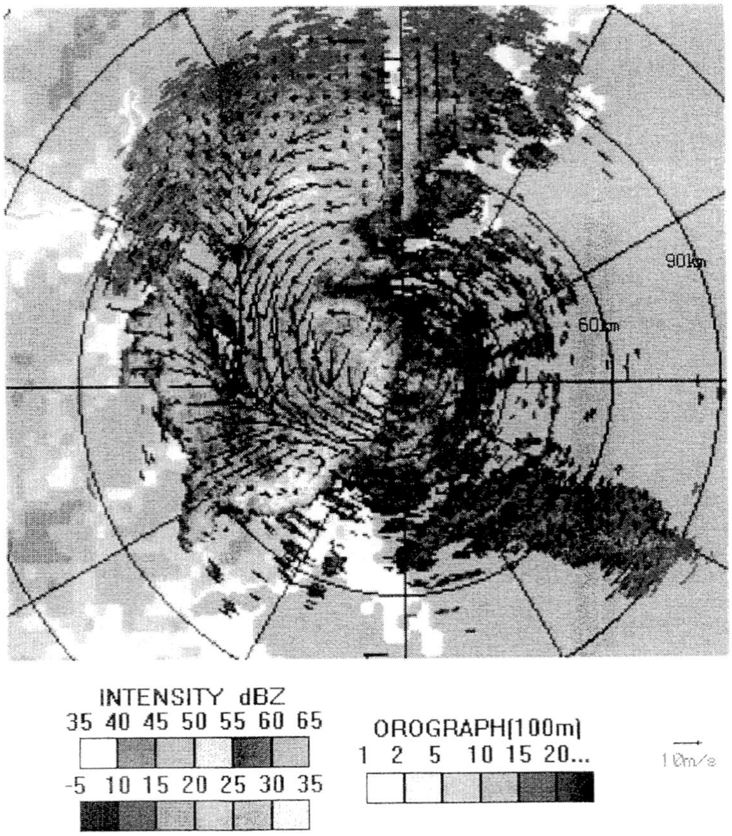

Fig.4b The image of retrieved two-dimensional wind field with intensity echo

Fujian Changle 1998.6.9 20:35

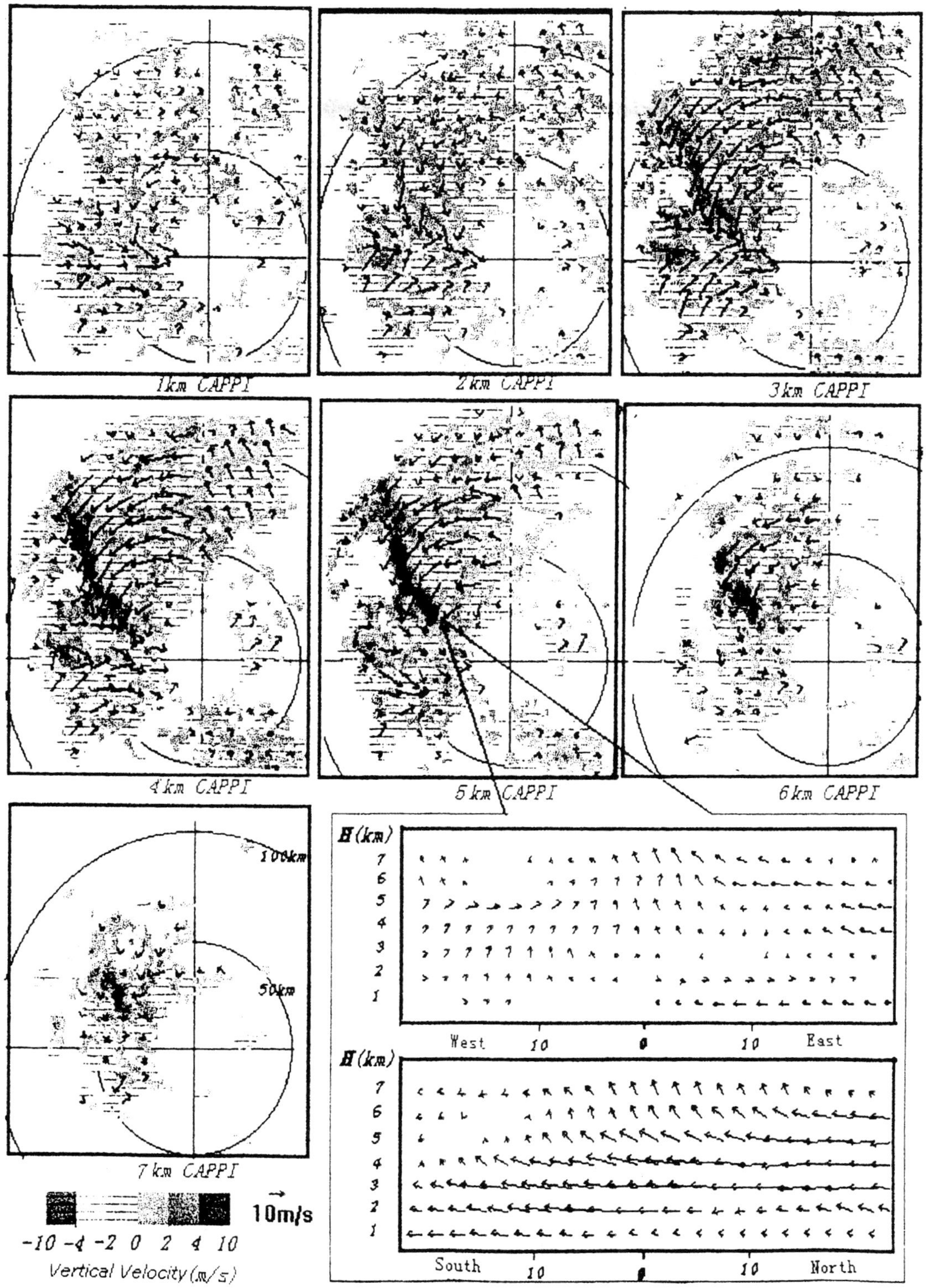

Fig.5 The Image of three-dimensional Retrieved Wind Field

Fujian Changle 1998.6.9 20:16

THE RESULTS OF DIFFERENT CUMULUS PARAMETERIZATIONS IN THE RAINSTORM PROCESS SIMULATION

Yuen Iek Chong Hao I Pan Ku Chi Meng António Viseu
Macao Meteorological and Geophysical Services

ABSTRACT

After the onset of the South China Sea Monsoon, a large area of rainstorm affected South China coastal region on 24 May 1998. Although in Macao, the rainstorm process only last from 1700UTC 23 to 1000UTC 24, the precipitation recorded, 100.4mm, was the maximum daily rainfall of that month. The red rainstorm warnings were then issued.

The Fifth-Generation NCAR / Penn State Mesoscale Modeling System, MM5, was introduced to the Macao Meteorological and Geophysical Services, and used to simulate this rainstorm process. The data of SCSMEX (South China Sea Monsoon Experiment) in 1998 was used as initial and boundary conditions for the model. Cumulus Parameterization Schemes were then applied to the simulation. As the simulative results were compared to the observation, the overall outcome shows that, in this case, the Anthes-Kuo scheme gives rather accurate descriptions of rainfall in Macao and the area over the Pearl River Estuary region.

1. INTRODUCTION

In order to be more accurate in predicting rainstorms, the Fifth-Generation NCAR / Penn State Mesoscale Modeling System, MM5, was introduced to the Macao Meteorological and Geophysical Services in 1997. Before using it operationally, necessary adjustments of the parameters have to be made in order to make it applicable in Macao and its adjacent area. Cumulus Parameterizations Scheme is one of them, which represents the subgrid-scale precipitation processes in meso-scale systems.

The analyzed surface and upper air data from SCSMEX (South China Sea Monsoon Experiment) in 1998 was used as initial and boundary conditions to the model. The simulation results from different Cumulus Parameterizations Schemes (CPS) were then compared with the observation. The main objective is to find out the most appropriate CPS for Macao and its adjacent areas.

2. REVIEW OF THE RAINSTORM PROCESS

The onset of South China Sea Monsoon on late May 1998 indicated the beginning of rainy season in South China region. A cold front associated with a large area of cloud band laid across the East Asia region from 23 to 24 May. It extended from northeast to southwest and established a strong convective system around the South China coastal region. More than 100mm accumulated precipitation was recorded during the rainstorm process (Fig. 1a). From 0000UTC to 0700UTC 24, the convective system continued to develop in the Pearl River Estuary (PRE) and its adjacent regions. The highest precipitation recorded from one of the land station was 135mm. The frontal system kept moving southeastward and reached the northern region of South China Sea (SCS) around 1200UTC on 24.

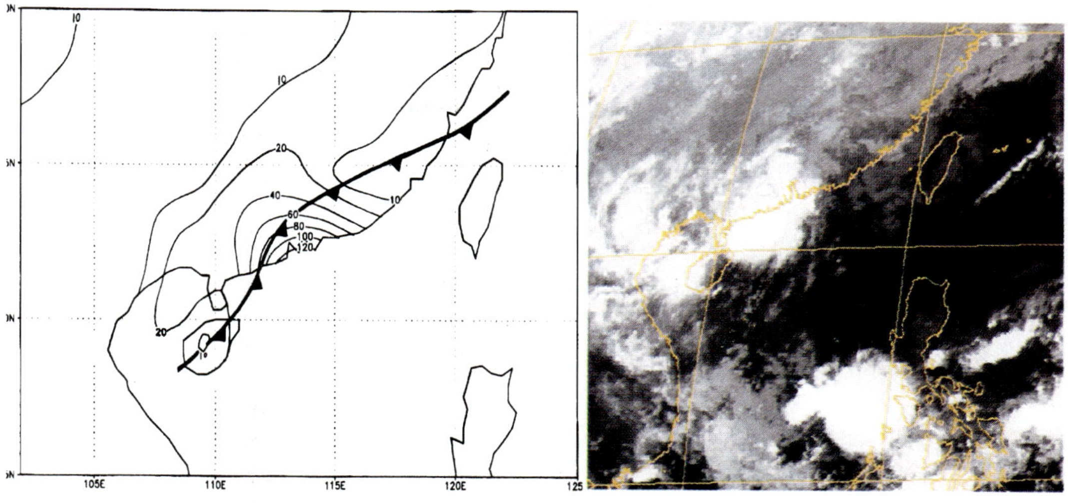

(a) (b)

Fig. 1. (a) Cold front's position at 0000UTC 24 and the accumulated rainfall (mm) from 1200UTC 23 to 1200UTC 24 over South China region, (b) GMS-5 IR image at 0035UTC 24 May 1998.

Based on the 6-hourly accumulated precipitation (not shown), the northwest of Canton and the Beibu Wan were shown to be the two major regions of rainfall at 1800UTC 23, with 51mm/6hr and 38mm/6hr rainfall respectively. At 0000UTC 24, the rainstorm over the Beibu Wan started to weaken while the rainstorm over the northern part of Canton moved southeastward to the PRE region with 53mm/6hr rainfall. Then it moved slower and developed eastward, which caused the Pearl River Estuary region to experience the maximum 6 hourly rainfall of 84mm at 0600UTC 24. As it continued to move southward to the northern region of SCS in the next 6 hours, the rainfall recorded on land decreased rapidly to 15mm/6hr. On the other hand, satellite images (Fig. 1b) showed that some rainstorms remained developing over the central region of SCS, i.e. west of Philippines.

Macao is located at the west coast of the Pearl River Estuary. Totally 100.4mm of rainfall was recorded during the rainstorm from 1700UTC 23 to 1000UTC 24 May. Maximum hourly rainfall of 46mm/hr was marked at 0300UTC 24 May 1998.

3. WEATHER REVIEW

A rainstorm process is always integrated by various weather systems at different levels: The southward moving cold front was the main cause of the rainstorm. It created large area of convective instability, which triggered the establishment and development of the meso-scale cyclone system (MCS) at the lower level. Fig. 2a shows that a MCS was formed at the north of the Beibu Wan at about 1200UTC 23 and moved southeastward over PRE region around 1200UTC 24. During that time, cold dry northeasterly air from the north met with the warm moist southwesterly air from the south and strengthened rapidly, thus intensified the convergence of air mass and moisture.

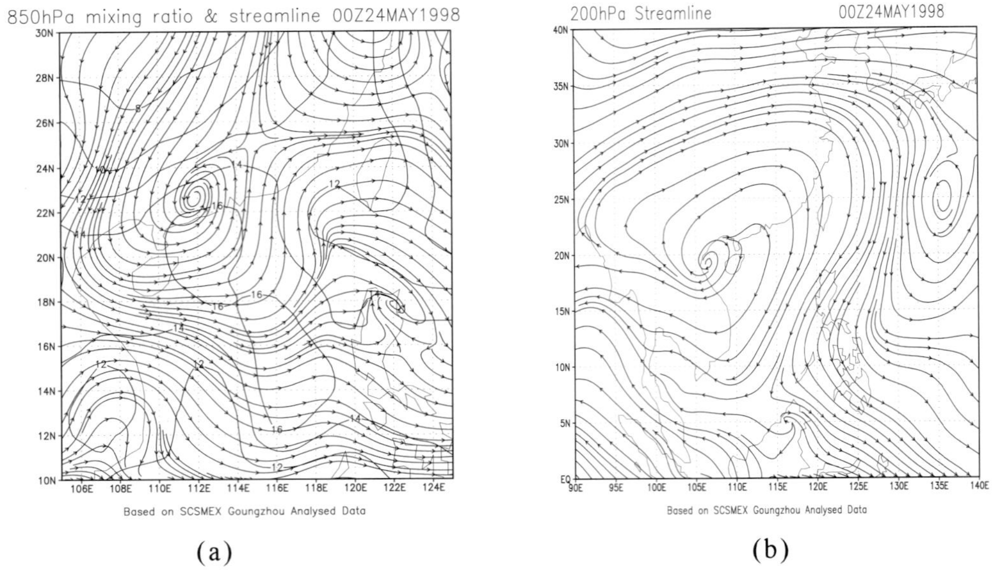

Fig. 2. Analyzed features at 0000UTC 24. (a) moisture (solid) and streamline fields at 850hPa, (b) streamline fields at 200hPa.

A 500hPa trough developing from Central China to Southern China moved southeastward to be ahead of the temperature trough, thus, increased the baroclinic instability of the South China region (not shown). At upper level, a large anticyclone at 200hPa provided an upper air divergence condition for the development of the convective system. During the rainstorm process, the Southeast Asia Anticyclone persisted over the East Asia area at 200hPa (Fig. 2b). The center of the anticyclone located at the Beibu Wan with its ridge extended northeastward across the South China area. As the ridge moved southward to the South China coastal region between 0000UTC to 0600UTC 24, it provided a good condition for the development of deep convective system from low-level disturbance. A deep-developing convective system was established under the situation of the warm-moist air convergence at lower level, divergence at upper level and the strengthening of baroclinic instability at mid-level.

4. A BRIEF INTRODUCTION TO THE MODELING SYSTEM

The PSU/NCAR MM5 modeling system used in this study is a 3-dimensional, limited-area, non-hydrostatic model with 23 levels at the terrain-following σ vertical coordinate. There are 65x81 horizontal grid points with 54km resolution centered at 20°N, 115°E by Mercator projection. The explicit cloud scheme was used to simulate the non-convective resolved-scale precipitation.

The grid-point data from SCSMEX 1998 were used as the initial and boundary conditions with 1°x1° horizontal resolution and 11 vertical levels from 1000hPa to 10hPa. The initial time for the simulation was 1200UTC 23. Four kinds of CPS were selected from the MM5: 1. Anthes-Kuo (named AK), 2. Grell (named GR), 3. Kain-Fritsch (named KF), 4.Betts-Miller (named BM).

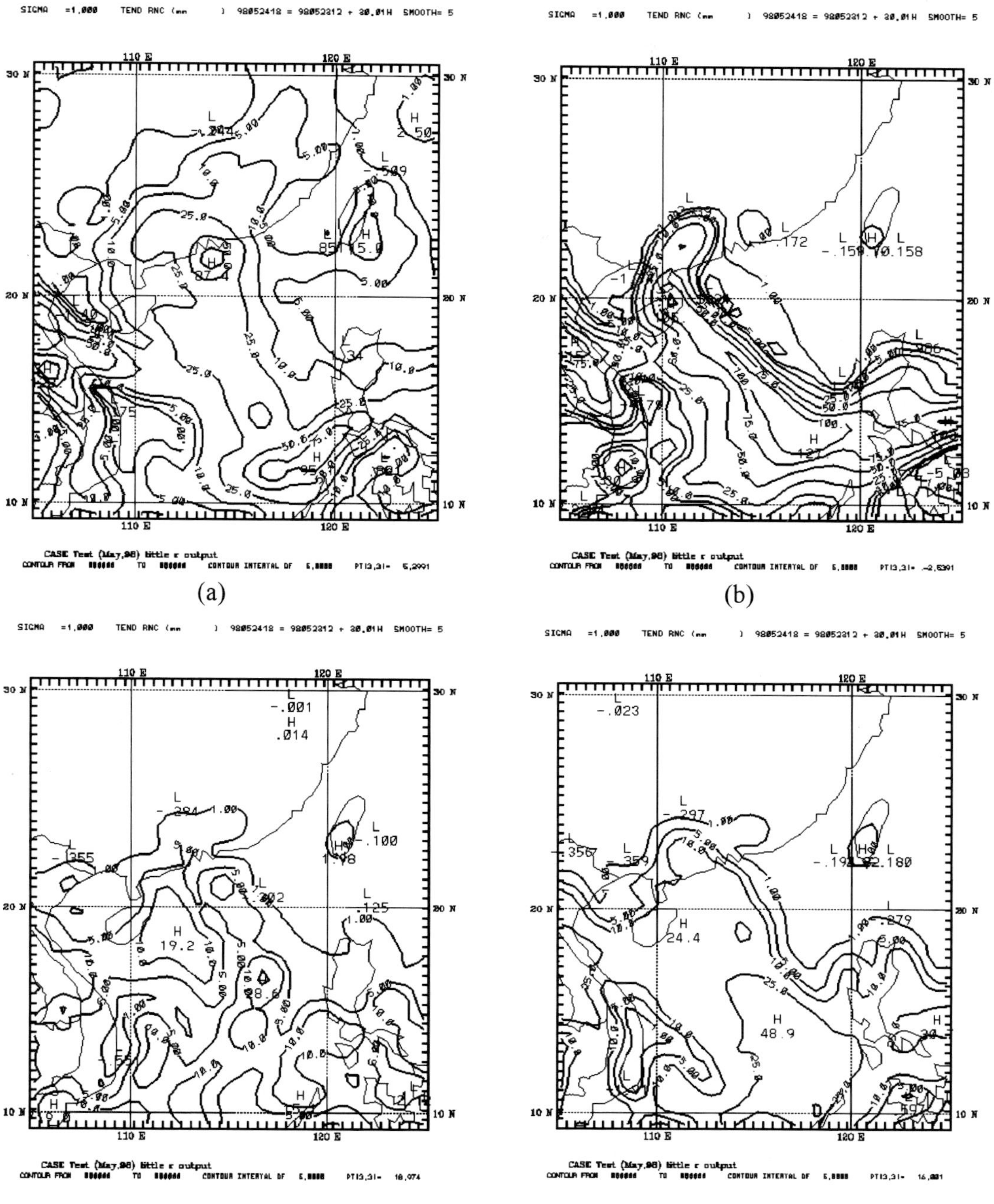

Fig. 3. Simulated convective rainfall in subgrid scale accumulated from 1200UTC 23 to 1200UTC 24. Negative value came from the graphic package. (a) AK, (b) BM, (c) GR and (d) KF.

5. RESULTS

(a) Simulation of convective rain

Fig. 3a shows results from AK scheme. The precipitation forms at 1700UTC over the northwest part of Canton moving southeastward and extends to the PRE region at 2100UTC 23. It strengthens afterwards with the maximum intensity of the precipitation up to 13mm/hr at 0600UTC 24. Then it moves to the SCS at 1300UTC 24 and decays. At 1900UTC 24, it moves northeastward to the in-land again centered at the east coast of PRE. During the rainy periods from 1200UTC 23 to 1800UTC 24, the accumulated precipitation from the model is 87mm over the PRE region. It is obviously shown that the evolution, intensity variation tendency and the location of precipitation areas are closed to the observation except the northward in-land movement. On the other hand, the simulation shows a delay in the evolution. The occurrence of intensification of rainfall and its southward movement are 3 hours later than the observation.

In BM scheme, a weak rainfall center forms at 1600UTC 23 over the northwest part of Canton (Fig. 3b) which develops and moves slowly southwestward. Then it intensifies and extends toward the Hainan Island at 2100UTC 23, crossing the SCS and combines with the rain band from the west of Philippines, thus, forming a large area of precipitation over the central region of SCS. The BM scheme has simulated the initial position of the precipitation, but it appeared to have a smoother evolution process and more southward movement than the observation. There is only 1mm rainfall produced over the PRE region.

In GR scheme (Fig. 3c), scattered convective rainfalls are simulated over the SCS at 1400UTC 23. Rainfall over the west of Canton occurs at 1900UTC 23. It hardly develops and moves southward to the northern part of SCS, forming a rain band from northwest to southeast and then moves eastward. During the rainy periods from 1200UTC 23 to 1800UTC 24, the maximum accumulated rainfall situates at the central region of the SCS and to the west of Philippines, with 29mm of rainfall. Two other maximum precipitation areas were both locate at the northern part of SCS with up to 19mm of rainfall. Only a few rainfalls are produced over the PRE region.

The KF scheme (Fig. 3d) is the latest one to initiate the rainfall. Precipitation is simulated around the Hainan Island and the northern part of SCS at 2000UTC 23. Rainfall over the northwest of Canton forms at 0300UTC 24 and extends southward to the PRE region at 0800UTC 24 with less than 10mm precipitation. The maximum rainfall is centered at the central SCS. Therefore, no intense convective rainfalls are simulated over the PRE region either by GR or KF scheme.

(b) Simulation of non-convective rain

In AK scheme (Fig. 4a), the non-convective rain center is simulated at 1700UTC 23, over the northwest part of Canton province and extends eastward with a total precipitation of 12mm from 1200UTC 23 to 1800UTC 24. Another rain center forms at 0500UTC 24 over Zhejiang province with a total rainfall of 20mm. The precipitation area that occur at 0200UTC 24 in BM scheme (Fig. 4b) is similar to the rainfall occurs over the northwest of Canton in AK scheme. The rain belt advances eastward with maximum accumulated rainfall of 47mm. Another rain center with 28mm of rainfall forms over Zhejiang at 1500UTC 24.

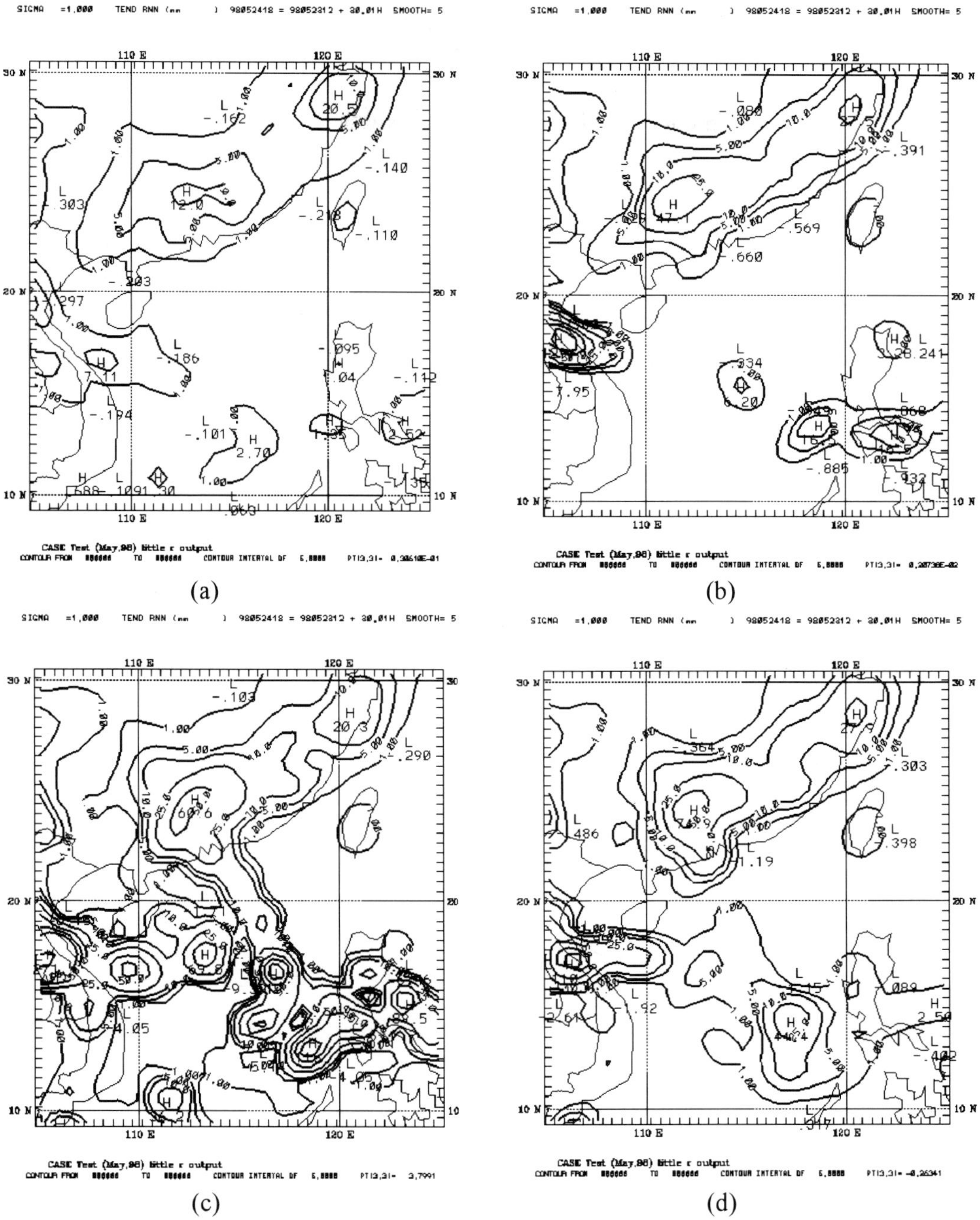

Fig. 4. Simulated non-convective rainfall in grid scale accumulated from 1200UTC 23 to 1200UTC 24. Negative value came from the graphic package. (a) AK, (b) BM, (c) GR and (d) KF.

The rainfall areas form over northwest of Canton and proceed northeastward in GR and KF scheme on 2300UTC and 2000UTC 23 respectively. The total precipitation is 60mm in GR and 75mm in KF. From 0700UTC 23 to 1200UTC 24, a total rainfall of 20mm is simulated over the PRE region and Zhejiang in GR scheme (Fig. 4c). There are several maximum rainfall areas over the SCS. The highest record is found over the west of

Philippines with 113mm of rainfall.

The rain over northwest of Canton in KF scheme also extends southward to PRE region while it advances northeastward with 20mm of rainfall (Fig. 4d). At 1500UTC 24, 28mm of total rainfall is simulated over Zhejiang. However, the rainfall area over west of Philippines experiences a precipitation of 44mm.

(c) Simulation result of AK scheme in different levels

At 925hPa (Fig. 5a), a cyclone is centered at Beibu Wan and creates a convergence zone from northeast to southwest lying across the South China region. It proceeds to the PRE region at 0600UTC 24 as the cyclone moves southeastward. The cyclone finally decays at 1800UTC 24. Fig. 5b shows a meso-scale cyclone locates at the southwest of South China region at 850hPa. Its westerly air stream at the southern part of the cyclone enhances to form a low-level jet with maximum windspeed of 20kt at 0600UTC 24. The rain band over the PRE region is enhanced by the transportation of moisture integrated from the southerly airstream to the eastern part of the cyclone. The South China region is in front of a trough in low level at 700hPa (not shown). It develops at 0600UTC 24 into a convergence zone over the PRE region and then forms a vortex at 1200UTC 24.

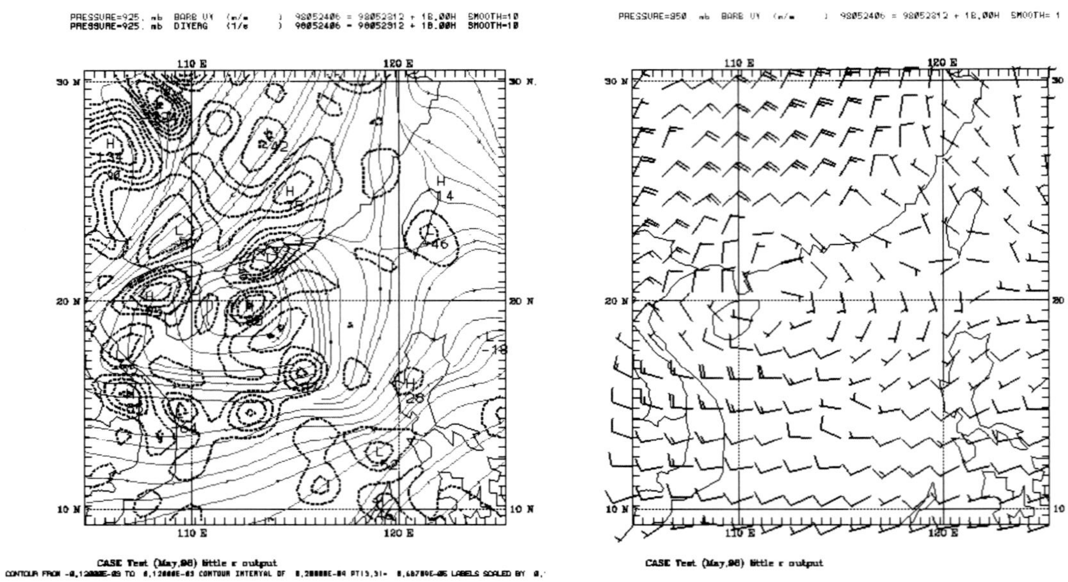

(a) (b)

Fig. 5. Features simulated at 0600UTC 24 by AK scheme. (a) streamline (solid) and divergence (dash) at 925hPa, (b) wind barb at 850hPa.

A 500hPa trough develops over South China region at 0600UTC 24 (Fig. 5c) and weakens at 0000UTC 25. At upper level (200hPa), an anticyclone, which centered at the northern Vietnam, covers a large area of Southeast Asia at 1200UTC 23. The South China coastal regions are affected by the northerly air stream, which came from the eastern part of the anticyclone. The maximum divergence centres at the northeast of the anticyclone and moves southeastward to the PRE region with the anticyclone. The peak divergence period occurs at 0600UTC 24 (Fig. 5d) and decays at 1800UTC 24.

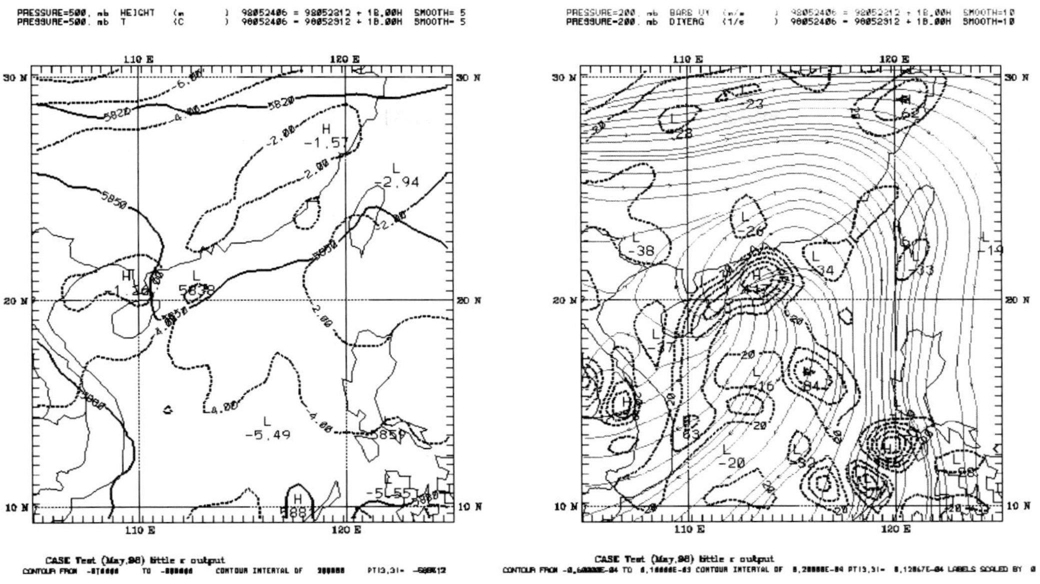

(c) (d)

Fig. 5. Features simulated at 0600UTC 24 by AK scheme. (c) geopotential height (solid) and isotherms (dash) at 500hPa, (d) streamline (solid) and divergence (dash) at 200hPa.

In divergence fields, the precipitation area corresponds to the divergence centre at 200hPa (Fig. 5d) and the convergence area in lower level. The convergence centre at 925hPa coincides with the rainfall area (Fig. 5a). As it moves to the PRE region at 0600UTC 24, the intensified precipitation is at its maximum.

Comparing the simulated rainfall with the features in different levels, the convective rain in sub-grid scale is mainly located at the northeast of the low-level meso-scale cyclone. This result coincides with the study of Shou (Shou 1997) on asymmetric structure of the meso-scale cyclone. Shou claimed that the distribution of rainfall in the cyclone was asymmetric. It is located in the southern and northern wind shears and the centre of the low-level cyclone. In this case, the rainfall is mainly centred at the northern wind shear area. In conclusion, the rainfall develops rapidly due to the conjoint conditions of the convergence at 925hPa, the moisture transportation of low-level jet at 850hPa, the vortex at 700hPa and the formation of disturbance at 500hPa as well as the divergence at 200hPa.

6. DISCUSSION

(a) Convective rain and Non convective rain

From the AK scheme, it is obviously shown that the convective rain is centered at the northern part of SCS and the South China Coastal region while the non-convective rain in northern part of Canton and Zhejiang province. The result is closed to observation. The distribution of convective and non-convective rain in BM scheme is similar to the AK's, but with no rain simulate over the PRE region. In GR scheme, aside from the simulation of the convective rain over the SCS area, the non-convective rain is also found over the central part of SCS and South China region. Moreover, the non-convective rainfall to the west of Philippines reaches up to 113mm. This is obviously contradicted with the observation. KF scheme shows similar result as the GR, only that the amount of convective rain and non-convective rain are both underestimated with maximum rainfall of 49mm and 44mm respectively. Both the GR and KF have the tendency of underestimating the convective rain (Wang and Nelson 1997).

AK and BM schemes in convective rain (Fig. 3) as well as BM, GR and KF schemes in non-convective rain (Fig.4) show that there is more than 100mm rainfall simulated over the central Vietnam. However, the satellite image taken during the raining period (Fig.1b) demonstrated that only low-level clouds existed occasionally. The high-rainfall simulations over that area are not reliable since no convective cloud bands were found over central Vietnam.

(b) Rainfall location

Based on the simulation of total precipitation, all the schemes, except AK scheme, show that the rainfall areas shifted northward over the South China region. (Fig.6). On the other hand, all the four CPSs can simulate the convective rain over the central region of SCS. Since some swirls existed from the middle to low level over the tropical area, the development of convective rain will accelerate if suitable large-scale environment is provided. It is shown that the four schemes in this simulation can predict the convective rain over the tropical ocean, and the non-convective rain that caused by the large scale upward movement during the passage of the frontal system (Liao and Wang 1998). Among the four CPSs, only the AK scheme can describe the precipitation located in the zone between the sea and land area.

(c) Rainfall evolution

The distribution of rain simulates indicated that AK scheme is similar to the observation with facts as follow: The daily maximum rainfall is located over the PRE. The rain belt moves southeastward and the convection rain is centered at the South China coastal regions and SCS area. And the non-convection rain is mainly over the Zhejiang region. Nevertheless, there are also signs of under-estimation of the total amount of precipitation together with the delay of the movement and development of the rain band.

7. CONCLUSIONS

The summer and winter monsoon cause great variations of weather situation in South China coastal region. Weather forecast becomes difficult during the transition period. The case in this paper is a typical case of precipitation occurs during the first rainy season in South China. The southwesterly monsoon brings warm and humid airstream to the South China coastal region. When there is a weak cold front proceeding southward, the conjoint constituent of the mid level to upper air trough associate with the MCS at lower level, thus, creates unstable conditions. The energy from the warm humid air at low level is released and thus intensive convection will be developed and persisted.

Forecast of precipitation is one of the most difficult issues in numerical weather prediction. And one of the main difficulties is the description of the convection rain in sub-grid scale which can be referred to as the CPS problem (Olson et al. 1995). Many studies tried to compare the performance of different CPSs. Their results varied among different regions, cases and seasons (Wang and Seaman 1997). The simulation in this case over South China coastal region shows that the AK scheme gives better description in precipitation over the PRE region and thus provides a reference for the appropriate choice of CPS. However, more cases from different seasons should be examined in order to determine the usage of CPS in MM5 for Macao and its adjacent areas.

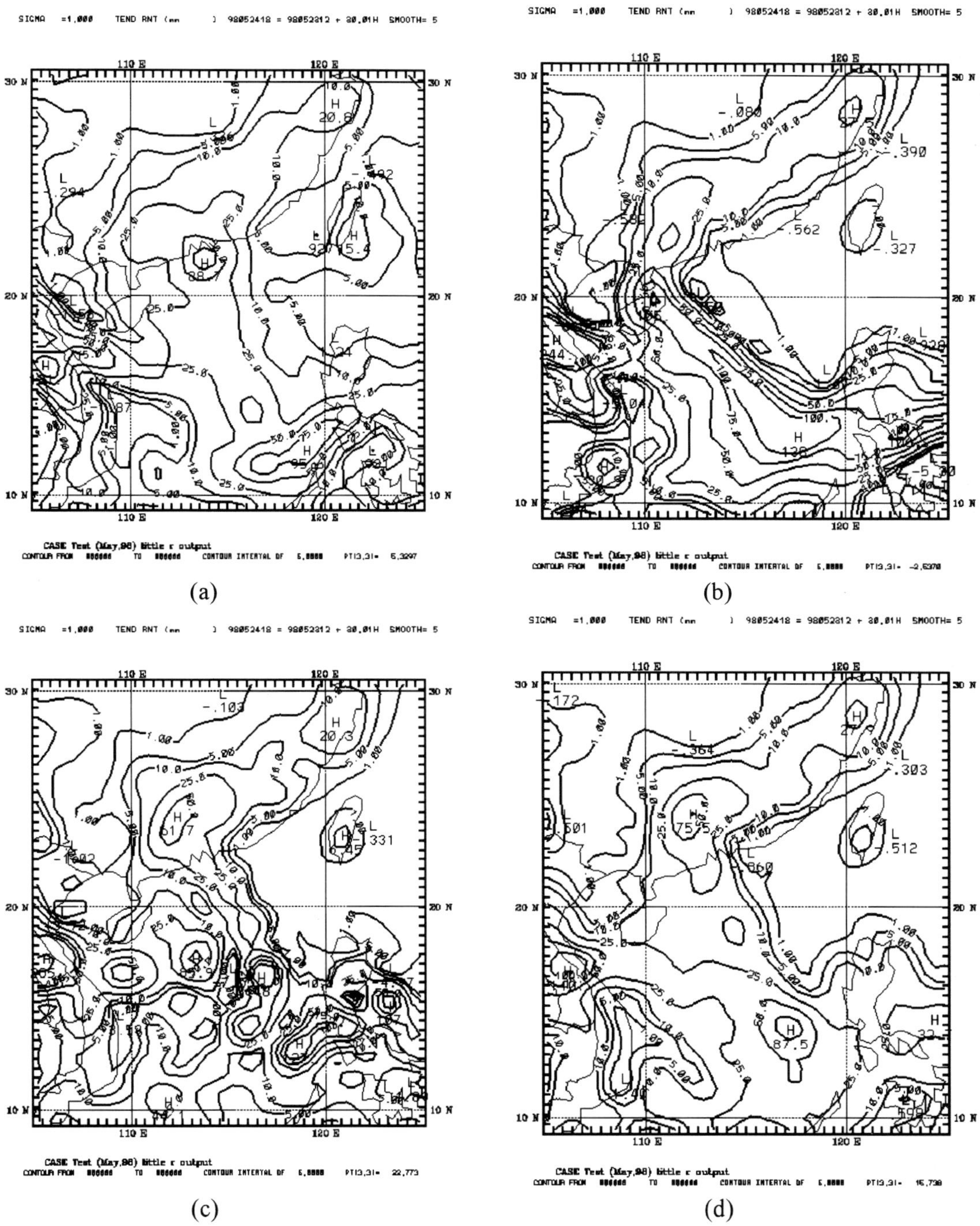

Fig. 6. Simulated total rainfall from 1200UTC 23 to 1200UTC 24 May 1998. Negative value came from the graphic package. (a) AK, (b) BM, (c) GR and (d) KF.

REFERENCES

Dong, He, Wei, and Pan, 1998: Numerical simulation of one mesoscale heavy precipitation process, *scientia meteorologica sinica*, Jiangsu Meteor. Soc., **18**, 4, 339-347.

He, Shan, and Wang, 1996: An experiment in the tropical numerical weather prediction with Betts-Miller convective scheme, *New technical research in typhoon and rainstorm NWP*, China Meteor. Press, 38-44.

Liao, and Wang, 1988: *Theory and Applications of NWP*. China Meteor. Press.

Mesoscale and Microscale Meteorology Division, National Center for Atmospheric Research, July 1997: *PSU/NCAR Mesoscale Modeling System Tutorial Class Notes: MM5 Modeling System Version 2.*

Olson, D. A., N.W. Junker, and B. Korty, 1995: Evaluation of 33 years of quantitative precipitation forecasting at the NMC, *Weather Forecasting*, **10**, 498-511.

Shou, Li, Lin, and Li, 1997: Analysis of the meso- β scale to a rainstorm process in Central China, *Journal on Meteor. Application*, **1**,5(3).

Wang, Wei, and Nelson L. Seaman, 1997: A Comparison Study of Convective Parameterization Schemes in a Mesoscale Model, *Monthly Weather Review*, **125**, 252-278.

~ Torrential Rain ~

THE SYSTEM DISTRIBUTION AND THE CIRCULATION FEATURES OF THE RAINSTORMS DURING THE SOUTH CHINA RAINSTORM EXPERIMENT IN 1998

Wang Yongming, Su Baixing and Chang Yue
Guangzhou Central Meteorological Observatory, Guangzhou 510080

ABSTRACT

Based on the data from South China Rainstorm Experiment and South China Sea Monsoon Experiment in 1998, we analyzed the features and variation rules of the physical quantities of large-scale, synoptic-scale and meso-scale systems of six rainstorms, which had happened during the experiments period. The counter-Hadley cells was found having established during the rainstorm periods in some degree. The system distribution of the MCCs in South China was also given in this study.

Key words: rainstorm experiment, feature of synoptic system, counter-Hadley cell, MCC in South China

1. INTRODUCTION

During the period of South China Sea Monsoon Experiment and South China rainstorm Experiment from May to June in 1998, there were six rainstorm processes in the experiment region and its vicinity. Two of them occurred in the Pearl River Delta, two others in the north of Pearl River Delta and in the central and northern part of Guangdong Province, and the remaining two in the west of Pearl River Delta and in the southwest of Guangdong Province. Among these, the three heavier processes lasted for 16 to 48 h, with 24-hour rainfall about 400 mm. Except that rainstorm processes in the last ten days of June were intensively observed on the surface level, for the other five experiments upper-air observation were carried out four times each day and extra data were added at 975 hPa, 950 hPa and 900 hPa, and the surface observation were done every hour. The spatial distance of the automatic surface meteorological observing stations in the delta region was about 10 to 40 km, about 20km on average. Based on these observations, we analyzed the different scale weather systems and their activity in the rainstorms from the angle of synoptic meteorology, and from what we could understand about the features of these systems and their distribution. Because the automatic surface meteorological observing stations were dense, surface analysis was confined to mesoscale systems in the delta region as well as surface frontal systems. The calculations used in this paper are the assimilation data from Guangzhou Regional Meteorological Centre.

2. THE FEATURES OF THE LARGE SCALE AND SYNOPTIC SCALE SYSTEMS IN THE MIDDLE-UPPER LEVEL

In 1998, the subtropical high was strong and shaped like a belt from May to June, extending from western Pacific through the South China Sea and the coast of South China into the Indochina Peninsula, with its ridge swaying between 8° N and 15° N. Among the six rainstorm processes, the 5880-gpm contour came near the coast of South China two times, and to the middle and north of Guangdong Province four times. So Guangdong Province lay to the northern edge of subtropical high.

In the rainstorm processes, the middle latitude was controlled mainly by westerly waves on 500 hPa, with two types of trough activity: one was the 'step' trough that was formed from Hexi corridor to Sichuan-Guangxi region, the other was the shear line in the east of the

Plateau. The former appeared three times, and the latter emerged two times, while both appeared together once. Southern trough activity occurred during all the above rainstorm events. But only during two cases did the southern troughs go into Guangdong Province. In addition, a weak shear line lasted for a short time in the northwest of South China Sea during one of the cases. During the other three cases, the westerly troughs only came close to Guangdong Province, and Guangdong Province lay under the southwesterly flow ahead of the trough.

During the rainstorms, the centre of the South Asia high went from the southern part of Indochina Peninsula to its northern part and to the Beibu Gulf. Although the centre of South Asia high went into the south of plateau in the last ten days of June, during the rainstorm period on 23 to 25 June, a sub-centre of the South Asia high or a weak ridge established in the northwest of South China Sea and in the southwest of Guangdong Province. The rainfall region appeared in the north, northeast, or near its sub-centre of South Asia high.

3. THE ACTIVITIES OF SHEAR LINE AND VORTEXES AT THE BOUNDARY LAYER

In the six rainstorms, five had shear lines in South China and north of South China Sea and some of the shear lines acted with low vortexes. The shear lines could be divided into three types. One was the southward, which went southward from south of Yangtze River into South China, and then swayed in the south-middle of Guangdong Province or moved southwards into the sea. Being influenced by the dynamical effect of the cold high, which had transformed before and moved eastwards into the sea, the south or southeast current prevailed in the north of South China Sea and the south-middle of Guangdong Province between 1000 and 975 hPa, and the southwest current prevailed in the southern edge of the shear line in 950~850hPa. The second was the coastal type—a shear line swayed in the Beibu Gulf and the north of South China Sea near 20°N, the south-middle part of Guangdong Province was controlled by the south-southeast current which was to the north of the shear line. The third was the compound type. It was combined by two kinds of shear line above, there was a shear line in the region from Beibu Gulf to the north of South China Sea and there was another in the south of Yangtze River. The shear line in south of Yangtze River appeared in the direction of north-east to southwest when it moved southwards, its west section moving into the south of Guangdong Province and converging in Beibu gulf region with the shear line from the north of South China Sea. At the same time, the southeast current prevailed in the south- middle part of Guangdong Province below 900 hPa, but the southwest current dominated at 850 hPa. Among these rainstorms, the five above had shear lines, but the remaining one apparently had no activity of shear line. The southwest current affected South China and only a very weak shear line was in the west coast of Guangdong Province at 975 to 950 hPa for a short time.

During these periods of heavy rainfall, five rainstorms contained mesoscale vortexes in different intensity. Some of them had corresponding circulations below 850 hPa. Some of them had shallow circulation only at some altitudes in the boundary layer. For example, on 9 June at 0000 to 1200UTC, a low vortex appeared only in 975 to 950 hPa, and at 0000UTC on 4 June, it was at 950 to 850 hPa. The horizontal scale of these mesoscale vortices was 100 to 400km. These vortices were related closely to the shear line. As the southward shear line appeared, the discontinual or shallow vortices always formed on it, moving southward with it and going eastwards into Guangdong Province and they were baroclinicity obviously (Su et al., 1999). As the coastal shear line appeared, the vortices formed on the shear line went along it from Beibu Gulf through Leizhou Peninsula into the sea area in the west of

Guangdong Province or the region outside the Pearl River Estuary. Apart from baroclinicity, the vortices were also a cold one sometimes (Su et al., 1999). As the compounding shear line appeared, the low vortices, which was formed from the shear lines, moved from Beibu Gulf to the sea area in the west of Guangdong Province, when the shear line in the north of South China Sea disappeared and the west section of the shear line, which had come from the south of Yangtze River into the South China, extended into the vortex. So this kind of vortex was of obvious baroclinicity and its scale was larger than the common one.

4. THE ACTIVITY OF THE SURFACE FRONTAL AND DISTURBANCE

During heavy rainfall periods, cold air was weak but active. All six rainstorms had close relations with frontal activity. Among these fronts, there were five that went into Guangdong Province. The last one, which had frontogenesis in the south of Yangtze River, only moved southwards to the north of South China. Guangdong Province lay in the warm area far from the front. Among the five fronts that had moved into Guangdong Province, some swayed in the middle or north of Guangdong Province, some went on moving southwards to the coast, two had occluded in the south-middle of Guangdong Province and the point of the occlusion were near the Pearl River Firth.

All of the rainstorms had disturbances on surface (Su et al., 1999), which acted as eddy, mesoscale shear line and mesoscale trough, whose horizontal scales were 15 to 200 km and were very shallow only existing near the surface. The processes of disturbances consisted of generation, development and weakening, with life cycle of 3 to 39 h. It was difficult to find them with the routine way if there were no surface observations in density. As the main rainfall area at the end of May was not in the Pear River Delta, and without abundant data, we only analyzed one disturbance. But on average each of the five other rainstorms had 5.8 disturbances, with one having 6 disturbances at the most. Because that the disturbances always separated or converged, there were always 2 to 4 disturbances at one time. Besides moving and swaying, they could also rotate, and the heavy rain occurred in the right forepart of the disturbance movement.

5. THE ESTABLISHMENT OF COUNTER-HADLEY CELL

From the cross-section of the composite stream line of ω' (to increase the effect, set $\omega'=3\omega$) and the meridional velocity v along the longitude passing through the rainstorm We found that during the prime period of the rainstorm, a meridional counter-Hadley cell established in the area between the location of rainstorm and equator. There were three rainstorm processes having the same pattern including the rainstorm on 8 to 9 June. The cross-section diagram along 114° E passing through the rainstorm region shows that before the rainstorm began, the upper troposphere was mainly controlled by a descending current, the upper level current flew southwards, and the low level current went up to the north of 25 ° N, thus the circulation was in disorder (Fig. 1a). When the rainstorm began, the upper air over the rainstorm region had changed to ascending motion. The current ascended to 500 to 100 hPa and then went down southwards, the current above 200 hPa went across the equator , and that below 200 hPa descended in 5° N to 25° N and divided into two branches under 700 hPa, one flew to the equator and the other went to rainstorm region northwards. At this time, the meridional counter-Hadley cell had established in the region of 5 to 25° N. Accompanying with the strengthening of rainfall and the intensifying of the ascending motion, the circulation reached to its prime period and centred in 10° N to 20° N, 500 hPa to 300 hPa (Fig. 1b). Sometimes there were two centres, with the northern one near 20° N below

500 hPa. When the ascending current in the rainstorm region weakened, the passage of the descending current in 5° N to 10° N was destroyed. The counter-Hadley cell disappeared gradually (Fig. 1c) and rainstorm ceased. There was one rainstorm process different from others. It appeared on 24 May, when monsoon just set off in the south of South China Sea. There were convective monsoon clouds over the South China Sea between 7° N to 12° N. The upper air current to the south of 15° N over the South China Sea showed disorder, thus the counter-Hadley cell was not obvious. On 25 May, monsoon convective clouds developed vigorously, and the strong upward current appeared near 10° N, at 0000UTC the upward motion replaced that in the rainfall region, and then the torrential rain ended (Fig. 2). The establishment of the counter-Hadley cell made it clear that the South China rainstorm had close relation to the circulation of low latitude. The vapour and energy at the low level in the low latitude input to the rainstorm region, and the upper air current in the region exhalation to the low latitude. Thus the circulations were contributing to the occurring of rainstorm.

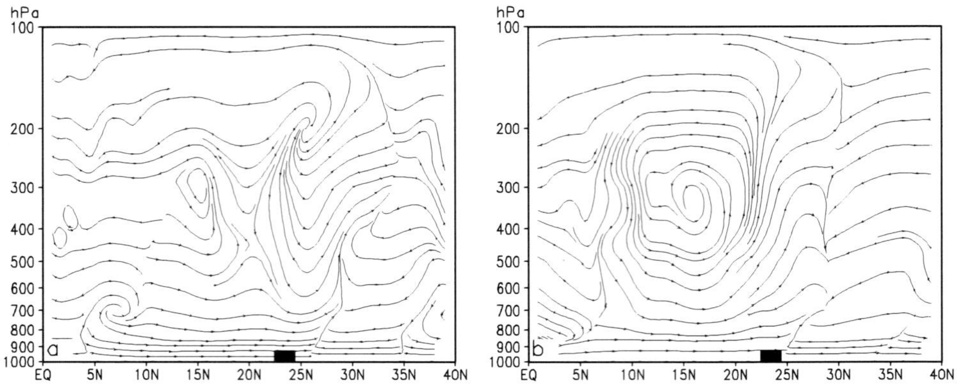

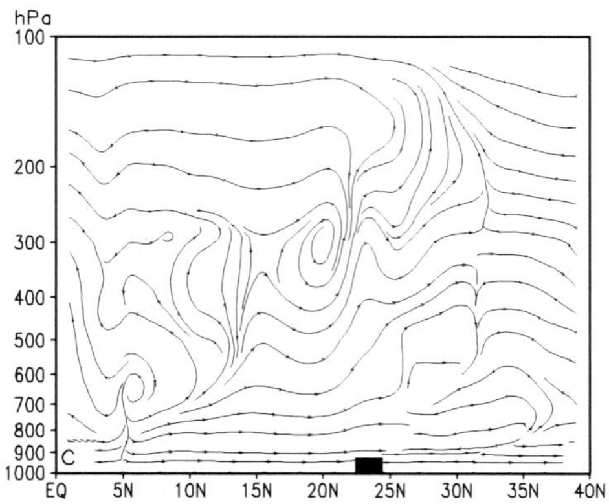

Fig. 1. Vertical cross-section of the resultant of ω' and v along 114° E (ω' =3ω, the black part is the region of the rainstorm) (a) at 1200UTC 8 June, 1998 (before the beginning time) (b) at 0000UTC 9 June (in the prime) (c) at 1800UTC 9 June (the ending time).

6. SOUTH CHINA MCC AND ITS SYSTEMATIC DISTRIBUTION

Among the six rainstorms, five had MCCs, which were added up to 10 altogether. There was one MCC at least and three MCCs at most (Table 1). Of cloud top area $T_{BB} \leqslant -50°$ C, the horizontal scale were 50 km to 500 km. The scales at the beginning time were small (but the rainfall was intensive) only between 5 km and 150 km, and in the prime period they were 200 km to 500 km. The life cycle of MCC was 9 to 28 h. The MCC, which caused the rainstorm, mainly acted in the south of Guangdong Province and the coastal region, and sometimes only half of a cloud cluster reached the land.

The South China MCC developed before the frontal face, near the front and even in the warm area without front, but in the surface layer, there all were shallow disturbances. Between 1000 and 850 hPa, some MCCs had shear line. Besides, there were always short time baroclinic mesoscale vortices, which had different depth. But there was also one MCC relating to southwest current and weak shear line in the low layer. So the South China MCC was baroclinic in the low layer, which was different from that of warm core structure in the middle-low layer as pointed out by Qiqiang He (Yang et al., 1991). The South China MCC had no fixed position relative to the meso scale vortex in the boundary layer, maybe they appeared to the east, south and north of the low vortex, sometimes even in the centre of it. MCCs also appeared to the north or near the shear line. They always had southwest jet in the low levels, but mainly had southeast current in 1000 hPa to 975 hPa. The MCC usually occurred in the southwest current in front of the westerly trough in 700 hPa to 400 hPa, some short and weak small troughs moving into this area. The South China MCC's cyclonic circulation was weak, and the altitude which had cyclonic circulation was low, the upper level of MCC was to the northern edge of the South Asia high or near the sub-centre of the high, which had anticyclonic circulation. Under this kind of allocation, South China MCC located in a region in which the low layer was convergent, the middle-low layer was of positive vorticity, and the upper layer was divergent and of negative vorticity. In addition, the whole troposphere was controlled by the ascending current, which composed an important part of the meridional counter-Hadley cell. The system distribution of the South China MCC and the feature of the circulation were given (Fig. 3).

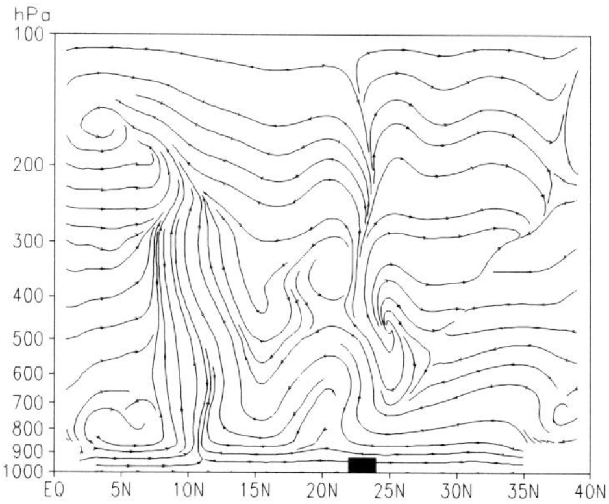

Fig. 2. Vertical cross-section of the resultant of ω' and v along 113 ° E at 0000UTC 25 May (ω' =3ω. the black part is the former region of the rainstorm).

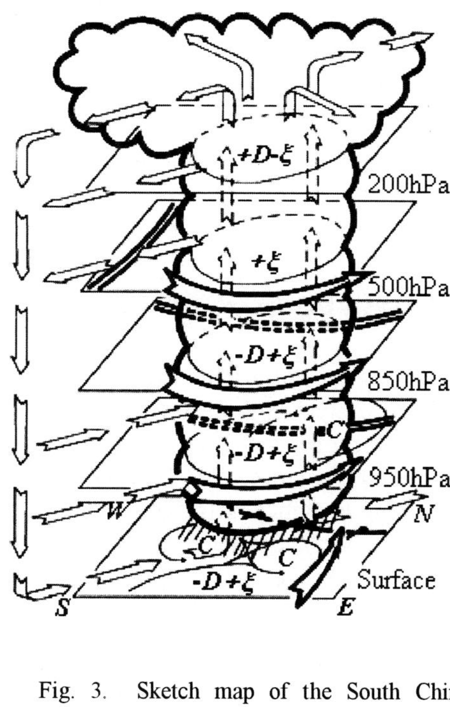

Fig. 3. Sketch map of the South China MCC's system distribution and circulation (the double line is trough line or shear line, the arrow is dominant flow or jet current in the corresponding altitude).

Table 1. The systems' distribution and the features of South China MCC during the period of the trail in 1998.

Rainstorm number	Ordinal number	Beginning time	Life time (hour)	T_{BB} ≤-50°C (scale:km)	Surface system	The boundary layer system	Middle level system	Relative position to the South Asia high
2	1	08 p.m. 23 May	19	150~500	disturbance before the front	low vorticity	southwest current in the front of the trough	to the north
3	2	02 a.m. 30 May	11	100~200	disturbance before the occluded front	shear line low vorticity	small shear line	same as the above
	3	05 p.m. 30 May	22	100~250	same as the above	same as the above	same as the above	
4	4	05 p.m. 04 June	7	150~200	same as the above	shallow low vorticity	small southern trough SW current	to the north north-east
	6	02 p.m. 04 June	10	200~500	same as the above	low vorticity	same as the above	same as the above
5	7	05 p.m. 08 June	16	50~400	disturbance before the front	shear line low vorticity	SW current in front of the trough	to the north north-east
	8	08 p.m. 08 June	28	100~400	same as the above	same as the above	same as the above	same as the above
6	9	06 a.m. 23 June	9	50~200	disturbance in the warm area	weak shear of the south west current	SW current in front of trough	near the sub-centre
	10	00 a.m. 23 June	9	50~350	same as the above	same as the above	same as the above	same as the above
	11	08 p.m. 24 June	16	150~400	same as the above	same as the above	same as the above	same as the above

Note: The first number of rainstorm was not listed because the process had no MCC.

7. CONCLUDING REMARKS

(1) During the experiment period, all six rainstorms are located to the northern edge of the zonal subtropical high, South Asia high or near the sub-centre of the South Asia high. The westerly trough ('step' trough, southern trough) from its upstream drew near or went into Guangdong Province. In the five rainstorms, there were shear lines and fronts acting in Guangdong Province or the coastal region.
(2) In the boundary layer, most of the rainstorms had mesoscale low vortexes or mesoscale shear lines. On the surface, there were some disturbances in the warm sector ahead of front or near the front. Below 975 hPa, the southeast current prevailed in the middle and south of Guangdong Province, and the level of 950 to 850 hPa was controlled by southwest current.
(3) When the rainstorm was in the prime period, the ascending current in the rainstorm, the descending current near the equator, the current flew to the equator in the upper level and the current to the rainstorm region in the low level constituted the meridional counter-Hadley cell. When the cell disappeared, the rainstorm was over, which indicated the close relation of the South China rainstorm and the circulation in the low latitude.

(4) The horizontal scale of South China MCC is 50 km to 500 km, the altitude of cyclonic circulation was low, it appeared as a mesoscale low vortex, mesoscale shear line or disturbance in the boundary layer, and it had obvious baroclinicity. In the upper layer the MCC was near the anticyclonic system. Therefore the convergence (or positive vorticity) regions were in the low levels, the divergence and negative vorticity regions were in the upper levels. The ascending current controlled the whole layer. This ascending current is an important part of the counter-Hadley cell established in the period of the rainstorm.

REFERENCES

Su Baixing, Wang Yongming, and Chang Yue, 1999: The analysis of the features of the mesoscale systems at boundary layer in two rainstorm processes, *Rain gust. Damage*, **1**, 42-48. (In Chinese)

Yang Guoxing, He Qiqiang, and Lu Hancheng, 1991: Mesoscale Meteorology, *Beijing: Meteorological Press*, 122-143. (in Chinese)

A SYNOPTIC STUDY OF HEAVY RAINFALL OVER TAIWAN

Chia-Chung Yu
Department of Earth Sciences, National Taiwan Normal University, and
Meteorological Application and Development Foundation, Taipei
E-Mail: masf-123@m2.dj.net.tw

ABSTRACT

When Taiwan is not directly under typhoon circulation and not in Meiyu, extremely heavy rainfall exceeding 130 mm per day may occasionally occur over the northeast region during the northeast monsoon period, especially from October to December, and over the southwest region during the southwest monsoon period. In this study, effort is made to find out the key factors that cause such heavy rainfalls, based on synoptic maps as well as other data. It is found that flow confluence and orographic lifting play very important roles in producing the heavy rainfalls over northern, southern or central Taiwan. The confluent flows that cause the heavy rainfalls over northern Taiwan mostly result from certain combinations of a continental cold high and a tropical depression. The confluent flows that cause the heavy rainfalls over southwestern Taiwan depend on the position and orientation of the subtropical high, typhoon activity and the movement of the Yangtze-River low. Several such synoptic patterns are identified.

1. INTRODUCTION

In Taiwan, due to the geographic location and the NNE-SSW oriented Central Mountain Range, rainfalls generally occur in the northeastern region during the winter monsoon season and in the southwestern region during the summer monsoon season. Under some special synoptic situations, daily rainfall can well exceed 100 mm. The most well known such situations are typhoon passages in summer and fall and Meiyu in May and June.

There are other not-so-well-known synoptic situations in which daily rainfall may reach hundreds of millimetres. Such situations are the most troubling to weather forecasters. Heavy rainfalls can cause floods threatening lives and causing severe property damages. It is therefore important to understand such causes of heavy rainfalls and thereby to improve the forecast skill in order to reduce the potential damage.

2. DATA AND CASE SELECTION

There have been many studies on heavy rainfalls over Taiwan due to typhoons or Meiyu-related activities. In the present study, we only select heavy rainfall events that are not a direct part of a typhoon or a Meiyu front (referred to hereafter as selected heavy rainfall, SHR). Here the term 'heavy rainfall' is defined according to the Central Weather Bureau, as rainfall accumulation no less than 130 mm in a day and during which the hourly rate reaches 15mm or more for at least one hour. The purpose of this study is to find out under what synoptic situations the SHRs occurred, in what scales and frequencies, at where and why.

The data used in this study are a) Taiwan rainfall records 1985-1998, b) surface and upper-air charts, c) satellite cloud images and d) soundings. All the data are obtained from the Central Weather Bureau.

3. RESULTS

The SHRs in northeastern Taiwan occurred mostly between October and December. The station that recorded the highest number of SHR is at Suao, located on the southern edge of the fan-shaped watershed of the Lanyang River that faces the western Pacific. Obviously, the particular location makes Suao the most favorable place for orographically forced convection in northeasterly flow. In the fall the seasonal averaged number of SHR is about four at Suao but the year to year variation is extreme large, from zero occurrence in 1993 to 16 in 1998. The frequency of SHR becomes much less frequent at other stations.

In central and southern Taiwan, the SHRs occurred mostly in summer. Over the 1985-1998 14-year period the station at Tainan recorded 22 SHRs between July and September, 1.5 per season on the average. Again, the inter-annual variations are large, from zero in 1993 and 1998 to five in 1988 and 1992. In the same period, the number of SHR events at Kaohsiung, Chiayi and Taichung are 15, 12 and 11 respectively.

What did cause the SHR? After examining the flow patterns over Taiwan in all the selected cases, a common feature is found among them. Namely, the low-level flow over Taiwan was confluent or convergent. The flow confluence or convergence combining with the local orographic forcing produced strong ascending motion of the moisture-rich air and hence resulted in heavy rainfall. The confluent easterly flows for the fall SHRs and the confluent southerly flows for the summer SHRs are schematically shown in Fig. 1a and 1b respectively. The confluent westerly flow for the summer SHRs is just a 90-degree rotation of the flow in Fig. 1b (not shown)

Although heavy-rainfall events are of mesoscale at most, it is the large-scale flow field that provides the breeding environment for them to develop. The synoptic patterns that produce confluent flows in the fall SHR events are of the following types:

(a) Fall type A (FA)

The synoptic pattern of FA consists of a mid-latitude or high-latitude cold high over China, a tropical depression or Typhoon over the South China Sea or the Philippines and a front running westward toward Taiwan. In the late fall, the continental cold high pressure is well established with low-level northeasterly winds generally prevailing over the East China Sea and Taiwan. The front separating the easterlies from the tropical flow is generally not very active unless the flow on either side of the front is reinforced. The existence of a typhoon or tropical depression over the Philippines or the South China Sea brings the moisture-rich southeasterly flow to Taiwan, thereby, increase the flow convergence and reactivates the front. The uplifting of the easterlies by local topography further enhances the ascending motion, excites strong convection and therefore brings localized heavy downpours.

For example, Fig. 2 shows the surface map at 00UTC, November 15, 1996, a typical example of Type FA. Taiwan was under low-level easterly confluent flow and a stationary front between the continental cold high and the tropical depression over the South China Sea. The daily rainfalls at Suao and Keelung exceeded 211 mm and 167 mm respectively. The rainfall at Ilan well exceeded 50 mm per day but did not reach the criteria of heavy rainfall because of its location. The SHR lasted for two days. See Table 1a for rainfall at the northern stations between November 14 and 16, 1996.

Tables 1b to 1d show the daily rainfall distributions over northern Taiwan in three other time periods, 17-28 November 1998, 9-10 October 1996 and 4-6 November 1995 respectively. During these time periods, Taiwan was under synoptic patterns of the same type, with only considerable large shift of the center of the cold high. Particularly worthy of attention is Table 1b. Between November 17 and 28, 1998, two tropical depressions, one after the other, moved westward passing the Philippines into the South China Sea and concurrently there were oscillations within the high pressure region over China. Heavy rainfall lasted for several days at Suao. These cases suggest that the tropical system is perhaps the most critical element in the type FA pattern. This is indeed supported by other evidence. In the fall of 1998 and in the October and November of 1996, the tropical activities in the region from the South China Sea to the Philippines were comparatively the most active in the data period. The corresponding numbers of SHR occurrences at Suao in these months were 7, 5, 4, 5 and 6 respectively. In the late fall and early winter of 1997, the same tropical region was the most inactive in the data period and correspondingly there was no heavy rainfall at all over northern Taiwan.

SHRs are found in other types of synoptic pattern without tropical depression or typhoon over the above-mentioned area. Such patterns are:

(b) Fall type B (FB)

The synoptic pattern of FB consists of a breaking-off high pressure moving toward the Yellow Sea and a cold front in the east extending toward Taiwan. Due to the breaking-off high pressure, the flow over Taiwan is generally easterly. Flow confluence generally exists over Taiwan despite of the up-slope ridge and the down-slope trough over the east and west sides of the Central Mountain Range respectively. Specific humidity over Taiwan is higher than usual due to the longer travel of air over the sea. The front near Taiwan is active due to the eastward movement of the breaking-off high. Local orographic lifting combined with the general flow convergence over the northern Taiwan produces localized convection on the windward slopes particularly at Suao.

Fig. 3 shows the surface map at 00UTC 16 November 1992, for an example of this type. Taiwan was in the convergent easterly flow and under the tail of the cold front. The daily rainfall at Suao was well over 500 mm, but the rainfalls at the near-by stations Ilan and Hualian were only 54 mm and 118.5 mm respectively (see Table 2). Clearly, orographic effect dominated the rainfall distribution.

(c) Fall type C (FC) or Winter type (W)

The synoptic pattern of FC (or W) consists of a dominating pressure ridge extending southwestward from mid-latitude to the Gulf of Tongking and confluent easterly flow prevailing over Taiwan. The most critical features of this pattern are the strong pressure gradient and the E-W running isobars over Taiwan and its vicinity. The easterly flow is moisture-rich and generally confluence over Taiwan. Again, the orographic lifting of the strong easterly flow combined with the general flow convergence over the northern Taiwan produces localized convection on the windward slopes particularly at Suao.

Fig. 4a and 4b show the surface map at 00UTC, November 19, 1994 as an example in the late fall and the surface map at 00UTC, January 30, 1991, as an example in the early winter, respectively. The two maps are generally similar with only minor differences. The daily rainfalls at Suao in the two cases are comparable and heavy, slightly above 291 mm and 242 mm respectively. The amounts of daily rainfall at the other stations were far below the criteria of heavy rainfall in both cases (see Table 3a and b).

The synoptic patterns that produce confluent flows in the summer SHR events are:

(a) Summer type A (SA)

The synoptic pattern of SA is composed of a pronounced frontal trough extending from Japan through the East China Sea to southern China, an accompanying wind-shear zone separating the northeast flow and the southwest flow up to 700 hPa level and a southwesterly low-level jet close to Taiwan. The strong elongated frontal trough brings broad and strong southerly flow to Taiwan. The existence of the low-level jet and its relationship with the heavy rainfall appear similar to the Meiyu cases studied by Chen and Yu 1988.

As an example, Fig. 5 shows the surface map at 00UTC, July 5, 1992. The stations in southern Taiwan recorded 200 to 350 mm daily rainfalls. In the next few days the storm activity that produced the heavy rainfalls advanced northward up to central Taiwan (Table 4).

(b) Summer type B (SB)

Type SB consists of an eastward moving Yangtze-River low and a southwestward extension of the subtropical high. This arrangement brings the most moisture abundant air into a potentially unstable environment. Deep convection triggered by orographic lifting, dumps heavy rain over most of the central and southern Taiwan. For example, Fig. 6 shows the surface map at 00UTC, July 1, 1997. Taiwan was in the southerly confluent flow. The region of heavy rainfall extended northward up to Chiayi (see Table 5)

(c) Summer type C (SC)

Type SC consists of a northward-moving typhoon over the East China Sea and a relative high pressure over the Philippines. This arrangement brings high-humidity convergent westerly flow to Taiwan

Example: Fig. 7 shows the surface map at 00UTC, August 8, 1997. Typhoon Tina was over the East China Sea brought confluent westerlies to Taiwan. Heavy rainfalls were recorded at southern stations (see Table 6).

(d) Summer type D (SD)

In this Type, a typhoon over the northern East China Sea interacting with another typhoon no far apart or a downgraded tropical cyclone inland. The two jointly form a wide area of low pressure, which brings confluent westerlies to Taiwan similar to that of Type SC.

Fig. 8 shows the surface map at 00UTC, August 12, 1994. Typhoon Doug left Taiwan on August 9 after dumping heavy rainfall over the windward slopes (see Table 7A). It began to interact with Typhoon Ellie on August 11 and moved southwestward toward east China. The westerly flow on the southern fringe produces heavy rainfall over the central and southern Taiwan, particularly over the southern part (see Table 7B). It is noted that over southern Taiwan the rainfall due to the convergent westerly flow resulted from the interaction between Typhoon Ellie and (later downgraded) Typhoon Doug is much more intense than the rainfall during the passage of Typhoon Doug (see Table 7A).

4. CONCLUSIONS

The above results show:

(a) Orographic lifting and the confluence of moisture-rich flow are the two most important factors in producing the heavy rainfalls over Taiwan.

(b) Most heavy rainfalls over northeastern Taiwan are associated with confluent easterly flow between a mid-latitude high over China and a tropical low near the Philippines or the South China Sea. The tropical low pressure system in this region ensures the availability of moisture-rich air over Taiwan.

(c) Most heavy rainfalls over southwest Taiwan are produced by the lifting of the confluent moisture-rich southerly, southwesterly and westerly flow brought in by a low pressure system in the East China Sea or the coastal region.

The synoptic patterns identified in this work may be helpful in operational forecast. It is noted that the number of heavy rainfall events has a pronounced year-to-year fluctuation and that tropical cyclones constitute an important component of most of the SHR-producing patterns. The inter-annual variation of the heavy rainfall may be related to that of tropical cyclones. This is a topic that warrants further study.

ACKNOWLEDGMENTS. This work is a part of the research effort by the author as consultant to the Central Weather Bureau. The author wishes to thank Mr. S.-L. Shieh, Director General of the Central Weather Bureau, for continual encouragement, Dr. L. Peng for helping in improving the manuscript, and Mr. S.-T. Wang for valuable suggestions. Thanks are due to Mr. H.-L. Chung for drawing the figures and Miss P.-H. Chung for typing the tables.

REFERENCES

Chen, G. T.-J., and C.-C.Yu, 1988: Study of low-level jet and extremely heavy rainfall over northern Taiwan in the Mei-yu season. *Mon. Wea. Rev.*, **116**, 884-871.

Table 1. Daily Rainfall (mm) over Northeastern Taiwan (Type FA).

a). Nov. 14-16, 1996.

Date	KEELUNG	ILAN	SUAO
14	33.5	38.5	113.5
15	**169.5**	69.5	**211.5**
16	70.0	92.5	**156.5**

b). Nov. 17-28, 1998.

Date	KEELUNG	ILAN	SUAO
17	38	9	24
18	48	39	98
19	72	64	129
20	57	68	**157**
21	71	95	**162**
22	28	23	**130**
23	31	37	3
24	49	39	**191**
25	33	46	58
26	12	47	105
27	121	**266**	**301**
28	5	0.6	8

c). Oct. 9-12, 1996.

Date	KEELUNG	ILAN	SUAO
9	17.5	14.7	**237.0**
10	75.0	71.5	**299.2**
11	41.5	41.0	18.4
12	41.5	60.5	**151.1**

d). Nov. 4-6, 1995.

Date	KEELUNG	ILAN	SUAO	HUALIEN
4	12.0	33	53.5	5.0
5	35	**244.5**	**259.3**	**173.5**
6	7	70.0	94.3	0

Table 2：Daily Rainfall (mm) over Northeastern Taiwan, Nov. 13-17, 1992 (Type FB).

Date	KEELUNG	SUAO	ILAN	HUALIEN
13	33.7	59.8	32.2	1.0
14	40.0	129.0	46.0	16.5
15	21.5	67.5	26.6	38.0
16	53.5	**538.5**	54.0	118.5
17	41.5	**136.0**	66.3	44.0

Table 3：Daily Rainfall (mm) over Northeastern Taiwan (Type FC).

a). Dec. 17-19, 1994.

Date	KEELUNG	ILAN	SUAO	HUALIEN
17	7.5	10.2	14.1	21.5
18	28.0	33.5	**127.6**	28.0
19	11.0	80.0	**291.4**	75.5

b). Jan. 26-31, 1991.

Date	KEELUNG	ILAN	SUAO	HUALIEN
26	37.1	32.9	116.8	0.5
27	21.9	27.4	62.4	0.9
28	55.7	98.1	**152.5**	46.1
29	56.2	31.1	**198.1**	2.6
30	42.7	30.4	**242.2**	0.4
31	20.5	6.9	84.4	0.1

Table 4 : Daily Rainfall (mm) over Central and Southwestern Taiwan, Jul. 4-8, 1992 (Type SA).

Date	TAI-CHUNG	WUCHI	SUN MOON LAKE	YUSHAN	ALISHN	CHIAYI	TAINAN	KAO-HSIUNG	PING-TUNG (S)	PING TUNG (N)
4	24	10	16	56	95	27	137	90	100	103
5	11	7	27	97	115	57	**215**	**221**	**350**	**330**
6	38	0.5	37	**183**	**181**	84.5	10	1	4	11
7	**160**	52	60	115	**160**	86	**139**	14	29	29
8	16	8	21	31	41	**188**	127	23	28	32

Table 5 : Daily Rainfall (mm) over Central and Southwestern Taiwan, Jun. 30-Jul. 1, 1997 (Type SB).

Date	TAI-CHUNG	YUSHAN	ALISHAN	CHIAYI	TAINAN	KAO-HSIUNG	PING-TUNG(S)	PING-TUNG(N)	HENG-CHUN
6/30	1.1	2.5	32.0	1.5	8.5	103.2	69	82	86.5
7/1	59.2	62.5	**171.5**	**216.5**	**293.5**	115.7	**167**	**173**	108.0

Table 6 : Daily Rainfall (mm) over Central and Southwestern Taiwan, Aug. 6-9, 1997 (Type SC).

Date	HSIN-CHU	TAI-CHUNG	WUCHI	SUN MOON LAKE	YU-SHAN	ALI-SHAN	CHIAYI	TAI-NAN	KAO-HSIUNG	PING-TUNG (S)	PING-TUNG (N)
6	0	0.2	0	5	11	8	58	3	84	45	44
7	0	**151**	24	11	12	39	**216**	35	**131**	**494**	**533**
8	0	9	12	37	36	70	37	**132**	93	**207**	**191**
9	0	46	5	17	28	59	114	77	10	68	88

Table 7 : Daily Rainfall (mm) over Northern, Central and Southern Taiwan, August 8-15, 1994.

a). During the Passage of Typhoon DOUG.

Date	TAIPEI	HSINCHU	TAI-CHUNG	ALI-SHAN	CHIAYI	TAINAN	KAO-HSIUNG	PING-TUNG(N)	PING-TUNG(S)
8	109	190	**236**	**545**	23	25	31	86	72
9	4	0	15	**339**	18	42	26	88	64

b). During the Fujiwara Interaction between Typhoon DOUG and Typhoon ELLIE (Type SD).

Date	TAIPEI	HSIN-CHU	TAI-CHUNG	ALISHAN	CHIAYI	TAINAN	KAO-HSIUNG	PING-TUNG(N)	PING-TUNG(S)
10	0	0.8	105	152	104	105	10	12	17
11	0	0	31	56	**224**	66	61	**166**	**140**
12	0	0	14	34	96	**291**	**361**	**173**	**166**
13	0	0	-	14	-	118	**283**	31	27
14	4	14	-	1	30	109	**156**	93	62
15	86	0	-	52	43	96	34	118	95

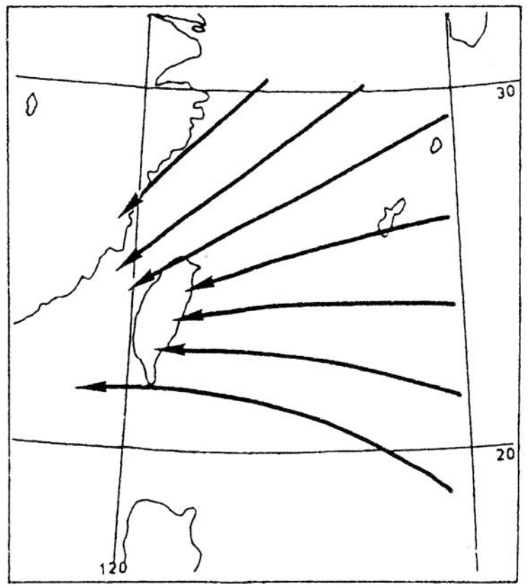

Fig. 1a. Schematic sketch of easterly confluent flow.

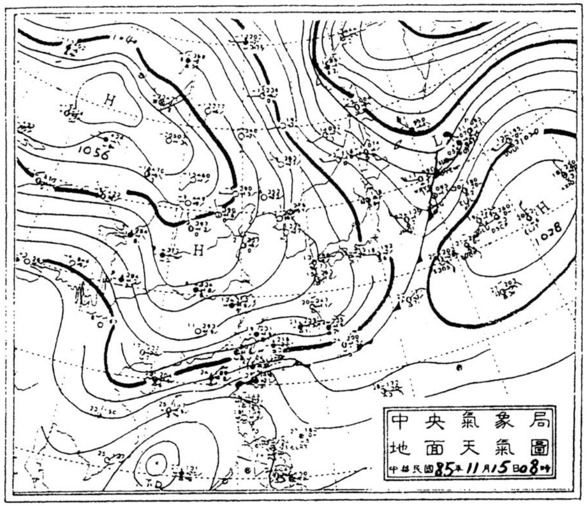

Fig. 2. Example of Type FA, surface map 00Z Nov. 15, 1996.

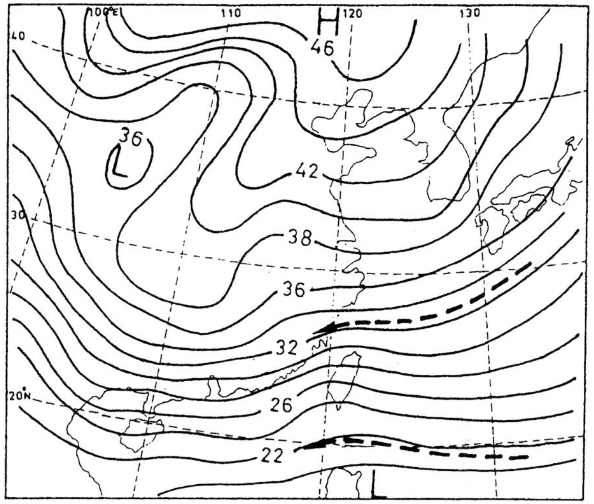

Fig. 4a. Example of Type FC, surface map 00Z Dec. 19, 1994.

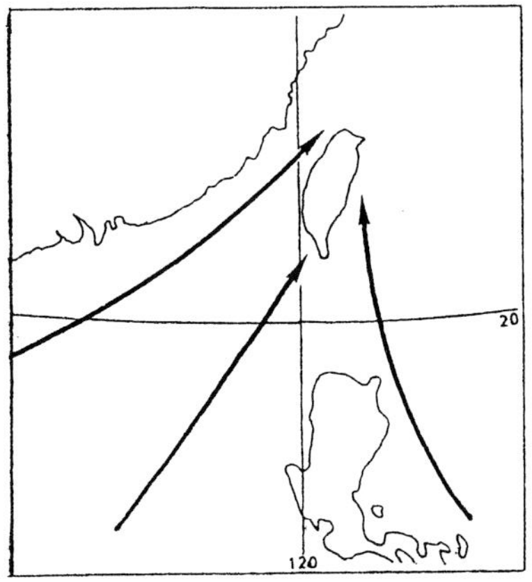

Fig. 1b. Schematic sketch of southerly confluent flow.

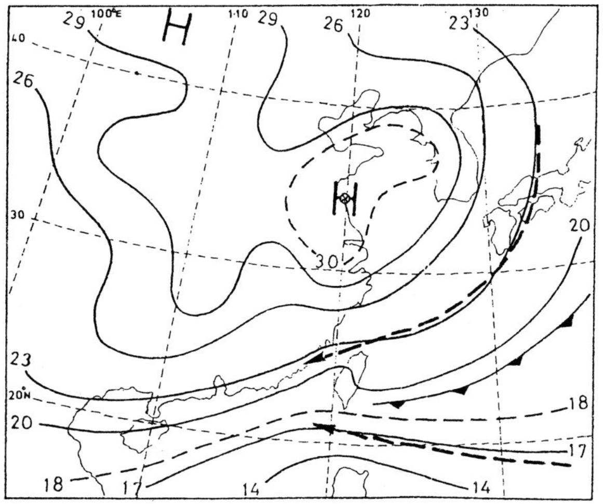

Fig. 3. Example of Type FB, surface map 00Z Nov. 16, 1992.

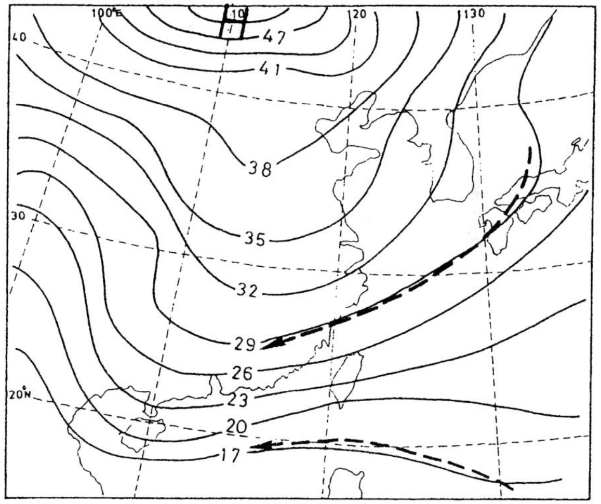

Fig. 4b. Example of Type W, surface map 00Z Jan. 30, 1991.

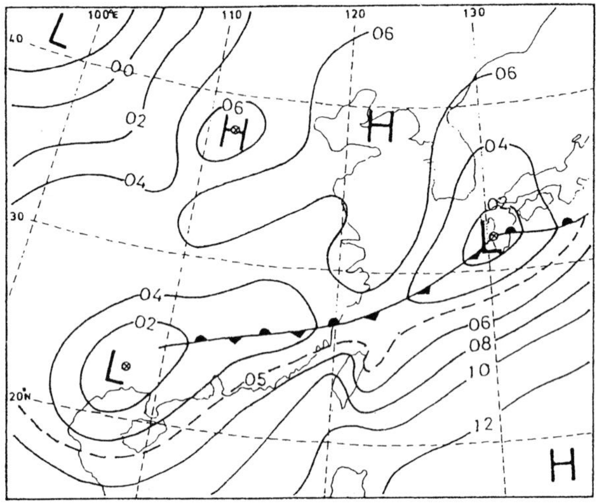

Fig. 5. Example of Type SA, surface map
00Z Jul. 5, 1992.

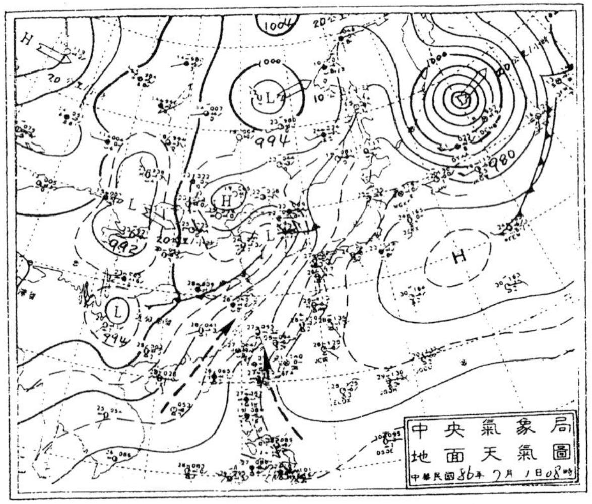

Fig. 6. Example of Type SB, surface map
00Z Jul. 1, 1997.

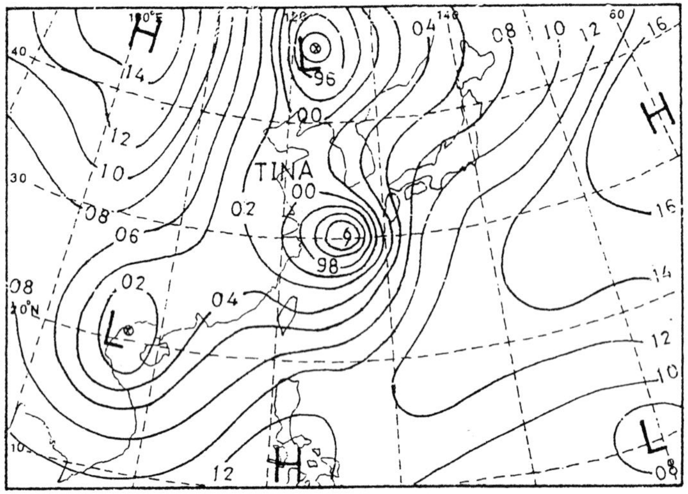

Fig. 7. Example of Type SC, surface map
00Z Aug. 8, 1997.

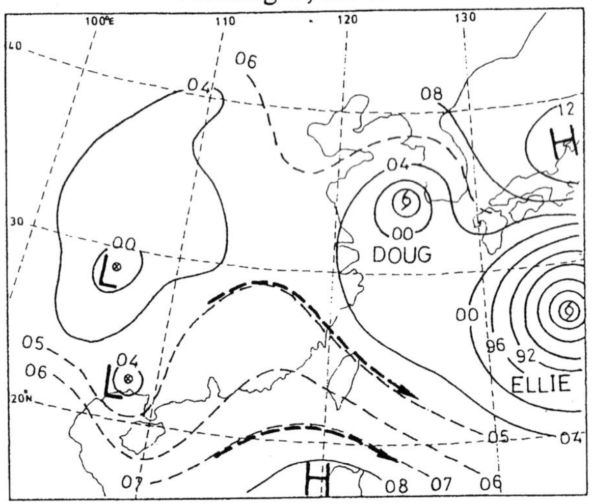

Fig. 8. Example of Type SD, surface map
00Z Aug. 12, 1994.

SHORT-TERM RAINFALL FORECASTS USING THE REGIONAL SPECTRAL MODEL (RSM): PERFORMANCE AND VERIFICATION

W. K. Wong, H. Lam, C. C. Lam and C. K. Chow

Hong Kong Observatory, 134A Nathan Road, Kowloon, Hong Kong

ABSTRACT

Heavy rain during spring and summer in Hong Kong often causes considerable disruptions in social and economical activities. The Hong Kong Observatory (HKO) is running a high resolution Regional Spectral Model (RSM) to improve short-term rainfall forecasts. For operational use at the HKO, if rainfall forecasts over model grid points covering Hong Kong territory exceed certain threshold values, model rainfall alerts will be issued to the forecasters.

The RSM system, which is adapted from the Japan Meteorological Agency (JMA), implements physical initialization to incorporate hourly rainfall amount based on (1): rainfall analysis using radar reflectivity against raingauge data and (2): cloud cover and cloud top temperature from Geostationary Meteorological Satellite. Numerical simulations for past heavy rainstorm cases using 20-km horizontal resolution reveal that the model can give a significant improvement over the 60-km resolution in prediction of rainfall in the initial hours of forecasts.

In this paper, we will discuss the model performance in rainfall forecasts for April – June 1999. Objective verification of rainfall forecasts using scores like Probability of Detection (POD) and False Alarm Ratio (FAR) will be presented. Model performance will also be illustrated using rainstorm cases. Rainfall amount, spatial distribution and its temporal trend during forecast hours will be compared with actual observations.

1. INTRODUCTION

The Hong Kong Observatory (HKO) is operating a high-resolution numerical weather prediction (NWP) model to facilitate rainstorm forecasts. During spring and summer, persistent and heavy rain associated with the low-pressure surface trough, monsoon or tropical cyclones often lead to social and economical disruptions. If heavy rain over a certain amount has been fallen or is expected to fall generally over Hong Kong, the HKO will issue rainstorm signal to alert the public about the occurrence of heavy rain. The rainstorm signal system is composed of three levels of warning according to the amount of rainfall: the AMBER, RED and BLACK with their respective threshold of 30, 50 and 70 millimetres of rainfall in an hour. If a rainstorm system develops rapidly, typically in 0-3 hours in the vicinity of Hong Kong, forecasters can base on the synoptic scenario depicted from rapidly update automatic weather station observations and sequences of radar or satellite imageries to issue the warning to the public. In longer term (more than 3 hours), the NWP products will become important for forecasters' reference to predict the occurrence of heavy rain. If a considerable amount of surface rainfall is predicted in model output, forecasters will be alerted to monitor closely the local or synoptic development during that period.

The direct model output data for predicting evolution of meteorological elements and surface rainfall are usually displayed as sets of graphics ready for forecasters. Moreover, to speed up the model interpretation process, an automatic weather forecast scheme is also devised to produce a text-based forecast for the Hong Kong area summarizing all key features and spatial distribution of model data prediction. For instance, if certain elements in model predictions exceed their warning thresholds, thunderstorm warning or rainstorm warnings can then be issued. In this connection, the automatic forecast scheme can provide a baseline for forecasters to add on their experience to produce a better forecast.

The purpose of this paper is to discuss the performance of Regional Spectral Model (RSM) in rainfall forecasts for April – June 1999. The model configurations will be outlined in Section 2 and the automatic forecast scheme for model rainstorm warning signal will also be briefly described. The score of rainfall forecast in April to June using RSM of 60-km horizontal resolution will be given in Section 3. Some rainstorm cases will be highlighted to illustrate the improvement of rainfall prediction due to increase in model spatial resolution. A summary will be given in Section 4.

It is noted that in mentioning the time for a rainfall event, it is often referred to the local time (HKT), which is 8 hours ahead of the universal time (UTC).

2. MODEL DESCRIPTION AND AUTOMATIC WEATHER FORECAST SCHEME

The HKO began to run a 60-km version of RSM in 1998 to simulate synoptic to mesoscale phenomena and to provide guidance for heavy rain forecast. The development of RSM in the HKO was based on the Regional Spectral Model of the Japan Meteorological Agency (JMA). Details of the model are described in NPD/JMA (1997). It provides 3-hourly forecasts up to a total of 48-hours twice a day. The prediction output from the previous run provides the initial condition for the current forecast. The boundary conditions are obtained from the Global Spectral Model (GSM) of JMA. Though the model simulated well on synoptic feature for evolution of atmospheric and thermodynamic fields in some rainstorm cases, the mesoscale features usually could not be clearly depicted and the rainfall amount was less than actual, especially when the heavy rain event happened in the first few forecast hours. To improve model performance, a higher resolution version, with 20-km spatial grid size, was then developed. This higher resolution version uses radar derived rainfall data for physical initialization to better initialize the moisture field so as to alleviate the model spin-up problem for forecasting rainfall. It was illustrated in a simulation of rainstorm case (Lam and Wong, 1999) that this higher resolution model can improve the rainfall forecast within the first 12 hours.

The 20-km version has been on trial run since April 1999. Due to limitation of computer resource, the model is configured to provide 12 hour forecast at hourly intervals. It is nested inside the 60-km RSM with its domain enlarged to get the boundary conditions. Fig. 1 shows the domain for the two models. The 20-km RSM runs eight times a day while the 60-km RSM runs four times. The Nowcasting Division of the HKO routinely prepares rainfall analysis data and the data are used in hourly "pre-runs" during the period prior to initial hours in both models. In particular, the rainfall data are inserted three times in 20-km RSM while only once for 60-km model (Fig. 2).

The direct model output will be processed through an automatic scheme to generate a text-based forecast for Hong Kong. Forecast elements over grids in Hong Kong region are used (Fig. 3). The scheme produces forecast information on wind, state of sky, precipitation type, visibility, temperature, relative humidity and warnings for thunderstorms and rainstorms. For example, hourly rainfall predicted by 20-km RSM over 25 grids around Hong Kong are used to calculate the rainstorm signal according to the following criteria:

Table 1. Criteria for Rainstorm Warning Forecast.

Warning	**AMBER**	**RED**	**BLACK**
Hourly rainfall (RF)	10 mm ≤ RF < 16 mm	16 mm ≤ RF < 22 mm	RF ≥ 22 mm

These criteria are derived from calibration of model output with actual rainfall observations.

3. RAINFALL FORECAST PERFORMANCE AND VERIFICATION

The 20-km RSM has been semi-operational since April 1999. During April and June 1999, there were several rainstorm events that brought considerable amount of rain over Hong Kong. There were also two tropical cyclones affecting Hong Kong during the period. Typhoon Leo (9902) necessitated the hoisting of No. 8 Northeast Gale or Storm Signal on 2 May and that was the earliest No. 8 signal in a year since 1946. Typhoon Maggie (9903) hit Hong Kong directly on 7 June and the Increasing Gale or Storm Signal No. 9 was hoisted during the early hours on that day. Performance of rainfall forecast during the tropical cyclone period is highly affected by the accuracy of forecast track of the tropical cyclone. Asymmetric typhoon bogus is used routinely in the model for better track forecast. The forecast track of Typhoon Leo using 60-km RSM will be discussed to illustrate certain capability of the model to forecast the track for tropical cyclones.

In verifying rainfall forecast by 60-km RSM, rainfall amount forecasted at HKO grid point will be compared with actual records at HKO. The actual total rainfall on a day is taken from the amount of 24-hour rainfall recorded since mid-night. The DAY 1 (D+1) forecast rainfall is the total amount at HKO grid during 2300 HKT (1500 UTC) on the same analysis date to the same hour on the following day. According to rainfall verification scheme in HKO, the rain events are classified in five categories as tabulated below:

Table 2. Classification of daily rainfall amount in verifying 60-km RSM rainfall forecast.

Type	No rain	Light	Moderate	Heavy	Very heavy
Daily rainfall amount (mm)	0	<10	10-25	25-50	50-100

If the forecast score is calculated from the number of correct forecasts within one adjacent category over the total number of samples, the rainfall forecast scores based on 0000 UTC and 1200 UTC analyses in these three months is summarized in the following Table 3. Some of the rainstorm cases that are not predicted correctly will be discussed in below sections.

Table 3. DAY 1 Rainfall Forecast Score for 60-km RSM.

Score / %	April	May	June
0000 UTC	83.3	93.5	90.0
1200 UTC	86.2	93.5	90.0

a. Rainfall Forecast Performance in April

The first AMBER rainstorm event occurred on 12 April due to the influence of passage of a low-pressure trough. The AMBER rainstorm warning was hoisted from 0115 HKT to 0800 HKT. Rain was the heaviest over Hong Kong Island and eastern part of Lantau Island. However, both 0000 and 1200 UTC forecasts of the 60-km RSM on 11 April predicted the total rainfall in the following day was less than 10 mm. In fact, the rainstorm system started developing significantly after mid-night. As the physical initialization process has to ingest the radar derived rainfall data for three hours prior to analysis time, the forecast by the 20-km RSM with base time at 1500 UTC 11 April still did not predict much rainfall during this period. However, the rainfall predicted by 20-km RSM run at 1800 UTC was much higher than previous run after the radar derived rainfall data were ingested (Fig. 4).

During the last few days on April, the local weather was affected by the Typhoon Leo (9902). The rainfall predicted by the model during the period did not agreed well with the actual rainfall recorded. It is found that the forecast tracks were biased westward to the best tracks, resulting in error of rainfall forecast. This bias error was gradually corrected when Leo was closer to Hong Kong. In particular, based on the forecast with initial time at 0000 UTC on 1 May (Fig. 5), the error in 24 hour and 36 hour forecast positions from the best-track positions were 70 and 68 km respectively. The DAY 1 forecast rainfall was about 4 mm at HKO grid while the maximum value among the 9 model grid points for warnings was about 11 mm. In comparison, the actual HKO rainfall recorded on 2 May was 9.6 mm. The northwest turning of track for Typhoon Leo on 1 May was examined through various set-up of model simulation in Lam et.al. (1999). It was concluded that the success of the RSM in predicting the northwest turning was attributed to the use of asymmetric bogussing profile in the analysis and the fact that the forecast weakening of vortex strength of Typhoon Leo.

b. Rainfall Forecast Performance in May

An AMBER rainstorm warning signal was hoisted during 0440 – 1345 HKT on 19 May as a result of the passage of a trough of low pressure over south China coast on 18 May. The largest amount of rainfall recorded at HKO was observed during 0400 – 0800 HKT, with a maximum hourly rainfall of over 25 mm ending at 0500 HKT recorded at HKO. Based on 0000 and 1200 UTC 60-km RSM runs on 18 May, the 3-hourly rainfall forecast ending at 0500 and 0800 HKT did not exceed 5 mm (Fig. 6).

The rainfall amount predicted by successive runs using 20-km RSM was closer to actual rainfall records. It is evident from the radar imageries (Fig. 7) that the rainstorm was due to an intensifying band of echoes extended from western coast of Guangdong. With physical initialization using these data, the amount of rainfall forecast during early morning of 19 May were successively adjusted and exceeded the model AMBER rainstorm warning threshold (1-hr RF > 10 mm). In forecast using analysis at 1500 UTC, the maximum rainfall among all 25 warning grid points at 0600 HKT was 10.5 mm. In the following run using analysis at 1800 UTC, the threshold was reached as early as 0500 HKT and this was the time when the actual AMBER warning signal was issued to public (Fig. 8). In this rain case, the series of rain forecasts and alert generated by the 20-km model provided forewarning of the rain in time to the forecasters.

c. Rainfall Forecast Performance in June

The direct hit of Typhoon Maggie (9903) resulted in over 60 mm of daily rainfall at HKO on 7 June. As the Typhoon Maggie was initially moving over the western North Pacific where it was close to the boundary region of RSM, the error of track forecast was quite substantial. When it took a northwest course and entered the South China Sea, the forecast tracks by 60-km RSM were closer to the best track. Moreover, there was still some southward bias found in model runs on 6 June and this led to only moderate level of DAY 1 forecast rainfall that were less than actual records.

There was another heavy rainfall event on 21 June due to the influence of an unstable southerly air stream after the northward passage of low-pressure surface trough. AMBER rainstorm warning was sued by HKO between 0900 HKT and 1040 HKT. The 0000 UTC model run on 20 June predicted nil DAY 1 forecast rainfall while the 1200 UTC run forecasted a moderate level. The successive cycle runs for 20-km RSM out-performed the rainfall forecast by 60-km RSM in this case. It is shown in Fig. 9 that the rainfall predicted during the early hours on 21 June was gradually corrected by a successive update of 20-km RSM forecasts.

4. DISCUSSION AND SUMMARY

The performance of rainfall forecast during April to June 1999 using 60-km RSM is discussed. The rainfall forecast results for May is the best among the three months in this study. In predicting the *moderate* rainfall for DAY 1, the Probability of Detection (POD) is 0.5 and the False Alarm Ratio (FAR) is also 0.5. To forecast the *heavy* rain event on DAY 1, the POD can only be 0.33 while FAR reaches 0.67.

It is the objective to improve the rainfall forecast performance by increasing the model resolution to 20 km and using the most current rainfall data from radar observations to initialize the model. With 3-hourly update cycle run, the model can analyze the rapid development of rainstorm system to improve the rainfall forecast. It is found that 20-km RSM can simulate well the rain amount and its trends. In forecasting the AMBER rainstorm cases in April to June, the average POD is over 0.45 but the FAR is also very large (>0.9) due to the high sensitivity of model initial moisture content to the radar rainfall distribution. It is found that once the surface rainfall is generated through physical initialization in the model, the rainfall field can persist for several subsequent runs even after the rain event has been over, resulting in the large number of false alarms.

It is an on-going work to tune the model parameters to reduce the false alarms in the forecast of heavy rain events. The current thresholds for issuance of rainstorm and other warnings in the automatic forecast are based on previous rainstorm case studies. Therefore, these limits will be revised upon more rainstorm cases are analyzed.

ACKNOWLEDGEMENTS. The Regional Spectral Model (RSM) was adapted from Japan Meteorological Agency (JMA). We thank JMA for their support in the development of RSM in Hong Kong.

REFERENCES

Lam, C.C., and W. K. Wong, 1999: Numerical Simulation of the Heavy Rainstorm of 9 June 1998: the Impact of Physical Initialization, *Third International Scientific Conference on the Global Energy and Water Cycle. 16-19 June 1999, Beijing, China.*

Lam, C.C., W.K.Wong, and H. Lam, 1999: Forecasting the track of Typhoon Leo (9902) over the South China Sea, *to be submitted.*

Matsumura, T., I. Takano, K. Aonashi, and T. Nitta, 1995: Improvement of spin-up of precipitation calculation with use of observed rainfall in the initialization scheme. *J. Meteorol. Soc. Japan,* **73**, 353-368.

NPD/JMA, 1997: *Outline of the operational numerical weather prediction at Japan Meteorological Agency, March 1997.*

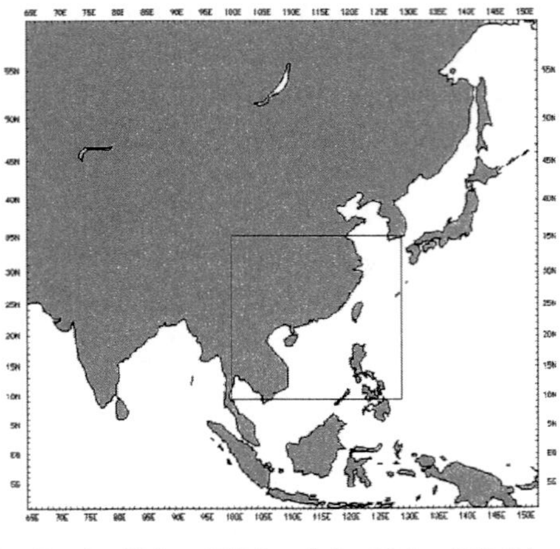

Fig. 1. Domains for the 60-km RSM and the 20-km RSM (small square box).

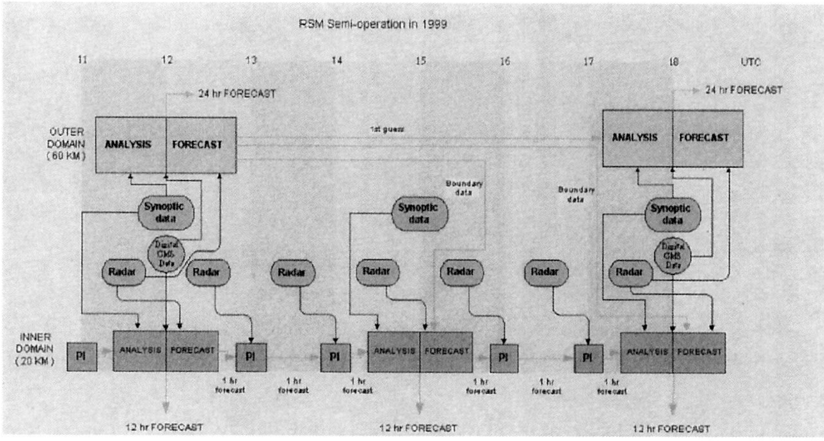

Fig. 2. Data flow for the 60-km and 20-km RSM.

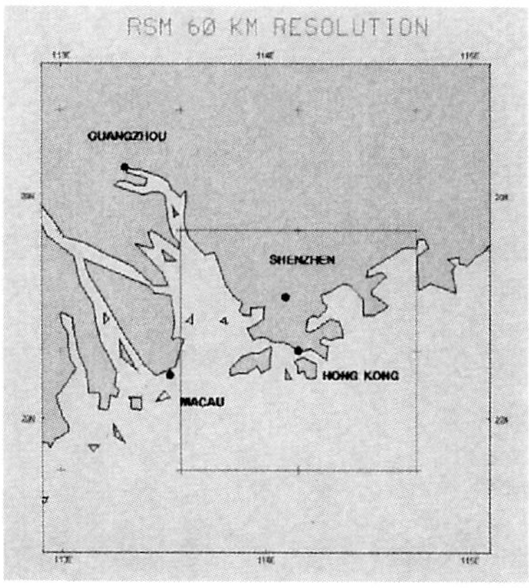

Fig. 3a. Model grid data in the area indicated are used in generating warnings for the 60-km RSM for HK.

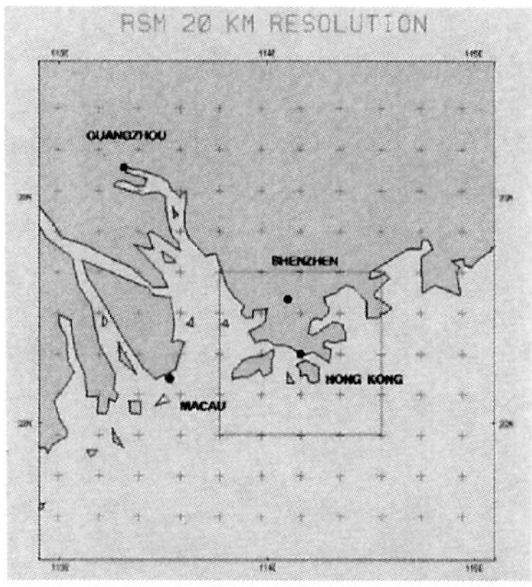

Fig. 3b. Model grid data (totally 25 points) are used to produce warning forecast for the 20-km RSM for HK.

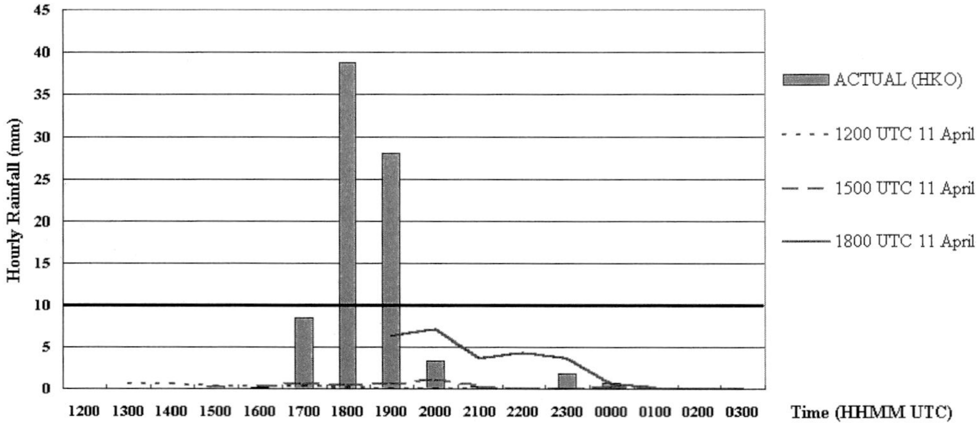

Fig. 4. Comparison of actual HKO hourly rainfall against hourly forecast rainfall predicted by consecutive 20-km RSM run of initial time at 1200, 1500, 1800 UTC 11 April. The horizontal thick solid line shows the threshold to issue AMBER rainstorm warning by 20-km RSM.

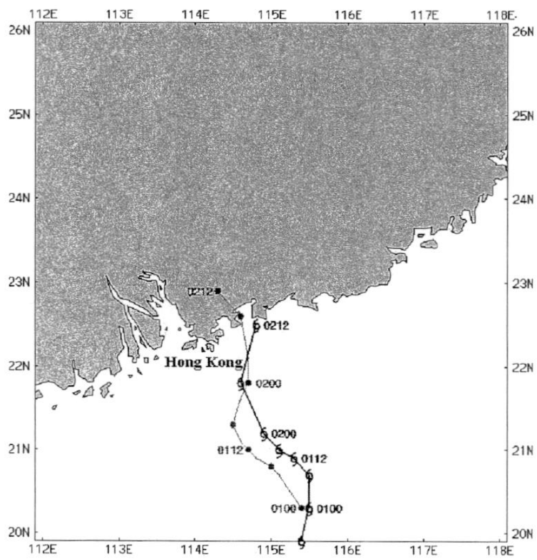

Fig. 5. Forecast tracks against HKO best track (marks are spaced in 6-hour interval) for Typhoon Leo (9902) using 60-km RSM based on analysis at 0000 UTC 1 May, 1999.

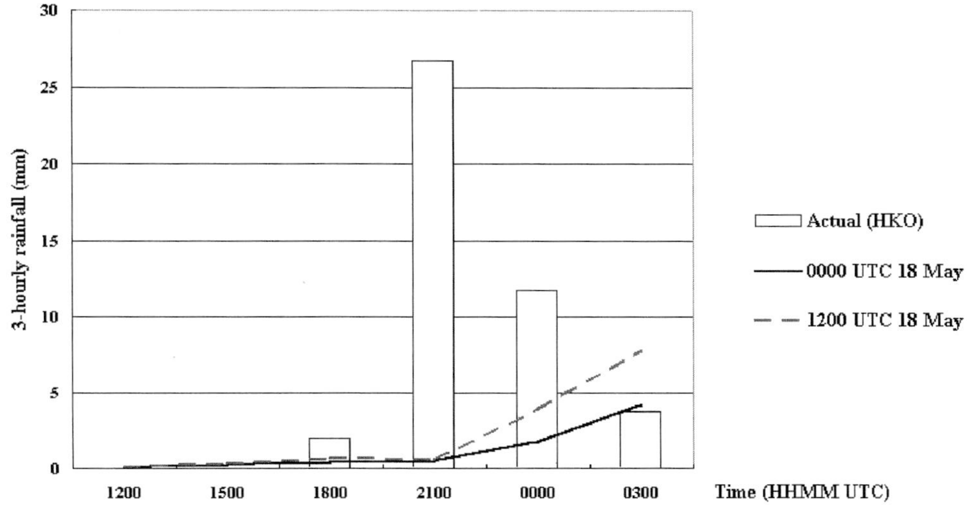

Fig. 6. 3-hourly rainfall records at HKO against the 3-hourly forecast rainfall by 60-km RSM during 1200 UTC 18 May to 0300 UTC 19 May. The forecast rainfall data are based on model runs with initial hours at 0000 and 1200 UTC on 18 May.

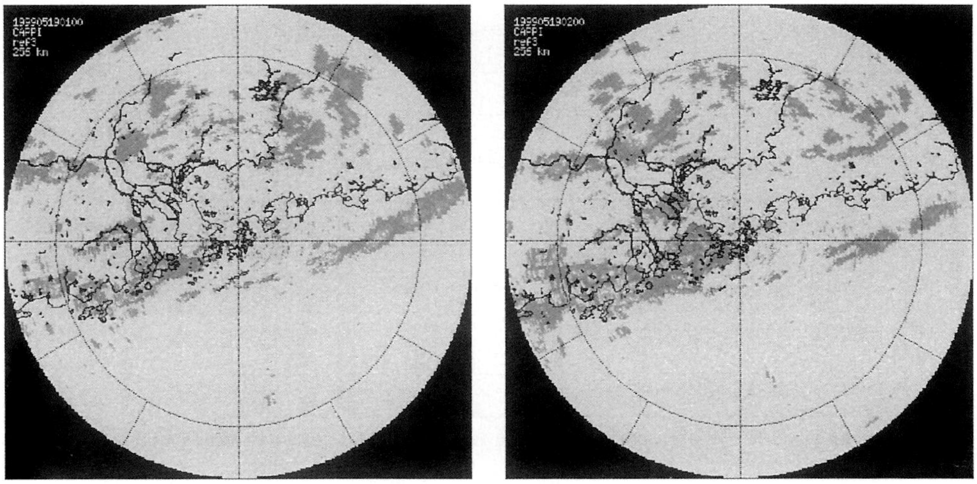

Fig. 7. Hourly radar reflectivity CAPPI (at 3 km) imageries during 0100 – 0200 HKT on 19 May 1999.

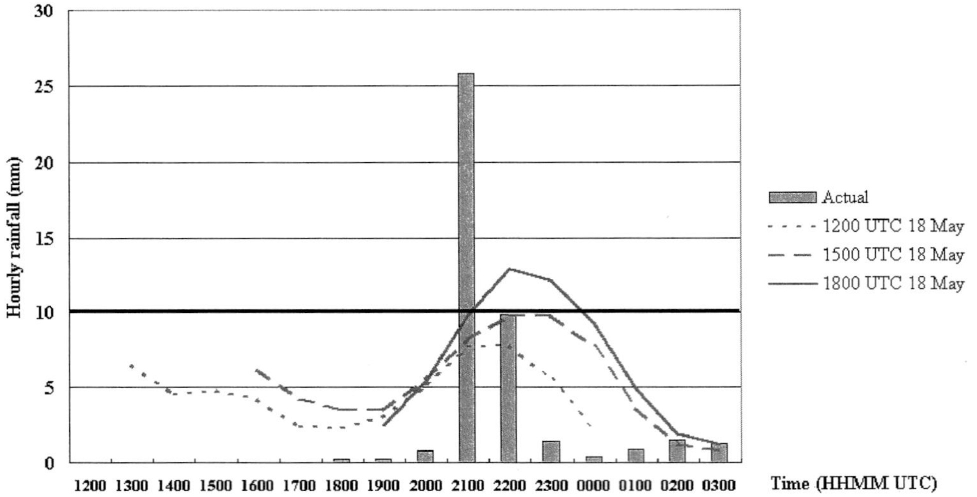

Fig. 8. Hourly rainfall recorded at HKO against the forecast rainfall by 20-km RSM run with initial time at 1200, 1500 and 1800 UTC on 18 May 1999.

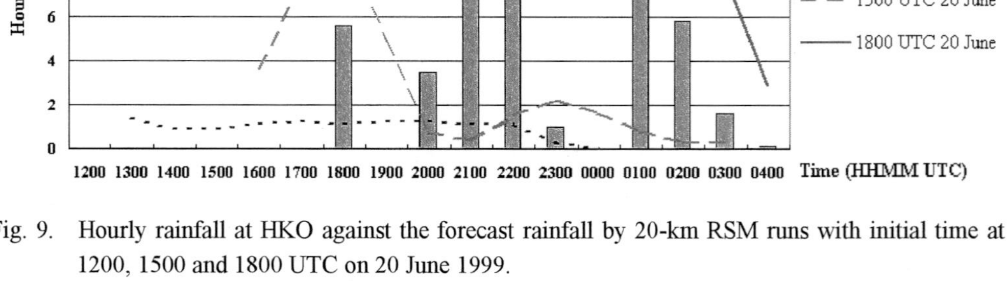

Fig. 9. Hourly rainfall at HKO against the forecast rainfall by 20-km RSM runs with initial time at 1200, 1500 and 1800 UTC on 20 June 1999.

STUDY OF HEAVY PRECIPITATION IN TAIWAN DURING THE MEI-YU SEASON

Koung-Ying Liu Pei-Chen Chang
Department of Atmospheric Science
Chinese Culture University, Taipei

ABSTRACT

In this paper, the EC/TOGA Advanced 1.125° x 1.125° analysis data for May and June of 1992, 1993, 1995 and 1996 are used to investigate the heavy rainfall during the Mei-Yu season in Taiwan. With a criteria of 100 mm per day for at least two surface stations and three automatic rain gauges of the Central Weather Bureau, twelve heavy-rain cases are selected. For the first part of this study, a composite analysis based on nine cases is carried out (three cases in 1994 are not included). The composite analysis shows that a weak baroclinic zone associated with the Mei-Yu front forms a northern boundary to the southwesterly flow and the embedded mesoscale convective systems. As the result of the interaction between the tropical and extratropical weather systems, the convectively ageostrophic winds not only increase the Mei-Yu frontal intensity but also strengthen the low-level jet (LLJ). The LLJ plays an important role in the development of Mei-Yu frontal systems through the water vapor supply and vertical motion enhancement. A vertical cross-section analysis reveals the intensification of the Mei-Yu front and its transition from an "anafront" to a "katafront". The heavy rainfall event is closely associated with this transformation.

In the second part of this study, nine cases are simulated with the PSU/NCAR Mesoscale Model MM5 and the simulation results show that the composite warm moist flow mainly comes from the Bay of Bengal. But the moisture flux comes from the sea west of Philippines as the return flow of the subtropical Pacific High. Another branch of moisture flux is the cross-equatorial flow from the west of Australia. The numerical sensitivity experiments show that as the latent heating term is turned off or the relative humidity field over the Bay of Bengal is reduced to forty percent, the LLJ and the corresponding heavy rainfall does not exist.

Key words: Mei-Yu front, heavy precipitation.

1. INTRODUCTION

The development of heavy rainfall events during the Mei-Yu season is one of the main focuses for the study of local weather hazards in Taiwan. Chi (1969) and Wang (1970) performed climatological investigations of the rainfall phenomena during Taiwan's Mei-Yu season. Later in 1977, the mean structure and the synoptic characteristics of the Mei-Yu frontal systems in Taiwan were studied (Chen and Tsay, 1977; Chen,1983; 1979; Liu, 1981). Then, the "528" flush flood event in the northern Taiwan occurred on 28 May 1981 (Liao et al, 1982; Liu, 1982) and the TAMEX experiment arose from it (Kuo and Chen, 1990). In the following 10 years of post-TAMEX period, many articles related to heavy-rainfall research are published and these papers contributed a lot to the remarkable improvement of the forecasting skill for heavy rainfalls.

Generally speaking, the Mei-Yu phenomenon in Taiwan is not as stable as that along with the Yangtze River. The complexity is due to the interaction between the different scales of atmospheric motion and the combined influence of the tropical and extratropical airflow and/or weather systems (Chen, 1991; Liu and Shyu, 1992), and also the orographic effect (Hong, 1992; Jou, 1994). Many problems still remain unsolved and, therefore, further investigations (even another field experiment) are needed.

In this study, twelve cases in the years of 1992, 1993, 1995 and 1996 are studied through conventional and modeling composite analysis. All the selected cases fit the rainfall amount criteria of 100 mm per day for at least two surface stations and three automatic rain gauges of the Central Weather Bureau. Using the EC/TOGA advanced $1.125^\circ \times 1.125^\circ$ analysis data provided by the Subtropical Data Bank (sponsored by the National Science Council in Taiwan), a composite study based on nine cases is carried out. Furthermore, in order to investigate the importance of the water vapor flux and latent heat release, one case is simulated using the PSU/NCAR Mesoscale Model MM5. The simulation shows that the latent heat release associated with the maximum water vapor flux plays an important and unique role in the production of LLJ and the high rainfall for the Mei-Yu frontal system.

2. RESULTS OF THE COMPOSITE ANALYSIS

Using the EC/TOGA analysis dataset mentioned above, we analyzed these synoptic features: (1) the position of Mei-Yu front, i.e., the position of 340°K contour line of equivalent potential temperature at 850 hPa (Saito, 1966), (2) the 850- and 700-hPa troughs, (3) the ridge of the Pacific High at 500 hPa, (4) LLJ and the maximum axis of the water vapor flux, and (5) the vertical structure of the Mei-Yu front. Fig. 1 is the composite chart of the major synoptic-scale meteorological features. In this composite chart, each symbol represents the mean position of the meteorological feature among the nine cases. Principal findings from this composite chart are listed as follows.

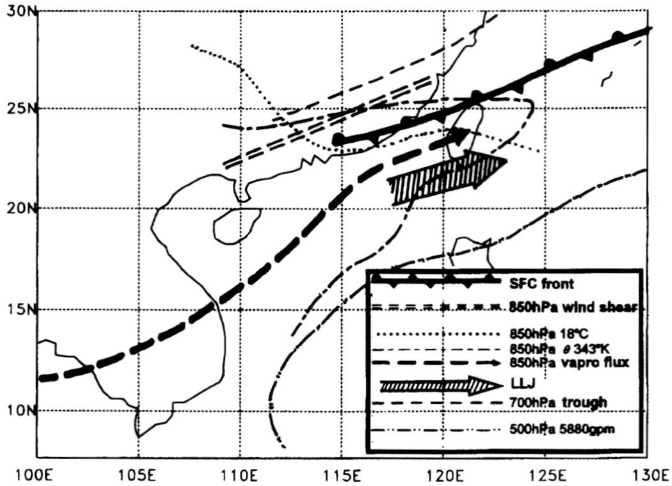

Fig. 1. Major synoptic features on the composite chart of 9 heavy precipitation events.

(1) The Mei-Yu front is located at the northern tip of Taiwan. It extends in the northeast-southwest orientation and forms a boundary of the southwesterly flow and the embedded mesoscale convective systems.

(2) An up-stream tilt of the trough occurs only in the lower troposphere, i.e., from the surface to 700 hPa. This reveals that the Mei-Yu front is a shallow weather system with weak baroclinicity.

(3) On the 850-hPa level, a warm tongue as indicated by the 343°K potential-temperature contour line is over Taiwan and a cold tongue associated with the modified polar continental air mass extends from the East China Sea to the central China. Along with the warm tongue is the axis of the maximum water vapor flux coming from the Bay of Bengal.

(4) There is a strong southwesterly wind, i.e., the low-level jet (LLJ), in front of the Mei-Yu front. The mean axis of the LLJ extends from the South China Sea to the southern part of Taiwan. The warm tongue and the axis of the maximum water vapor flux are on the positive (cyclonic) shear side of the LLJ.

(5) Fig. 2 illustrates the composite 850-hPa level wind field. It shows that the LLJ over Taiwan is really a convergence of two branches of air streams. One branch comes from the Bay of Bengal. The other branch is a combination of the returning flow of the subtropical Pacific High and the cross-equatorial flow from the sea west of Australia. Krishnamurti (1983; personal communication) pointed that the tropical disturbances embedded in the trade wind can be steered by the returning flow (southwesterly wind) of the Pacific High, as they pass over the ridge of that subtropical high pressure. These tropical disturbances move northeastward first and then follow the converging flow into the western coast of Taiwan. There is another source of moisture and relative vorticity (see Fig. 3) which supports the positive advection of water vapor and, hence, deep convection and the resulting heavy rainfalls.

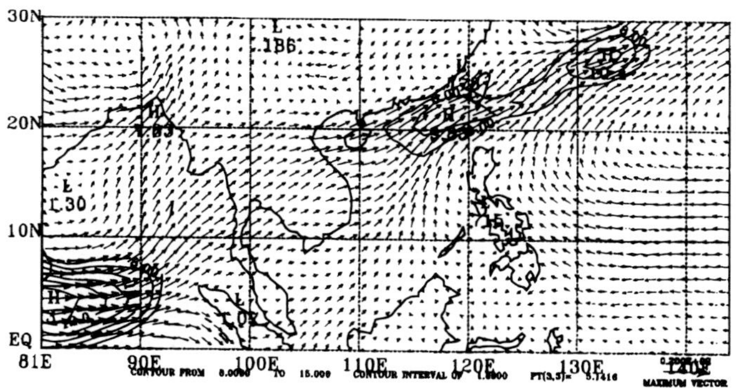

Fig. 2. Same as Fig.1 but for the 850hPa wind field.

(6) As Fig. 3 indicates the water vapor flux (also the maximum axis of equivalent potential temperature) comes along with the LLJ and the returning flow from the Pacific High also converges with the LLJ, a positive feedback mechanism between the convectively ageostrophic wind and latent heat release could stimulate the development of mesoscale convective systems and then producing heavy rainfalls in the vicinity of Taiwan. Model results (see below) confirm this positive response derived from the composite analysis.

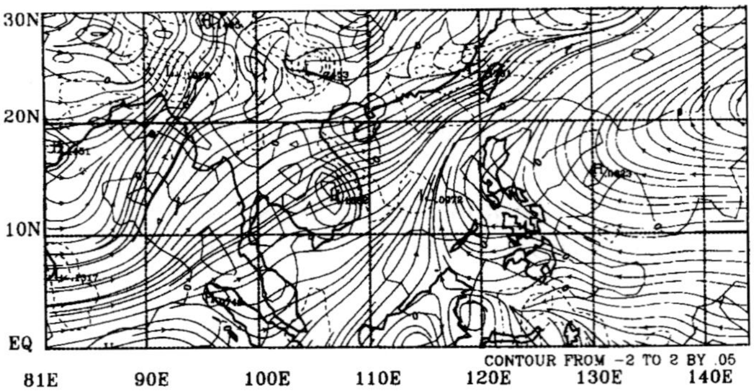

Fig. 3. Same as Fig.1 but for 850hPa divergence of specific humidity (Dq) (solid line with or without a center represents $Dq>0$, dotted line is negative) and vapor streamlines.

(7) The 500-hPa subtropical ridge extends southwestward from the Pacific High to the north of Philippines.

As mentioned above, the Mei-Yu front is a shallow system with weak baroclinicity. However, during the period of heavy rainfalls in Taiwan area, frontogenesis process continues and the horizontal gradient of equivalent potential temperature in the Mei-Yu frontal zone increases rapidly (see Fig. 4). Furthermore, as indicated by the vertical cross-section analysis of the frontal secondary circulation, the Mei-Yu front in this period changes from an "anafront" to a "katafront". The heavy rainfalls occur during this transition (see below).

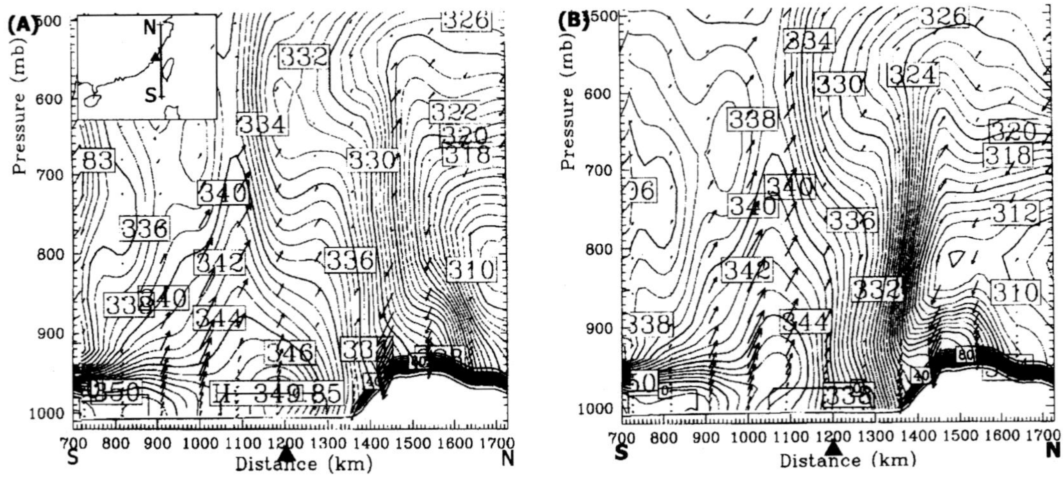

Fig. 4. Cross-section analysis (▲ marked the north tip of Taiwan) of θ_e and convective ageostrophic wind and 1200UTC May 3, 1994 (A), and 3 hours later (B). The inserted figure shows the reference coordinate.

3. NUMERICAL SIMULATION RESULTS

Previous studies indicate that PSU/NCAR Mesoscale Model (MM5) is a reliable model to study the Mei-Yu frontal system and the associated heavy precipitation (Yang et al., 2000). Therefore, the MM5 model (Version 2.11) is used to simulate the structure of Mei-Yu front and the relation between the LLJ and the moisture flux. First, the vertical structure of Mei-Yu front of 3 May 1994 is investigated. The model initial field is 0000 UTC of 1 May 1994. Heavy rainfall terminated 60 hours after initialization. At that time the Mei-Yu front is located over the north of Taiwan (see Fig. 4), as represented by the 340°K contour line of equivalent potential temperature (θ_e) at 850 hPa. In Fig. 4 of a north-south (N-S) cross-section analysis, the isolines with 1°K interval represent θ_e contour lines, and arrows represent the convectively ageostrophic wind. The thick black belt is the mainland China topography. The simulation results in Fig. 4 can be summarized as follows:

(1) As indicated in the synoptic composite, the Mei-Yu front is a shallow system (up to about 700 hPa) with weak baroclinicity. However, during the merging period (60 h to 63 h after initialization) of tropical and extratropical weather systems, heavy rainfall events and frontogenesis process proceed, and the $\nabla\theta_e$ in the frontal zone increases rapidly.

(2) The LLJ plays an important role in the development of heavy rain systems in two folds. First, LLJ takes the warm moist air from the Tropics into the system and builds up the warm tongue over Taiwan. Second, it produces the convectively ageostrophic wind. Along the Mei-Yu front, warm moist air goes up and tilts toward the cold side, and the cold air goes down and deflects to the warm side (see Fig.4). Trough the latent heat release, temperature gradient in the warm side is also increased. Both factors are very important to the frontogenesis process and the production of heavy precipitation.

(3) Comparing Figs. 4a and 4b, it is clear that the Mei-Yu front not only propagates southward during that period, but also transforms from an "anafront" to a weak "katafront" (as indicated by the weak downward motion below 800 hPa at the leading edge of the cold air).

Second, the contributions of the moisture flux and the latent heat release are also analyzed. Fig. 5 illustrates the simulated 850-hPa wind field at 0000 UTC of 27 May 1993. The low-level jet, as indicated by the isotaches, is over Taiwan in a NE-SW orientation. Using the same initial data, we make two sensitivity experiments, one without latent heat release and the other with the specific humidity field over the Bay of Bengal reduced to forty percent. When we turn off the latent heat, there is no LLJ (see upper-left inserted figure in Fig. 5), and the strength of vertical motion and the $\nabla\theta_e$ in the frontal zone is reduced (figure omitted). When the specific humidity over the Bay of Bengal is reduced to 40, the LLJ originally extending from that area to Taiwan is not in existence. On the other hand, the part of LLJ from the south is not influenced. Therefore, the maximum wind occurs to the north of Taiwan (see lower-right inserted figure in Fig. 5). The results indicate that there is a positive feedback mechanism between the latent heat release and the LLJ.

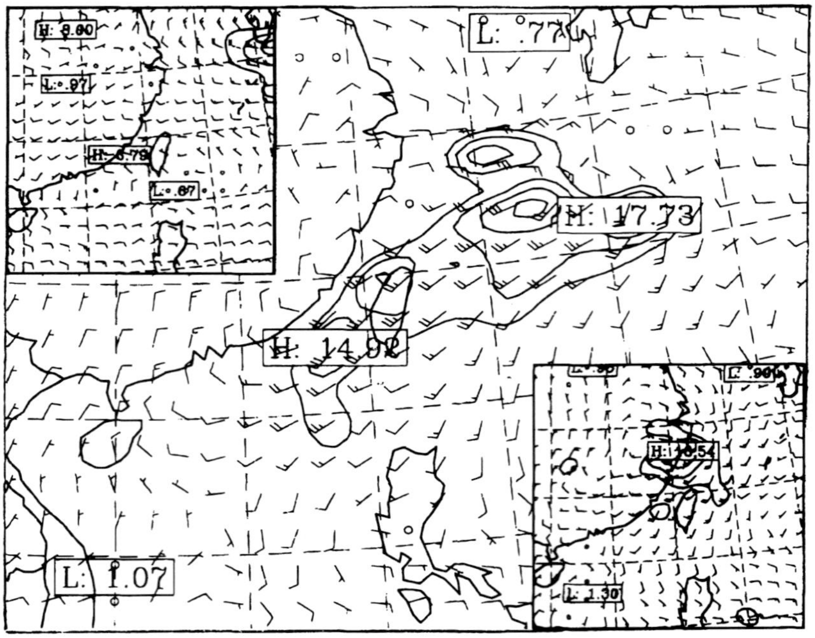

Fig. 5. 850hPa model output wind field (for future information see text).

4. CONCLUDING REMARKS

This study indicates that the heavy rainfall in Taiwan during May and June is a result of the combined influence of the tropical and extratropical meteorological systems. The two systems are the Mei-Yu front and the LLJ (or the lower-tropospheric water vapor flux). As these two systems merge into one, heavy precipitation systems form over Taiwan. The Mei-Yu front is a shallow system with weak baroclinicity. During the period of heavy rainfall, the Mei-Yu front reveals a transformation from an "anafront" to "katafront". There is a positive feedback mechanism between latent heat release and the LLJ. Strong upward motion and heavy rainfalls occur at the leading edge of the Mei-Yu front. Radar observations as analyzed by Liu and Huang (1998) show similar results.

ACKNOWLEDGEMENTS. The authors wish to express their appreciation to those who contributed to this and three previous preceding conferences. Their hard work not only makes the conferences in success but also builds up bridges over seas. Thanks are also to Dr. Yang for helpful comments. Without those people this study cannot be in such a success.

REFERENCES

Chen, G. T.-J., and C. Y.-Tsay, 1977: A detailed analysis of a case of Mei-Yu system in the vicinity of Taiwan. *Tech. Rep. Mei-Yu-001*, Dept. of Atmos. Sci., Natl. Taiwan University.

Chen, G. T.-J., 1983: Observational aspects of the Meiyu phenomenon in substropical China. *J. Meteor. Soc. Japan*, **61**.

Chen, G. T-J., 1991: Mesoscale features observed in the Taiwan Mei-Yu season. *J. Meteor. Soc. Japan*, **70**.

Chi, K. H., 1969: On the Rainfall distribution of Taiwan. *Meteorological Bulletin*. CWB, Taipei (in Chinese with English abstract).

Hong, S. H., 1992: On orographic influence on precipitation. *Proceedings of the Intl. Conference on East Asia and Western Pacific Meteor. And Climate*. Edited by Kyle, W. J., and C. P. Chang. World Scientific.

Jou, B. J. D., 1994: Mountain-originated mesoscale precipitation system in Northern Taiwan: A case study of 21 June 1991. *TAO*,**5**.

Kuo, Y. H., and G. T. J. Chen, 1990: The Taiwan Area Mesoscale Experiment (TAMEX). An overview. *Bull. Amer. Soci.*, **7**.

Liao, S. Y., et al., 1982: A study of mesoscale disturbance in Southern China and the vicinity. *Tech. Rep. NSC71-0202-M008-05*, Natl. Central University (in Chinese with English abstract).

Liu, K. Y., 1981: On the extreme weathers of Taiwan during the Mei-Yu season. *Tech. Rep., NSC-69n-0202-05(02)* Weather Central, CAF (in Chinese with English abstract).

Liu, K. Y., 1982: The relationship of the「528」heavy rain event and mesoscale weather system. *Quarterly Journal of Meteorology*, **91** (in Chinese with English abstract).

Liu, K. Y., and T. Y. Shyu, 1992: Heavy rainfall under the combined influence of the tropical and Mei-Yu frontal system. *Proceedings of the Intl. Conference on East Asia and Western Pacific Meteor. And Climate*. Edited by Kyle, W. J., and C. P. Chang. World Scientific.

Liu, K. Y., and Huang W. H., 1998: Radar climatology and related heavy rainfall during the Mei-Yu seasons in Southwest Part of Taiwan. *Hwa Kang Journal of Sci.*, Vol. **15**, Chinese Culture Univ.(in Chinese with English abstract).

Saito, N., 1966: A preliminary study of summer monsoon of Southern and Eastern Asia. *J. Met. Soc. Japan.*, **44**.

Wang, S. T., 1970: On the Mei-Yu of Taiwan. *Quarterly Journal of Meteorology*, **44**, (in Chinese with English abstract).

Yang, M.-J., F.-C. Chien and M.-D. Cheng, 2000: Precipitation parameterizations in a simulated Mei-Yu front. *Terr. Atmos. Oceanic Sci.*, in press (June issue).

PRELIMINARY ANALYSIS OF GROUND-BASED GPS OBSERVATIONS IN HUANAN TORRENTIAL RAINFALL EXPERIMENT*

XU Baoxiang ZHOU Xiuji LI Yanxing(1)
HE Ping WANG Rong SONG Chunmei
Chinese Academy of Meteorological Science, Beijing 100081

1. INTRODUCTION

It is well known that water vapor plays a pivotal role in energy conversion in weather processes of all temporal and spatial scales. Therefore, water vapor content is a very important physical parameter, especially in short-term forecasting and nowcasting. Being a kind of effective greenhouse gas, water vapor is even more important in climatic research. In fact, atmospheric water vapor itself is an index for atmospheric warming as warmer air has the capacity to retain more water vapor.

Our ability to measure regional or global distribution of water vapor is quite limited. The traditional method relies on global upper-air sounding system. Balloons equipped with sensors are launched twice daily at upper-air sounding station and water vapor distribution with height can be obtained. Unfortunately, the temporal and spatial distributions of such data are inadequate for research purpose. Although satellite-based techniques have made significant progress, difficulties still exist in the lower troposphere.

Using ground-based GPS to measure the satellite signal delay, we can retrieve atmospheric integrated precipitable water vapor (IPWV). Through the GPS network, the quest to attain real-time or near-real-time monitoring of the distribution and variation of atmospheric water vapor looks more promising. This paper attempts to assess GPS's potential in meteorological applications by means of analyzing the data observed during the Huanan Torrential Rainfall Experiment.

2. EXPERIMENT DESIGN AND DATA PROCESSING

From May to June 1998, Huanan Torrential Rainfall Experiment was conducted in Guangdong and Fujian in southern China. Two continuously operating GPS stations were located in Shantou and Yangjiang of Guangdong from 5 May to 26 June, lasting a total of 53 days. Sampling was done every 15 minutes and the lowest elevation angle of the GPS satellites for data collection was 15°. Routine meteorological observations, such as temperature, pressure, moisture and wind, were made in situ at the two stations. In collaboration with the experiment, GPS tracking stations in Haikou and Tianjin also took simultaneous observations.

Data processing of GPS observations includes: (1) computation of the accurate

---

* Supported by the special project of the National Science and Technology Plan (95-special-03) and the National Key Basic Research and Development Project (G 1998040909)

(1) The First Crustal Deformation Monitoring Center, Tianjin 300180

geocentric coordinates of the Shantou and Yangjiang stations: benchmark technique was adopted by selecting the International Global Reference Frame (IGRF96) as the reference frame and choosing nine GPS tracking stations (Shanghai, Haikou, Wuhan, Tianjin and Lhasa in China; TSKB and Usuda in Japan; IRKT in Russia and Kitab in Uzbekistan) as fiducial points; (2) determination of the zenith wet delay of each observation at Shantou and Yangjiang: the effect caused by ionospheric delay was removed using double-frequency receivers at both stations; wet delay had a good correlation with surface pressure and could be obtained with accuracy down to millimeter; and from the relationship between wet delay and IPWV, the IPWV at the corresponding observation time was derived.

3. MAJOR RESULTS

(1) IPWV values deduced from upper-air sounding data are slightly different from those obtained from GPS at the same station. The latter is systematically biased towards lower values. However, the trends of evolution with time and the variation in amplitudes are in good agreement. Fig. 1 shows the comparison of results from the Shantou station from 5 May to 3 June.
(2) During the 53-day experiment, GPS-derived IPWV values vary in time with obvious regularity. IPWV values increase sharply before rain and decrease rapidly as rain weakens and ceases, clearly illustrating the accumulation and release processes of water vapor during the weather events. Fig. 2 is a comprehensive comparison of GPS-derived IPWV values with the rainfall histogram recorded at Yangjiang from 23 May to 3 June. While showing a good correspondence, any quantitative relationships that exist between the two will need to be studied further.
(3) Further analyses also show the correspondence between GPS-derived IPWV data and hourly accumulated rainfall values. Rain events usually happen at high content of water vapor. Furthermore, heavy rain is also associated with wild fluctuations in IPWV values (Fig. 2).
(4) Water vapor computed from sounding data is the result accumulated along the path of the launched sounding balloon. Because the balloon follows the wind, its path must differ significantly from the zenith path for the retrieved GPS signals. This will obviously have an effect on interpretation consistency when the results are compared. In particular, more sophisticated work should be done to further determine and improve the accuracy in retrieval methodologies for water vapor information.
(5) IPWV values retrieved from GPS correlate very well with water vapor flux deduced from upper-air sounding. Fig. 2 shows the computed results of the corresponding rainfall processes. Water vapor flux illustrated here is the integrated variability of water vapor over the station (including all altitude layers and all directions). Water vapor flux on individual altitude layer should be considered further and this requires the vertical distribution of atmospheric water vapor for comparison with the water vapor flux at different altitudes.

4. DISCUSSION

(1) Analyses have shown that ground-based GPS observation can provide reliable information on atmospheric IPWV. With continuous (every 15 minutes) coverage, it is

potentially useful for studying the source and movement of water vapor. If a surface GPS network can be set up, such information is likely to play an important role in short-term weather forecasting and nowcasting.

(2) For better use of surface GPS observation data of high temporal and spatial resolutions, we may consider the possibility of retrieving the vertical distribution of water vapor and combining it with in situ sounding data to compute the convergence and divergence parameters on each layer. This also opens the way for applying ground-based GPS techniques in numerical weather prediction.

Acknowledgements. Under the enthusiastic support of the Meteorological Bureaus of Guangdong and Fujian, especially with the contribution of the Meteorological Bureaus of Shantou and Yiangjiang, the Huanan Torrential Rainfall Field Experiment was completed successfully. Ms. HUANG Xingyuan did a lot of work for this publication. The authors would like to express their cordial thanks.

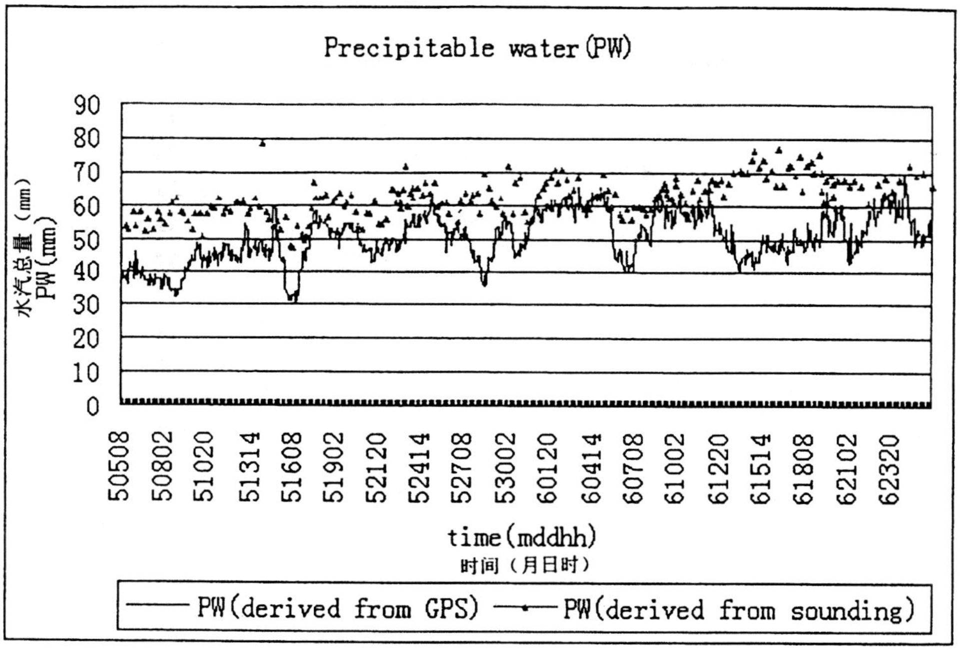

图1 1998年5月5日08时(50508)至6月23日20时(62320)汕头站由GPS和无线电探空观测分别求出的水汽总量之对照

Fig. 1. The comparison of atmospheric Precipitable Water(PW) retrieved from GPS and rawinsonde observations respectively at shantou during 08:00 May 5 to 20:00 June 23, 1998

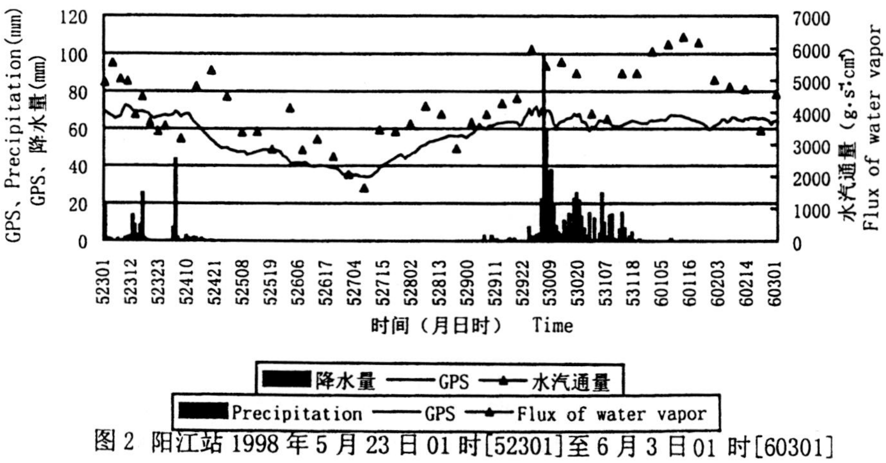

图2 阳江站1998年5月23日01时[52301]至6月3日01时[60301] GPS反演水汽总量与水汽通量及地面降水量的对照（北京时）

Fig. 2. The correspondence of atmospheric precipitable water by GPS, the flux of water vapor and precipitation, yangjiang, 5,23,01 [52301]—6,3,01[60301], 1998.

PRELIMINARY PERFORMANCE EVALUATION OF A RAINSTORM NOWCASTING SYSTEM

Edwin S.T. Lai and P. W. Li
Hong Kong Observatory, 134A Nathan Road, Kowloon, Hong Kong

ABSTRACT

A rainstorm nowcasting system SWIRLS (Short-range Warning of Intense Rainstorms in Localized Systems) is currently under development at the Hong Kong Observatory. Phase I of SWIRLS went through operational trial in 1998 and was officially put into service in April 1999 (Li *et al.* 2000). Movement and intensity change of individual radar echo are analyzed by TREC (Tracking Radar Echoes by Correlation). Extrapolation based on the respective TREC vector provides the forecast echo position and intensity, but development of the latter has the option of being constrained by an idealized profile. Meanwhile, making use of the dense raingauge data over Hong Kong, SWIRLS adjusts the reflectivity-rainfall (Z-R) relationship in real time. Combining the echo forecast and the time-adjusted Z-R relationship, SWIRLS generates short-range quantitative precipitation forecasts (QPF) over Hong Kong every 6 minutes, i.e. time interval of radar scans.

QPFs by SWIRLS during several heavy rain episodes in 1998 and 1999 are objectively evaluated using common performance measures such as NAP (No Alarm Probability), FAR (False Alarm Rate), Critical Success Index (CSI) and Heidke Skill Index (HSI). Results from this study help to: (a) assess the reliability and robustness of SWIRLS QPFs; (b) reveal strength and weaknesses in SWIRLS QPFs for forecasters' guidance and reference in operating the rainstorm warning system; (c) focus researchers' attention to rain characteristics peculiar to Hong Kong, hence pointing the way forward in SWIRLS development work.

1. INTRODUCTION

In Hong Kong, heavy rain in the summer is usually associated with the southwest monsoon, monsoon troughs and tropical cyclones. A three-tiered Rainstorm Warning System (RWS) is in operation based on hourly rainfall rate of 30 mm (amber warning), 50 mm (red warning) and 70 mm (black warning) as recorded by 10 raingauges or more over Hong Kong. While perfect matching of predicted and actual rainfall distribution over a small area like Hong Kong (in the order of 50 x 50 km) remains an ambitious goal, operational forecasting systems such as SWIRLS should at least demonstrate a minimum level of skill and capability for forecasters operating the RWS.

A host of performance measures, namely NAP (No Alarm Probability), FAR (False Alarm Rate), Critical Success Index (CSI) and Heidke Skill Index (HSI) are applied to the critical rainfall criteria operationally used in the RWS. Owing to the rarity of black warning cases during 1998 and 1999, this evaluation exercise will focus on the amber (30 mm per hour) and red (50 mm per hour) criteria. But as we would also like to know how SWIRLS performs in leading up to the amber warning, an additional level at 20 mm per hour, say pre-amber threshold, will also be assessed in this study. Additional assessment is also made for criteria as applied to 5 gauges instead of the operational prerequisite of 10 gauges. The idea is again to see how effective SWIRLS is in the build-up stages of heavy rain events.

2. ECHO MOVEMENT BY TREC

Movement of individual radar echo between two consecutive CAPPI scans at 6-minute interval is derived using correlation method, i.e. TREC – Tracking Radar Echoes by Correlation. Cross-correlation technique was first used to determine the movement of the entire storm cluster (Hilst and Russo, 1960; Kessler and Russo, 1963; Crane, 1979; Bjerkaas and Forsyth, 1980). With the availability of higher resolution radar data, this technique was extended to retrieve the individual motion vectors of embedded storm cells (Rinehart and Garvey, 1978; Rinehart, 1979; Smythe and Zrnic, 1983; Tuttle and Foote, 1990).

The main advantage of using correlation method is that it requires only information from one single radar at two different observation times. Although the derived motion field is essentially two-dimensional in the horizontal and does not include strong vertical motion normally expected in convective cloud systems, the approximation should be good enough given that the effect of horizontal advection should be more pronounced if only short-term motion is considered. Once the motion vectors are computed, the entire echo field can be extrapolated to produce a short-term (say, one-hour) forecast field.

TREC method is applied to two successive CAPPI reflectivity fields at 1-km level at 6-minute interval. The first field (at time T_1) is divided into a number of equally sized "boxes" or two-dimensional arrays of pixels. Computation can be repeated for all possible arrays found at T_2 to determine which array results in the highest correlation, and the centre of this second array will be the end point of motion vector (i.e. TREC vector). In practice, to reduce computation time, a search area centred on the first array is prescribed. The radar reflectivity field has a resolution of 480 x 480 pixels. The arrays are spaced 5 pixels apart, giving a TREC vector field on a 93 x 93 grid. The equivalent spatial resolution is of the order of 1 to 5 km depending on the range of scan.

Before the whole vector field is analyzed objectively, each motion vector has to pass a preliminary consistency check. If a TREC vector deviates more than 25 degrees from the local mean of its neighbourhood (which is calculated from the set of 25 vectors located around it), it will be replaced by this mean vector. In this way, the "noisiness" as a result of sole consideration of maximum correlation in deriving the motion vector can be reduced at an early stage. To produce a continuous gridded vector field, a three-pass objective analysis using modified Cressman weighting function is then applied (Cressman 1959). Results have shown that TREC is able to reproduce realistic wind fields in severe weather systems. For example, the perturbed flow in the westerlies was well depicted in the travelling squall line on 2 May 1998 (Fig. 1). TREC has also demonstrated its usefulness in radar eye fix and timing of gale onset during the close passage of tropical cyclones, e.g. Typhoon York on 16 September 1999 (Fig. 2).

The forecast radar reflectivity pattern is obtained by extrapolation based on the smoothed TREC vectors. Temporal integration makes use of the two-time-level semi-Lagrangian scheme (Staniforth and Cote, 1991). As the vector field is computed from consecutive radar CAPPI images six minutes apart, the integration scheme can output forecasts in 6-minute interval, and each successive forecast pattern is then used for the next step of integration.

3. ECHO INTENSITY CHANGE

The next problem relates to growth or decay in echo size and intensity. TREC winds provide a basis to estimate the extent of growth and decay associated with each echo pixel.

As the TREC vector traces the trajectory of individual echo, the difference in echo intensity along the flow between two successive radar scans can also be deduced. This information can be suitably included in the forecast if so desired.

Keeping things relatively simple to start with, there are basically three approaches in predicting short-term echo intensity changes: (a) assuming no change as from the latest scan; (b) assuming linear extrapolation from the latest scans; (c) imposing an idealized intensity profile upon the intensity trends as observed from the latest scans.

Option (a) may seem simplistic at times, but given the complexity of the problem, it has the advantage of providing forecasts with good continuity, or even reasonable estimates in some cases should rain generation and degeneration balance out on the broader scale. Option (b) in general gives reasonable extrapolation within the first hour. But given the short life cycle of convective cells, linear extrapolation will soon become unrealistic and rainfall forecasts based on such an assumption will soon become disproportionately large. Option (c) provides a possible, though crude, remedy by imposing an idealized empirical intensity profile in the Gaussian form. SWIRLS operationally generates both 1-hour and 2-hour rainfall forecasts assuming no intensity change, i.e. option (a), as well as based on forecast intensity change deduced from option (c).

4. RADAR AND RAINGAUGE RAINFALL RE-ANALYSIS

Conventionally, the Marshall-Palmer relationship (Marshall *et al.* 1955) is used in converting echo intensity (i.e. reflectivity, Z) to rainfall, R. But it is well known that the relationship is much generalized and does not apply well in many cases. To obtain a better estimate, SWIRLS makes use of raingauge data over Hong Kong to calibrate radar reflectivity in real time. In other words, the Z-R relationship will be adjusted in time as the rain event unfolds.

To optimize the parameters a and b in the conventional $Z = aR^b$ relationship, radar reflectivity is correlated every five minutes with the rainfall recorded by the raingauges underneath. The adjusted Z-R relationship is used subsequently in the forecast module, based on TREC winds, to convert the forecast radar reflectivity into rainfall over the raingauge positions. The procedures adopted largely follow Zawadzki *et al.* (1986). An example of SWIRLS forecast during the passage of Typhoon Maggie on 7 June 1999 is shown in Fig. 3.

When there are not enough reporting radar-raingauge pairs, such as when a rainstorm is still outside Hong Kong, either the standard Marshall-Palmer relationship ($a = 200$, $b = 1.6$) or the Z-R relationship from the previous rain episode will be used as first guess. As more radar-raingauge pairs become available, the Z-R relationship will be updated accordingly.

5. QPF VERIFICATION

Using notations of Y = Yes (rainfall criterion reached) and N = No (rainfall criterion not reached), we consider the four possible outcomes of YY, NY, YN and NN where LHS denotes forecast conditions and RHS denotes actual conditions (e.g. YN will translate as forecast criterion reached but actual criterion not reached). Then the four performance measures used are defined as:

NAP (No Alarm Probability) = NY / (YY+NY)
FAR (False Alarm Rate) = YN / (YY+YN)
CSI (Critical Success Index) = YY / (YY+NY+YN)
HSI (Heidke Skill Index) = [(YY+NN) – F] / [(YY+NY+YN+NN) – F]

In HSI, F can be any low-skill reference techniques such as persistence or climatology. In our study, F is a random forecast given by:

F = [(YY+NY)*(YY+YN)+(YN+NN)*(NY+NN)] / (YY+NY+YN+NN).

In both NAP and FAR, it is both desirable to have small values. In CSI, higher values mean better performance. In HSI, we want to stay on the positive side to beat the low-skill alternative.

There were no Black Rainstorm cases in Hong Kong in 1999 and most of the heavy rain during the year was due to visiting tropical cyclones (TC). Verification is carried out for Maggie, Sam, York and Cam, i.e. "TCs in 1999", each over a 4-hour period. A total of 143 forecasts have been made (the maximum of 160 forecasts not reached due to some data losses in the cases of Sam and York). Despite the shortage of good trough cases in 1999, we are also interested in obtaining a preliminary evaluation on how SWIRLS performs in the more volatile monsoon trough situations as compared to tropical cyclone rain. As such, results of a prolonged heavy rain event arising from a monsoon trough on "9 June 1998", from a total of 80 forecasts over an 8-hour period, are also shown alongside.

Results of NAP, FAR, CSI and HSI are tabulated in Tables 1, 2, 3, 4 respectively. In each table, there are two sets of scores: (a) for thresholds of pre-amber (20 mm per hour), amber (30 mm per hour) and red (50 mm per hour) reached at 10 raingauges; and (b) for thresholds of pre-amber (20 mm per hour), amber (30 mm per hour) and red (50 mm per hour) reached at 5 raingauges. The intention here is two-fold: firstly to evaluate how sensitive the results are to the density of raingauges, and secondly to see how effective SWIRLS is in the rain build-up towards the operational requirement of 10 raingauges.

For each set of scores in (a) and (b), verification was carried out for QPF made with the assumption of no intensity changes (i.e. "No"), as well as for QPF with imposed intensity profiles (i.e. "Yes"; Section 3 refers).

The key results from Tables 1 to 4 are extracted and summarized in Section 6 and observations discussed in terms of SWIRLS performance and development.

6. DISCUSSION

SWIRLS is found to be most useful at the amber threshold level of 30 mm per hour. Of all the "TCs in 1999" scores in Tables 1 - 4, "amber" comes out on top in 11 categories out of the maximum possible 16, compared with 4 for pre-amber and 2 for red (one of which is a joint top with amber). Comparing "pre-amber", "amber" and "red" for TCs in 1999, the best scores over the operational requirement of 10 gauges are:

	Intensity changes "No"	Intensity changes "Yes"
NAP	Pre-amber	Amber
FAR	Red	Amber
CSI	Pre-amber	Amber
HSI	Pre-amber	Amber and Red

While judgment on "red" is reserved owing to comparatively fewer cases, it is obvious that the "no intensity changes" option holds an advantage at the pre-amber stage of lesser rain. If, however, the raingauge requirement is scaled down from 10 to 5, the "no intensity changes" option loses its advantage in the pre-amber stage (only coming out on top in terms of NAP) and the best scores become very much dominated by "amber". Such observations indicate that the overall design objective of SWIRLS pitching at the amber level has more or less been met.

However, for forecasters' decision-making relating to the Rainstorm Warning System revolving around the critical amber and red stages, verification results suggest that the "with intensity changes" option is likely to be more reliable. Introducing intensity changes in QPF will significantly reduce NAP at the expense of only a slight increase in FAR. But ideally, the absolute values in FAR need to be reduced further in the next phases of SWIRLS.

In terms of CSI and HSI, amber warning at the operational 10 gauges prerequisite based on forecast intensity change peaks at the values of 0.33 and 0.29 respectively. In the build-up towards "amber", an alternative way to gain some forecast lead time is to keep count of the number of gauges reaching the amber criterion. If a build-up stage of 5 gauges is inserted, the corresponding CSI and HSI values are as high as 0.54 and 0.48 respectively. Focusing on the amber scores in Tables 1 - 4, it can be shown that reducing the number of gauges to from 10 to 5 gives overwhelmingly better results. The better scores between "10" and "5" gauges for TCs in 1999 in different categories are tabulated below:

	Intensity changes "No"	Intensity changes "Yes"
NAP	"5"	"5"
FAR	"5"	"5"
CSI	"5"	"5"
HSI	"5"	"5"

Preliminary comparisons are made between tropical cyclone rain and monsoon trough rain as represented by the 9 June 1998 case. The TC scores are found to be better than the corresponding trough scores in the majority of cases. One possible explanation is that heavy rain associated with tropical cyclone is very much the result of rain concentration by advective processes, mechanisms that currently are well handled by SWIRLS.

Score differences after incorporating intensity changes in the QPF for "amber" and "red" at 10 raingauges are shown as follows, with improvement (negative values in NAP and FAR and positive values in CSI and HSI) highlighted in italic:

	Amber	Red
NAP	***-0.73(TC)*** vs ***-0.93(trough)***	***-0.55(TC)*** vs ***-0.17(trough)***
FAR	***-0.09(TC)*** vs ***-0.62(trough)***	+0.73(TC) vs +0.80(trough)
CSI	***+0.22(TC)*** vs ***+0.60(trough)***	***+0.22(TC)*** vs ***+0.10(trough)***
HSI	***+0.24(TC)*** vs ***+0.31(trough)***	***+0.29(TC)*** vs ***+0.12(trough)***

Introducing intensity changes in the QPF generally gives better results except for the FAR scores in the red warnings. Once again, it substantiates the earlier observation that imposing an idealized intensity profile is better than just advecting the rain area assuming no intensity changes. However, one interesting aspect to note is that the improvement is less noticeable in the trough situation for the really intense cases of red warnings. It is obvious that enhanced convection under such circumstances are controlled by mechanisms (e.g. terrain effect, localized convergence on the smaller scales, merging cells, etc.) that cannot be adequately generalized by means of typical intensity profiles. Further development work in SWIRLS on these aspects is required. Tracking of echo groups, pattern recognition, terrain correction, and real-time intensity profile fitting are just some of the approaches currently under investigation.

While we hope to compile more verification statistics as SWIRLS gains further field experience, particularly "red" and "black" rainstorm cases for more meaningful interpretation, the preliminary results obtained so far are already quite useful for operational guidance and research reference. Notwithstanding the best of theories and numerical techniques, there will always be a place for the engineering approach in the design of a QPF nowcasting system tailor-made for the specific locale, both in terms of rain characteristics and operational warning requirement. For example, CSI scores for the VSRF system in JMA ranges between 0.2 and 0.4 for rainfall rate of 10 mm per hour (JMA 1998). If verification for this criterion is run on SWIRLS, the CSI score will presumably be quite unsatisfactory. But in Hong Kong, forecasters are not so concerned about rain of 10 mm per hour anyway! And if systematic verification can guide us towards the optimal level and areas of interest, so much the better.

REFERENCES

Bjerkaas, C.L., and D.E. Forsyth, 1980: Operational test of a three-dimensional echo tracking program. *Preprints 19th Conf. Radar Meteorology* Miami Beach, Amer. Meteor. Soc., 244-247.

Crane, R.K., 1979: Automatic cell detection and tracking. *IEEE Trans. Geosci.Electron.,* **GE-17**, 250-262.

Cressman, G.P., 1959: An operational objective analysis system. *Mon. Wea. Rev.,* **87**, 367-374.

Hilst, G.R., and J.A. Russo, Jr., 1960: An objective extrapolation technique for semi-conservative fields with an application to radar patterns. Tech. Memo. No. **3**, Travelers Weather Research Centre, Hartford, CT.

Japan Meteorological Agency (JMA), Forecast Department, Forecast Division, 1998: *Analysis and Very-Short-Range Forecasting of Precipitation.*

Kessler, E., III, and J.A. Russo, Jr., 1963: Statistical properties of weather echoes, *Preprints 10th Weather Radar Conf.,* Washington, DC, Amer. Meteor. Soc., 25-33.

Li, P.W., W.K. Wong, K.Y. Chan, and S.T. Lai, 2000: SWIRLS – An Evolving Nowcasting System. *Hong Kong Observatory Tech. Note,* **No. 100** (to be published).

Marshall, J.S., W. Hitschfeld, and K.L.S. Gunn, 1955: Advances in radar weather. *Adv. Geophys.* **2**, 1-56.

Rinehart, R.E., 1979: *Internal storm motions from a single non-Doppler weather radar.* NCAR/TN-146+STR, 262pp.

______, and E.T. Garvey, 1978: Three-dimensional storm motion detection by conventional weather radar. *Nature,* **273,** 287-289.

Smythe, G.R., and D.S. Zrnic, 1983: Correlation analysis of Doppler radar data and retrieval of the horizontal wind. *J. Climate and Appl. Meteor.,* **22**, 297-311.

Staniforth, A., and J. Cote, 1991: Semi-Lagrangian integration schemes for atmospheric models – a review. *Mon. Wea. Rev.*, **119**, 2206-2223.

Tuttle, J.D., and G.B. Foote, 1990: Determination of the Boundary Layer Airflow from a Single Doppler Radar. *J. Atmos. Ocean. Tech.,* **7**, 218-232.

Zawadzki, I., C. Desrochers, and E. Torlaschi, 1986: A Radar-raingauge Comparison. Preprints, *23rd Conf. on Radar Meteorology*, Snowmass, Colorado, Amer. Meteor. Soc., 121-124.

Table 1. NAP (No Alarm Probability).

(a)

10 gauges reaching criteria		Pre-amber ≧20 mm per hour		Amber ≧30 mm per hour		Red ≧50 mm per hour	
Intensity Change	No	TCs in 1999	0.58	TCs in 1999	0.83	TCs in 1999	1.00
		9 June 1998	0.78	9 June 1998	1.00	9 June 1998	1.00
	Yes	TCs in 1999	0.16	TCs in 1999	0.10	TCs in 1999	0.45
		9 June 1998	0.00	9 June 1998	0.07	9 June 1998	0.83

(b)

5 gauges reaching criteria		Pre-amber ≧20 mm per hour		Amber ≧30 mm per hour		Red ≧50 mm per hour	
Intensity Change	No	TCs in 1999	0.50	TCs in 1999	0.60	TCs in 1999	1.00
		9 June 1998	0.75	9 June 1998	1.00	9 June 1998	1.00
	Yes	TCs in 1999	0.20	TCs in 1999	0.04	TCs in 1999	0.38
		9 June 1998	0.08	9 June 1998	0.00	9 June 1998	0.85

Table 2. FAR (False Alarm Rate).

(a)

10 gauges reaching criteria		Pre-amber ≧20 mm per hour		Amber ≧30 mm per hour		Red ≧50 mm per hour	
Intensity Change	No	TCs in 1999	0.70	TCs in 1999	0.75	TCs in 1999	0.00
		9 June 1998	0.90	9 June 1998	1.00	9 June 1998	0.00
	Yes	TCs in 1999	0.69	TCs in 1999	0.66	TCs in 1999	0.73
		9 June 1998	0.88	9 June 1998	0.38	9 June 1998	0.80

(b)

5 gauges reaching criteria		Pre-amber ≧20 mm per hour		Amber ≧30 mm per hour		Red ≧50 mm per hour	
Intensity Change	No	TCs in 1999	0.77	TCs in 1999	0.44	TCs in 1999	1.00
		9 June 1998	0.92	9 June 1998	1.00	9 June 1998	0.00
	Yes	TCs in 1999	0.75	TCs in 1999	0.45	TCs in 1999	0.84
		9 June 1998	0.85	9 June 1998	0.73	9 June 1998	0.69

Table 3. CSI (Critical Success Index).

(a)

10 gauges reaching criteria		Pre-amber ≧20 mm per hour		Amber ≧30 mm per hour		Red ≧50 mm per hour	
Intensity Change	No	TCs in 1999	0.21	TCs in 1999	0.11	TCs in 1999	0.00
		9 June 1998	0.07	9 June 1998	0.00	9 June 1998	0.00
	Yes	TCs in 1999	0.29	TCs in 1999	0.33	TCs in 1999	0.22
		9 June 1998	0.12	9 June 1998	0.60	9 June 1998	0.10

(b)

5 gauges reaching criteria		Pre-amber ≧20 mm per hour		Amber ≧30 mm per hour		Red ≧50 mm per hour	
Intensity Change	No	TCs in 1999	0.19	TCs in 1999	0.31	TCs in 1999	0.00
		9 June 1998	0.06	9 June 1998	0.00	9 June 1998	0.00
	Yes	TCs in 1999	0.24	TCs in 1999	0.54	TCs in 1999	0.15
		9 June 1998	0.14	9 June 1998	0.27	9 June 1998	0.11

Table 4. HSI (Heidke Skill Index).

(a)

10 gauges reaching criteria		Pre-amber ≧20 mm per hour		Amber ≧30 mm per hour		Red ≧50 mm per hour	
Intensity Change	No	TCs in 1999	0.12	TCs in 1999	0.05	TCs in 1999	0.00
		9 June 1998	-0.02	9 June 1998	-0.03	9 June 1998	0.00
	Yes	TCs in 1999	0.20	TCs in 1999	0.29	TCs in 1999	0.29
		9 June 1998	0.02	9 June 1998	0.28	9 June 1998	0.12

(b)

5 gauges reaching criteria		Pre-amber ≧20 mm per hour		Amber ≧30 mm per hour		Red ≧50 mm per hour	
Intensity Change	No	TCs in 1999	0.04	TCs in 1999	0.27	TCs in 1999	-0.05
		9 June 1998	-0.14	9 June 1998	-0.07	9 June 1998	0.00
	Yes	TCs in 1999	0.09	TCs in 1999	0.48	TCs in 1999	0.13
		9 June 1998	-0.01	9 June 1998	0.08	9 June 1998	-0.09

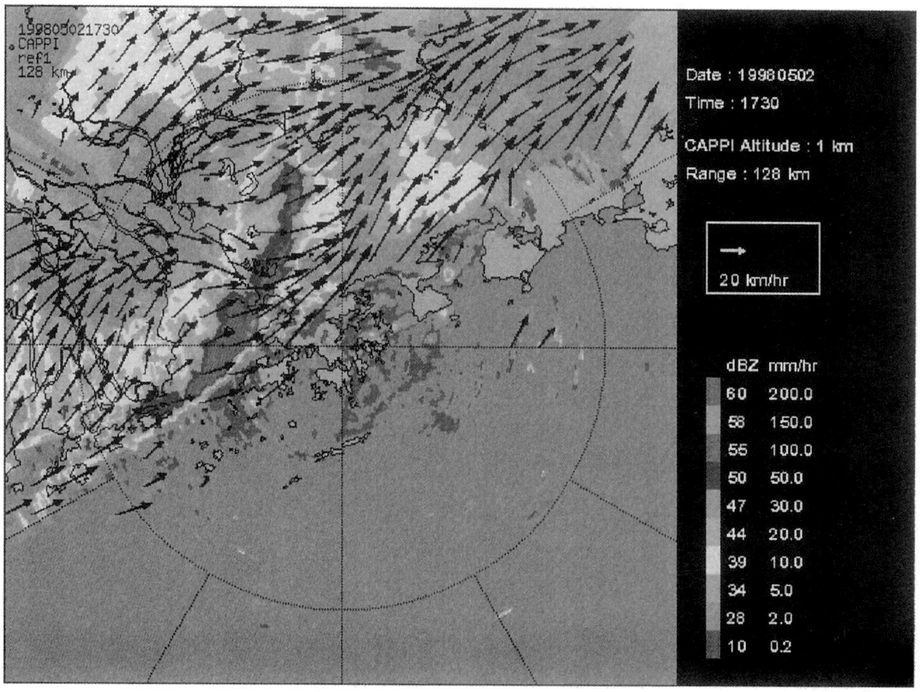

Fig. 1. TREC analysis of a squall line approaching Hong Kong from the northwest at 17:30 HKT on 2 May 1998.

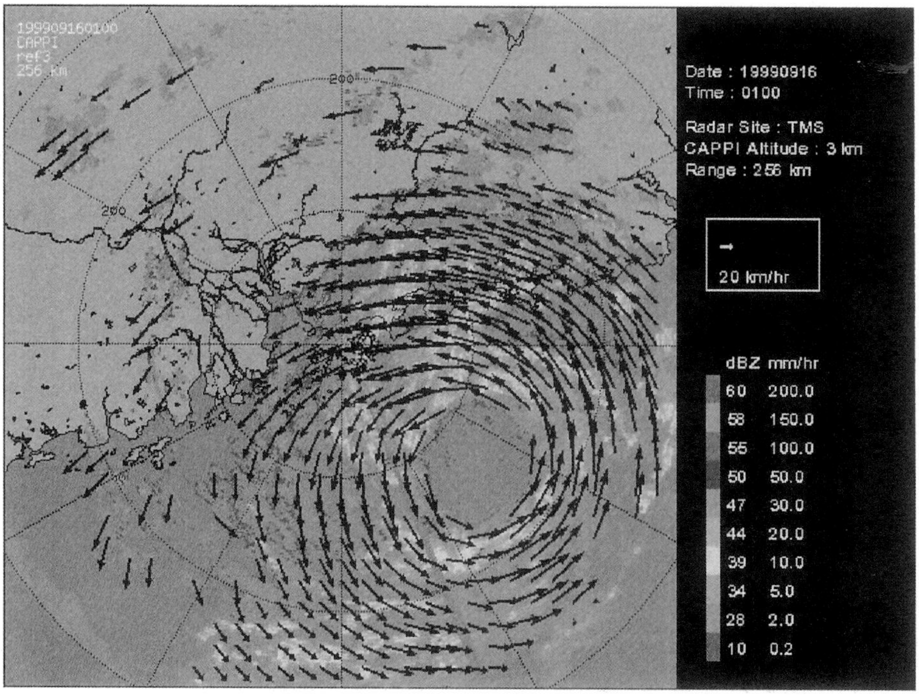

Fig. 2. TREC analysis of Typhoon York just to the southeast of Hong Kong at 01:00 HKT on 16 September 1999.

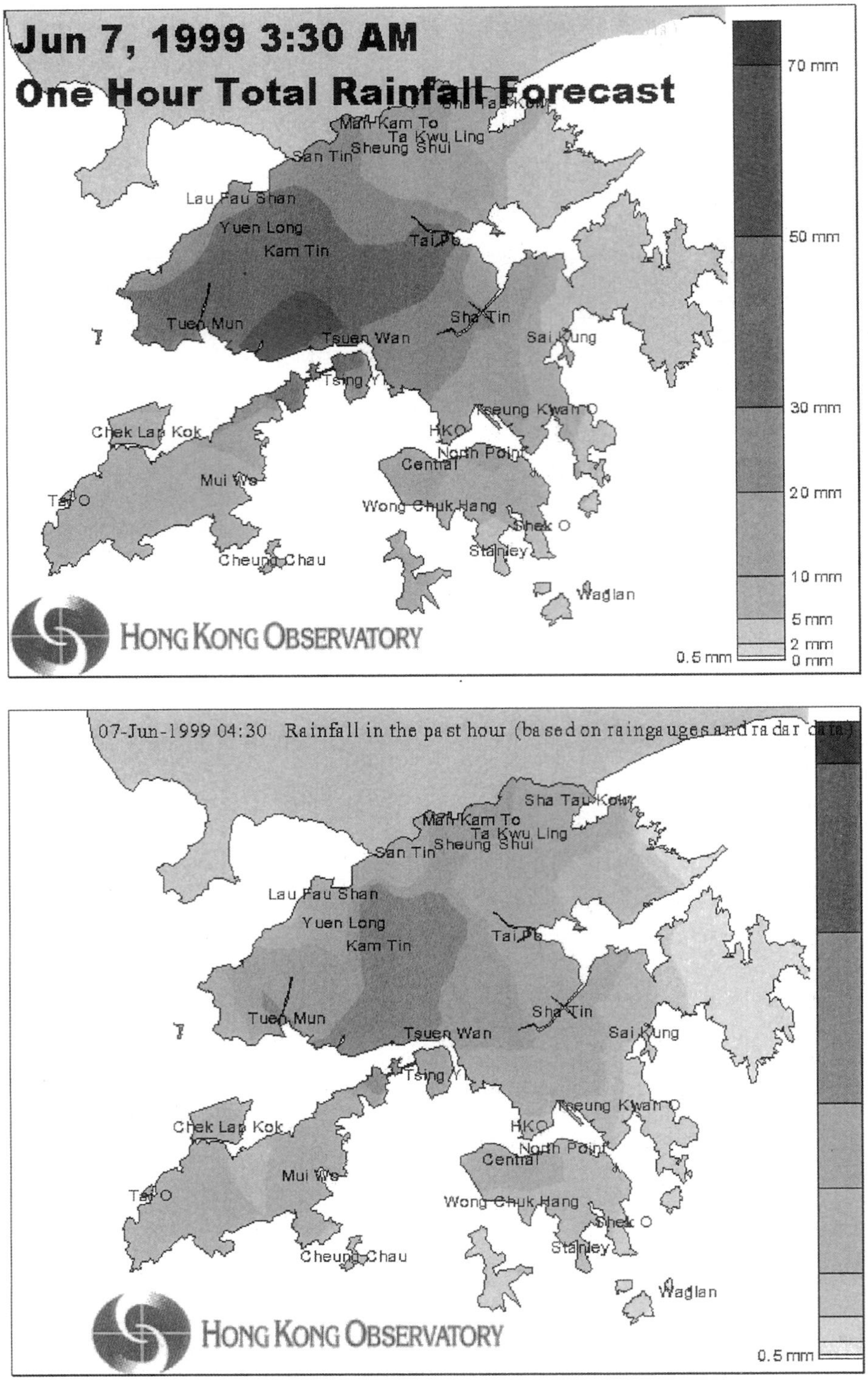

Fig. 3. 1-hour rainfall forecast by SWIRLS for Typhoon Maggie issued at 3:30 HKT on 7 June 1999 (upper panel) and the corresponding one hour rainfall accumulation reported by the raingauges from 3:30 HKT to 4:30 HKT on 7 June 1999 (lower panel).

~ Air Pollution ~

EMISSION INVENTORY OF CIVIL AIRCRAFT OVER TAIWAN

Po-Hsiung Lin, Wei-So Sun, and Jen-Ping Chen
Department of Atmospheric Sciences, National Taiwan University

ABSTRACT

The inventory of aircraft emissions over the Taipei Flight Information Region (FIR) has been developed with 1997 as the base year. We first analyze the spatial and temporal distributions of flight frequency base on the official civil aviation monthly reports, airline schedules, pilots' on air reports. Fuel consumption and trace-gas emissions are then calculated by using the engine performance database of the International Civil Aviation Organization (ICAO). Compared with the official records, the amount of fuel consumption estimated in this study is within 0.5% accuracy. The emission inventory is built for different flight types (domestic or international), aircraft types (commercial jets, cargo jets or propeller aircraft), as well as the state of aircraft operation (idle, takeoff, climb-out, cruise and approach/landing). Engine types, number of engine per aircraft, as well as engine performance at different operation states are all considered in estimating the emission rates.

The emission rates of six trace species (NO_x, CO, CO_2, Sulphur, HC and H_2O) are estimated for major flight routes and airports, and further separated into three height levels (ground, near surface, and cruising altitudes). To perceive their relative importance, we took the ratios of the annual emissions from aircraft over the Taipei FIR with those from ground transportation in Taiwan. The most significant ratios are 14.9% for NO_x, 9.7% for Sulphur, 0.87% for CO, and 0.43% for HC. Over 80% of the NO_x were emitted into the upper troposphere, whereas most of the CO and HC were produced on the surface. A further comparison shows that the amount of aircraft-produced NO_x over the Taiwan area is about 70% of those produced over China mainland.

1. INTRODUCTION

There has been increasing attention on the impact of aircraft emissions on global environment, such as the emissions of greenhouse gases, depletion of stratospheric ozone, production of tropospheric ozone, and the formation of cirrus-like contrail clouds. All these factors are relevant to the change of radiation budgets in the atmosphere-earth system. For example, anthropogenic carbon dioxide (CO_2) may have contributed to about 2 Wm^{-2} radiative forcing in 1990, and one-tenth of which is resulted from aviation emissions (IPCC, 1996). The radiative effect from contrails is of similar magnitude (IPCC, 1999). World passenger traffic on scheduled airlines has doubled during the past ten years (IATA, 1996). At the same time, aviation fuel consumption has reached about 3% of the fuel used worldwide. Such fast growth has promoted on not only the current status but also future prospect of aviation impacts. Scenarios of global aviation emissions have been studied and compared among National Aeronautics and Space Administration of the United Sates (NASA; Baughchum et al., 1996), the Deutsches Zentrum für Luft- und Raumfahrt (DLR) and European Abatement of Nuisances Caused by Air Transport/European Commission (ANCAT/EC; Gardner, 1998). They estimated a 4% average growth rate of aviation fuel consumption during 1992 to 2015. The Intergovernmental Panel on Climate Change (IPCC, 1999) summarized the conclusions of these projects for policy-makers in helping them realize current scientific understanding of aviation effect on global atmosphere.

The projects mentioned above have provided a global view on the effect of aviation emissions. But one should not ignore the importance of aircraft emissions because the effect on a local or regional scale is much stronger due to the relatively short lifetime of all emission products except water vapor (H_2O) and CO_2. According to the report of Air Transport Action Group (ATAG, 1997), the air traffic growth rates in East Asia and the western Pacific region is 10.1% between 1985 and 1995 and is expected to be about 7.4% between 1995 and 2010. These rates are about twice of the global average. In Taiwan, the annual growth rate between 1991 and 1996 is over 20% (CAA, 1998), also much higher than the world average of 5% per year. Furthermore, as indicated by the key position in the East Asia/western Pacific region shown in Fig. 1, the Taipei Flight Information Region (FIR) hosts several busiest international and transcontinental flight routes in Asia. The route between Hong Kong and Taipei alone carries more than one-third of the flights in Asia (IATA, 1996). One can thus expect a high local impact from the aircraft emission over Taiwan. We therefore intend to estimate the inventory of aircraft emissions over the Taiwan area in order to provide crucial information for atmospheric modeling and government policy-making.

We first investigate the spatial (horizontal and vertical) and temporal (annual, seasonal, weekly, and diurnal) distributions of both domestic and international flights over the Taipei FIR. The average emissions from a single aircraft are then estimated by considering the fuel flow rate and emission indices of different engine types for different operation states (i.e., idle, takeoff, climb-out, cruise, and approach/landing) using data set from International Civil Aviation Organization (ICAO). Combining the flight information and emission rates, we obtain the emission rates of several trace gases along different flight corridors (both domestic and international) and airports over the Taipei FIR. Finally, these emission inventories are compared with those of surface transportation as well as the aircraft emissions over China mainland.

2. METHODOLOGY AND DATA SOURCES

Ideally, the emission during one complete flight operation can be obtained according to the equation:

$$e_i = n\sum_{k=1}^{5}(q_{ik} \cdot f_k \cdot t_k), \tag{1}$$

where $\boldsymbol{e}$ is the amount of emission (g), $\boldsymbol{i}$ is a chemical species, $\boldsymbol{n}$ is the number of engines per aircraft, $\boldsymbol{k}$ is the engine operation mode (idle, takeoff, climb-out, cruise, approach/landing), $\boldsymbol{q}$ is the emission index (mass of pollutant produced by burning a unit mass of fuel, in g kg^{-1}), $\boldsymbol{f}$ is the fuel flow rate (kg s^{-1}), and $\boldsymbol{t}$ is the operated time (s). There are seven engine operation modes in one landing and takeoff cycle (LTO): taxi-out, takeoff, climb-out, cruise, approach, landing and taxi-in (ICAO, 1995). The taxi-out and taxi-in modes are generally combined into one mode called "idle." The approach and landing modes are also combined into a single mode because of their similar emission indices. Except the "cruise" mode, all other engine modes are defined as operations at heights below 1,000 feet. Table 1 lists the mode-specific engine power setting and operation time except for the cruise mode, for which the operation time depends on individual flight schedule and cruise speed. The emission factor $\boldsymbol{e}$ and the fuel flow rate $\boldsymbol{f}$ in equation (1) are engine-dependent parameters.

Equation (1) gives the complete aviation emission calculation for one flight operation between two airports (i.e., one LTO cycle). In reality, however, it is difficult to trace the operation specifics of every single flight (such as the aircraft and engine types, engine performance, payload, operation time of different engine modes, type of fuel used, fuel consumption, etc.). A compromised solution is to use averaged values for the parameters in equation (1). This can be done first by classifying aircraft into passenger jet, cargo jet, propeller aircraft, and military aircraft. Equation (1) is then modified as:

$$E_{ij} = \sum_{j=1}^{3}\left(N_j \sum_{k=1}^{5}(Q_{ijk} \cdot F_{jk} \cdot T_{jk})\right), \tag{2}$$

where E is the average amount of emission (g), $\boldsymbol{j}$ is the aircraft type, $\boldsymbol{N}$ is the average number of engines, $\boldsymbol{Q}$ is the average emission index (g kg^{-1}), $\boldsymbol{F}$ is the average fuel flow rate (kg s^{-1}), and $\boldsymbol{T}$ is the average operated time (s). In this study, we did not estimate the emission by military aircraft, so $\boldsymbol{j} = 1$ to 3 refers to passenger jet, cargo jet and propeller aircraft.

Five data sources are used in deriving the above factors: (a) monthly report of the Civil Aeronautics Administration (CAA); (b) weekly timetable of domestic airlines; (c) pilot's way-point report; and (d) ICAO engine emissions data bank; and (e) emission content of propeller aircraft engine provided by a local airline. The CAA monthly report contains the operating information of all airports in Taiwan, including LTO cycles, passenger and cargo flight frequency, passenger and cargo loads, aircraft types, etc. From this data set we obtained the engine types and the average number of engine per aircraft (*N*) for the Taipei FIR (Table 3). In the weekly domestic airline timetable one can find not only the time schedule of the domestic flights but also the aircraft types used in all domestic airports. Combined with the CAA monthly report, one can obtain the specifics about domestic traffic, including the ratio between jets and propeller aircraft as well as the annual fuel consumption.

The pilot's way-point report is a communication record required by the Air Traffic Services Division in CAA when aircraft is flying above 10,000 feet and passing through specific locations in the route (the way-points; solid round points in Fig. 2). The main contents are reporting time, departure time and airport, destination airport, current altitude and speed, as well as previous and next way-points. An interactive software interface is developed for extracting records from the way-point database according to specified criteria. From this database we obtained certain information about international flights, such as the cruise speed, flight frequency in a certain route, the distribution of flight altitudes, and whether the flight is departing/arriving or just passing through.

The ICAO engine emissions data bank contains information relevant to jet aircraft emissions. From this data bank we obtain the factors $\boldsymbol{Q}$ and $\boldsymbol{F}$ of 6 chemical species (index $\boldsymbol{i}$) for two jet aircraft types (index $\boldsymbol{j}$) under different engine operation modes (index $\boldsymbol{k}$). Since the early 1970s, engine manufacturers had started measuring emissions to demonstrate compliance with regulations developed for the airport vicinity. ICAO established emission indices for several pollutants based on measurements performed on 86 types of engines (ICAO, 1995), and 16 of them are commonly used on the commercial aircraft operated in Taipei FIR according to the CAA monthly report. We therefore derived the mean emission rates of these 16 engine types in Table 2a. However, the ICAO data bank does not provide the factors $\boldsymbol{Q}$ and $\boldsymbol{F}$ for either the cruise mode of jet aircraft or propeller aircraft. After consulting with some airline pilots, we found it reasonable to assume that the engine output (and thus the factors $\boldsymbol{Q}$ and $\boldsymbol{F}$) during the cruise mode can be taken as the mean of those of the climb-out and approach/landing modes. The values we obtained this way turn out to be consistent with those of Brasseur et al. (1998). Note that the emission index (Q) of CO_2, sulphur (S, mainly in the form SO_2) and H_2O does not depend on the engine operation mode but the fuel type. From the ICAO data bank and other supporting references (eg. Zhao and Turco, 1995; Brasseur et al., 1998), we obtained the values of 3.14 kg kg^{-1}, 0.4 g kg^{-1}, and 1.23 kg kg^{-1} for CO_2, S and H_2O, respectively, for common aircraft fuels. As for the propeller aircraft, we were able to obtain the mixing ratio of various gases in the engine exhaust from a maintenance manual provided by the TranAsia (Fushing) Airways. The mixing ratios are then converted into emission indices by taking CO_2 as the reference species (Table 2b).

The ICAO engine emissions data bank also provides the average operation time for all but the cruise mode in Table 1. For the cruise mode the average operation time is estimated by combining the average cruise speed with the route length (distance between two airports or way-points shown in Fig. 3). The averaged cruise speed for subsonic jets is 250 m s^{-1} with a 16 m s^{-1} standard deviation, and 130 m s^{-1} for propeller craft (IATAG, 1998). From the way-point reports in the Taipei FIR we derived an average cruise speed of 248.7 m s^{-1}, which confirm with the IATAG report (1998). Although the actual cruise time depends on weather conditions, control tower operations and aircraft mechanical conditions, they can be regarded as constants on a long-term average.

3. TEMPORAL AND SPATIAL DISTRIBUTIONS FLIGHTS IN TAIPEI FIR

To obtain a detailed data set for regional aircraft emissions, we need to know the temporal and spatial variations of flight frequencies in the airports and in flight corridors over the Taipei FIR. We first present the flight frequency in four time formats: yearly, monthly, daily, and hourly. Fig. 4a shows the annual frequency of domestic and international air traffic in the Taipei FIR from 1988 to 1999 according to the CAA monthly reports. Between 1988 and 1997, the domestic flight frequency increased from about 130,000 LTO per year to near 580,000 LTO per year, almost doubling every 5 years. However, the fast growing trend subsided significantly after 1997 probably because of over-expansion and several flight-safety incidences. During the same period of time, the international flights increased steadily but with a slower pace, from 47000 LTO in 1988 to 135000 LTO in 1997. The domestic to international flight ratio went up from 2.8 in 1988 to 4.6 in 1996, then drop back to 3.7 in 1998. To facilitate discussions, we choose the 1997 as the baseline year for estimating the temporal and spatial variations of aircraft emissions.

From the CAA monthly reports we also obtained the monthly air traffic frequency (LTO cycles) over the Taipei FIR. As shown in Fig. 4b, there is no significant seasonal variation for both the domestic and international flights. The situation in most part of the northern hemisphere is quite different. For instance, in the North Atlantic region the flight frequency is about 15% higher during July and 15% lower during January compared with the annual average (IPCC, 1999). Such seasonality also exists over the North Pacific region but with much lower amplitude of $\pm 5\%$. Since the global mean of the monthly variation is similar to that in the North Pacific, the seasonality in other regions must be less obvious or even opposite to that in the North Atlantic region. So, it is not surprising that the flight frequencies over the Taipei FIR and in the southern hemisphere essentially have no seasonality.

Daily and diurnal variations in flight frequency do exist over the Taipei FIR according to the weekly domestic airline timetable. As shown in Fig. 5a, the LTO of domestic flights peak on Sunday, rises slightly from the weekend and comes down on Monday. The LTO of international flights, on the other hand, is slightly higher on Sunday, Friday and Tuesday than the other days. Fig. 5b shows the diurnal variations of LTO. Most of the domestic flights begin their operation after 6:00 a.m. The flight frequency increases rather quickly and reaches a plateau of about 140 LTO per hour between 9:00 a.m. and 3 p.m. After about 5 p.m., the flights diminish gradually and finally close all operation by 11:00 p.m. The international flights operate similarly to the domestic flights but with a somewhat longer duration. It had slight flat pattern during 8:00 AM to 11:00 PM. It is believed that the airline schedule is associated with the social activities, the airport operation and the vision reason of aircraft piloting.

For vertical distribution of civil aviation in Taipei FIR, the ELA and HLG way-points in the data source (c) were selected for discussion. In general situation, an aircraft must keep the cruise height until it is 70 miles (126 kilometres) within the airport. ELA, locating on Route-A1 is the southwestern boundary way-point between Taipei and Hong Kong FIRs. All the flights passing ELA are international traffic and the aircraft are expected to be in the upper air (Fig. 6a). In contrast, HLG locates on the overlap point of Route-A1 and Route-W4, and the latter route is the busiest domestic corridor. Three main cruise levels are clearly identified into the heights of 6 to 7, 9 to 10 and 11 to 12 kilometres at the HLG way-point (Fig. 6b). In general, the international flight is not forced to give on-air report above HLG except it is a non-stop flight and penetrates northerly using W4 route. These international flights fly above 9 km height and more. The lowest cruise height (6-7km) is used by the northern-directional domestic flights which will decrease their altitudes soon to approach the Song-Shang airport which is close to the APU way-point in Fig. 2.

Data sources (b) and (c) had been extracted from the domestic and international route-used frequency in Fig. 7 and Fig. 8 for analyzing the spatial distribution of civil aviation in Taipei FIR. For domestic air traffic, then maximum percentage (19.1%) of route-used frequency occurred over northwestern Taiwan (between HLG and APU way-points). The percentage decreases southerly along the western coastline and central mountain region to 11.7% over southwestern Taiwan. In summary, the aviation emission on the western land is 81% and 7% along the eastern coastline. There was 8% amount across Taiwan Strait (Makung, Chinmen and Machu transportation) and 4% to Lan-Yu and Green Island.

For international traffic, Fig. 8 gives the quantitative value of route-used frequency among the way-points. [ELA-APU and APU-BLN (Route A1) which penetrated southwestern to northeastern Taipei FIR had the biggest ratio (47%) in Taipei FIR. APU way-point and the routes-crossing point over Taipei City had a 64% frequency of upper air traffic. Another routes-crossing way-point, HCN over Kenting National Park, had a 36 % frequency of upper air traffic. There was 44.6% of the international flights penetrating the southwestern boundary of Taipei FIR, and 24.8% crossing the Taiwan Strait into the mainland. 34.7% of the international flights penetrated the northeastern boundary to North Pacific and North America, and the south and east boundaries had 15.3% and 5.4% traffic respectively. The only unknown traffic weighting in Taipei FIR region is the northwestern area which is controlled by another centre. Meanwhile, the ratio between the number of the over-flight commercial jet and the number of the jet which was landing or takeoff commercial jet from Taipei FIR was 35 % from the diagnose of the way-point reports. We had the condition to search the flight which passes any way-point on the Taipei FIR boundary and their destinations or starting places are not one of the airports in Taipei FIR. These direct penetrated crafts kept their cruise height around 10 or 12 kilometres above the surface showed in Fig. 6. This information emphasizes the fact of aviation emissions in the upper air.

4. AIRCRAFT EMISSIONS IN TAIPEI FIR

Equation (1) in section two gives the complete aviation emission calculation of one flight through the LTO cycle and cruise period between two airports. But in reality, no data is available to trace each flight with its actual operation including the payload, the time spent, the fuel burned, the engine performance and so on. A compromised solution is to modify equation (1) into a simpler form:

$$E = \sum_{j=1}^{3} \left(N_j \sum_{i=1}^{5} (R_{ij} \cdot T_{ij}) \right) \quad (2)$$

where E is again the amount of emission (g), N_j is the average number of engines in three different types of aircraft such as the commercial jet, cargo and propeller aircraft. R is emission rate (g/s) and T is operation time (s). The index i is the five operation modes (idle, take-off, climb-out, cruise, approach) reduced from eight operation modes in equation (1) to fit the existing items of engine emissions in the ICAO dataset. Cruise mode was defined in this study as the averaged status of climb-out and approach modes' emissions, except the time variable. The time for cruise mode was calculated through the geographic distance between two airports or way-points. The averaged cruise speed for subsonic jets is 250 ms^{-1} with a 16 ms^{-1} standard deviation, and 130 ms^{-1} for propeller craft (IATAG, 1998).

The three parameters in the above equation were acquired from data sources in section two and had been organized into look-up tables. Table 1 lists the T values and Table 2 summarizes the emitted rate for six emissions from three different aircraft types (index j) under five (index i) engine modes. From the studying of the CAA monthly report, sixteen engine types operated in Taipei FIR had been found in the ICAO data bank for engine exhaust emissions. The means for the performance of these engines were dealt in the Table 2, and the emitted rates are consistent to other literature (Brasseur et al., 1998). Table 3 provides the mean and deviation of N_j for the different types of aircraft. The deviation is caused by the fact that a subsonic aircraft may own two to four engines.

Six types of propeller passenger aircraft are used in Taipei FIR. All of them are the two-engine types and the ATR-72 aircraft operated by the TransAsia Airways is the biggest one for medium distance transportation between local cities, Penghu and Kinmen islands. The ratio of ATR-72 operation in the domestic air traffic is 9.5%. The reason for concerning propeller craft's contribution has been arisen due to its LTO amount was more than the commercial jet in 1997, although its operation time is shorter and emission distribution is limited along the routes between Taiwan and small islands. Unfortunately, only one engine for the AI®ATR propeller aircraft, like ATR-72 aircraft, was registered in the ICAO exhaust dataset. Lacking engine emission from propeller craft, we used the emission rate R of ATR-72 aircraft to represent other kinds of propeller crafts operated in Taipei FIR.

Cargo jets operated in Taipei FIR occupied 11% of the international air traffic, but such aircraft are equipped with more engines (3.42) and heavier payloads. Baughcum et al.

(1996) estimated the deviation caused by payload factor from 0.8 % to 7.7%. There is no available information to trace the payload in each cargo flight and its flight schedule, but the total amount of cargo flight and their engine types can be found in the monthly CAA report.

The generations of aircraft engine fall into three main categories. They are:

(a) natural products of fossil fuel combustion, such as CO_2, H_2O and nitrogen oxides (NO_x),
(b) inefficient combustion products of carbon monoxide (CO), HC and S, and
(c) fuel related emissions consisting of sulphur compounds and other trace species.

ICAO selected four pollutants (NO_x, S, CO, HC) to list in the ICAO data bank of engine exhaust emissions. The emitted CO_2 and H_2O were calculated through the amount of fuel burned and their constant exhaust coefficients.

Table 4 and Table 5 summarizes the annual amount of six emissions in 1997 with equation (2), and its characteristics will be discussed more under the following issues:

(a) domestic and international traffic
The amount of all the emissions from international traffic was nearly 1.3 to 1.5 times more than that of domestic traffic, but HC and CO emissions had 0.86 and 0.3 times individually. The reason is that more HC and CO are produced under an ill-burned condition, e.g. the idle and take-off modes. Sixteen domestic airports provided higher LTO numbers. In contrast, the time duration of the international traffic is longer and altitude higher in cruise mode, and only two airports in Taipei FIR are for migration landing and take-off.

(b) ground level, near-surface and cruise level
The ground level was represented by the idle and take-off engine modes and the climb-out and approach modes were treated as the near-surface level emission. The reason is that civil aircraft run the take-off procedure with full power setting to lift the craft to 35 feet above the ground, and then enter the climb-out procedure up to 1,000 feet with 85% power-setting. For CO and HC, the emissions on the ground level were 1.7 and 3.1 times of that on the cruise level, the emissions on the near-surface level were less than 4% of the total. On the other hand, over 80% of S, NO_X, CO_2 and H_2O emitted into the cruise level (Table 6).

(c) different craft types
For CO_2, cargo jets contributed 8% of the emission to international traffic. This result is consistent with the study of Baughcum et al. (1996), who estimated the Boeing 747 Cargo emission over North Pacific. The emissions, except CO , emitted from propeller crafts were less than 0.1 times than subsonic jet crafts had. The ratio for CO emission was near equal each other.

5. SUMMARIES AND DISCUSSION

Four data sources, three flight datasets and one engine exhaust emission dataset, had been analyzed to outline the spatial and time distribution of civil aviation emissions of 1997 in Taipei FIR area. The annual amount of six emissions was provided quantitatively for ground level, near-surface and cruise level. The contributions from different aircraft types were shown, too. The error of the emission inventory rises from the averaged process of the performances of these sixteen engines and the engine number in equation (2). Time constant assumed in four engine modes in one LTO process and using one propeller engine type representing all propeller crafts is not reasonable at all. In order to check the bias of the emission inventory, the amount of fuel burned annually was estimated in Table 2 as well as in the flight schedules in dataset (a) to (c). We compared it with the independent official record (personal contact with Chinese Petroleum Corporation). Our calculation (1,778 kilotons per year) was 99.7% of the civil aviation fuel burned in Taiwan. This objective reference gives a high confidence in our results.

The emission ratio between civil aviation and land transportation over Taiwan was listed in Table 7. Combined with the result of Table 4 and Table 5, NO_X emitted into the cruise level (6-12 km) had 12% of the amount emitted from the land traffic. Meanwhile, the annual NOx emission in 1997 from all possible sources over Taiwan was about 750 kilotons, predicted by the Environmental Protection Administration (EPA). So the NOx emission from civil aviation might make a 4% contribution in the country.

Ma and Chou (1999) estimated that NO_x emission of civil aviation over China mainland during 1997 to 1998 is 44.6 kilotons/year. NO_x emission inventory in this study is 31.6 kilotons/year, a ratio of 70% between Taipei FIR and entire China mainland. They mentioned that about 78% NOx emission occurred at the altitude band of 9 to 12 km, which is less than the value of 80.3% in this study. We believe that a 2% difference between each other is caused by two reasons. One is the calculation on near-surface level where the component from climb-out and approach modes in our work is. The other is the calculation of the domestic crafts which cruise height is near 6 km level, not up to 9 km height.

The consideration for the exhaust modes of five engines in equation (2) is the specialty in our inventory. But our inventory has not countered the military aviation contribution yet. IPCC (1999) estimated that the ratio of the worldwide civil and military aviation is 82:18 in 1992, and it will decrease to 93:7 in 2015. This information provides us a reference for the component of military emission.

ACKNOWLEDGEMENT. This work was supported by the Grant National Sciences Council NSC87-2621-P-117-001. The authors are grateful to the data provider, Civil Aeronautics Administration.

REFERENCES

ATAG, 1997: *Asia/Pacific Air Traffic Growth & Constraints.* Air Transport Action Group, Geneva, Switzerland, 50pp.

Baughcum, S. L., S. C. Henderson, T. G. Tritz, and D. C. Pickett, 1996: *Scheduled civil aircraft emission inventories for 1992: database development and analysis.* NASA-CR-4700, National Aeronautics and Space Administration, Langley Research Center, Hampton, VA, USA, 196pp.

Brasseur, G. P., R. A. Cox, D. Hauglustaine, I. Isaksen, J. Lelieveld, D. H. Lister, R. Sausen, U. Schumann, A. Wahner, and P. Wiesen, 1998: European scientific assessment of the atmospheric effects of aircraft emissions. *Atmos.Environ.*, **32**, 2329-2418.

CAA, 1998: *The monthly statistics report of commerical aircraft transport.* Civil Aeronautics Administration, 164pp (in chinese)

Gardner, R. M., (ed.), 1998: *ANCAT/EC2 aircraft emissions inventories for 1991/1992 and 2015: Final Report.* EUR-18179, ANCAT/EC Working Group, ISBN-92-828-1924-6, 84pp.

IATA, 1996: *World Air Transportation Statistics.* International Air Transport Association, Geneva, Switzerland, 40th ed., 111pp.

IATAG, 1998: *International Airline Guide 98-99.* EPS Aviation Publishers, 388pp.

ICAO, 1995: *ICAO Engine Exhaust Emissions Databank, 1st Edition-1995.* ICAO-9646-AN/943, International Civil Aviation Organization, Montreal, Quebec, Canada, 53pp.

IPCC, 1996: *Climate change 1995: Impacts, Adaptations and Mitigation of Climate Change: Scientific Technical Analysis. Contribution of Working Group II to the Second Assessment Report of the Intergovernmental Panel on Climate Change* [Watson, R.T., N.C. Zinyowera, and R.H. Moss (eds.)]. Cambridge University Press, Cambridge, United Kingdom and New York, 880 pp.

IPCC, 1999: *Aviation and the Global Atmosphere.* [Penner, J. E., D. H. Lister, D. J. Ggriggs, D. J. Dokken, and M. Mcfarland (ed.)]. Cambridge University Press, 373pp.

Ma, J., and X. Zhou, 1999: Developing of a three-dimensional inventory of aircraft NO_x emissions over China. *Atmos.Environ.* **34**, 389-396.

Zhao, J., and R. P. Turco, 1995: Nucleation simulations in the wake of a jet aircraft in stratospheric flight. *J. Aerosol Sci.*, **26**, 779-795.

Table 1. Engine modes and performances defined in ICAO exhaust emissions data bank.

Mode	Rated Output (%)	Time (minutes)
Idle	7	26.0
Take-off	100	0.7
Climb-out	85	2.2
Approach	30	4.0

Table 2. The emitted rates (g/s) for differential aircraft styles (commercial jet- cargo jet-propeller sequence) in five engine operations.

	HC	CO	NO_X	S	CO_2	H_2O
	0.68	4.22	0.73	0.07	533.8	209.1
Idle	3.10	9.56	0.78	0.08	628.0	246.0
	0.19	11.7	0.21	0.06	440.0	170.0
	0.20	0.9	50.6	0.75	5871.8	2300.1
Take-off	0.17	0.46	80.9	0.98	7661.6	3001.2
	0.039	0.878	2.15	0.06	440.0	170.0
	0.18	0.85	31.6	0.60	4741.4	1857.3
Climb-out	0.16	0.38	49.2	0.80	6280.0	2460.0
	0.039	0.585	1.96	0.05	408.0	160.0
	0.17	1.41	5.10	0.20	1601.4	627.3
Approach	0.35	2.48	6.37	0.26	2041.0	799.5
	0.039	0.585	0.9	0.05	408.0	160.0
	0.24	1.68	15.60	0.40	3171.4	1242.3
Cruise	0.41	2.66	22.88	0.53	4176.2	1635.9
	0.039	0.585	1.43	0.05	408.0	160.0

Table 3. The averaged engine numbers and their deviation in the parentheses of differential aircraft types in Taipei FIR.

Commercial jet	Cargo jet	Propeller craft
2.36 (0.7)	3.42 (0.5)	2 (0)

Table 4. Annual amount of six emissions from the international commercial jet and cargo jet which are represented in the parentheses.

	HC tons/year	CO tons/year	NO_X tons/year	S tons/year	CO_2 kilotons/year	H_2O kilotons/year
Ground	148.0 (130)	918.8 (399.5)	457.7 (123.6)	19.48 (4.5)	152.9 (34.8)	59.9 (13.6)
Near surface	9.7 (2.8)	63.7 (17.3)	764.6 (214.0)	18.23 (4.5)	143.1 (35.3)	56.1 (13.8)
Cruise level	280.2 (45.3)	1906.9 (294.0)	17459.7 (2554.0)	444.6 (60.4)	3483.1 (473.2)	1366.74 (185.5)
Sub-total	438.2 (178.1)	2889.4 (710.8)	18682.0 (2891.6)	482.3 (69.4)	3779.1 (543.3)	1482.7 (212.9)
Total	616.3	3600.2	15447.3	551.7	4322.4	1695.6

Table 5. Annual amount of six emissions from the domestic commercial jet and propeller aircraft which are represented in the parentheses.

	HC tons/year	CO tons/year	NO_X tons/year	S tons/year	CO_2 kilotons/year	H_2O kilotons/year
Ground	354.5 (101)	2200.9 (6216)	1095.5 (142)	45.02 (29.21)	353.5 (239.6)	138.6 (92.7)
Near surface	23.0 (4.93)	152.4 (74)	1830.3 (161)	42.18 (6.32)	330.8 (51.8)	129.5 (20.1)
Cruise level	97.2 (16.8)	668.5 (252)	6192.3 (615)	160.32 (21.25)	1258.5 (176.4)	493.0 (68.9)
Sub-total	474.7 (122.73)	3021.8 (6542)	9118.1 (918)	247.52 (56.78)	1942.8 (467.8)	761.1 (181.7)
Total	597.43	9563.8	10036.1	304.3	2410.6	942.8

Table 6. The relative percentages of emissions amount in three vertical levels.

	HC	CO	NO_X	S	CO_2	H_2O
Ground	60.4	73.9	5.6	11.5	11.5	11.5
Near surface	3.4	1.9	9.6	8.2	8.2	8.2
Cruise level	36.2	24.2	84.8	80.3	80.3	80.3

Table 7. The ratio of emissions amount between civil aviation and land transportation in 1997 over Taiwan.

HC	CO	NO_X	S
0.43%	0.87%	14.9%	9.7%

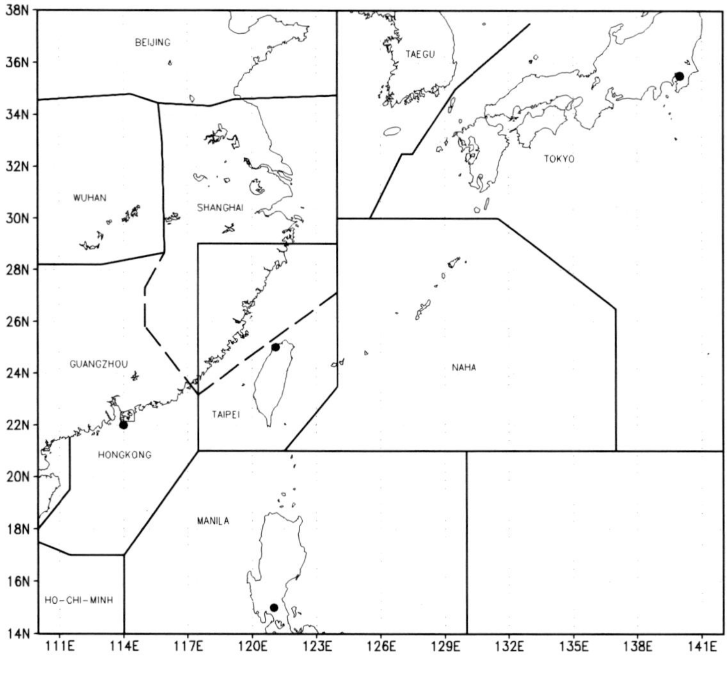

Fig. 1. The geography of Taipei Flight Information Region (FIR) in the western Pacific.

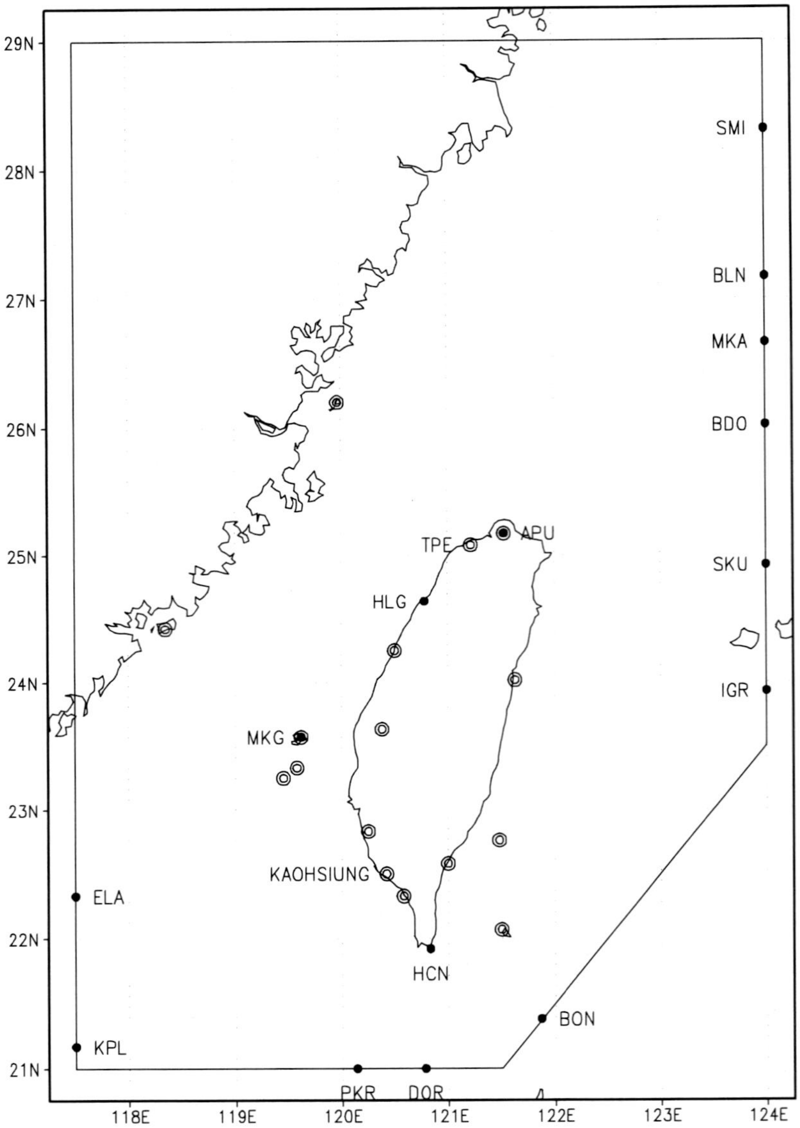

Fig. 2. The way-points (solid circle) and airports (empty circle) in Taipei FIR.

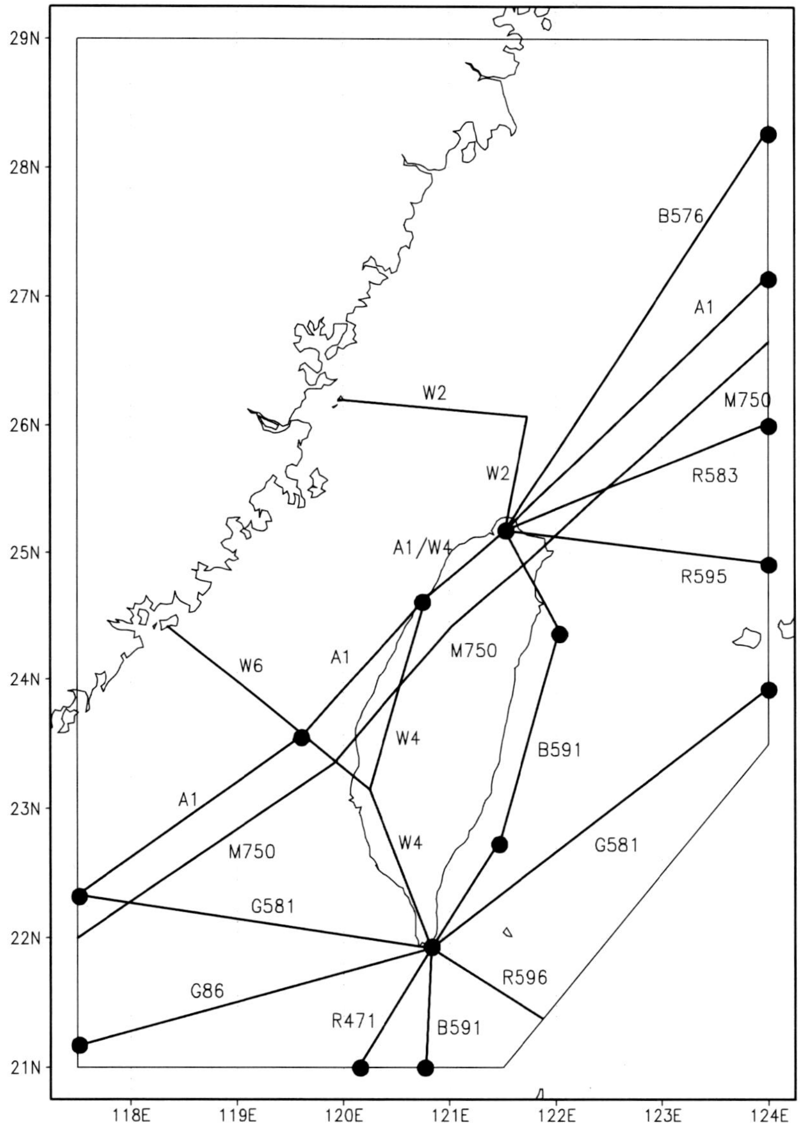

Fig. 3. Flight routes in Taipei FIR. The route names are marked besides the lines.

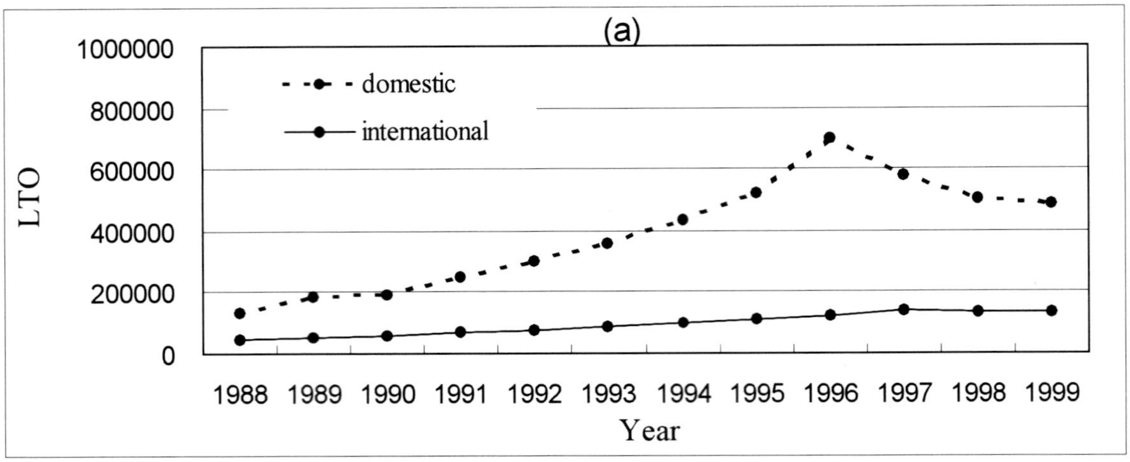

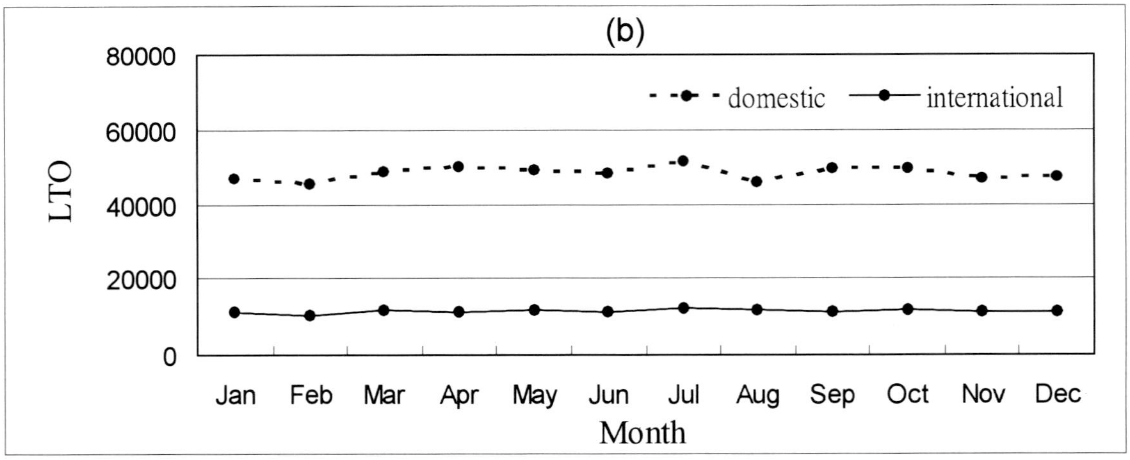

Fig. 4. (a) the annual variation of domestic (dash line) and international (solid line) civil air traffic. (b) same as (a), but is the monthly variation in 1997. LTO is the number of the civil aircraft landing or takeoff.

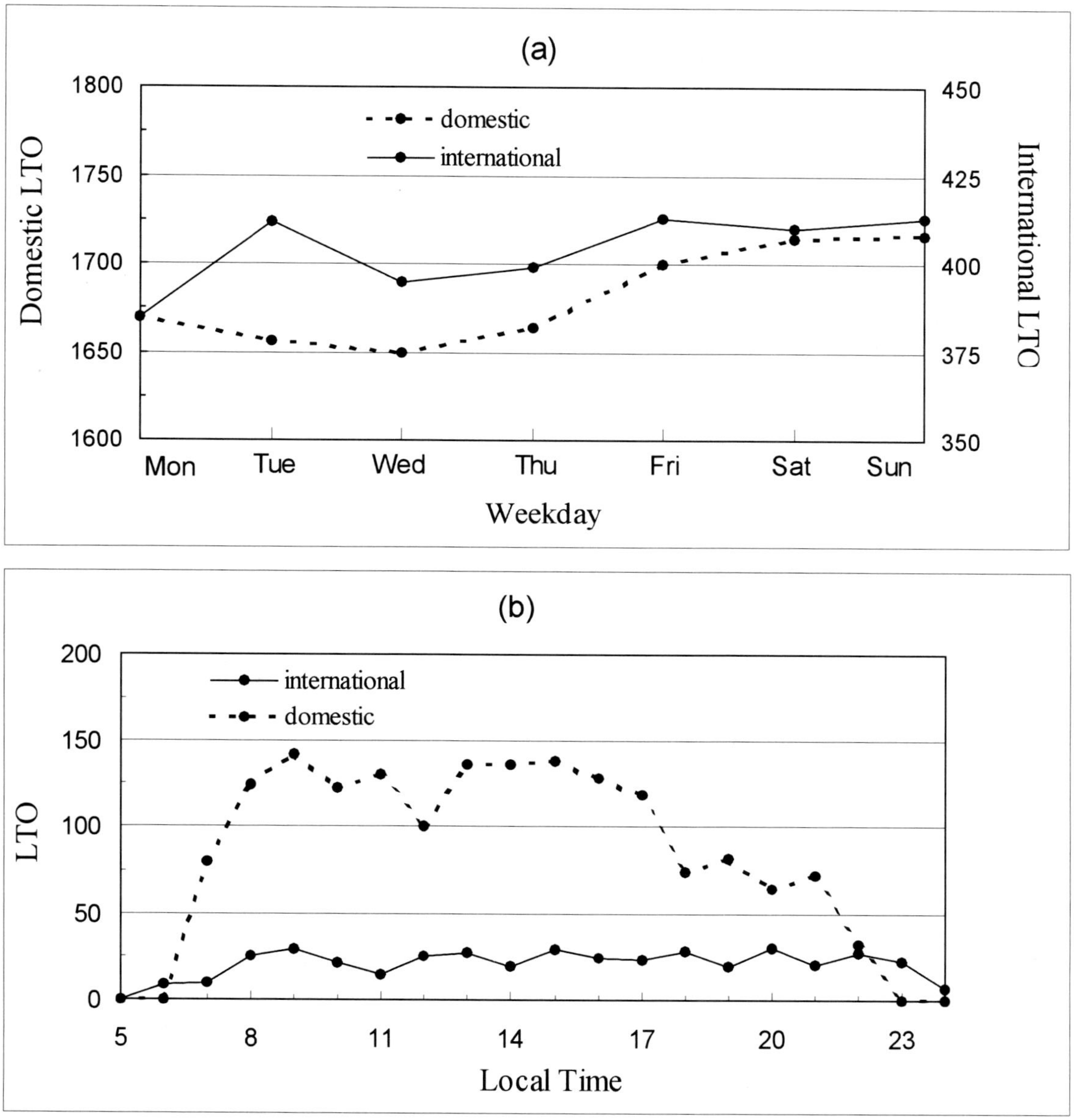

Fig. 5. (a) same as Fig. (4a), but is the weakly variation in 1997. (b) same as Fig. (5a), but is the hourly variation in 1997. LTO is the number of the civil aircraft landing or takeoff.

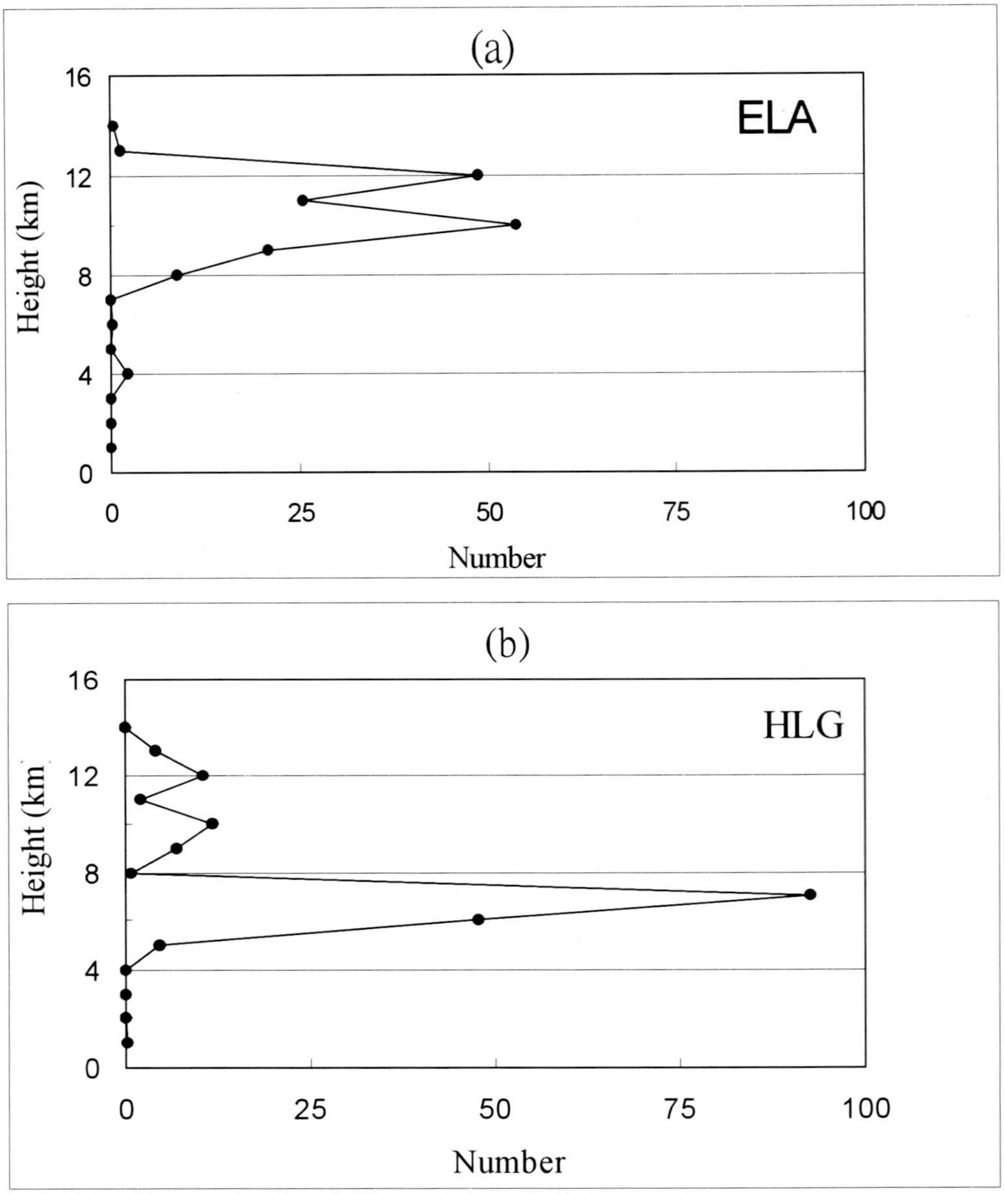

Fig. 6. Vertical distribution of civil aviation at ELA and HLG way-points. The geo-location of these two way-points are shown in Fig. 2. The value in the x-coordinate is the daily mean of the airplanes number passing above these way-points.

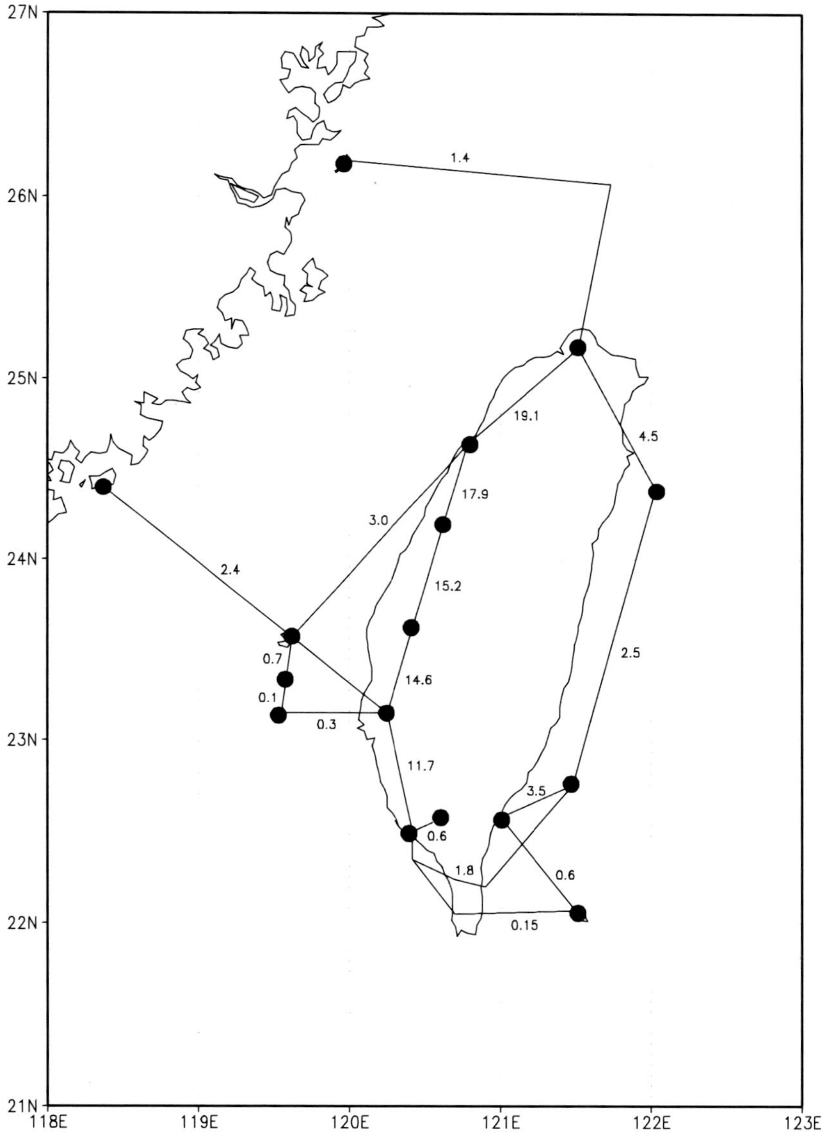

Fig. 7. Route-frequency of domestic air traffic in 1997 (Unit: %).

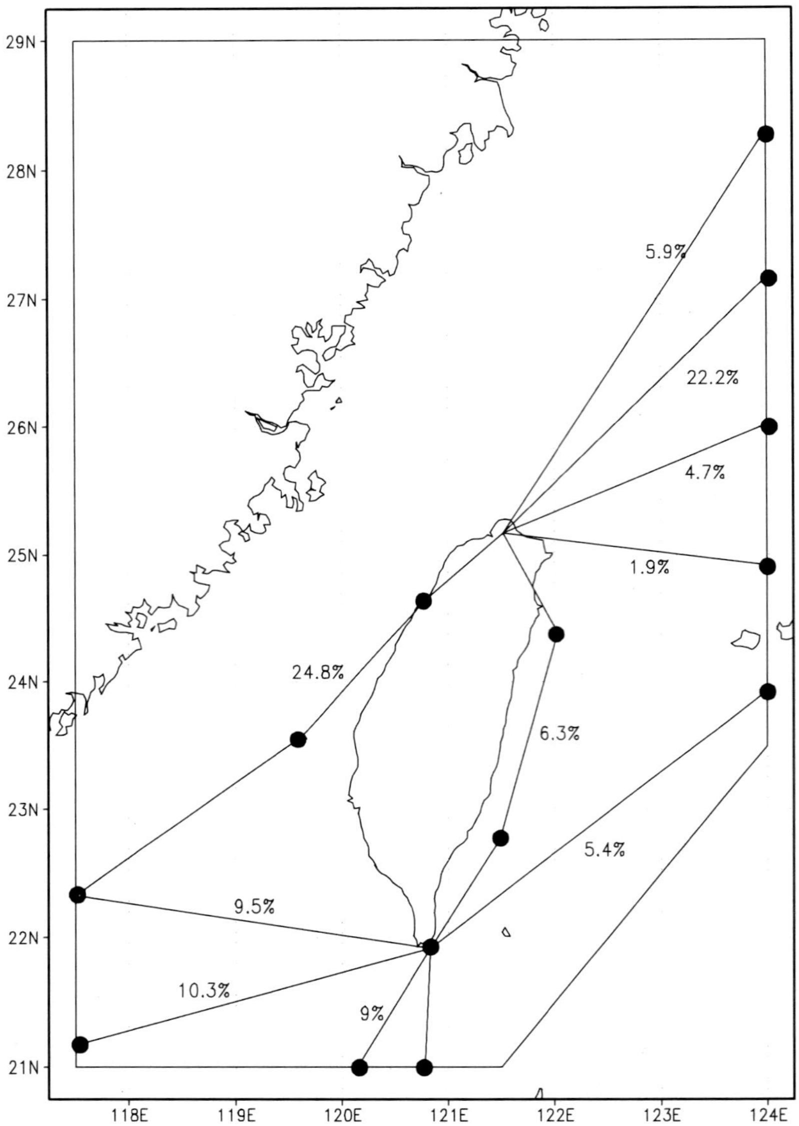

Fig. 8. same as Fig. 7, but is for the international air traffic.

NUMERICAL SIMULATION OF THE SEASONAL VARIATION OF TROPOSPHERIC OZONE AND FUTURE SCENARIO FORECAST IN CHINA

Yang Jian*, Li Weiliang and Zhou Xiuji
Chinese Academy of Meteorological Sciences, Beijing 100081

ABSTRACT

An updated version of the Regional Acid Deposition Model (RADM) driven by meteorological fields derived from the Chinese Regional Climate Model (CRegCM) is used to simulate seasonal variation of tropospheric ozone over eastern China. The results show that:

(1) Peak O_3 concentration area moves from southern to northern China in response to the shifting of solar perpendicular incidence point from south to north.

(2) In eastern China, the highest O_3 month-average concentration occurs in July, the lowest in January, and intermediate values in April and October. The pattern mainly depends on solar radiation, the O_3 concentration precursors of NO_X and NMHC, and the ratio of NMHC/NO_X.

(3) Diurnal variations of O_3 over eastern China are well defined. O_3 concentration increases with the rising sun, reaching its maximum at noon and then falling off. The range of variation is highest in summer (40×10^{-9} in volume fraction) and lowest in winter (20×10^{-9} in volume fraction).

(4) Diurnal variations of O_3 over western China are not as clear. The range of variation is less than 10×10^{-9} in volume fraction.

(5) NO_X concentration increases as its emission increases. In January, surface O_3 decreases as NO_X increases. In July, O_3 generally increases over the country. In April and October, O_3 decreases in northeastern China, northern China and Sichuan province but increases in other areas.

Key words: numerical simulation, seasonal variation, tropospheric ozone, China

1. INTRODUCTION

Although ozone is a component of trace gases, it is important to the global or regional environmental ecosystem and climate change. Tropospheric ozone is an important active oxidant (Ramanathan et al. 1985; and WMO 1990) in photochemistry and is a prerequisite in the production of hydroxy radicals which control the chemical cycles of many reduced gases in the atmosphere (WMO, 1992). On the other hand, tropospheric ozone is one of major air pollutants that, in high concentration, can endanger health, affect surface ecosystem (Reich et al. 1985) and damage plants (Coffin et al. 1977). In addition, because of its absorption properties in the infrared spectrum near 9.6μm, tropospheric ozone is an important greenhouse gas. Perturbation of tropospheric ozone, particularly in the upper troposphere, may have significant climatic impacts (Fishman et al. 1979). Most of the researches done on tropospheric ozone have focused on case studies. Seasonal variations of tropospheric ozone, especially surface ozone, are usually analyzed using observed data (e.g. Walter et al. 1989). In this paper, we use a three-dimensional Eulerian acid deposition model and a Chinese regional climate model to simulate and analyze seasonal variations of tropospheric ozone over China.

---

* Now Ph.D. student at Institute of Atmospheric Physics, CAS

2. MODEL DESCRIPTION

To simulate seasonal variations of tropospheric ozone over China, two independent models with one-way coupling are used: a regional chemical model (Chang and Brost 1987), driven by meteorological fields derived from regional climate model (Gong and Li 1996).

(a) Chinese Regional Climate Model

Based on the regional climate model version 1 (RegCM1) of NCAR, Li et al. 1997 has developed a Chinese regional climate model (CRegCM) to predict the regional climate change over China. In the adaptation processes, changes have been made to the cumulus convective parameterization scheme as well as the lateral boundary and horizontal diffusion. More details can be found in Gong and Li 1996. The model successfully simulated the seasonal surface air temperature and precipitation over China (Li et al. 1997). The proven capability of the CRegCM in simulating the regional climate of China provides the basis for running the regional chemical model. CRegCM's boundary conditions and initial field are provided by NCAR CCM1.

(b) Chemical Model

An updated version of the Regional Acid Deposition Model (RADM) of Chang and Brost 1987 is used in this paper. The model is an Eulerian transport, transformation, and removal model that sub-divides the atmosphere over China into 14 vertical levels, each with 44 (North-South) $\times$49 (East-West) horizontal grid cells approximately 100 $\times$100 km^2 in size. The model top is at 10 hPa. The center of grid is 101.5°E, 36.5°N. The model solves a set of chemical species conservation equations,

$$\frac{\partial C}{\partial t} = -\nabla . VC + \nabla .(K_e \nabla C) + P_{chm} - L_{chm} + E + \left(\frac{\partial C}{\partial t}\right)_{clouds} + \left(\frac{\partial C}{\partial t}\right)_{dry} \qquad (1)$$

where C is the species volume mixing ratio, V is the three-dimensional velocity vector at each grid point in the model domain, K_e is the eddy diffusivity used to parameterize the subgrid scale fluxes of trace species due to non-cloudy turbulent motions, P_{chm} and L_{chm} are the production and loss rates due to chemical interactions, E is the emission rate, $(\partial C/\partial t)_{clouds}$ is the time rate of change of concentration due to cloud effects (including subgrid scale vertical re-distribution, aqueous chemical interactions and scavenging), and $(\partial C/\partial t)_{dry}$ represents the change in concentration due to dry deposition.

Equation (1) is numerically solved with an operator-splitting technique. Horizontal and vertical transport uses a time step of Δt_{ad}=150s, followed by diffusion and dry deposition computations with the same time step. Gas phase chemistry is solved with a variable chemical time step Δt_c which is often considerably smaller than Δt_{ad}.

In our simulations, only anthropogenic emissions were considered. Anthropogenic NO_X and SO_2 emissions in China were based on the annually-averaged 1° $\times$1° gridded inventories of Bai 1996. Data in surrounding areas were taken from Kato and Akimoto 1987. The emission data of other species were derived from their ratio with NO_X (Brost 1988). All emission data were interpolated into the lowest model layer grids.

3. TEST DESIGN

(a) Control Test

Four model runs focussing on January, April, July and October were used to represent the four seasons. One-month prediction by CRegCM was run with initial fields from CCM1. Hourly outputs from CRegCM were used to run RADM. RADM was run for the same period to simulate ozone variations.

(b) Sensitivity Test

According to the economic development of different regions, future changes in emission rates were prescribed. We assumed that the emissions in Hebei, Shandong, Jiangsu, Zhejiang, Fujian and Guangdong provinces would increase by 30%. In the northeastern part, Shanxi, Henan, Anhui, Hunan, Guangxi, Sichuan (east of Chengdu) and Shaanxi (east of Xian) provinces, emissions increase was set at 50%. In all other areas, emissions increase was taken to be 100%. RADM was re-run and changes in O_3 and NO_X were computed.

4. RESULTS AND ANALYSIS

(a) Control Test

Figs. 1 a, b, c and d show the distributions of model-calculated monthly averaged NO_X in the surface layer for January, April, July and October respectively. Figs. 2 a, b, c and d give the observed surface NO_2 distributions for the same months. From Fig. 1, surface NO_X concentration was found to be higher in January and lower in July. Higher NO_X in January was mainly due to relatively small consumption as a result of less wet deposition and lower temperatures. But the distribution patterns showed very little variations over the four seasons. High NO_X in the southern part of northeastern China, northern China and the Yangtze delta matched the emission pattern. Observation data in Fig. 2 showed similar patterns except that the high center was not as obvious. Magnitudes from model results were generally smaller than observed. This was probably due to the omission of nature emission in model simulation as only anthropogenic sources were considered. In rural areas, nature emission constituted a large part of the total emission. Another possible cause for the differences could be attributed to the fact that most of the observation data were obtained near cities.

Table 1 is the comparison of model-calculated and observed concentrations of NO_X at Longfeng Shan, Lin An, Wanliguan and Qingdao. Simulation results in pollution areas were better than those for rural areas. For example, the absolute value of mean NO_X differences in Qingdao was 16 $\times 10^{-9}$ in volume fraction and the relative difference was only 49%; but in Wanliguan, although the absolute difference was only 0.05 $\times 10^{-9}$ in volume fraction, the relative difference was as high as 226%. Model simulation was therefore less effective in rural areas. Possible causes could be due to unrealistic emissions in rural areas and the observation data being highly sensitive to environmental factors.

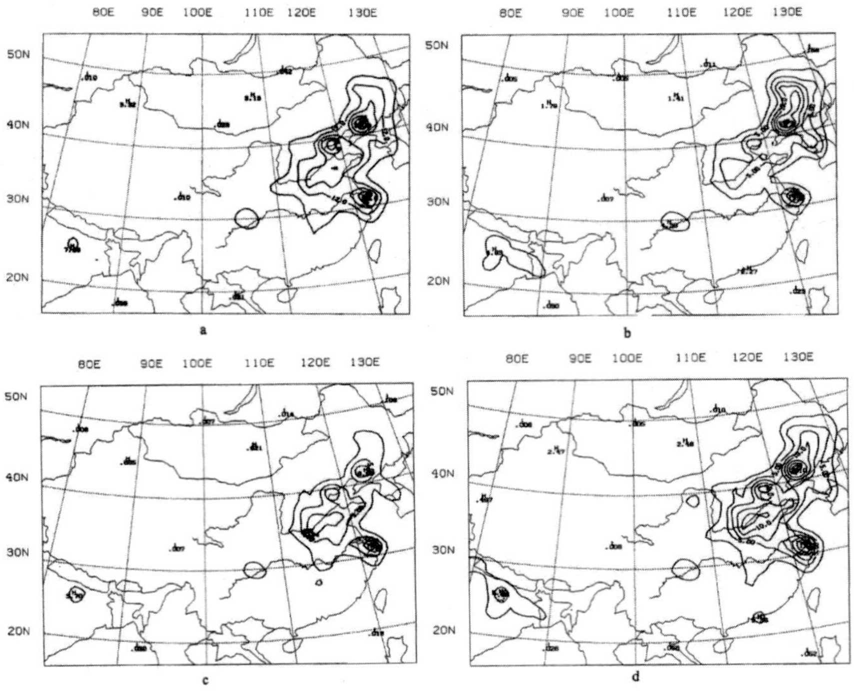

Fig. 1. Model-calculated monthly averaged concentration of surface NO_X (unit: 10^{-9} in volume fraction) - (a) January; (b) April; (c) July, (d) October.

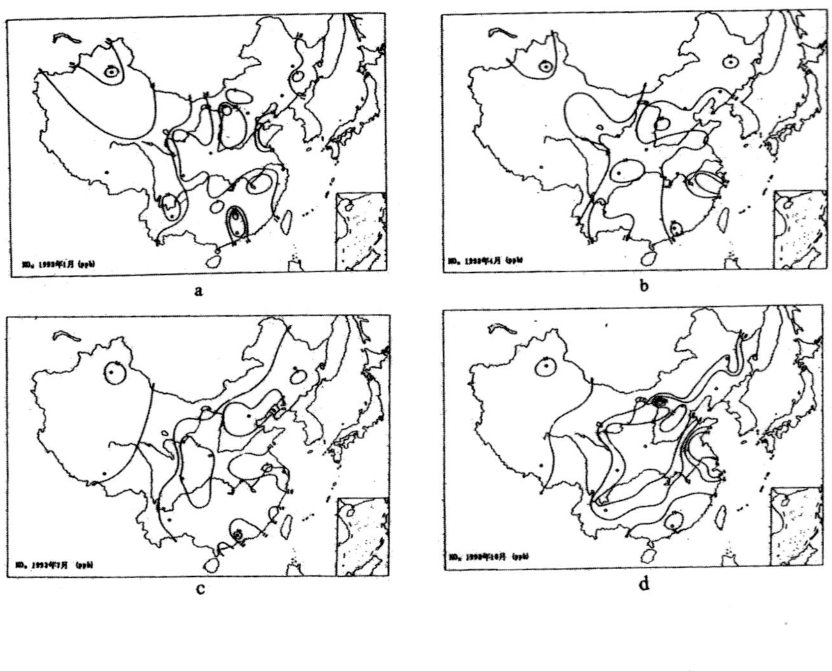

Fig. 2. Observed monthly averaged concentration of surface NO_2 (unit: 10^{-9} in volume fraction) - (a) January; (b) April; (c) July, (d) October.

Table 2 is the comparison of model-calculated and observed O_3 concentrations at the same four locations. In general, O_3 simulations were better than NO_X. Except in Wanliguan, the simulated O_3 values were always lower than the observed values when the simulated NO_X were higher than the observed values. In Wanliguan, it was interesting to note that for the January run, although there was a significant difference between the simulated and observed NO_X concentrations, the O_3 difference was only 3 $\times 10^{-9}$ in volume fraction; whereas for the July run, the difference of NO_X was only 0.01 $\times 10^{-9}$ in volume fraction but the difference of O_3 was as large as 20 $\times 10^{-9}$ in volume fraction. The seasonal and daily variations in rural areas were very small, indicating that the O_3 concentrations in rural areas were less affected by photochemical processes.

Table 1. Model-calculated (Mod.) and Observed (Obs.) Monthly Averaged NO_X Concentrations in Four Sites (unit: 10^{-9} in volume fraction).

	Longfeng Shan		Lin An		Wanliguan		Qingdao	
Month	Mod.	Obs.	Mod.	Obs.	Mod	Obs.	Mod.	Obs.
1	7.7	4.8	17.1	2.4	0.17	0.02	33.4	49.4
4	3.6	2.2	4.9	9.2	0.06	0.03	10.3	37.1
7	0.7	.2(Aug.)	5.2	0.7	0.04	0.03	10.3	24.7
10	4.0	1.2	11.2	15.7	0.05	0.04	16.7	24.7

Table 2. Model-calculated (Mod.) and Observed (Obs.) Monthly Averages O_3 Concentrations in Four Sites (unit: 10^{-9} in volume fraction).

	Longfeng		Lin An		Wanliguan		Qingdao	
Month	Mod.		Mod.	Obs.	Mod.	Obs.	Mod.	Obs.
1	19.4	27.2	21.7	35.2	46.0	43.5	10.5	18.7
4	32.7	37.4	40.8	33.0	44.2	52.4	33.5	23.4
7	45.4	27.2	45.2	21.5	41.3	61.7	33.1	23.4(Aug.)
10	30.7	41.7	34.5	56.9	45.8	42.5	22.0	37.4

From the comparison against observation data, we found that the model had the ability to simulate the seasonal variations of O_3 over China. Fig. 3 presents the model-calculated spatial distribution of surface (0 – 76 m) O_3 in the four seasons. It shows high ozone concentration over the Tibetan Plateau in January, April and October with mean values exceeding 50 x 10^{-9} in volume fraction. But in July, the high concentration zone was not apparent and the magnitude was smaller. This did not agree with the observations made by Yan et al. 1996. One possible reason could be that we did not alter the emission in our model according to the seasons and during the summer there was higher ozone consumption. In addition, our model did not consider the injection of ozone from the stratosphere. All these might have led to lower surface ozone values over the Tibetan Plateau.

Seasonal variations of surface O_3 over eastern China (east of 110°E) were observed. High concentration zone appeared in southern China (south of 30°N) in January (Fig. 3a) and moved north in April in response to the shifting of solar perpendicular incidence point. In northern and northeastern China where the surface ozone concentration had been low, an increasing trend was observed (Fig. 3b). By July, the high ozone areas in northern and northeastern China were quite distinct (Fig. 3c). The ozone distribution became less latitude-dependent and corresponded more to the distribution of NO_X. In October, the surface ozone values in northern China decreased as the solar perpendicular incidence point migrated to the south (Fig. 3d). During the winter (January), high ozone concentration re-emerged in southern China and the one-year cycle was completed. From the seasonal variations of ozone, it was shown that fluctuations in solar radiation, the induced changes in temperatures, humidity and precipitation were the main factors controlling the distribution of surface ozone.

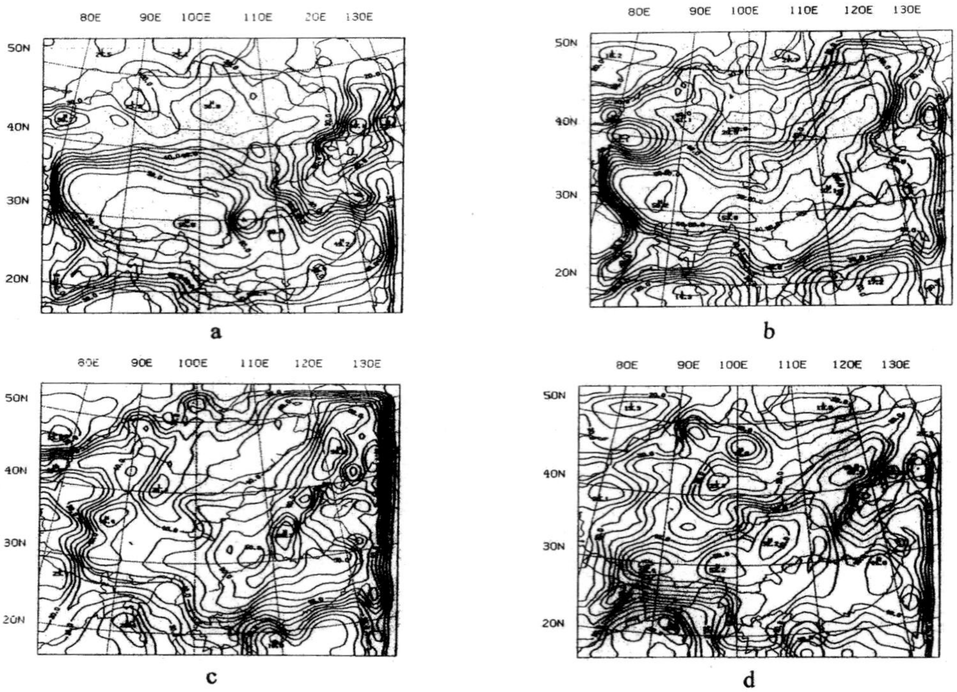

Fig. 3. Model-calculated monthly averaged concentration of surface O_3 (unit: 10^{-9} in volume fraction) - (a) January; (b) April; (c) July, (d) October.

NO_X was the main O_3 precursor. Its variation would greatly affect ozone concentration. By analyzing the observation data in Lin An, Luo and Zhou 1994 showed that there was a non-linear relationship between NO_X and O_3. When the NO_X concentration was higher than 3×10^{-9} in volume fraction, O_3 concentration would decrease with increasing NO_X, and *vice versa*. Fig. 4 shows the relationship between surface NO_X and O_3 for the January run. It was found that when NO_X concentration was above 5×10^{-9} in volume fraction, O_3 concentration decreased with increasing NO_X. In January (Fig. 1a), the low surface O_3 concentrations in northeastern and northern China, where the NO_X level was almost higher than 10×10^{-9} in volume fraction (Fig. 3), were due to the high NO_X level in these areas.

NMHC was another main O_3 precursor. Its variation could also greatly affect O_3 concentration. Liu et al. 1996 pointed out that there was a non-linear relationship between $NMHC/NO_X$ and O_3. At low NO_X concentration, the variation of $NMHC/NO_X$ had little effect on O_3 concentration; but at high NO_X, an increase in the ratio would lead to increased O_3 concentration. In polluted areas, O_3 could be produced from photochemical reaction under suitable meteorological condition ($NMHC + NO_X + h\nu + M \rightarrow O_3 +$ other products). $O(^3P)$ was produced by NO_2 photolysis through absorbing UV radiation energy. It then combined with O_2 to form ozone, the main production mechanism for ozone. But NO would react with ozone rapidly, consuming ozone and inhibiting its accumulation. However, NMHC could turn NO into NO_2, and ozone production would be enhanced through increasing NO_2 concentration and reduced NO. So under certain $NMHC/NO_X$ ratios, an increase in NMHC could increase O_3 production. Fig. 5 shows the simulated ratio of $NMHC/NO_X$ in July. The ratio over the Tibetan Plateau was very high with averaged values exceeding 100 (in Fig. 5, all ratios above 100 are set as 100). The observed $NMHC/NO_X$ ratio at Wanliguan was as high as 1000. This shows that natural NMHC emission over Tibetan Plateau was higher than anthropogenic emission. The ratio in eastern China was between 10 and 20. The seasonal difference between July and January (July minus January) for $NMHC/NO_X$ and O_3 are given in Figs. 6 a and b respectively. As shown in Fig. 6a, the $NMHC/NO_X$ ratio north of the Yangtze River was higher in July than that in January (positive in figure). However, south of the Yangtze River, the ratio was higher in January than that in July (negative in figure). The ozone differences in Fig. 6b had features similar to the $NMHC/NO_X$ patterns, again suggesting that the $NMHC/NO_X$ ratio had important influence on ozone production.

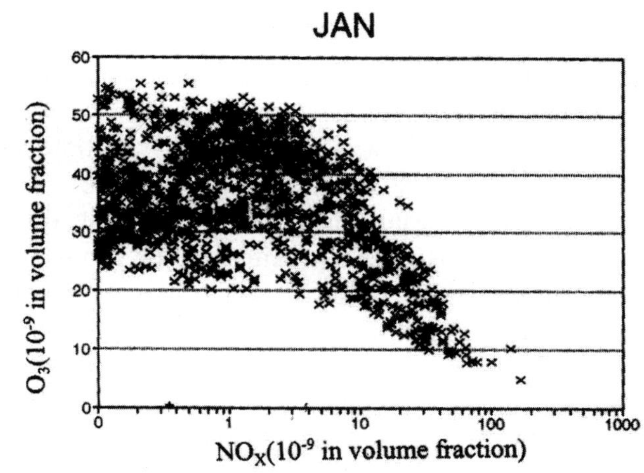

Fig. 4. Model-calculated relationship between surface O_3 and NO_X for monthly averaged concentrations in January (unit: 10^{-9} in volume fraction).

Figs. 7a, b, c and d are the distribution of monthly averaged ozone concentrations at 500 hPa in January, April, July and October respectively. It was shown that ozone distribution was mainly determined by large-scale wind field with an ozone maximum appearing in northeastern China in all four months. In January and April, there is a high ozone concentration over Tibet Plateau, but in July and October, there is a low ozone concentration.

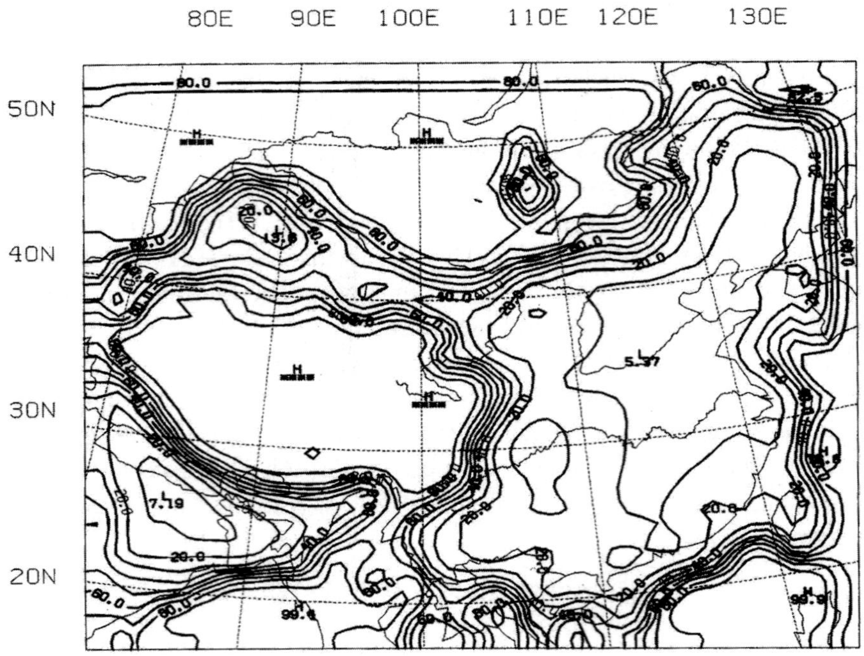

Fig. 5. Model-calculated monthly averaged ratio of $NMHC/NO_X$ in July.

Fig. 8 is the model-calculated surface ozone diurnal variations in Lin An, Qingdao, Longfeng Shan and Wanliguan. The abscissa is Beijing time. The diurnal variations revealed a typical photochemical feature in all sites except at Wanliguan. The ozone concentrations increased with the rising sun, peaking at noon and then decreasing gradually towards the lowest level before sunrise. Diurnal ranges were large in Lin An and Qingdao, located in the mid latitudes, with a peak value of 40×10^{-9} in volume fraction in the summer. In Longfeng Shan at higher latitudes, the diurnal range was relatively small, only 20×10^{-9} in volume fraction in the summer. Variations in diurnal ranges between high and low latitudes were due to the different intensity levels of solar radiation.

In Wanliguan, the diurnal ozone variation was complex. In fact, the diurnal as well as the seasonal variations were both relatively small. The smallest range of about 5×10^{-9} in volume fraction occurred in winter. With the exception of January, ozone concentration no longer had its peak at midday but instead took on a decreasing trend, especially in July. This suggested ozone consumption through photochemical processes. Luo et al. 1996 studied the photochemical processes focussing on Wanliguan by using a box model and found that there was a net consumption of O_3 at the rate of 0.5×10^{-9} in volume fraction per day. Our simulation results showed that there were two peaks of high surface ozone values in July at 00 and 11 hour (Beijing time) and low surface ozone values at 14-15 hour (Beijing time). The results were similar to the observations from Tang et al. 1996.

time). The results were similar to the observations from Tang et al. 1996.

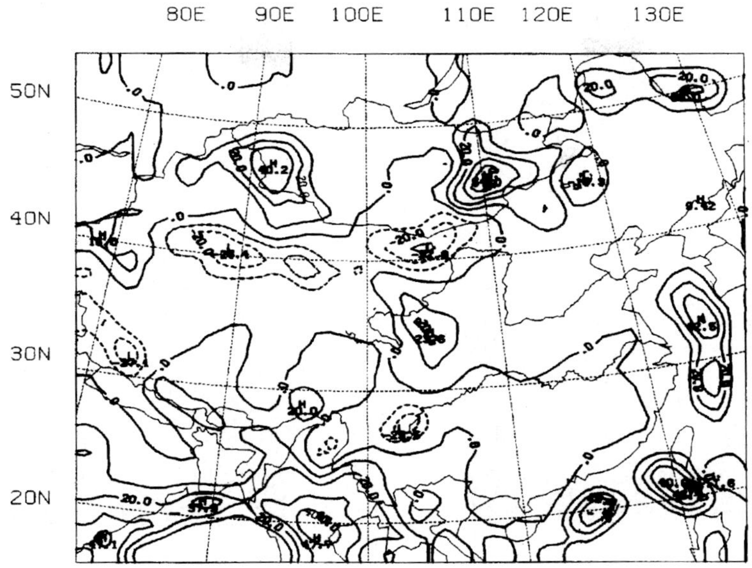

a

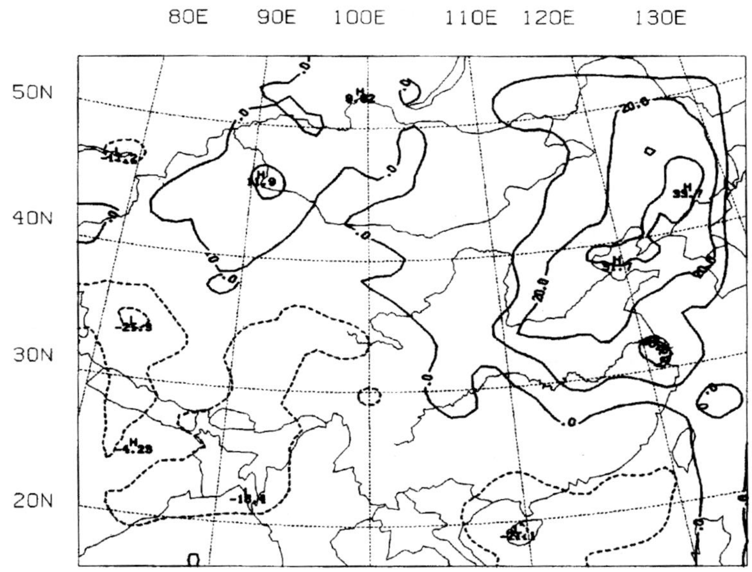

b

Fig. 6. Model-calculated difference of monthly averaged surface concentration (July run minus January run) – (a) NMHC/NO_X ; (b) O_3.

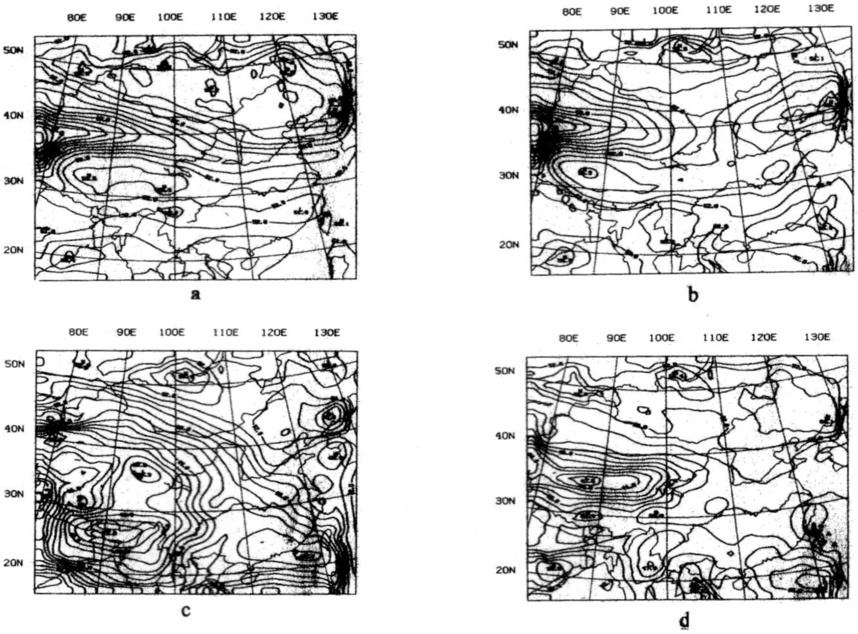

Fig. 7. Model-calculated monthly averaged ozone concentration at 500 hPa (unit: 10^{-9} in volume fraction) - (a) January; (b) April; (c) July, (d) October.

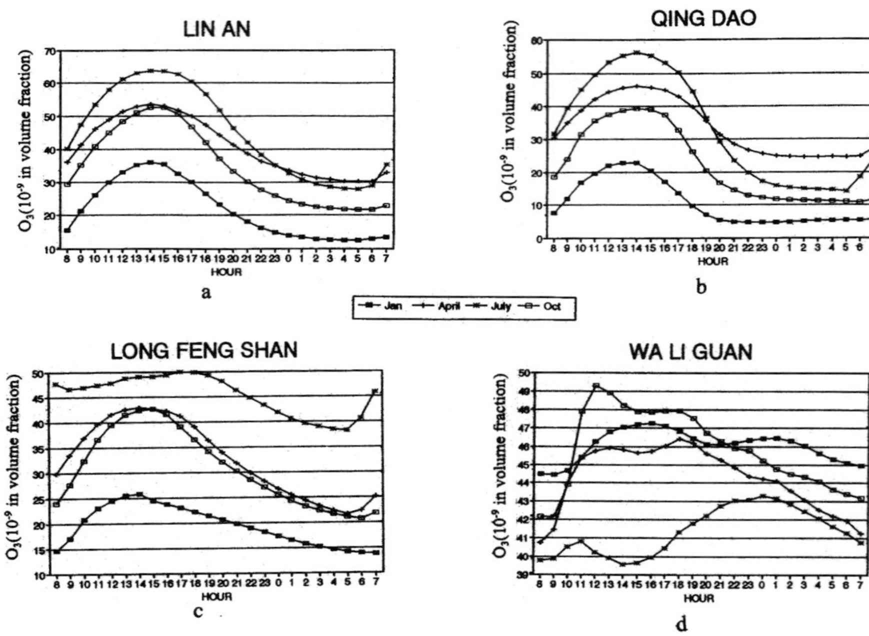

Fig. 8. Model-calculated monthly averaged diurnal variations of surface ozone (unit: 10^{-9} in volume fraction) – (a) Lin An; (b) Qingdao; (c) Longfeng Shan; and (d) Wanliguan.

Figure 9 shows the model-calculated diurnal variations of monthly averaged O_3, NO_2 and NO in July at Lin An. The trends were opposite between O_3 and NO_2. At midday when O_3 reached its highest level, NO_2 reached its lowest level. This illustrates the main process of O_3 production and the reaction mechanisms are given as follows:
NO and O formed through NO_2 photolysis and ozone by the combination reaction of O and O_2, expressed as:

$$NO_2 + h\nu\ (\lambda < 0.4\mu m) \rightarrow NO + O(^3P) \quad (2)$$
$$O(^3P) + O_2 + M \rightarrow O_3 + M \quad (3)$$

But O_3 is also consumed by its reaction with NO:

$$O_3 + NO \rightarrow NO_2 + O_2 \quad (4)$$

So the net O_3 production is zero. But if NO is converted into NO_2 by other oxidants (see the following reactions), there will then be a net production of O_3 :

$$RO_2 + NO \rightarrow NO_2 + RO \quad (5)$$
$$HO_2 + NO \rightarrow NO_2 + OH \quad (6)$$

where RO_2 and HO_2 are the products of the oxidized VOC, CH_4 and CO.

The low daytime NO_2 concentration in Fig. 9 was mainly due to the following consumption reactions:

$$HO + NO_2 + M \rightarrow HNO_3 + M \quad (7)$$
$$HO_2 + NO_2 \rightarrow HNO_4 \quad (8)$$
$$CH_3COO_2 + NO_2 + M \rightarrow PAN + M \quad (9)$$

At night, photolysis mechanisms were not at work and other radicals also had low levels, leading to NO_2 accumulation.

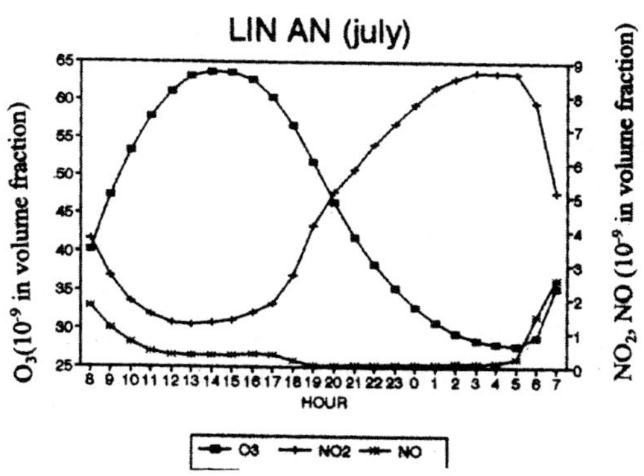

Fig. 9. Model-calculated diurnal variations of monthly averaged surface O_3, NO_2 and NO in July at Lin An (unit: 10^{-9} in volume fraction).

(b) Sensitivity Test

Fig. 10 shows the monthly averaged changes of surface NO_X (sensitivity test minus control test) and the variations in NO_X concentration over the four seasons. The concentration maximum in eastern China had the highest value of 17 $\times 10^{-9}$ in volume fraction in January and the lowest value of 5 $\times 10^{-9}$ in volume fraction in July. The smaller increase in NO_X in July was mainly due to stronger photochemical reactions and higher humidity in summer.

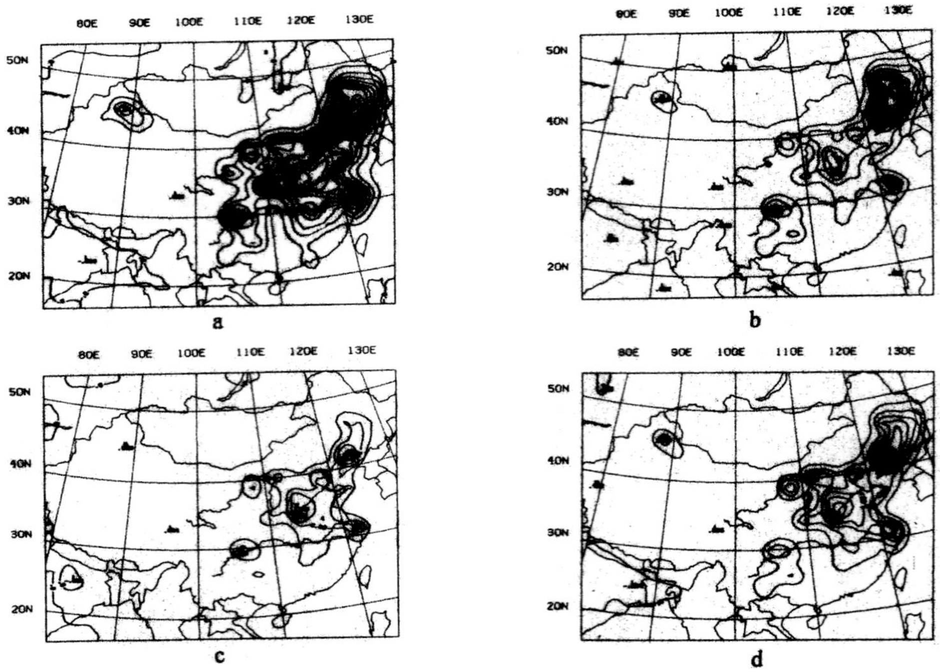

Fig. 10. Changes in monthly averaged surface NO_X (sensitivity test minus control test; unit: 10^{-9} in volume fraction) - (a) January; (b) April; (c) July, (d) October.

The monthly averaged changes of surface O_3 (sensitivity test minus control test) in Fig. 11 were more complex than those of NO_X. There were regional differences in the variations as well as seasonal differences within the same area. In January, surface O_3 was reduced in most parts of eastern China. The largest reduction observed was 5 $\times 10^{-9}$ in volume fraction. Over western China, increases were rather insignificant. In April, surface O_3 decreased in northern and northeastern China, Sichuan basin and the northern part of Xinjiang but increased to the south of the Yangtze River. Surface O_3 increased everywhere in July, with the highest value of 4.5 $\times 10^{-9}$ in volume fraction located in Hubei and Hunan provinces. In October, the distribution of O_3 variations was similar to that of April except in the Yangtze delta where O_3 decreased. In a non-linear relationship between O_3 and NO_X, an increase in surface O_3 generally corresponded to a decrease in NO_X while a decrease in O_3 corresponded to an increase in NO_X.

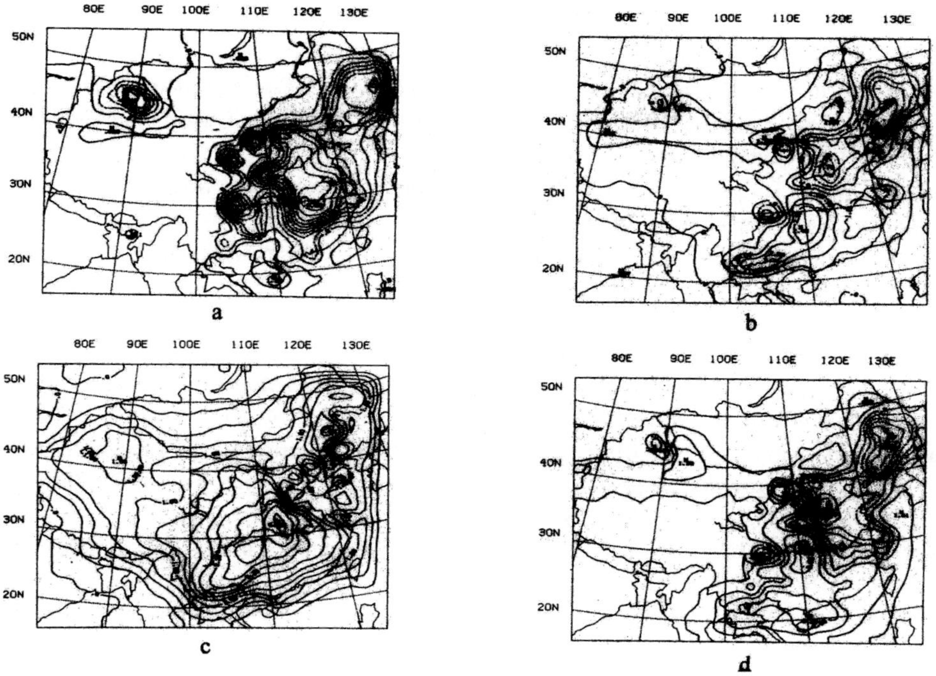

Fig. 11. Changes in monthly averaged surface O_3 (sensitivity test minus control test; unit: 10^{-9} in volume fraction) - (a) January; (b) April; (c) July, (d) October.

5. SUMMARY

In this study, the seasonal variations of ozone over China were simulated using RADM and CRegCM. Temporal and spatial variations of ozone were well depicted. The ozone maximum moved from south to north in response to the change of solar perpendicular incidence point from south to north, and *vice versa*. Monthly averaged ozone concentration over eastern China was highest in July and lowest in January, with intermediate values in April and October. The patterns mainly depended on the variations in solar radiation intensity, NO_X, NMHC and NMHC/NO_X. Diurnal variation of surface ozone was prominent over eastern China. Ozone concentration increased with the rising sun, peaking at noon and decreasing towards its lowest level before sunrise. The diurnal range of surface ozone over eastern China was lowest in winter (about 20×10^{-9} in volume fraction) and highest in summer (about 40×10^{-9} in volume fraction). However, in western China, the diurnal variation was not as obvious with a range of no more than 10×10^{-9} in volume fraction. NO_X concentrations increased as emissions increased. Due to higher NO_X values in January, surface ozone was reduced as a result of increased emissions. In July, ozone increased as a whole. In April and October, ozone decreased in Sichuan province, northern and northeastern China, but increased elsewhere, showing a non-linear relationship between NO_X and O_3.

REFERENCES

Bai Naibin, 1996: 1° ×1° grid emission of CO_2 , SO_2 and NO_X in China mainland area. *The Ozone Change in China Area and Its Impact on Climate and Environment (I)*. Zhou Xiuji (Eds.), China Meteorology Press, 145–150 (in Chinese).

Brost, R.A., 1988: Three-dimensional modeling transport of chemical species from continents to the Atlantic Ocean. *Tellus,* **40**(B), 358–379.

Chang, J.S., and R.A. Brost, 1987: A three–dimensional Eulerian acid deposition model: physical concepts and formulation. *J. Geophys. Res.*, **92**, 1468–1470.

Coffin, D.L., and H.E. Stokinger, 1977: *Biological effects of air pollutants, Air Pollution.* A.C. Stern (Eds.), II, 231–260.

Fishman, J., S. Soloman, and P.J. Crutzen, 1979: Observations and theoretical evidence in support of a significant in–situ photochemical source of tropospheric ozone. *Tellus,* **31**, 432–446.

Gong Wei, and Li Weiliang, 1996: The regional climate model for assessing the climate effect of ozone change. *The Ozone Change in China Area and Its Impact on Climate and Environment (I),* Zhou Xiuji (Eds.), China Meteorology Press, 251–266 (in Chinese).

Kato, N., and H. Akimoto, 1987: Anthropogenic emission of SO_2 and NO_X in Asia emission inventories. *Atmos. Environ.,* **26**, 2997–3017.

Li Weiliang, et al., 1997: Simulation of regional climate over China with Chinese regional climate model. *Acta Meteor. Sinica,* **3**, 307-319.

Liu, S. C., et al., 1996: Model study of tropospheric trace species distributions PEM-West A. *J. Geophys. Res.*, **101**, 2073-2085.

Luo, C., J. Tang, and X.J. Zhou, 1996: Measurements and model simulations of surface ozone variation at different background concentration of the precursors. *Acta Meteor. Sinica,* **Vol. 10**, No. **2**, 215-227.

Luo, C., and X.J. Zhou, 1994: A regional model study of the variation and distributions of ozone and its precursors on Eastern Asia and West Pacific Ocean Regions. *Acta Meteor. Sinica*, **Vol. 8**, No.**2**, 195–202.

Ramanathan, V., et al., 1985: Trace gas trends and their potential role in climate change. *J. Geophys. Res.,* **90**, 5545–5566.

Reich, P.B., et al., 1985: Ambient levels of ozone reduce net photosynthesis in trees and crop species. *Science,* **230,** 566–570.

Tang Jie, and X.S. Li, et al., 1996: Observation and initial result of surface ozone in Wang Li Guan Mountain, Qing Hai province. *The Ozone Change in China Area and Its Impact on Climate and Environment (I),* Zhou Xiuji (Eds.), China Meteorology Press, 19-29 (in Chinese).

Walter, E., et al., 1989: Surface ozone: trend details, seasonal variations, and interpretation. *J. Geophys. Res.* **94,** 1828-1829.

WMO, 1990: *Review of the contribution of ozone and other minor gases to atmospheric radiation and their possible effect on global climate change*. Report No.**10**, Geneva, Switzerland.

WMO, 1992: *Scientific assessment of ozone depletion: 1991, World Meteorological Organization Global Ozone Research and Monitoring Project*. Report No.**25**, Geneva.

Yan, P., et al., 1996: The distribution feature of O_3, NO_X, SO_2 in China rural area. *The Ozone Change in China Area and Its Impact on Climate and Environment (I),* Zhou Xiuji (Eds.), China Meteorology Press, 1-9 (in Chinese).

AUTHOR INDEX